Differential Equations with *Maple V* ®

SECOND EDITION

Differential Equations with *Maple V* ®

SECOND EDITION

Martha L. Abell
James P. Braselton

ACADEMIC PRESS

A Harcourt Science and
Technology Company

ACADEMIC PRESS
San Diego   San Francisco   New York   Boston
London   Sydney   Tokyo

ACADEMIC PRESS
A Harcourt Science and Technology Company
525 B Street, Suite 1900, San Diego, CA 92101-4495, USA
http://www.apnet.com

Academic Press
24–28 Oval Road, London NW1 7DX, UK
http://www.hbuk.co.uk/ap/

Library of Congress Catalog Card Number: 99-64632
International Standard Book Number: 0-12-041560-7

Printed in the United States of America
99 00 01 02 03   IP   9 8 7 6 5 4 3 2 1

Contents

Preface

Maple's diversity makes it particularly well suited to performing many calculations encountered when solving many ordinary and partial differential equations. In some cases, Maple's built-in functions can immediately solve a differential equation by providing an explicit, implicit, or numerical solution; in other cases, Maple can be used to perform the calculations encountered when solving a differential equation. Because one goal of elementary differential equations courses is to introduce the student to basic methods and algorithms and have the student gain proficiency in them, nearly every topic covered in *Differential Equations with Maple V*, Second Edition, includes typical examples solved by traditional methods and examples solved using Maple. Consequently, we feel that we have addressed one issue frequently encountered when implementing computer-assisted instruction. In addition, *Differential Equations with Maple V*, Second Edition, uses Maple to establish well-known algorithms for solving differential equations.

Taking advantage of Release 5 of Maple, this edition of *Differential Equations with Maple V* introduces the fundamental concepts of differential equations as encountered in typical introductory courses in ordinary and partial differential equations and uses Maple to solve typical problems of interest to students, instructors, and scientists. Other features to help make *Differential Equations with Maple V*, Second Edition, as easy to use and as useful as possible include the following:

1. **Applications.** New applications, many of which are documented by references, from a variety of fields, especially biology, physics, and engineering, are included throughout the text.

2. **Getting Started.** The Appendix provides a brief introduction to Maple, including discussions about entering and evaluating commands, loading packages, and taking advantages of Maple's extensive help facilities. Appropriate references to *The Maple Book* are included as well.

3. **Release 5 Compatibility.** All examples illustrated in the book were completed using Release 5 of Maple. Although most computations can continue to be carried out with earlier versions of Maple, such as Releases 3 and 4, we have taken advantage of the new features in Release 5 as much as possible.

4. **Detailed Table of Contents.** The table of contents includes all chapter, section, and subsection headings. By using it along with the comprehensive index, we hope that users will be able to locate information quickly and easily.

5. **Comprehensive Index.** In the index, mathematical examples and applications are listed by topic, or name, as well as commands along with frequently used options: particular mathematical examples as well as examples illustrating how to use frequently used commands are easy to locate. In addition, commands in the index are cross-referenced with frequently used options. Functions available in the various packages are cross-referenced both by package and alphabetically.

6. **CD-ROM.** All Maple input that appears in *Differential Equations with Maple V*, Second Edition, is included on the CD-ROM packaged with the text.

Differential Equations with Maple V, Second Edition, may be used as a handbook that addresses some ways to use Maple for computations of explicit, implicit, numerical, or graphical solutions of a variety of differential equations and as a supplement for courses in ordinary and/or partial differential equations. The content, examples, and applications will be appreciated by students, faculty, scientists, and other professionals who use differential equations.

Of course, we must express our appreciate to those who assisted in this project. We would like to express appreciation to our editor, Robert Ross, and our production editor, Vanessa Gerhard, at Academic Press for providing a pleasant environment in which to work. In addition, Waterloo Maple Software, especially Ben Friedman, has been most helpful in providing us with up-to-date information about Maple V. Finally, we thank those close to us, especially Imogene Abell, Lori Braselton, Ada Braselton, and Martha "Mattie" Braselton, for enduring with us the pressures of meeting a deadline and for graciously accepting our demanding work schedules. We certainly could not have completed this task without their care and understanding.

M. L. Abell
J. P. Braselton
Statesboro, Georgia

CHAPTER 1

Introduction to Differential Equations

The purpose of *Differential Equations with Maple V* is twofold. First, we introduce and discuss in a very standard manner all topics typically covered in an undergraduate course in ordinary differential equations as well as some supplementary topics, such as Laplace transforms, Fourier series, and partial differential equations, that are not. Second, we illustrate how Maple is used to enhance the study of differential equations not only by eliminating the computational difficulties but also by overcoming the visual limitations associated with the solutions of differential equations. In each chapter, we first briefly present the material in a manner similar to that most in differential equations texts and then illustrate how Maple can be used to solve typical problems. For example, in Chapter 2, we introduce the topic of first-order equations. We first show how to solve the problems by hand and then show how Maple can be used to perform the same solution procedures. Finally, we illustrate how commands such as `dsolve` can be used to solve some equations directly. On the other hand, in Chapter 3 we discuss some applications of first-order equations. Because we are experienced and understand the methods of solution covered in Chapter 2, we make use of `dsolve` and similar commands to obtain solutions. In doing so, we are able to emphasize the applications themselves as opposed to becoming bogged down in calculations.

The advantages of using Maple in the study of differential equations are numerous, but perhaps the most useful is that of being able to produce the graphics associated with solutions of differential equations. This is particularly beneficial in the discussion of applications because many physical situations are modeled with differential equations. For example, we will see that the motion of a pendulum can be modeled by a differential equation. When we solve the problem of the motion of a pendulum, we use technology to actually watch the pendulum move. The same is true for the motion of a mass attached to the end of a spring as well as many other problems. In having this ability, the study of differential equations becomes much more meaningful as well as interesting.

If you are a beginning Maple V user and especially new to Version 5, the Appendix at the end of the text contains an introduction to Maple V including discussions about entering and evaluating commands, loading miscellaneous library functions and packages, and taking advantage of Maple's extensive help facility, especially those new to Release 5.

Although Chapter 1 is short, the vocabulary introduced will be used throughout the text. Consequently, even though, to a large extent, it may be read quickly, subsequent chapters will take advantage of the terminology and techniques discussed here.

1.1 Definitions and Concepts

We begin our study of differential equations by explaining what a differential equation is.

**Definition
Differential
Equation**

A **differential equation** is an equation that contains the derivative or differentials of one or more dependent variables with respect to one or more independent variables. If the equation contains only ordinary derivatives (of one or more dependent variables) with respect to a single independent

EXAMPLE 1: Determine which of the following are examples of ordinary differential equations:

(a) $\frac{dy}{dx} = \frac{x^2}{y^2 \cos y}$,   (b) $\frac{dy}{dx} + \frac{du}{dx} = u + x^2 y$,   (c) $(y-1)dx + x\cos y\, dy = 1$,   and   (d) $x^2 y'' + xy' + (x^2 - n^2)y = 0$.

SOLUTION: All of the equations are ordinary differential equations. Notice that the equation in part (c) includes differentials.

∎

If the equation contains partial derivatives of one or more dependent variables, then the equation is called a **partial differential equation**.

EXAMPLE 2: Determine which of the following are examples of partial differential equations:

(a) $u \frac{\partial u}{\partial t} = \frac{\partial u}{\partial x}$; (b) $u u_x + u = u_{yy}$; (c) $\frac{\partial^2 u}{\partial x^2} + \frac{\partial^2 u}{\partial y^2} = 0$; (d) $\frac{\partial^2 u}{\partial t^2} = \frac{\partial^2 u}{\partial x^2}$ and (e) $\frac{\partial u}{\partial t} = \frac{\partial^2 u}{\partial x^2}$.

SOLUTION: All of these equations are partial differential equations. In fact, the equations in parts (c), (d), and (e) are well known and called **Laplace's equation**, the **wave equation**, and the **heat equation**, respectively.

■

Generally, given a differential equation, our goal in this course will most often be to construct a solution or a numerical approximation of the solution. The first level of classification, distinguishing ordinary and partial differential equations, was discussed earlier. We extend this classification system with the following definition.

Definition **Order**	The **order** of a differential equation is the order of the highest order derivative appearing in the equation.

EXAMPLE 3: Determine the order of each of the following differential equations:

(a) $\frac{dy}{dx} = \frac{x^2}{y^2 \cos y}$, (b) $u_{xx} + u_{yy} = 0$, (c) $\left(\frac{dy}{dx} \right)^4 = y + x$, and (d) $y^3 + \frac{dy}{dx} = 1$.

SOLUTION: (a) The order of this equation is one because it includes only one first-order derivative, $\frac{dy}{dx}$.

(b) This equation is classified as second order because the highest order derivatives, both u_{xx} representing $\frac{\partial^2 u}{\partial x^2}$, and u_{yy}, representing $\frac{\partial^2 u}{\partial y^2}$, are of order two. Hence, Laplace's equation is a second-order partial differential equation.

(c) This is a first-order equation because the highest order derivative is the first derivative. Raising that derivative to the fourth power does not affect the order of the equation. The expressions

$$\left(\frac{dy}{dx} \right)^4 \quad \text{and} \quad \frac{d^4 y}{dx^4}$$

do not represent the same quantities: $\left(\frac{dy}{dx} \right)^4$ represents the derivative of y with respect to x, $\frac{dy}{dx}$, raised to the fourth power; $\frac{d^4 y}{dx^4}$ represents the fourth derivative of y with respect to x.

(d) Again, we have a first-order equation, because the highest order derivative is the first derivative.

■

The next level of classification is based on the following definition.

Definition **Linear** **Differential** **Equation**	An ordinary differential equation (of order n) is **linear** if it is of the form $$a_n(x)\frac{d^n y}{dx^n} + a_{n-1}(x)\frac{d^{n-1}y}{dx^{n-1}} + \cdots + a_2(x)\frac{d^2 y}{dx^2} + a_1(x)\frac{dy}{dx} + a_0(x)y = f(x),$$ where the functions $a_i(x), i = 0, 1, \ldots, n$, and $f(x)$ are given and $a_n(x)$ is not the zero function.

If the equation does not meet the requirements of this definition, then the equation is said to be **nonlinear**. (A similar classification is followed for partial differential equations. In this case, the coefficients in a linear partial differential equation are functions of the independent variables.)

EXAMPLE 4: Determine which of the following differential equations are linear: (a) $dy/dx = x^3$, (b) $\frac{d^2 u}{dx^2} + u = e^x$, (c) $(y - 1)dx + x\cos y\,dy = 1$, (d) $\frac{d^3 y}{dx^3} + y\frac{dy}{dx} = x$, (e) $\frac{dy}{dx} + x^2 y = x$, (f) $\frac{d^2 x}{dt^2} + \sin x = 0$, (g) $u_{xx} + yu_y = 0$, and (h) $u_{xx} + uu_y = 0$.

SOLUTION: (a) This equation is linear, because the nonlinear term x^3 is the function $f(x)$ of the independent variable in the proceeding general formula.

(b) This equation is also linear. Using u as the dependent variable name does not affect the linearity.

(c) Solving for dy/dx, we have $\frac{dy}{dx} = \frac{1-y}{x\cos y}$. Because the right-hand side of this equation includes a nonlinear function of y, the equation is nonlinear (in y). However, if we solve for dx/dy, we find that

$$\frac{dx}{dy} = \frac{\cos y}{1 - y}x.$$

This equation is linear in the variable x. Notice that we take the dependent variable to be x and the independent variable to be y in this equation.

(d) The coefficient of the term dy/dx is y and, thus, is not a function of x. Hence, this equation is nonlinear.

(e) This equation is linear. The term x^2 is merely the coefficient function.

(f) This equation, known as the **pendulum equation** because it models the motion of a pendulum, is nonlinear because it involves a nonlinear function of x, the dependent variable in this case. (t is the independent variable.) This function is $\sin x$.

(g) This partial differential equation is linear, because the coefficient of u_y is a function of one of the independent variables.

(h) In this case, there is a product of u and one of its derivatives. Therefore, the equation is nonlinear.

■

In the same manner that we consider systems of equations in algebra, we can also consider systems of differential equations. For example, if x and y represent functions of t, we will learn to solve the **system of linear equations**

$$\begin{cases} x' = ax + by \\ y' = cx + dy \end{cases},$$

where a, b, c, and d represent constants and differentiation is with respect to t. We will see that systems of differential equations arise naturally in many physical situations that are modeled with more than one equation and involve more than one dependent variable.

1.2 Solutions of Differential Equations

When faced with a differential equation, our goal is frequently, but not always, to determine solutions to the equation.

Definition
Solution

> A **solution** to the nth-order ordinary differential equation
>
> $$F\left(x, y, y', y,'' \ldots, y^{(n)}\right) = 0$$
>
> on the interval $a < x < b$ is a function $\phi(x)$ that is continuous on $a < x < b$ and has all the derivatives present in the differential equation such that
>
> $$F\left(x, \phi, \phi', \phi'', \ldots, \phi^{(n)}\right) = 0$$
>
> on $a < x < b$.

In later chapters, we will discuss methods for solving differential equations. Here, in order to understand what is meant to be a solution, we either give both the equation and a solution and then verify the solution or use Maple to solve equations directly.

EXAMPLE 1: Verify that the given function is a solution to the corresponding differential equation:
(a) $dy/dx = 3y$, $y(x) = e^{3x}$; (b) $d^2u/dx^2 + 16u = 0$, $u(x) = \cos 4x$; and (c) $y'' + 2y' + y = 0$, $y(x) = xe^{-x}$.

SOLUTION: (a) Differentiating y we have $dy/dx = 3e^{3x}$, so substitution yields

$$\frac{dy}{dx} = 3y$$

$$3e^{3x} = 3e^{3x}.$$

(b) Two derivatives are required in this case: $du/dx = -4\sin 4x$ and $d^2u/dx^2 = -16\cos 4x$. Therefore,

$$\frac{d^2u}{dx^2} + 16u = -16\cos 4x + 16\cos 4x = 0.$$

(c) In this case, we illustrate how to use Maple. (If you are a beginning Maple user, see the **Appendix** for help getting started with Maple.) After defining y,

```
> y:=x->x*exp(-x);
```

$$y := x \rightarrow xe^{(-x)}$$

we use `diff` to compute $y' = e^{-x} - xe^{-x}$, naming the resulting output `dy`.

```
> dy:=diff(y(x),x);
```

$$dy := e^{(-x)} - xe^{(-x)}$$

Similarly, we use `diff` to compute $y'' = -2e^{-x} + xe^{-x}$, naming the resulting output `d2y`.

```
> d2y:=diff(y(x),x$2);
```

$$d2y := -2e^{(-x)} + xe^{(-x)}$$

Finally, we compute $y'' + 2y' + y = 2e^{-x} + 2(e^{-x} - xe^{-x}) + xe^{-x} = 0.$

```
> d2y+2*dy+y(x);
```

$$0$$

We graph this solution with `plot`. Entering

```
> plot(y(x),x=-1..1);
```

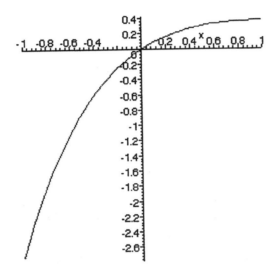

graphs $y(x) = xe^{-x}$ on the interval $[-1,1]$.

∎

   In the previous example, the solution is given as a function $y(x)$ of the independent variable. In these cases, the solution is said to be **explicit**. In solving some differential equations, however, we can only find an equation involving x and y that the solution satisfies. In this case, the solution is said to be **implicit**.

EXAMPLE 2: Verify that the given implicit function satisfies the differential equation.

Function: $2x^2 + y^2 - 2xy + 5x = 0$

Differential equation: $\dfrac{dy}{dx} = \dfrac{2y - 4x - 5}{2y - 2x}$

SOLUTION: We first use implicit differentiation to compute the derivative of $2x^2 + y^2 - 2xy + 5x = 0$:

$$4x + 2y\frac{dy}{dx} - 2x\frac{dy}{dx} - 2y + 5 = 0$$

$$\frac{dy}{dx}(2y - 2x) = 2y - 4x - 5$$

$$\frac{dy}{dx} = \frac{2y - 4x - 5}{2y - 2x}.$$

Hence, the given implicit solution satisfies the differential equation $\frac{dy}{dx} = \frac{2y-4x-5}{2y-2x}$.
We also illustrate how to use Maple to differentiate $2x^2 + y^2 - 2xy + 5x = 0$
implicitly. After clearing all prior definitions of x and y, if any, by entering
$x := 'x' : y := 'y'$: we use D to differentiate $2x^2 + y^2 - 2xy + 5x = 0$, naming the
result step_1.

> **x:='x':y:='y':**

step_1:=D(2*x^2+y^2-2*x*y+5*x=0);

$$step\_1 := 4D(x)x + 2D(y)y - 2D(x)y - 2xD(y) + 5D(x) = 0$$

We then replace each occurrence of D(x) in step_1 by 1 with subs, naming the
result step_2. The symbol D(y) appearing in the result represents $y' = \frac{dy}{dx}$;
step_2 represents the equation $4x + 2y\frac{dy}{dx} - 2x\frac{dy}{dx} - 2y + 5 = 0$.

> **step_2:=subs(D(x)=1,step_1);**

$$step\_2 := 4x + 2D(y)y - 2y - 2xD(y) + 5 = 0$$

Finally, we obtain the derivative by solving step_2 for D(y) with solve.

> **solve(step_2,D(y));**

$$\frac{1}{2}\frac{4x - 2y + 5}{-y + x}$$

Hence, the given implicit solution satisfies the differential equation $\frac{dy}{dx} = \frac{2y-4x-5}{2y-2x}$.
The solution (an ellipse) is graphed using implicitplot, which is contained in
the **plots** package. First, we load the **plots** package by entering

> **with(plots):**

Note that the commands contained in the **plots** package are not displayed
because a colon (:) is included at the end of the command instead of a semicolon
(;). If a semicolon had been included, a list of the commands contained in the
plots package would have been returned.
Then, entering

> **implicitplot(2*x^2+y^2-2*x*y+5*x=0,x=-7..2,y=-7..2);**

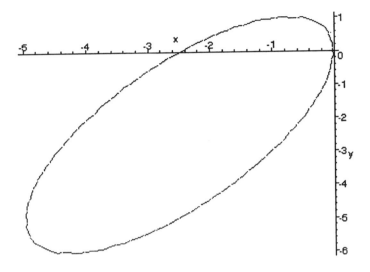

graphs the equation $2x^2 + y^2 - 2xy + 5x = 0$ on the rectangle $[-7,2] \times [-7,2]$.

■

Most differential equations have more than one solution. We illustrate this property in the following examples, where we use the Maple command `dsolve` to solve the indicated equations. Generally, the command

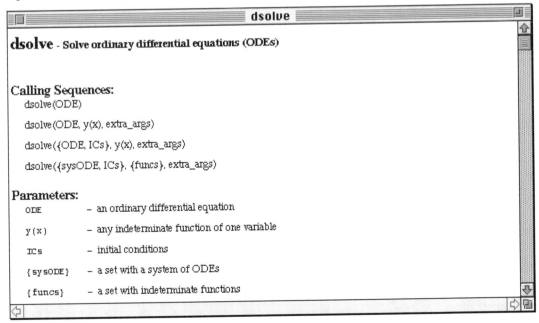

```
dsolve(F(x,y(x),diff(y(x),x),...,diff(y(x),x$n))=0,y(x))
```

attempts to solve the differential equation $F(x, y, y', \ldots, y^{(n)}) = 0$ for y. Detailed help regarding `dsolve` is obtained by entering `?dsolve`, which contains a complete description of the `dsolve` command, a discussion of its various options, and several examples, as illustrated in the previous screen shot.

EXAMPLE 3: Verify that the differential equation $y' = -y \cos x$ has infinitely many solutions.

SOLUTION: We use `dsolve` to solve the first-order linear equation and name the result sol. We interpret the result to mean that if C is any number, a solution to the equation is $y = Ce^{-\sin x}$. Thus, the equation has infinitely many solutions.

```
> sol:=dsolve(diff(y(x),x)=-y(x)*cos(x),y(x));
```

$$sol := y(x) = \_C1e^{(-\sin(x))}$$

We graph several solutions with `plot`.

```
> toplot:={seq(subs(_C1=i,rhs(sol)),i=-5..5)}:
plot(toplot,x=0..4*Pi,color=BLACK);
```

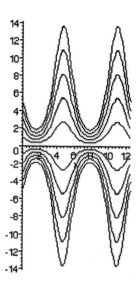

EXAMPLE 4: Verify that $y'' + y = 0$ has infinitely many solutions.

SOLUTION: We use `dsolve` to solve this second-order linear equation and name the resulting solution `sol`. We interpret the result to mean that if c_1 and c_2 are any constants, $y = c_1 \cos x + c_2 \sin x$ is a solution of $y'' + y = 0$.

```
> sol:=dsolve(diff(y(x),x$2)+y(x)=0,y(x));
```

$$sol := y(x) = \_C1 \cos(x) + \_C2 \sin(x)$$

In particular, this result indicates that $y = c \cos x$ is a solution to $y'' + y = 0$ for any value of c (set $c_2 = 0$) and that $y = c \sin x$ is a is a solution to $y'' + y = 0$ for any value of c (set $c_1 = 0$).

Some of the members of the family of solutions are graphed with `plot`. First, we use `seq` to generate a set of 11 functions obtained by replacing c in $y = c \cos x$ by $-2.5, -2, -1.5, \ldots, 1.5, 2$, and 2.5, naming the resulting set `toplot1`, and then a set of 11 functions obtained by replacing c in $y = c \sin x$ by $-2.5, -2, -1.5, \ldots, 1.5, 2$, and 2.5, naming the resulting set `toplot2`.

```
> ivals:=seq(.5*i,i=-5..5):
toplot1:={seq(subs({_C1=i,_C2=0},rhs(sol)),
        i=ivals)}:
toplot2:={seq(subs({_C1=0,_C2=i},rhs(sol)),
        i=ivals)}:
```

Then, the set of functions `toplot1` and `toplot2` are graphed with `plot` for $0 \leq x \leq 4\pi$. Neither graph is displayed as it is generated because we include a colon at the end of each command.

```
> A:=array(1..2):
A[1]:=plot(toplot1,x=0..4*Pi,color=BLACK):
A[2]:=plot(toplot2,x=0..4*Pi,color=BLACK):
```

Instead, we show the graphs side by side using `display`, which is contained in the **plots** package.

```
> with(plots):
display(A);
```

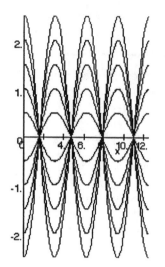 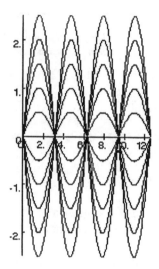

■

1.3 Initial and Boundary Value Problems

In many applications, we are given not only a differential equation to solve but also one or more conditions that must be satisfied by the solution(s) as well. For example, suppose that we want to find an antiderivative of the function $f(x) = 3x^2 - 4x$. Then, we solve the differential equation $dy/dx = 3x^2 - 4x$ by integrating:

$$dy/dx = 3x^2 - 4x \Rightarrow y = \int (3x^2 - 4x)dx \Rightarrow y = x^3 - 2x^2 + C.$$

```
> int(3*x^2-4*x,x);
```

$$x^3 - 2x^2$$

Because the solution involves an arbitrary constant and all solutions to the equation can be obtained from it, we call this a **general solution**. On the other hand, if we want to find a solution that passes through the point (1,4), we must find a solution that satisfies the **auxiliary condition** $y(1) = 4$. Substitution into $y = x^3 - 2x^2 + C$ yields

$$y(1) = (1)^3 - 2(1)^2 + C = 4 \Rightarrow C = 5.$$

Therefore, *the* member of the family of solutions $y(x) = x^3 - 2x^2 + C$ that satisfies $y(1) = 4$ is $y(x) = x^3 - 2x^2 + 5$.

The following commands illustrate how to graph some members of the family of solutions by substituting various values of C into the general solution. We also graph the solution to the problem

$$\begin{cases} dy/dx = 3x^2 - 4x \\ \quad y(1) = 4 \end{cases}.$$

First, we use `seq` to generate a table of functions $x^3 - 2x^2 + C$ for $C = -10, -8, \ldots, 8, 10$, naming the the resulting set of function `toplot`. Note that we use c to represent C to avoid conflict with the built-in symbol C. The set of functions `toplot` is not displayed (for length reasons) because a colon (:) is included at the end of the command. Using `nops`, we see that `toplot` contains 11 functions.

```
> cvals:=seq(2*i,i=-5..5):

toplot:={seq(x^3-2*x^2+c,c=cvals)}:

nops(toplot);
```

$$11$$

To graph the 11 functions contained in `toplot`, we use `plot`.

```
> plot(toplot,x=-2..3,view=[-2..3,-15..15]);

plot(x^3-2*x^2+5,x=-2..3,view=[-2..3,-15..15]);
```

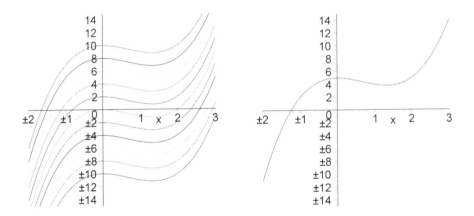

Notice that this first-order equation requires one auxiliary condition to eliminate the unknown coefficient in the general solution. Frequently, the independent variable in a problem is t, which usually represents time. Therefore, we call the auxiliary condition of a first-order equation the **initial condition**, because it indicates the initial value (at $t = t_0$) of the dependent variable. Problems that involve an initial condition are called **initial-value problems**.

EXAMPLE 1: Consider the first-order equation

$$\frac{dv}{dt} = 32 - v,$$

which is solved to determine the velocity at time t, $v(t)$, of an object of mass $m = 1$ subjected to air resistance equivalent to the instantaneous velocity of the object. If the initial velocity of the object is $v(0) = 0$, determine the solution that satisfies this initial condition.

SOLUTION: A general solution to this equation is found to be $v(t) = 32 + ce^{-t}$, where c is a constant, with dsolve.

```
> gensol:=dsolve(diff(v(t),t)=32-v(t),v(t));
```

$$gensol := v(t) = 32 + e^{(-t)}\_C1$$

Substituting into the general solution, we have $v(0) = 32 + c = 0$. Hence, $c = -32$, and the solution to the initial-value problem is $v(t) = 32 - 32e^{-t}$. dsolve can be used to solve this initial-value problem as well.

```
> gensol:=dsolve({diff(v(t),t)=32-v(t),v(0)=0},v(t));
```

$$genol := v(t) = 32 - 32e^{(-t)}$$

■

If dsolve cannot find an exact solution to an initial-value problem or if numerical results are desired, dsolve together with the numeric option

```
dsolve({diff(y(x),x)=f(x,y(x)),y(x0)=y0},y(x),numeric)
```

attempts to find a numerical solution to the initial-value problem $\{y' = f(x,y), y(x_0) = y_0\}$.

EXAMPLE 2: Graph the solution to the initial-value problem $\{y' = \sin x^2, y(0) = 0\}$ on the interval $[0, 10]$. Evaluate $y(5)$.

SOLUTION: In this case, we see that dsolve is able to solve the initial-value problem, although the result is given in terms of the FresnelS function.

```
> exactsol:=dsolve({diff(y(x),x)=sin(x^2),y(0)=0},y(x));
```

$$exactsol := y(x) = \frac{1}{2}\sqrt{2}\sqrt{\pi}; \text{FresnelS}\left(\frac{\sqrt{2}x}{\sqrt{\pi;}}\right)$$

Here is Maple's description of the `FresnelS` function.

```
> ?Fresnels
```

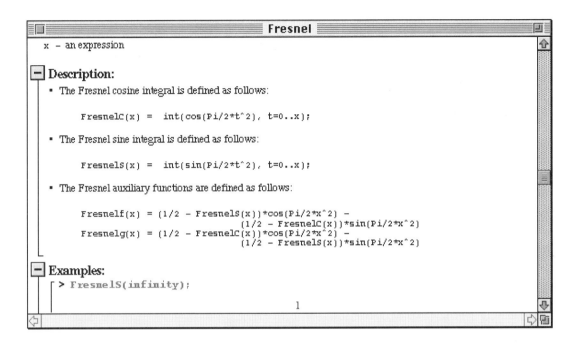

Using `dsolve` together with the `numeric` option, we obtain a numerical solution to the initial-value problem.

```
> numsol:=dsolve({diff(y(x),x)=sin(x^2),y(0)=0},
      y(x),numeric);
```

$$numsol := \mathbf{proc}(rkf45\_x) \dots \mathbf{end}$$

which we graph with `odeplot`. Note that `odeplot` is contained in the **plots** package, so we first load the commands contained in the **plots** package by entering `with(plots)`.

```
> with(plots):
```

Thus, entering

```
> odeplot(numsol,[x,y(x)],0..10);
```

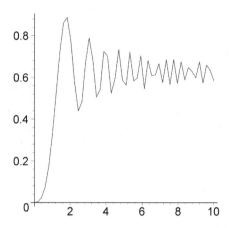

graphs x versus y for $0 \le x \le 10$.
The value of $y(5)$ is found by evaluating `numsol`

```
> numsol(5);
```

$$[x = 5, y(x) = 0.5279173072879902]$$

and indicates that $y(5) \approx 0.5279$.

■

Because first-order equations involve a single auxiliary condition, which is usually referred to as an initial condition, we use the following examples to distinguish between **initial** and **boundary value** problems that involve higher order equations.

EXAMPLE 3: Consider the second-order differential equation $x'' + x = 0$, which models the motion of a mass with $m = 1$ attached to the end of a spring with spring constant $k = 1$, where $x(t)$ represents the displacement of the mass from the equilibrium position $x = 0$ at time t. A general solution of this differential equation is found to be $x(t) = A \cos t + B \sin t$, where A and B are arbitrary constants, with `dsolve`.

```
> sol:=dsolve(diff(x(t),t$2)+x(t)=0,x(t));
```

$$sol := x(t) = \_C1 \sin(t) + \_C2 \cos(t)$$

Because this is a second-order equation, we need two auxiliary conditions to determine the two unknown constants. Suppose that the initial displacement of the mass is $x(0) = 0$ and the initial velocity is $x'(0) = 1$. This is an **initial-value problem** because we have two auxiliary conditions given at the same value of t, namely $t = 0$. Use these initial conditions to determine the solution of this problem.

SOLUTION: Because we need the first derivative of the general solution, we calculate $x'(t) = -A \sin t + B \cos t$. Substitution yields $x(0) = A = 0$ and $x'(0) = B = 1$. Hence, the solution is $x(t) = \sin t$. dsolve can solve this initial-value problem as well.

```
> dsolve({diff(x(t),t$2)+x(t)=0,
      x(0)=0,D(x)(0)=1},x(t));
```

$$x(t) = \sin(t)$$

■

EXAMPLE 4: The shape of a bendable beam of length 1 unit that is subjected to a compressive force at one end is described by the graph of the solution $y(x)$ of the differential equation $\frac{d^2y}{dx^2} + \frac{\pi^2}{4}y = 0$, $0 < x < 1$. If the height of the beam above the x-axis is known at the endpoints $x = 0$ and $x = 1$ then we have a **boundary-value problem**. Use the boundary conditions $y(0) = 0$ and $y(1) = 2$ to find the shape of the beam.

SOLUTION: First, we use dsolve to find a general solution to the equation. The result indicates that a general solution is $y(x) = A \cos(\pi x/2) + B \sin(\pi x/2)$.

```
> dsolve(diff(y(x),x$2)+Pi^2/4*y(x)=0,y(x));
```

$$y(x) = \_C1 \sin\left(\frac{1}{2}\pi x\right) + \_C2 \cos\left(\frac{1}{2}\pi x\right)$$

Applying the condition $y(0) = 0$ to the general solution yields
$$y(0) = A \cos 0 + B \sin 0 = A = 0.$$

Similarly, $y(1) = 2$ indicates that
$$y(1) = B \sin(\pi/2) = B = 2,$$

so the solution to the boundary-value problem is $y(x) = 2\sin(\pi x/2)$, $0 < x < 1$. `dsolve` is also able to solve this boundary-value problem.

```
> dsolve({diff(y(x),x$2)+Pi^2/4*y(x)=0,
      y(0)=0,y(1)=2},y(x));
```

$$y(x) = 2\sin\left(\frac{1}{2}\pi x\right)$$

This function that describes the shape of the beam is graphed with `plot`.

```
> plot(2*sin(1/2*Pi*x),x=0..1);
```

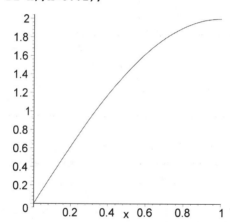

We will see that it is often impossible to find exact solutions of higher order nonlinear initial-value problems. In those cases, we can often use `dsolve` together with the numeric option to generate an accurate approximation of the solution or commands, such as `DEplot`, contained in the **DEtools** package to graph solutions.

```
>   ?DEtools
```

EXAMPLE 5: Rayleigh's equation is the nonlinear equation

$$x'' + \left(\frac{1}{3}(x')^2 - 1\right)x' + x = 0$$

and arises in the study of the motion of a violin string. Graph the solution to Rayleigh's equation on the interval [0,15] if (a) $x(0) = 1, x'(0) = 0$; (b) $x(0) = 0.1, x'(0) = 0$; and (c) $x(0) = 0, x'(0) = 1.9$.

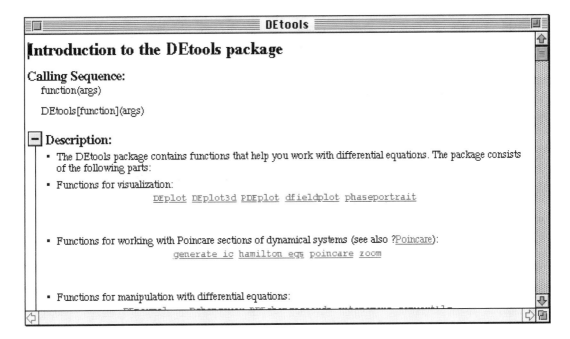

SOLUTION: After loading the **DEtools** package, we use DEplot to graph the solution to each initial-value problem.

```
> with (DEtools):
```

> [*DEnormal, DEplot, DEplot3d, DEplot_polygon, DFactor, Dchangevar, GCRD, LCLM, PDEchangecoords, RiemannPsols, abelsol, adjoint, autonomous, bernoullisol, buildsol, buildsym, canoni, chinisol, clairautsol, constcoeffsols, convertAlg, convertsys, dalembertsol, de2diffop, dfieldplot, diffop2de, eigenring, endomorphism_charpoly, equinv, eta_k, eulersols, exactsol, expsols, exterior_power, formal_sol, gen_exp, generate_ic, genhomosol, hamilton_eqs, indicialeq, infegen, integrate_sols, intfactor, kovacicsols, leftdivision, liesol, line_int, linearsol, matrixDE, matrix_riccati, moser_reduce, mult, newton_polygon, odeadvisor, odepde, parametricsol, phaseportrait, poincare, polysols, ratsols, reduceOrder, regular_parts, regularsp, riccati_system, riccatisol, rightdivision, separablesol, super_reduce, symgen, symmetric_power, symmetric_product, symtest, transinv, translate, untranslate, varparam, zoom*]

Thus, entering

```
> DEplot(diff(x(t),t$2)+
      (1/3*diff(x(t),t)^2-1)*diff(x(t),t)+x(t)=0,
          x(t),t=0..15,[[x(0)=1,D(x)(0)=0]],
               stepsize=0.1,linecolor=BLACK);
```

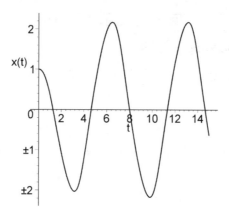

graphs the solution that satisfies $x(0) = 1, x'(0)$ for $0 \leq t \leq 15$, and entering

```
> DEplot(diff(x(t),t$2)+
        (1/3*diff(x(t),t)^2-1)*diff(x(t),t)+x(t)=0,
            x(t),t=0..15,
                [[x(0)=0.1,D(x)(0)=0],[x(0)=0,D(x)(0)=1.9]],
                stepsize=0.1,linecolor=[BLACK,GRAY]);
```

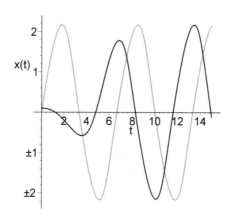

graphs the solutions that satisfy $x(0) = 0.1, x'(0) = 0$ (in black) and $x(0) = 0, x'(0) = 1.9$ (in gray).

■

1.4 Direction Fields

The geometric interpretation of solutions to first-order differential equations of the form $dy/dx = f(x, y)$ is important to the basic understanding of problems of this type. Suppose that a solution to this equation is a function $y = \psi(x)$, so a solution is the graph of the function ψ. Therefore, if (x,y) is a point on this graph, the slope of the tangent line is given by $f(x,y)$. A set of short line segments representing the tangent lines can be constructed for a large number of points. This collection of line segments is known as the **direction field** of the differential equation and provides a great deal of information concerning the behavior of the family of solutions. This is due to the fact that by determining the slope of the tangent line for a large number of points in the plane, the shape of the graphs of the solutions can be seen without actually having a formula for them. The direction field for a differential equation provides a geometric interpretation of the behavior of the solutions of the equation. Throughout this text, we will frequently display graphs of various solutions to a differential equation along with a graph of the direction field. Direction fields are generated with `DEplot` and `dfieldplot`, which are contained in the **DEtools** package. After loading the **DEtools** package, the commands

```
DEplot(diff(y(x),x)=f(x,y(x)),y(x),x=x0..x1,y=y0..y1)
```

and

```
dfieldplot(diff(y(x),x)=f(x,y(x)),y(x),x=x0..x1,y=y0..y1)
```

graph the direction field associated with $dy/dx = f(x, y)$ for $x_0 \leq x \leq x_1$ and $y_0 \leq y \leq y_1$.

EXAMPLE 1: Graph the direction field associated with the differential equation $dy/dx = e^{-x} - 2y$.

SOLUTION: Entering

```
> with(DEtools):

with(plots):
```

first loads the **DEtools** and **plots** packages and then graphs the direction field associated with the equation $dy/dx = e^{-x} - 2y$ for $-1/2 \leq x \leq 1$ and $-3/4 \leq y \leq 3/4$, naming the resulting graphics object p1. p1 is then displayed using the `display` command, which is contained in the **plots** package.

```
> p1:=DEplot(diff(y(x),x)=exp(-x)-2*y(x),
             y(x),x=-1/2..1,y=-3/4..3/4):

display(p1);
```

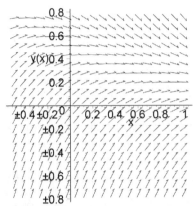

A general solution of this first-order linear equation is found to be $y = e^{-x} + Ce^{-2x}$ with dsolve. (Generally, Maple is able to find a general solution of first-order linear equations such as this with dsolve or the linearsol command, which is contained in the **DEtools** package.)

```
> y:='y':
diffyq:=diff(y(x),x)=exp(-x)-2*y(x):
gensol:=dsolve(diffyq,y(x));
```

$$gensol := y(x) = \mathbf{e}^{(-x)} + \mathbf{e}^{(-2x)}\_C1$$

At this point, we name $y(x)$ the result obtained in gensol with assign and then use seq and subs to define the set of seven functions, to_plot, obtained by replacing_C1 in $y(x)$ by i for $i = -3, -2, -1, 0, 1, 2$, and 3. This set of seven functions is then graphed on the interval $[-1/2,1]$ with plot and the resulting graphics object is named p2. p1 and p2 are then shown together with display.

```
> assign(gensol):
toplot:={seq(subs(_C1=i,y(x)),i=-3..3)}:
p2:=plot(toplot,x=-1/2..2,-1..1,color=BLACK):
display({p1,p2},view=[-1/2..1,-3/4..3/4]);
```

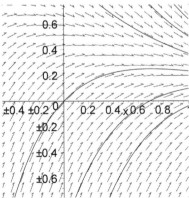

■

Maple allows us to graph solutions of equations and associated direction fields that would be nearly impossible by traditional methods, as shown in the following example.

EXAMPLE 2: Graph the direction field associated with the differential equation $\frac{dy}{dx} = \frac{\cos y - y \cos x}{x \sin y + \sin x - 1}$.

SOLUTION: We begin by finding a general solution of $\frac{dy}{dx} = \frac{\cos y - y \cos x}{x \sin y + \sin x - 1}$ with dsolve, naming the resulting output gensol.

```
> y:='y':
      gensol:=dsolve(diff(y(x),x)=(cos(y(x))-y(x)*cos(x))/
            (x*sin(y(x))+sin(x)-1),y(x));
```

$$gensol := \_C1 + y(x) \sin(x) - x \cos(y(x)) - y(x) = 0$$

Thus, a general solution of $\frac{dy}{dx} = \frac{\cos y - y \cos x}{x \sin y + \sin x - 1}$ is $y \sin x - x \cos y - y = C$. Next we graph the solution for various values of C and the direction field associated with $\frac{dy}{dx} = \frac{\cos y - y \cos x}{x \sin y + \sin x - 1}$. First, we note that the graph of $y \sin x - x \cos y - y = C$ for various values of C is the same as the graph of the level curves of $f(x, y) = y \sin x - x \cos y - y$. We define tograph to be $y \sin x - x \cos y - y$ by using lhs to extract the left-hand side of the equation gensol and then using subs to replace each occurrence of $y(x)$ by y and_C1 by 0.

```
> tograph:=subs({y(x)=y,_C1=0},lhs(gensol));
```

$$tograph := y \sin(x) - \cos(y)x - y$$

Then, after loading the **plots** package, we use contourplot, which is contained in the **plots** package, to graph several level curves of tograph on the rectangle $[0, 4\pi] \times [0, 4\pi]$.

```
> with(plots):
cplot:=contourplot(tograph,x=0..4*Pi,y=0..4*Pi,
      grid=[40,40],axes=NORMAL,color=BLACK):
display(cplot);
```

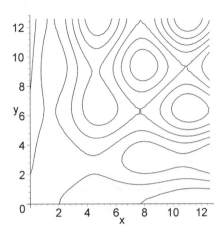

To graph the direction field associated with the equation, we use `dfieldplot` in the same manner as we used `DEplot` in the previous example. We name the direction field `dirfield` and display `cplot` and `dirfield` together with `display`.

```
> with(DEtools):
dirfield:=dfieldplot(diff(y(x),x)=(cos(y(x))-y(x)*cos(x))/
       (x*sin(y(x))+sin(x)-1),y(x),x=0..4*Pi,y=0..4*Pi):
display({cplot,dirfield});
```

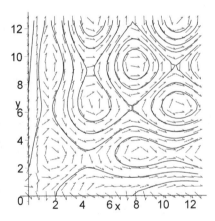

■

Maple is particularly useful in graphing the direction field associated with a system of equations. After the **DEtools** package has been loaded by entering `with(DEtools)`, the commands

```
DEplot([diff(x(t),t)=f(x(t),y(t)),
       diff(y(t),t)=g(x(t),y(t))],
   [x(t),y(t)],t=a..b,x=x0..x1,y=y0..y1)
```

or

```
dfieldplot([diff(x(t),t)=f(x(t),y(t)),
           diff(y(t),t)=g(x(t),y(t))],
    [x(t),y(t)],t=a..b,x=x0..x1,y=y0..y1)
```

graph the direction field associated with the system $\begin{cases} x' = f(x,y) \\ y' = g(x,y) \end{cases}$ for $x_0 \leq x \leq x_1$ and $y_0 \leq y \leq y_1$. If these commands are entered in the form

```
DEplot([diff(x(t),t)=f(x(t),y(t)),
       diff(y(t),t)=g(x(t),y(t))],
    [x(t),y(t)],t=a..b,inics,x=x0..x1,y=y0..y1)
```

or

```
dfieldplot([diff(x(t),t)=f(x(t),y(t)),
           diff(y(t),t)=g(x(t),y(t))],
    [x(t),y(t)],t=a..b,inics,x=x0..x1,y=y0..y1)
```

where `inics` is a list of initial conditions of the form

```
[[x(t0)=x0,y(t0)=y0],...,[x(tn)=xn,y(tn)=yn]]
```

the solutions that satisfy the conditions $x(t_0) = x_0, y(t_0) = y_0, \ldots, x(t_n) = x_n, y(t_n) = y_n$ for $a \leq t \leq b$ are graphed with the direction field, unless the option `arrows = NONE` is included in the `DEplot` or `dfieldplot` command, in which case the direction field is not displayed.

EXAMPLE 3 (Competing Species): Under certain assumptions the system of equations

$$\begin{cases} \frac{dx}{dt} = x(a - b_1 x - b_2 y) \\ \frac{dy}{dt} = y(c - d_1 x - d_2 y) \end{cases},$$

where a, b_1, b_2, c, d_1, and d_2 represent positive constants, can be used to model the population of two species, represented by $x(t)$ and $y(t)$, competing for a common food supply. Graph the direction field associated with the system if (a) $a = 1$, $b_1 = 2$, $b_2 = 1$, $c = 1$, $d_1 = 0.75$, and $d_2 = 2$; and (b) $a = 1$, $b_1 = 1$, $b_2 = 1$, $c = 0.67$, $d_1 = 0.75$, and $d_2 = 1$.

SOLUTION: After identifying $f(x, y) = x(a - b_1 x - b_2 y)$ and $g(x, y) = y(c - d_1 x - d_2 y)$, we define f and g.

```
> f:=(x,y)->x*(a-b[1]*x-b[2]*y);
g:=(x,y)->y*(c-d[1]*x-d[2]*y);
```

$$f := (x, y) \rightarrow x(a - b_1 x - b_2 y)$$

$$g := (x, y) \rightarrow y(c - d_1 x - d_2 y)$$

Then, for (a) we define $a = 1$, $b_1 = 2$, $b_2 = 1$, $c = 1$, $d_1 = 0.75$, and $d_2 = 2$; load the **DEtools** package; and graph the direction field associated with the system for $0 \leq x \leq 1$ and $0 \leq y \leq 1$ with DEplot. (Remember that if you have previously loaded the **DEtools** package during your current Maple session, you do not need to reload the **DEtools** package.)

```
> a:=1:b[1]:=2:b[2]:=1:c:=1:d[1]:=0.75:d[2]:=2:
> with(DEtools):
DEplot([diff(x(t),t)=f(x(t),y(t)),
        diff(y(t),t)=g(x(t),y(t))],
        [x(t),y(t)],t=0..10,x=0..1,y=0..1);
```

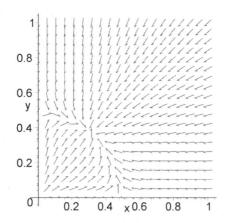

In this case, we see that both species appear to approach some equilibrium population. In fact, later we will see that this equilibrium population is obtained by solving the system of equations $\begin{cases} a - b_1 x - b_2 y = 0 \\ c - d_1 x - d_2 y = 0 \end{cases}$ for x and y.

For (b), we redefine $a = 1, b_1 = 1, b_2 = 1, c = 0.67, d_1 = 0.75$, and $d_2 = 1$ and use the `dfieldplot` command in the same way that we used the `DEplot` command in (a).

```
> a:=1: b[1]:=1: b[2]:=1:c:=0.67:d[1]:=0.75:d[2]:=1:
> dfieldplot([diff(x(t),t)=f(x(t),y(t)),
        diff(y(t),t)=g(x(t),y(t))],
            [x(t),y(t)],t=0..10,x=0..1,y=0..1);
```

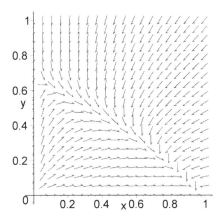

In this case, we see that it appears as though the species with population given by $y(t)$ eventually dies out while the species with population given by $x(t)$ eventually approaches some equilibrium population. Later, we will see that this is true and the equilibrium population of the species with population given by $x(t)$ will be found by computing the limit as $t \to +\infty$ of the solution to the differential equation $dx/dt = ax - b_1 x^2$.

■

Often, we can generate the direction field of a higher order equation by rewriting it as a system of first-order equations.

EXAMPLE 4 (Rayleigh's Equation): Write Rayleigh's equation

$$x'' + \left(\frac{1}{3}(x')^2 - 1\right)x' + x = 0$$

as a system of two first-order equations. Graph the direction field associated with the resulting system on the rectangle $[-4, 4] \times [-4, 4]$.

SOLUTION: We write Rayleigh's equation as a system by letting $y = x'$. Then,

$$y' = x'' = -\left(\frac{1}{3}(x')^2 - 1\right)x' - x = -\left(\frac{1}{3}y^2 - 1\right)y - x$$

so Rayleigh's equation is equivalent to the system

$$\begin{cases} x' = y \\ y' = -\left(\frac{1}{3}y^2 - 1\right)y - x \end{cases}.$$

The direction field associated with this system is then graphed with `DEplot`.

```
> with(DEtools):
DEplot([diff(x(t),t)=y(t)),
       diff(y(t),t)=-(1/3*y(t)^2-1)*y(t)-x(t)],
        [x(t),y(t)],t=0..15,x=-4..4,y=-4..4);
```

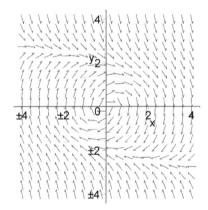

In the direction field, we see that solutions appear to tend to a closed curve, C. We can accurately approximate C. We use `DEplot` to graph the solutions to the equation if (a) $x(0) = 1, y(0) = 0$; (b) $x(0) = 0.1, y(0) = 0$; (c) $x(0) = 0, y(0) = 1.9$; and (d) $x(0) = -4, y(0) = 4$ for $0 \le t \le 15$. We see that the graph of solution (c) corresponds to C; the graphs of the other solutions all tend to C.

```
> DEplot([diff(x(t),t)=y(t),
       diff(y(t),t)=-(1/3*y(t)^2-1)*y(t)-x(t)],
        [x(t),y(t)],t=0..15,
        [[x(0)=1,y(0)=0],[x(0)=0.1,y(0)=0],
```

```
[x(0)=0,y(0)=1.9],[x(0)=-4,y(0)=4]],
x=-4..4,y=-4..4,linecolor=BLACK,
stepsize=0.1);
```

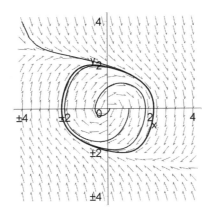

∎

CHAPTER 2

First-Order Ordinary Differential Equations

We will devote a considerable amount of time in this text to developing explicit, implicit, numerical, and graphical solutions of some differential equations. In this chapter we introduce frequently encountered forms of first-order ordinary differential equations and methods for constructing explicit, numerical and graphical solutions of them. Several of the equations along with the methods of solution discussed here will be used in other chapters of the text.

2.1 Theory of First-Order Equations: A Brief Discussion

In order to understand the types of first-order initial-value problems that have a unique solution, the following theorem is stated.[1]

---

[1] C Corduneanu, *Principles of Differential and Integral Equations*, Chelsea Publishing Co., New York (1977), pp. 19–24 or T. Apostol, *Mathematical Analysis* Second Edition, Addison-Wesley, Reading, MA (1974), p. 181.

Theorem

**Existence and
Uniqueness**

Consider the initial-value problem

$$\begin{cases} dy/dx = f(x,y) \\ \quad y(x_0) = y_0 \end{cases}.$$

If f and $\partial f/\partial y$ are continuous functions on the rectangular region R, $R = \{(x,y)|a<x<b, c<y<d\}$, containing the point (x_0, y_0), there exists an interval $|x - x_0|<h$ centered at x_0 on which there exists one and only one solution to the differential equation that satisfies the initial condition.

Frequently, we can use the command

```
dsolve({diff(y(x),x) = f(x,y(x)), y(x0) = y0},y(x))
```

to solve the initial-value problem $\{dy/dx = f(x,y), y(x_0) = y_0\}$;

```
dsolve({diff(y(x),x) = f(x,y(x))},y(x))
```

attempts to find a general solution of $dy/dx = f(x,y)$. Detailed help regarding dsolve is obtained by entering ?dsolve, which contains a complete description of the dsolve command, a discussion of its various options, and several examples. (See Section 1.2.)

EXAMPLE 1: Solve the initial-value problem

$$\begin{cases} dy/dx = x/y \\ \quad y(0) = 0 \end{cases}$$

Does this result contradict the existence and uniqueness theorem?

SOLUTION: This equation is solved with dsolve to determine the two solutions $y = \pm\sqrt{x^2 + C}$.

```
> exactsol:=dsolve(diff(y(x),x)=x/y(x),y(x));
```

$$exactsol := y(x) = \sqrt{x^2 + \_C1}, y(x) = -\sqrt{x^2 + \_C1}$$

When we include the implicit option in the dsolve command, Maple returns the family of solutions $y^2 - x^2 = C$.

```
> implicitsol:=dsolve(diff(y(x),x)=x/y(x),y(x),
      implicit);
```

$$implicitsol := y(x)^2 \quad - x^2 - \_C1 = 0$$

We note that the graph of $y^2 - x^2 = C$ for various values of C is the same as the graph of the level curves of $f(x, y) = y^2 - x^2$. Several members of this family are graphed with `contourplot`.

```
> tograph:=subs({y(x)=y,_C1=0},lhs(implicitsol));
```

$$tograph := y^2 - x^2$$

```
> with(plots):
contourplot(tograph,x=-1..1,y=-1..1,color=BLACK);
```

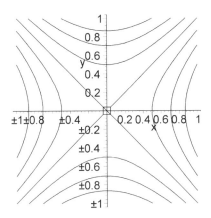

Application of the initial condition yields $0^2 - 0^2 = C$, so $C = 0$. Therefore, solutions that pass through (0,0) satisfy $y^2 - x^2 = 0$, so there are two solutions, $y = x$ and $y = -x$, that satisfy the differential equation and the initial condition. Although more than one solution satisfies this initial-value problem, the existence and uniqueness theorem is *not* contradicted because the function $f(x, y) = x/y$ is not continuous at the point (0,0): the requirements of the theorem are not met.

■

EXAMPLE 2: Verify that the initial-value problem $dy/dx = y, y(0) = 1$ has a unique solution.

SOLUTION: Notice that in this case, $f(x, y) = y$, $x_0 = 0$, and $y_0 = 1$. Hence, both f and $\partial f/\partial y$ are continuous on all rectangular regions containing the point $(x_0, y_0) = (0, 1)$. Therefore by the existence and uniqueness theorem, there exists a unique solution to the differential equation that satisfies the initial condition $y(0) = 1$.

We can verify this by solving the initial-value problem. The unique solution is $y = e^x$, which is computed with dsolve and then graphed with plot. Notice that the graph passes through the point (0,1), as required by the initial condition.

```
> sol:=dsolve({diff(y(x),x)=y(x),y(0)=1},y(x));
```

$$sol := y(x) = e^x$$

```
> assign(sol):
plot(y(x),x=-1..1);
```

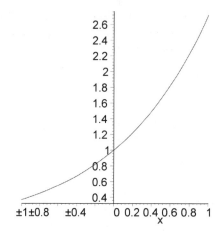

■

EXAMPLE 3: Show that the initial-value problem

$$\begin{cases} x\dfrac{dy}{dy} - y = x^2 \cos x \\ \qquad y(0) = 0 \end{cases}$$

has infinitely many solutions.

SOLUTION: Writing $x\dfrac{dy}{dx} - y = x^2 \cos x$ in the form $dy/dx = f(x,y)$ results in

$$\frac{dy}{dx} = \frac{x^2 \cos x + y}{x}$$

and because $\frac{x^2 \cos x + y}{x}$ is not continuous on an interval containing $x = 0$, the existence and uniqueness theorem does *not* guarantee the existence or uniqueness of a solution. In fact, using `dsolve`

```
> y:='y':
sol:=dsolve(x*diff(y(x),x)-y(x)=x^2*cos(x),y(x));
```

$$sol := \mathrm{y}(x) = x \sin(x) + x\_C1$$

we see that a general solution of the equation is $y = x \sin x + Cx$ and for every value of C, $y(0) = 0$.

We confirm this graphically by graphing several solutions. First, we use `assign` to name $y(x)$ the result obtained in sol and then use `seq` to define `toplot` to be a set of functions obtained by replacing the arbitrary constant in $y(x)$ by $-4, -3, \ldots, 3, 4$.

```
> assign(sol):
toplot:=seq(subs(_C1=i,y(x)),i=-4..4);
```

$$toplot := x \sin(x) - 4x, x \sin(x) - 3x, x \sin(x) - 2x, x \sin(x) - x,$$
$$x \sin(x), x \sin(x) + x, x \sin(x) + 2x, x \sin(x) + 3x, x \sin(x) + 4x$$

These functions are then graphed with `plot`.

```
> plot({toplot},x=-10..10,color=BLACK,
        view=[-10..10,-10..10]);
```

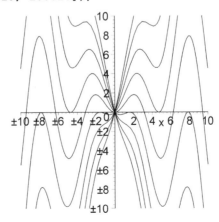

2.2 Separation of Variables

Definition **Separable Differential** **Equation**	A differential equation that can be written in the form $g(y)y' = f(x)$ or $g(y)dy = f(x)dx$ is called a **separable differential equation**.

Separable differential equations are solved by collecting all the terms involving y on one side of the equation, collecting all the terms involving x on the other side of the equation, and integrating:

$$g(y)dy = f(x)dx \Rightarrow \int g(y)dy = \int f(x)dx + C,$$

where C is a constant.

Often, separable equations can be solved with `dsolve` or with the `separablesol` command, which is contained in the **DEtools** package.

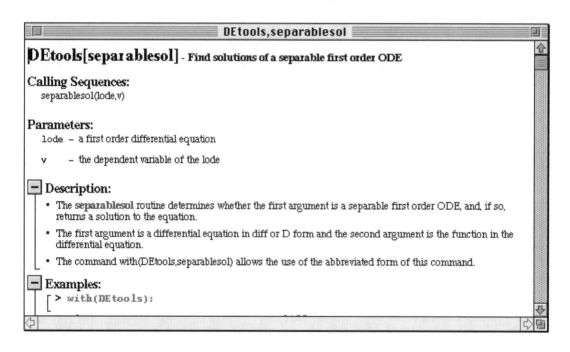

DEtools,separablesol

$\mathbb{D}$Etools[separablesol] - Find solutions of a separable first order ODE

Calling Sequences:
  separablesol(lode,v)

Parameters:
  lode – a first order differential equation

  v   – the dependent variable of the lode

Description:
- The **separablesol** routine determines whether the first argument is a separable first order ODE, and, if so, returns a solution to the equation.
- The first argument is a differential equation in diff or D form and the second argument is the function in the differential equation.
- The command with(DEtools,separablesol) allows the use of the abbreviated form of this command.

Examples:
  `> with(DEtools):`

EXAMPLE 1: Show that the equation

$$\frac{dy}{dx} = \frac{2y^{1/2} - 2y}{x}$$

is separable, and solve by separation of variables.

SOLUTION: The equation $\frac{dy}{dx} = \frac{2y^{1/2} - 2y}{x}$ is separable because it can be written in the form

$$\frac{dy}{2y^{1/2} - 2y} = \frac{dx}{x}.$$

To solve the equation, we integrate both sides and simplify. Observe that we can write this equation as

$$\int \frac{1}{2y^{1/2}} \frac{dy}{(1 - y^{1/2})} = \int \frac{dx}{x} + C_1.$$

To evaluate the integral on the left-hand side, let $u = (1 - y^{1/2})$ so $du = \frac{-dy}{2y^{1/2}}$. We then obtain

$$\int \frac{-du}{u} = \int \frac{dx}{x} + C_1$$

so that $-\ln|u| = \ln|x| + C_1$. Recall that $-\ln|u| = \ln\frac{1}{|u|}$, so we have

$$\ln\frac{1}{|u|} = \ln|x| + C_1.$$

Using Maple, we take advantage of the `changevar` command that is contained in the **student** package to perform the substitution.

```
> y:='y':
with(student):
step1:=changevar(1-sqrt(y)=u,
        Int(1/(2*y^(1/2)*(1-y^(1/2))),y), u);
```

$$step1 := \int -\frac{1}{u} du$$

The integral is then evaluated with `value`

```
> step2:=value(step1);
```

$$step2 := -\ln(u)$$

and we resubstitute $u = 1 - y^{1/2}$ into the result with `subs`.

```
> step3:=subs(u=1-sqrt(y),step2);
```

$$step3 := -\ln(1 - \sqrt{y})$$

The integral on the right-hand side of the equation is computed in the same way.

```
> step4:=int(1/x,x);
```

$$step4 := \ln(x)$$

Simplification yields

$$\frac{1}{|u|} = e^{\ln|x|+C_1} = C|x|$$

where $C = e^{C_1}$. Resubstituting, we find that $\frac{1}{|1-y^{1/2}|} = C|x|$ or

$$x = \pm \frac{1}{C(1 - y^{1/2})}$$

is a general solution of the equation $\frac{dy}{dx} = \frac{2y^{1/2} - 2y}{x}$. We obtain equivalent results with Maple by solving for either x or y.

```
> simplify(solve(step3=step4+c,x));
simplify(solve(step3=step4+c,y));
```

$$-\frac{e^{(-c)}}{-1 + \sqrt{y}}$$

$$2$$

$$\frac{(-1 + xe^c)^2 e^{(-2c)}}{x^2}$$

We obtain an equivalent result with dsolve. Entering

```
> gensol:=dsolve(diff(y(x),x)=(2*sqrt(y(x))-2*y(x))/x,
       y(x));
```

$$gensol := y(x) = \frac{(x + \_C1)^2}{x^2}$$

finds a general solution of the equation that is equivalent to the one we obtained by hand and names the result gensol. Alternatively, we can use separablesol, which is contained in the **DEtools** package, to find a slightly different form of the general solution.

```
> with(DEtools):
```

```
separablesol(diff(y(x),x)=(2*sqrt(y(x))-2*y(x))/x,

    y(x));
```

$$\{-\frac{1}{2}\ln(y(x)-1)+ \operatorname{arctanh}(\sqrt{y(x)}) - \ln(x) = \_C_1\}$$

To graph the solution for various values of _C1, which represents the arbitrary constant in the formula for the solution, we use seq, subs, and rhs to generate a set of functions obtained by replacing _C1 in the formula for the solution by i for each value of i in ivals = −2,?−1.75, ... , 1.75, and 2, naming the resulting set of functions toplot.

```
> ivals:=seq(-2+0.25*i,i=0..16):
```
```
toplot:={seq(subs(_C1=i,rhs(gensol)),i=ivals)}:
```

We then graph the set of functions toplot with plot.

```
> plot(toplot,x=-2..2,view=[-2..2,0..4],

    color=BLACK);
```

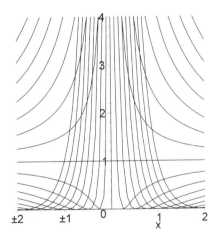

■

An initial-value problem involving a separable equation is solved through the following steps.

1. Find a general solution of the differential equation using separation of variables.
2. Use the initial condition to determine the unknown constant in the general solution.

EXAMPLE 2 Solve (a) $y \cos x\, dx - (1 + y^2)dy = 0, y(0) = 1$ and (b) the initial-value problem $y \cos x\, dx - (1 + y^2)dy = 0, y(0) = 1$.

SOLUTION: (a) Note that this equation can be rewritten as $\frac{dy}{dx} = \frac{y \cos x}{1+y^2}$. We first use dsolve to solve the solution. We are careful to include the argument, x, each time we type the dependent variable $y = y(x)$.

```
> sol1:=dsolve(diff(y(x),x)=y(x)*cos(x)/(1+y(x)^2),y(x));
```

$$sol1 := y(x) = -I\sqrt{-\text{LambertW}(e^{(2\sin(x)+2\_C1)})}$$

(Note that entering dsolve(D(y)(x)=y(x)*cos(x)/(1 + y(x)^2),y(x)) produces the same result.) In this case, we see that dsolve is able to solve the nonlinear equation, although the result contains the LambertW function. Given z, the Lambert W function returns the value of w that satisfies $z = we^w$. If we include the implicit option in the dsolve command, a more familiar form of the solution is found.

```
> sol2:=dsolve(D(y)(x)=y(x)*cos(x)/(1+y(x)^2),y(x),
       implicit);
```

$$sol2 := \sin(x) - \frac{1}{2}y(x)^2 - \ln(y(x)) + \_C1 = 0$$

We can also use Maple to implement the steps necessary to solve the equation by hand. We see that it is separable using odeadvisor

```
> odeadvisor(D(y)(x)=y(x)*cos(x)/(1+y(x)^2));
```

$$[\_separable]$$

and rewrite the equation as $\cos x\, dx = \frac{1+y^2}{y}dy$. To solve the equation, we must integrate both the left- and right-hand sides, which we do with int, naming the resulting output LHS and RHS, respectively.

```
> LHS:=int(cos(x),x);
RHS:=int((1+y^2)/y,y);
```

$$LHS := \sin(x)$$

$$RHS := \frac{1}{2}y^2 + \ln(y)$$

Alternatively, the separablesol command that is contained in the DEtools package, which is illustrated next, can be used to determine whether an equation is separable and, if so, solve it.

```
> with(DEtools):

sol2:=separablesol(
        D(y)(x)=y(x)*cos(x)/(1+y(x)^2),y(x));
```

$$sol2 := \left\{ \frac{1}{2}y(x)^2 + \ln(y(x)) - \sin(x) = \_C_1 \right\}$$

Therefore, a general solution to the equation is $\sin x = \ln|y| + \frac{1}{2}y^2 + C$. We now use contourplot to graph $\sin x = \ln|y| + \frac{1}{2}y^2 + C$ for various values of C by observing that the level curves of $\sin x - \ln|y| + \frac{1}{2}y^2$ correspond to the graph of $\sin x = \ln|y| + \frac{1}{2}y^2 + C$ for various values of C. First, we replace each occurrence of $y(x)$ in $\frac{1}{2}y(x)^2 + \ln(y(x)) - \sin(x)$ with y

```
toplot:=subs(y(x)=y,lhs(sol2[1]));
```

$$toplot := \frac{1}{2}y^2 + \ln(y) - \sin(x)$$

and then use contourplot to generate the graph.

```
> with(plots):

contourplot(toplot,x=0..10,y=0..10,
        color=BLACK,contours=10,grid=[60,60]);
```

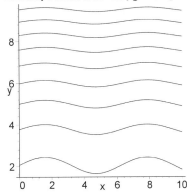

By substituting $y(0) = 1$ into this equation, we find that $C = 1/2$, so the implicit solution is given by $\sin x + \frac{1}{2} = \ln|y| + \frac{1}{2}y^2$.

```
> eval(subs({x=0,y=1},toplot));
```

$$\frac{1}{2}$$

■

APPLICATION

Kidney Dialysis

The primary purpose of the kidney is to remove waste products, such as urea, creatinine, and excess fluid, from blood. When kidneys are not working properly, wastes accumulate in the blood; when toxic levels are reached, death is certain. The leading causes of chronic kidney failure in the United States are hypertension (high blood pressure) and diabetes mellitus. In fact, one-quarter of all patients requiring **kidney dialysis** have diabetes. Fortunately, kidney dialysis removes waste products from the blood of patients with improperly working kidneys. During the hemodialysis process, the patient's blood is pumped through a **dialyzer**, usually at a rate of 1 to 3 deciliters per minute. The patient's blood is separated from the "cleaning fluid" by a semipermeable membrane, which permits wastes (but not blood cells) to diffuse to the cleaning fluid; the cleaning fluid contains some substances beneficial to the body, which diffuse to the blood. The cleaning fluid, called the **dialysate**, is flowing in the *opposite* direction to the blood, usually at a rate of 2 to 6 deciliters per minute. Waste products from the blood diffuse to the dialysate through the membrane at a rate proportional to the difference in concentration of the waste products in the blood and dialysate. If we let $u(x)$ represent the concentration of wastes in blood; $v(x)$ represent the concentration of wastes in the dialysate, where x is the distance along the dialyzer; Q_D represent the flow rate of the dialysate through the machine; and Q_B represent the flow rate of the blood through the machine, then

$$\begin{cases} Q_B u' = -k(u - v) \\ -Q_D v' = k(u - v) \end{cases},$$

where k is the proportionality constant.

If we let L denote the length of the dialyzer and the initial concentration of wastes in the blood is $u(0) = u_0$ while the initial concentration of wastes in the dialysate is $v(L) = 0$, then we must solve the initial-value problem

$$\begin{cases} Q_B u' = -k(u - v) \\ -Q_D v' = k(u - v) \\ u(0) = u_0, v(L) = 0 \end{cases}.$$

Solving the first equation for u' and the second equation for $-v'$, we obtain the equivalent system

$$\begin{cases} u' = -\frac{k}{Q_B}(u - v) \\ -v' = \frac{k}{Q_D}(u - v) \\ u(0) = u_0, v(L) = 0 \end{cases}.$$

Adding these two equations results in the linear equation in $u - v$,

$$u' - v' = -\frac{k}{Q_B}(u - v) + \frac{k}{Q_D}(u - v)(u - v)' = -\left(\frac{k}{Q_B} - \frac{k}{Q_D}\right)(u - v).$$

Let $\alpha = k/Q_B - k/Q_D$ and $y = u - v$. Then we must solve the separable equation $y' = -\alpha y$, which is done with dsolve, naming the resulting output step1.

> `step1:=dsolve(diff(y(x),x)=-alpha*y(x),y(x));`

$$step1 := y(x) = \_C1\ e^{(-\alpha x)}$$

> `y:=subs(_C1=c,rhs(step1));`

$$y := c\ e^{(-\alpha x)}$$

Using the facts that $u' = -\frac{k}{Q_B}(u - v) = -\frac{k}{Q_B}y$ and $u(0) = u_0$, we are able to use dsolve to find $u(x)$.

> `step2:=dsolve({diff(u(x),x)=-k/Q[b]*y,u(0)=u0},u(x));`

$$step2 := u(x) = \frac{k\,c\,e^{(-\alpha x)}}{Q_b\,\alpha} - \frac{k\,c - u0\,Q_b\,\alpha}{Q_b\,\alpha}$$

Because $y = u - v$, $v = u - y$. Consequently, because $v(L) = 0$ we are able to compute c.

> `leftside:=subs(x=L,rhs(step2)-y);`

$$leftside := \frac{k\,c\,e^{(-\alpha L)}}{Q_b\,\alpha} - \frac{k\,c - u0\,Q_b\,\alpha}{Q_b\,\alpha} - c\,e^{(-\alpha L)}$$

> `cval:=solve(leftside=0,c);`

$$cval := \frac{u0\,Qb\,\alpha}{k\,e^{(-\alpha L)} - k - e^{(-\alpha L)}\,Qb\,\alpha}$$

and determine u and v.

> `u:=simplify(subs(c=cval,rhs(step2)));`

$$u := -\frac{u0(k\,e^{(-\alpha x)} - k\,e^{(-\alpha L)} + e^{(-\alpha L)}\,Q_b\,\alpha)}{k\,e^{(-\alpha L)} - k - e^{(-\alpha L)}\,Q_b\,\alpha}$$

> `v:=subs(c=cval,u-y);`

$$v := -\frac{u0(k\,e^{(-\alpha x)} - k\,e^{(-\alpha L)} + e^{(-\alpha L)}\,Q_b\,\alpha)}{k\,e^{(-\alpha L)} - k - e^{(-\alpha L)}\,Q_b\,\alpha} + \frac{u0\,Q_b\,\alpha\,e^{(-\alpha x)}}{k\,e^{(-\alpha L)} - k - e^{(-\alpha L)}\,Q_b\,\alpha}$$

For example, in healthy adults, typical urea nitrogen levels are 11 to 23 milligrams per deciliter (1 deciliter = 100 milliliters), serum creatinine levels range from 0.6 to 1.2 milligrams per deciliter, and the total volume of blood is 4 to 5 liters (1 liter = 1000 milliliters).

Suppose that hemodialysis is performed on a patient with a urea nitrogen level of 34 mg/dl and serum creatinine level of 1.8 mg/dl using a dialyzer with $k = 2.25$ and $L = 1$. If the flow rate of blood, Q_B, is 2 dl/minute and the flow rate of the dialysate, Q_D, is 4 dl/minute, will the wastes in the patient's blood reach normal levels after dialysis is performed?

After defining the appropriate constants, we evaluate u and v.

```
> alpha:=k/Q[b]-k/Q[d]:

k:=2.25:

L:=1:

Q[b]:=2:

Q[d]:=4:

u0:=34+1.8:

u;

v;
```

$$50.06232707\ e^{(-0.5625000000\ x)} - 14.26232708$$

$$25.03116354\ e^{(-0.5625000000\ x)} - 14.26232708$$

and then graph u and v on the interval [0,1] with `plot`. Remember that the dialysate is moving in the direction *opposite* to the blood. Thus, we see from the graphs that as levels of waste in the blood decrease, levels of waste in the dialysate increase, and at the end of the dialysis procedure, levels of waste in the blood are within normal ranges.

```
> plot(u,x=0..1);

plot(v,x=0..1);
```

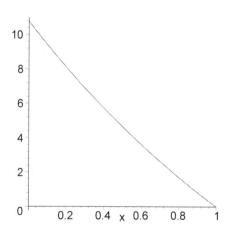

Typically, hemodialysis is performed 3 to 4 hours at a time three or four times per week. In some cases, a kidney transplant can free patients from the restrictions of dialysis. Of course, transplants have other risks not necessarily faced by those on dialysis; the number of available kidneys also affects the number of transplants performed. For example, in 1991 more than 130,000 patients were receiving dialysis while only 7000 kidney transplants had been performed.

Sources: D. N. Burghes and M. S. Borrie, *Modeling with Differential Equations*, Ellis Horwood Limited, pp. 41–45. Joyce M. Black and Esther Matassarin-Jacobs, *Luckman and Sorensen's Medical–Surgical Nursing: A Psychophysiologic Approach*, Fourth Edition, W. B. Saunders Company (1993), pp. 1509–1519, 1775–1808.

2.3 Homogeneous Equations

Definition
Homogeneous
Differential Equation

A differential equation that can be written in the form $M(x,y)dx + N(x,y)dy = 0$ where $M(tx,ty) = t^n M(x,y)$ and $N(tx,ty) = t^n N(x,y)$ is called a **homogeneous differential equation (of degree n)** .

It is a good exercise to show that an equation is homogeneous if we can write it in either of the forms $dy/dx = F(y/x)$ or $dy/dx = G(x/y)$.

EXAMPLE 1: Show that the equation $(x^2 + yx)dx - y^2 dy = 0$ is homogeneous.

SOLUTION: Let $M(x,y) = x^2 + yx$ and $N(x,y) = -y^2$. Because $M(tx,ty) = (tx)^2 + (ty)(tx) = t^2(x^2 + yx) = t^2 M(x,y)$ and $N(tx,ty) = -t^2 y^2 = t^2 N(x,y)$, the equation $(x^2 + yx)dx - y^2 dy = 0$ is homogeneous of degree two.

■

Homogeneous equations can be reduced to separable equations by either of the substitutions

$$y = ux \quad \text{or} \quad x = vy$$

Generally, use the substitution $y = ux$ if $N(x,y)$ is less complicated than $M(x,y)$ and use $x = vy$ if $M(x,y)$ is less complicated than $N(x,y)$. If a difficult integration problem is encountered after a substitution is made, try the other substitution to see if it yields an easier problem.

Often, the command `genhomosol`, which is contained in the **DEtools** package, can be used to find solutions to first-order homogeneous differential equations.

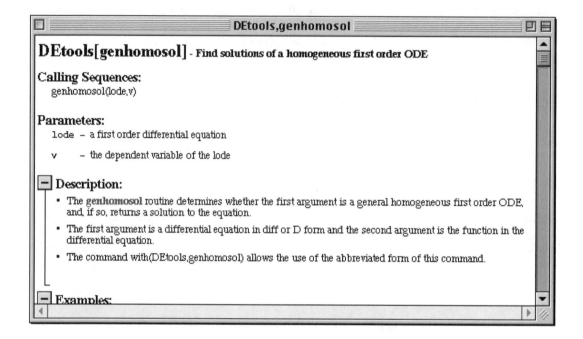

EXAMPLE 2: Solve the equation $(x^2 - y^2)dx + xydy = 0$.

SOLUTION: In this case, $M(x,y) = x^2 - y^2$ and $N(x,y) = xy$. Then $M(tx, ty) = t^2 M(x,y)$ and $N(tx, ty) = t^2 N(x,y)$, which means that $(x^2 - y^2)dx + xydy = 0$ is a homogeneous equation of degree two. In fact, after defining `diffeqn` to be $(x^2 - y^2)dx + xydy = 0$ and loading the **DEtools** package, we see that odeadvisor is able to determine that the equation is homogeneous as well.

```
> diffeqn:=x^2-y(x)^2+x*y(x)*diff(y(x),x)=0:

with(DEtools):

odeadvisor(diffeqn);
```

[[_homogeneous, class A], _rational, _Bernoulli]

Assume $x = vy$. Then, $dx = vdy + ydv$, and substituting into the equation and simplifying yields

$$0 = (x^2 - y^2)dx + xydy$$

$$= (v^2y^2 - y^2)(vdy + ydv) + vyydy$$

$$= v^3y^2dy - y^2vdy + v^2y^3dv - y^3dv + vy^2dy$$

$$= (v^3y^2 - y^2v + vy^2)dy + (v^2y^3 - y^3)dv$$

$$= y^2v^3dy + y^3(v^2 - 1)dv.$$

Dividing this equation by y^3v^3 yields the separable differential equation

$$\frac{dy}{y} + \frac{(v^2 - 1)dv}{v^3} = 0.$$

We solve this equation by rewriting it in the form

$$\frac{dy}{y} = \frac{(1 - v^2)dv}{v^3} = \left(\frac{1}{v^3} - \frac{1}{v}\right)dv$$

and integrating. This yields

$$\ln|y| = \frac{-2}{v^2} - \ln|v| + C_1$$

which can be simplified as $\ln|vy| = \frac{-2}{v^2} + C_1$, so

$$vy = Ce^{-2/v^2}, \quad \text{where } C = \pm e^{C_1}.$$

Because $x = vy$, $v = x/y$ and resubstituting into the preceding equation yields

$$x = Ce^{-2y^2/x^2}$$

as a general solution of the equation $(x^2 - y^2)dx + xydy = 0$.
We see that dsolve is able to solve the equation by finding exact solutions

```
> gensol:=dsolve(diffeqn,y(x));
```

$$gensol := y(x) = \sqrt{-2\ \ln(x) + \_C1}\ x, y(x) = -\sqrt{-2\ \ln(x) + \_C1}\ x$$

as well as an implicit solution when we include the implicit option in the dsolve command.

```
> implicitsol:=dsolve(diffeqn,y(x),implicit);
```

$$implicitsol := y(x)^2 + 2\ x^2\ \ln(x) - x^2\_C1 = 0$$

We can graph this implicit solution for various values of C by solving this equation for C

```
> tograph:=subs(y(x)=y,solve(implicitsol,_C1));
```

$$tograph := \frac{y^2 + 2\,x^2\ln(x)}{x^2}$$

and then noting that graphs of the equation $y^2 = x^2(C - 2\ln|x|)$ for various values of C are the same as the graphs of the level curves of the function $\frac{y^2+2x^2\ln|x|}{x^2}$. The `contourplot` command graphs several level curves $z = C$, C a constant, of the function $z = f(x, y)$. We may instruct Maple to graph the level curves of $z = f(x, y)$ for particular values of C by including the `contours` option.

For example, the level curves of $\frac{y^2+2x^2\ln|x|}{x^2}$ which intersect the x-axis at $x = 1, 2, \ldots, 9$, and 10 are the contours with values obtained by replacing each occurrence of y in $\frac{y^2+2x^2\ln|x|}{x^2}$ by 0 and x by 1, 2, ..., 9, and 10, which we do now with `seq` and `subs`, naming the resulting set of 10 numbers `contourvals`.

```
> contourvals:=[seq(subs({x=i,y=0},tograph),i=1..10)]:
```

Then, entering

```
> with(plots):
contourplot(tograph,x=0.1..10,y=-5..5,

        contours=contourvals,grid=[60,60],color=BLACK);
```

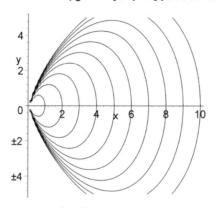

graphs several level curves of $\frac{y^2+2x^2\ln|x|}{x^2}$ for $0.01 \le x \le 10$ (we avoid $x = 0$ because $\frac{y^2+2x^2\ln|x|}{x^2}$ is undefined if $x = 0$) and $-5 \le y \le 5$. The option `contours = contourvals` instructs Maple to draw contours with values given in the list of numbers `contourvals`.

■

The next example illustrates how Maple can be used to help solve homogeneous equations.

EXAMPLE 3: Solve the equation $(x^{1/3}y^{2/3} + x)dx + (x^{2/3}y^{1/3} + y)dy = 0$.

SOLUTION: We begin by identifying $M(x,y) = x^{1/3}y^{2/3} + x$ and $N(x,y) = x^{2/3}y^{1/3} + y$ and then defining M and N.

```
> M:=(x,y)->x^(1/3)*y^(2/3)+x:

N:=(x,y)->x^(2/3)*y^(1/3)+y:
```

Next, we verify that the equation is homogeneous of degree 1 by computing and factoring $M(tx,ty)$ and $N(tx,ty)$ with `radsimp`. (Because Maple does not automatically simplify $(ab)^n = a^n b^n$ unless n is an integer, we must use `radsimp` to simplify the terms $(tx)^{1/3} = t^{1/3}x^{1/3}$, $(ty)^{1/3} = t^{1/3}y^{1/3}$, $(tx)^{2/3} = t^{2/3}x^{2/3}$, and $(ty)^{2/3} = t^{2/3}y^{2/3}$.)

```
> radsimp(M(t*x,t*y));

radsimp(N(t*x,t*y));
```

$$x^{\left(\frac{1}{3}\right)}t\left(y^{\left(\frac{2}{3}\right)} + x^{\left(\frac{2}{3}\right)}\right)$$

$$ty^{\left(\frac{1}{3}\right)}\left(y^{\left(\frac{2}{3}\right)} + x^{\left(\frac{2}{3}\right)}\right)$$

In this case, we see that `dsolve` computes a general solution of the equation, although the result is given in terms of an unevaluated integral.

```
> dsolve(M(x,y(x))+N(x,y(x))*diff(y(x),x)=0,y(x));
```

$$y(x) = \text{RootOf}\left(\ln(x) - \_C1 + \int^{\_Z} \frac{(\_ax)^{\left(\frac{1}{3}\right)}(x^2)^{\left(\frac{1}{3}\right)} + \_ax}{\_a(\_ax)^{\left(\frac{1}{3}\right)}(x^2)^{\left(\frac{1}{3}\right)} + \_a^2x + x^{\left(\frac{1}{3}\right)}(\_ax)^{\left(\frac{2}{3}\right)} + x} \, d\_a \right)$$

On the other hand, `genhomosol` can find a solution if we first assume that x and y are positive.

```
> with(DEtools):

assume(x>0,y>0):

genhomosol(M(x,y(x))+N(x,y(x))*diff(y(x),x)=0,
```

```
y(x));
```

$$\left\{ x\sim = \_C_1 x\sim \left(y\sim(x\sim)^5 x\sim^2 (y\sim(x\sim)x\sim^2)^{\left(\frac{1}{3}\right)} + y\sim(x\sim)^4 x\sim^4 \right. \right.$$

$$+ 2y\sim(x\sim)^4 (y\sim(x\sim)x\sim^2)^{\left(\frac{2}{3}\right)} x\sim^2 \sqrt{3} + y\sim(x\sim)^3 (y\sim(x\sim)x\sim^2)^{\left(\frac{1}{3}\right)} x\sim^4 \sqrt{3}$$

$$+ 4y\sim(x\sim)(y\sim(x\sim)x\sim^2)^{\left(\frac{1}{3}\right)} x\sim^6 - \sqrt{3}(y\sim(x\sim)x\sim^2)^{\left(\frac{2}{3}\right)} y\sim(x\sim)^3$$

$$- \sqrt{3}(y\sim(x\sim)x\sim^2)^{\left(\frac{1}{3}\right)} y\sim(x\sim)^2 x\sim^2 - 3(y\sim(x\sim)x\sim^2)^{\left(\frac{1}{3}\right)} x\sim^4$$

$$\left. + \sqrt{3}(y\sim(x\sim)x\sim^2)^{\left(\frac{2}{3}\right)} x\sim^6 + x\sim^8 \right)^{\left(\frac{1}{4}\right)} \left(y\sim(x\sim)^5 x\sim^2 (y\sim(x\sim)x\sim^2)^{\left(\frac{1}{3}\right)} \right.$$

$$+ 4y\sim(x\sim)^4 x\sim^4 + 6y\sim(x\sim)^2 (y\sim(x\sim)x\sim^2)^{\left(\frac{2}{3}\right)} x\sim^4$$

$$+ 4y\sim(x\sim)(y\sim(x\sim)x\sim^2)^{\left(\frac{1}{3}\right)} x\sim^6 + y\sim(x\sim)^4 (y\sim(x\sim)x\sim^2)^{\left(\frac{2}{3}\right)} x\sim^2 \sqrt{3}$$

$$+ 3\sqrt{3}y\sim(x\sim)x\sim^2)^{\left(\frac{1}{3}\right)} y\sim(x\sim)^2 x\sim^2 + 3\sqrt{3}y\sim(x\sim)^2 x\sim^6$$

$$\left. + \sqrt{3}(y\sim(x\sim)x\sim^2)^{\left(\frac{2}{3}\right)} x\sim^6 + x\sim^8 \right)^{\left(\frac{2}{3}\right)} \Big/ \left((y\sim(x\sim)^4 + x\sim^4)^{\left(\frac{1}{4}\right)} \right.$$

$$\left(y\sim(x\sim)(y\sim(x\sim)x\sim^2)^{\left(\frac{1}{3}\right)} + x\sim^2 \right)^3$$

$$\left. \left. \left(y\sim(x\sim)(y\sim(x\sim)x\sim^2)^{\left(\frac{1}{3}\right)} + \sqrt{3}(y\sim(x\sim)x\sim^2)^{\left(\frac{2}{3}\right)} + x\sim^2 \right) \right) \right\}$$

We now illustrate how Maple can be used to implement the steps we use to solve homogeneous equations. First, we let $x = vy$.

```
> x:='x':y:='y':

x:=v*y;
```

$$x := vy$$

We see that $D(x)$ represents dx. Similarly, $D(v)$ represents dv and $D(y)$ represents dy.

```
> D(x);
```

$$D(v)y + vD(y)$$

Next, we evaluate the equation with the substitution $x = vy$ and expand powers containing rational exponents with radsimp, naming the result output step_1.

```
> radsimp(step_1:=M(x,y)*D(x)+N(x,y)*D(y)=0);
```

$$step\_1 := y\left(yv^{\left(\frac{4}{3}\right)}D(v) + v^{\left(\frac{4}{3}\right)}D(y) + v\ yD(v) + v^2 D(y) + D(y)v^{\left(\frac{2}{3}\right)} + D(y)\right) = 0$$

To see that the equation in `step_1` is separable, we begin by using `collect` to collect together the terms containing `D(v)`, representing dv, and `D(y)`, representing dy.

```
> step_2:=collect(step_1,{D(v),D(y)});
```

$$step\_2 := y\left(v^{\left(\frac{4}{3}\right)} + v^2 + v^{\left(\frac{2}{3}\right)} + 1\right)D(y) + y\left(yv^{\left(\frac{4}{3}\right)} + vy\right)D(v) = 0$$

Observe that dividing both sides of this equation by $y^2\left(v^{4/3} + v^2 + v^{2/3} + 1\right)$ results in the separated equation

$$\frac{1}{y}dy + \frac{v^{1/3} + v}{v^{4/3} + v^2 + v^{2/3} + 1}dv = 0.$$

Alternatively, we can use `separablesol`, which is contained in the **DEtools** package, to solve this separable equation. First, we replace each occurrence of `D(y)` with `diff(y(v),v)`, each occurrence of `y` with `y(v)`, and each occurrence of `D(v)` with 1:

```
> step_3:=subs({D(y)=diff(y(v),v),y=y(v),D(v)=1},step_2);
```

$$step\_3 := y(v)\left(v^{\left(\frac{4}{3}\right)} + v^2 + v^{\left(\frac{2}{3}\right)} + 1\right)\left(\frac{\partial}{\partial v}y(v)\right) + y(v)\left(v^{\left(\frac{4}{3}\right)}y(v) + vy(v)\right) = 0$$

and then use `separablesol` to solve this equation in `step_4`.

```
> step_4:=separablesol(step_3,y(v));
```

$$step\_4 := \left\{\ln(y(v)) - \frac{1}{4}\ln\left(v^{\left(\frac{4}{3}\right)} + \sqrt{3}v^{\left(\frac{2}{3}\right)} + 1\right) - \frac{1}{4}\ln\left(v^{\left(\frac{4}{3}\right)} - \sqrt{3}v^{\left(\frac{2}{3}\right)} + 1\right)\right.$$
$$\left. + \frac{1}{4}\ln(v^4 + 1) + \frac{1}{2}\ln\left(v^{\left(\frac{4}{3}\right)} + 1\right) = \_C_1\right\}$$

The solution is then obtained by substituting $v = x/y$ into `step_4`.

```
> x:='x':y:='y':v:='v':
step_5:=subs({y(v)=y,v=x/y},step_4);
```

$$step\_5 := \left\{\ln(y) - \frac{1}{4}\ln\left(\left(\frac{x}{y}\right)^{\left(\frac{4}{3}\right)} + \sqrt{3}\left(\frac{x}{y}\right)^{\left(\frac{2}{3}\right)} + 1\right)\right.$$
$$\left. - \frac{1}{4}\ln\left(\left(\frac{x}{y}\right)^{\left(\frac{4}{3}\right)} - \sqrt{3}\left(\frac{x}{y}\right)^{\left(\frac{2}{3}\right)} + 1\right) + \frac{1}{4}\ln\left(\frac{x^4}{y^4} + 1\right) + \frac{1}{2}\ln\left(\left(\frac{x}{y}\right)^{\left(\frac{4}{3}\right)} + 1\right) = \_C_1\right\}$$

With the assumptions that x and y are positive, we can use `combine` to simplify this result considerably.

```
> assume(x>0,y>0):

step_6:=combine(simplify(step_5),ln);
```

$$step\_6 := \left\{ -\frac{1}{4}\ln\left(x\sim^{\left(\frac{4}{3}\right)}y\sim^{\left(\frac{2}{3}\right)} - \sqrt{3}x\sim^{\left(\frac{2}{3}\right)}y\sim^{\left(\frac{4}{3}\right)} + y\sim^2\right) \right.$$

$$\left. +\ln\left(\frac{(x\sim^4 + y\sim^4)^{\left(\frac{1}{4}\right)}\sqrt{x\sim^{\left(\frac{4}{3}\right)}y\sim^{\left(\frac{2}{3}\right)} + y\sim^2}}{\left(x\sim^{\left(\frac{4}{3}\right)}y\sim^{\left(\frac{2}{3}\right)} + \sqrt{3}x\sim^{\left(\frac{2}{3}\right)}y\sim^{\left(\frac{4}{3}\right)} + y\sim^2\right)^{\left(\frac{1}{4}\right)}}\right) = \_C_1 \right\}$$

■

APPLICATION

Models of Pursuit

Suppose that one object pursues another whose motion is known by a predetermined strategy. For example, suppose that an airplane is positioned at $B(1000,0)$ to fly to another airport A that is 1000 miles directly west of its position B, as illustrated in the following figure. Assume that the airplane aims toward A at all times. If the wind goes from south to north at a constant speed, w, and the airplane's speed in still air is b, determine conditions on b so that the airplane eventually arrives at A and describe its path.

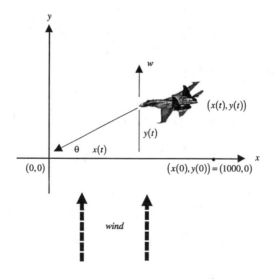

As described, the speed of the airplane, b, must be greater than the speed of the wind, w, $b > w$, in order for the plane to arrive at A. Observe that dx/dt describes the airplane's velocity in the x direction:

$$\frac{dx}{dt} = -b \cos \theta = \frac{-bx}{\sqrt{x^2 + y^2}},$$

because from right-triangle trigonometry we know that $\cos \theta =$ adjacent/hypotenuse $= x/\sqrt{x^2 + y^2}$. Similarly,

$$\frac{dy}{dt} = -b \sin \theta + w = \frac{-by}{\sqrt{x^2 + y^2}} + w$$

so

$$\frac{dy}{dx} = \frac{dy/dt}{dx/dt} = \frac{\frac{-by}{\sqrt{x^2+y^2}} + w}{\frac{-bx}{\sqrt{x^2+y^2}}} = \frac{by - w\sqrt{x^2 + y^2}}{bx}.$$

This is a homogeneous equation because it can be written in the form $dy/dx = F(y/x)$:

$$\frac{dy}{dx} = \frac{by - w\sqrt{x^2 + y^2}}{bx} = \frac{y}{x} - \frac{w}{b}\sqrt{1 + \left(\frac{y}{x}\right)^2}.$$

Therefore, we must solve the initial-value problem

$$\begin{cases} \frac{dy}{dx} = \frac{by - w\sqrt{x^2+y^2}}{bx} \\ y(1000) = 0 \end{cases}.$$

A general solution to the equation is found with `genhomosol`.

```
> with(DEtools):

assume(x>=0,b>0,w>0):

gensol:=genhomosol(diff(y(x),x)=
      (b*y(x)-w*sqrt(x^2+y(x)^2))/(b*x),y(x));
```

$$gensol := \left\{ x = \frac{\_C_1}{e^{\left(\frac{b \, \text{arcsinh}\left(\frac{y(x)}{x}\right)}{w}\right)}} \right\}$$

Alternatively, letting $y = ux$ and substituting into the equation results in the separable equation

$$\frac{dy}{dx} = \frac{du}{dx}x + u = \frac{bux - w\sqrt{x^2 + u^2 x^2}}{bx}$$

$$\frac{du}{dx}x + u = u - \frac{w}{b}\sqrt{1 + u^2}$$

$$\frac{1}{\sqrt{1 + u^2}} = -\frac{w}{b}\frac{1}{x}dx.$$

```
> x:='x':y:='y':

b:='b':w:='w':

y:=u*x:

step_1:=collect(radsimp(D(y)=
        (b*y-w*sqrt(x^2+y^2))/(b*x)),{D(x),D(u)});
```

$$step\_1 := D(u)x + uD(x) = \frac{bu - w\sqrt{1 + u^2}}{b}$$

```
> expand(step_1);
```

$$D(u)x + uD(x) = u - \frac{w\sqrt{1 + u^2}}{b}$$

Integrating the left-hand side of this equation yields $\int \frac{1}{\sqrt{1+u^2}} du = \ln|u + \sqrt{1 + u^2}| + C_1$

```
> left_int:=int(1/sqrt(1+u^2),u);
```

$$left\_int := \operatorname{arcsinh}(u)$$

```
> left_int:=convert(left_int,ln);
```

$$left\_int := \ln(u + \sqrt{1 + u^2})$$

and integrating the right results in $-\frac{w}{b}\int \frac{1}{x}dx = -\frac{w}{b}\ln|x| + C_2$. Note that absolute value bars are not necessary because x and y and, hence, u are nonnegative. Thus, $\ln(u + \sqrt{1 + u^2}) = -\frac{w}{b}\ln x + c$.

```
> right_int:=int(-w/(b*x),x)+c;
```

$$right\_int := -\frac{w\ln(x)}{b} + c$$

Because $y(1000) = 0$, $C = \frac{w}{b}\ln 1000$

```
> cval:=eval(solve(subs({x=1000,u=0},
      left_int=right_int),c));
```

$$cval := \frac{w\ln(1000)}{b}$$

and $\ln\left(u + \sqrt{1 + u^2}\right) = -\frac{w}{b}\ln x + \frac{w}{b}\ln 1000$.

```
> step_3:=subs(c=cval,left_int=right_int);
```

$$step\_3 := \ln(u + \sqrt{1 + u^2}) = -\frac{w\ln(x)}{b} + \frac{w\ln(1000)}{b}$$

which we solve for u:

$$\ln\left(u + \sqrt{1 + u^2}\right) = \ln\left(\frac{x}{1000}\right)^{-w/b}$$

$$u + \sqrt{1 + u^2} = \left(\frac{x}{1000}\right)^{-w/b}$$

$$\sqrt{1 + u^2} = \left(\frac{x}{1000}\right)^{-w/b} - u$$

$$1 + u^2 = \left(\frac{x}{1000}\right)^{-2w/b} - 2u\left(\frac{x}{1000}\right)^{-w/b} + u^2$$

$$2u\left(\frac{x}{1000}\right)^{-w/b} = \left(\frac{x}{1000}\right)^{-2w/b} - 1$$

$$u = \frac{1}{2}\left(\left(\frac{x}{1000}\right)^{-w/b} - \left(\frac{x}{1000}\right)^{w/b}\right).$$

```
> step_4:=solve(step_3,u);
```

$$step\_4 := \frac{1}{2}\frac{-1 + \left(e^{\left(\frac{w(-\ln(x)+\ln(1000))}{b}\right)}\right)^2}{e^{\left(\frac{w(-\ln(x)+\ln(1000))}{b}\right)}}$$

We solve for y by resubstituting $u = y/x$ and multiplying by x:

$$\frac{y}{x} = \frac{1}{2}\left(\left(\frac{x}{1000}\right)^{-w/b} - \left(\frac{x}{1000}\right)^{w/b}\right)$$

$$y = \frac{1}{2}x\left(\left(\frac{x}{1000}\right)^{-w/b} - \left(\frac{x}{1000}\right)^{w/b}\right).$$

```
> y:='y':
y:=x*step_4;
```

$$y := \frac{1}{2}\, x \frac{\left(-1 + \left(e^{\left(\frac{w(-\ln(x)+\ln(1000))}{b}\right)}\right)^2\right)}{e^{\left(\frac{w(-\ln(x)+\ln(1000))}{b}\right)}}$$

We graph y for various values of w/b by setting $b = 1$ and then using `seq` and `subs` to generate the value of y for $w = 0.25, 0.50, \ldots, 2.0$. These functions are then graphed with `plot`. Notice that the airplane never arrives at A if $w/b \geq 1$.

```
> b:=1:
wvals:=seq(0.25*i,i=1..8):
toplot:={seq(subs(w=i,y),i=wvals)};
plot(toplot,x=0..1000,color=BLACK,view=[0..1000,0..1000]);
```

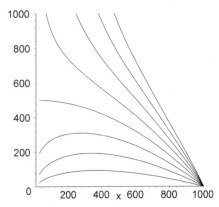

Sources: A particularly interesting and fun-to-read discussion of flight paths and models of pursuit can be found in *Differential Equations: A Modeling Perspective* by Robert L. Borrelli and Courtney S. Coleman and published by John Wiley & Sons, 1998.

2.4 Exact Equations

Definition
Exact Differential
Equation

A differential equation that can be written in the form

$$M(x, y)dx + N(x, y)dy = 0$$

where

$$M(x, y)dx + N(x, y)dy = \frac{\partial f}{\partial x}(x, y)dx + \frac{\partial f}{\partial y}(x, y)dy$$

for some function $f(x,y)$ is called an **exact differential equation**.

EXAMPLE 1: Show that the equation $2xy^3dx + (1 + 3x^2y^2)dy = 0$ is exact and that the equation $x^2ydx + 5xy^2dy = 0$ is not exact.

SOLUTION: Because $\frac{\partial}{\partial y}(2xy^3) = 6xy^2 = \frac{\partial}{\partial x}(1 + 3x^2y^2)$, the equation $2xy^3dx + (1 + 3x^2y^2)dy = 0$ is an exact equation. On the other hand, the equation $x^2ydx + 5xy^2dy = 0$ is not exact because $\frac{\partial}{\partial y}(x^2y) = x^2 \neq 5y^2 = \frac{\partial}{\partial x}(5xy^2)$. (However, $x^2ydx + 5xy^2dy = 0$ is separable.)

∎

If an equation is exact, we can find a function $f(x,y)$ such that $M(x,y) = \frac{\partial f}{\partial x}(x,y)$ and $N(x,y) = \frac{\partial f}{\partial y}(x,y)$.

1. Assume that $M(x,y) = \frac{\partial f}{\partial x}(x,y)$ and $N(x,y) = \frac{\partial f}{\partial y}(x,y)$.
2. Integrate $M(x,y)$ with respect to x. (Add an arbitrary function of y, $g(y)$.)
3. Differentiate the result in step 2 with respect to y and set the result equal to $N(x,y)$. Solve for $g'(y)$.
4. Integrate $g'(y)$ with respect to y to obtain an expression for $g(y)$. (There is no need to include an arbitrary constant.)
5. Substitute $g(y)$ into the result obtained in step 2 for $f(x,y)$.
6. A general solution is $f(x,y) = C$, where C is a constant.
7. Apply the initial condition if given.

Note: A similar algorithm can be stated so that in step 2, $N(x,y)$ is integrated with respect to y.

EXAMPLE 2: Solve $2x\sin ydx + (x^2\cos y - 1)dy = 0$ subject to $y(0) = 1/2$.

SOLUTION: The equation $2x\sin ydx + (x^2\cos y - 1)dy = 0$ is exact because

$$\frac{\partial}{\partial y}(2x\sin y) = 2x\cos y = \frac{\partial}{\partial y}(x^2\cos y - 1).$$

Alternatively, `odeadvisor` is able to determine that the equation is exact.

```
> with(DEtools:
diffeq:=2*x*sin(y(x))+x^2*cos(y-1)*diff(y(x),x)=0;
```

```
(x^2*cos(y(x))-1)*diff(y(x),x)=0:
odeadvisor(diffeq);
```

$$[\_exact]$$

Let $f(x,y)$ be a function with $\frac{\partial f}{\partial x}(x,y) = 2x \sin y$ and $\frac{\partial f}{\partial y}(x,y) = x^2 \cos y - 1$. Then, integrating $\frac{\partial f}{\partial x}(x,y)$ with respect to x yields

$$f(x,y) = \int 2x \sin y \, dx = x^2 \sin y + g(y).$$

Notice that the arbitrary function g of y serves as a "constant" of integration with respect to x. Because we have

$$\frac{\partial f}{\partial y}(x,y) = x^2 \cos y - 1$$

from the differential equation, and

$$\frac{\partial f}{\partial y}(x,y) = x^2 \cos y + g'(y)$$

from differentiation of $f(x,y)$ with respect to y,

$$g'(y) = -1.$$

Hence,

$$g(y) = -y + C_1.$$

Therefore,

$$f(x,y) = x^2 \sin y - y + C_1,$$

so a general solution of the exact equation is $x^2 \sin y - y + C_1 = C$. Simplifying, we have

$$x^2 \sin y - y = k,$$

where k is a constant. (Notice that we did not have to include the constant C_1 in calculating g because we combined it with the constant on the right-hand side of the equation in the general solution.) Because our solution requires that $y(0) = 1/2$, we must find the solution in the family of solutions that passes through the point $(0,1/2)$. Substituting these values of x and y into the general solution, we obtain $0^2 \sin(1/2) - 1/2 = k$, so $k = -1/2$, Therefore, the desired solution is $x^2 \sin y - y = 1/2$. We are able to use exactsol, which is contained in the **DEtools** package, or dsolve to find a general solution of the equation as well.

```
> sol1:=exactsol(diffeq,y(x));
```

$$sol1 := \{-x^2 \sin(y(x)) + y(x) = \_C_1\}$$

```
> sol2:=dsolve(diffeq,y(x));
```

$$sol2 := \_C1 + x^2 \sin(y(x)) - y(x) = 0$$

We can graph the solution for various values of the arbitrary constant by observing that the graph of $x^2 \sin y - y = k$ for various values of k is the same as the graph of the level curves of $f(x,y) = x^2 \sin y - y$, which are graphed with `contourplot`.

```
> gensol:=lhs(subs({_C1=0,y(x)=y},sol2));
```

$$gensol := x^2 \sin(y) - y$$

```
> with(plots):
contourplot(gensol,x=-5..5,y=-5..5,
```

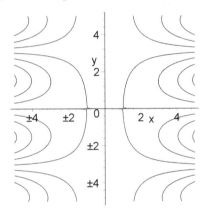

```
color=BLACK,grid=[70,70]);
```

Although `dsolve` can find the solution to the initial-value problem,

```
> sol3:=dsolve({diffeq,y(0)=1/2},y(x));
```

$$sol3 := y(x) = \text{RootOf}\left(\frac{1}{2} + x^2 \sin(\_Z) - \_Z\right)$$

it is easier to substitute the initial condition into the general solution found,

```
> step1:=subs({y(x)=1/2,x=0},sol2);
```

$$step1 := \_C1 - \frac{1}{2} = 0$$

solve for the constant,

```
> cval:=solve(step1);
```

$$cval := \frac{1}{2}$$

and substitute back into the general solution. The resulting equation is graphed with `implicitplot`, which is also contained in the **plots** package with the `contourplot` command. The option $grid = [70, 70]$ instructs Maple to sample 70 points in each of the x- and y-directions, helping ensure that the resulting graph appears smooth.

```
> tograph:=subs({_C1=cval,y(x)=y},sol2);
```

$$tograph := \frac{1}{2} + x^2 \sin(y) - y = 0$$

```
> implicitplot(tograph,x=-10..10,y=-10..10,
        color=BLACK,grid=[70,70]);
```

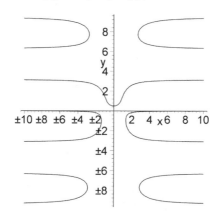

The following example illustrates how we can use Maple to assist us in carrying out the necessary steps encountered when solving an exact equation.

EXAMPLE 3: Solve $(2x - y^2 \sin(xy))dx + (\cos(xy) - xy \sin(xy))dy = 0$.

SOLUTION: We begin by identifying $M(x, y) = 2x - y^2 \sin(xy)$ and $N(x, y) = \cos(xy) - xy \sin(xy)$. We then define `capm`, corresponding to M, and `capn`, corresponding to N. We then see that the equation is exact because $\partial M / \partial y = \partial N / \partial x$, which is verified with `testeq`.

```
> capm:=(x,y)->2*x-y^2*sin(x*y):
capn:=(x,y)->cos(x*y)-x*y*sin(x*y):
testeq(diff(capm(x,y),y)=diff(capn(x,y),x));
```

$$true$$

Next, we compute $\int M(x,y)dx$ and add an arbitrary function of y, $g(y)$, to the result.

```
> f:=int(capm(x,y),x)+g(y);
```

$$f := x^2 + \cos(x\,y)y + g(y)$$

Differentiating f with respect to y gives us

```
> diff(f,y);
```

$$\cos(x\,y) - x\,y\sin(x\,y) + \left(\frac{\partial}{\partial y}g(y)\right)$$

and because we must have that $\frac{\partial f}{\partial y}(x,y) = N(x,y)$, we obtain the equation that we solve for $g'(y)$ with `solve`.

```
> solve(diff(f,y)=capn(x,y),diff(g(y),y));
```

$$0$$

Thus, $g(y)$ is a (real-valued) constant and a general solution of the equation is $x^2 + y\cos(xy) = C$. We can graph this general solution for various values of C by observing that the level curves of the function $x^2 + y\cos(xy)$ correspond to the graphs of the equation $x^2 + y\cos(xy) = C$ for various values of C.

```
> f:=subs(g(y)=0,f );
```

$$f := x^2 + \cos(x\,y)y$$

We now use `contourplot` to graph several level curves of $x^2 + y\cos(xy)$ on the rectangle $[0, 3\pi] \times [0, 3\pi]$.

```
> with(plots):
contourplot(f,x=0..3*Pi,y=0..3*Pi,
                grid=[70,70],color=BLACK);
```

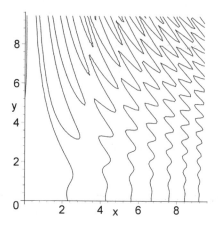

We see that `exactsol` or `dsolve` together with the `implicit` option is able to find an implicit solution of the equation after we rewrite it in the form $(2x - y^2 \sin(xy)) + (\cos(xy) - xy \sin(xy))y' = 0$.

```
> step1:=dsolve(capm(x,y(x))+
           capn(x,y(x))*diff(y(x),x)=0,y(x),implicit);
```

$$step1 := \_C1 + x^2 + \cos(x\,y(x))y(x) = 0$$

```
> with(DEtools):
altsol:=exactsol(capm(x,y(x))+
           capn(x,y(x))*diff(y(x),x)=0,y(x));
```

$$altsol := \{-x^2 - \cos(x\,y(x))y(x) = \_C_1\}$$

The solution obtained in `step1` can be graphed for various values of the arbitrary constant by first replacing each occurrence of _C1 in the left-hand side of `step1` by 0 and each occurrence of `y(x)` by y.

```
> f:=subs({_C1=0,y(x)=y},lhs(step1));
```

$$f := x^2 + \cos(x\,y)y$$

Then, entering `contourplot(f, x = 0..3* Pi, y = 0..3* Pi, grid = [70,70], color = BLACK)` produces exactly the same graph as obtained previously.

Alternatively, we can take advantage of the fact that **f** is a conservative vector field if there is a scalar field F satisfying $\mathbf{f} = \nabla F$. Thus, the equation $M(x,y)dx + N(x,y)dy = 0$ is exact if $\langle M(x,y), N(x,y) \rangle$ is a conservative vector field. We can determine whether **f** is a conservative vector field and, if so, its potential function F, with `potential`, which is contained in the `linalg` package. Thus, entering

```
> with(linalg):
potential([capm(x,y),capn(x,y)],[x,y],'F');
```

$$true$$

```
> F;
```

$$x^2 + \cos(x\,y)y$$

also shows that a general solution of the equation is $x^2 + y\cos(xy) = C$.

∎

Our last example illustrates how to solve an initial-value problem with `dsolve`.

EXAMPLE 4: Solve $(1 + 5x - y)dx - (x + 2y)dy = 0$ subject to the initial condition $y(0) = 1$.

SOLUTION: `dsolve` is successful in finding a general solution of the equation.

```
> gensol:=dsolve(1+5*x-y(x)-(x+2*y(x))*diff(y(x),x),y(x));
```

$$y(x) = \frac{1}{11} - \frac{1}{11}\frac{\frac{1}{2}(11\,x+2)\_C1 - \frac{1}{2}\sqrt{11(11\,x+2)^2\_C1^2 + 2}}{\_C1}$$

```
> implicitsol:=dsolve(1+5*x-y(x)-(x+2*y(x))*diff(y(x),x),
              y(x),implicit);
```

$$implicitsol := -\frac{1}{2}\ln((-5(11\,x+2)^2$$
$$- 2(11\,x+2)(-11\,y(x)+1) + 2(-11\,y(x)+1)^2)/$$
$$(11\,x+2)^2) - \ln(11\,x+2) - \_C1 = 0$$

In fact, we are able to use `dsolve` to solve the initial-value problem.

```
sol:=dsolve({1+5*x-y(x)-(x+2*y(x))*diff(y(x),x),y(0)=1},y(x));
```

$$sol := y(x) = \frac{1}{11} - \frac{2}{11}\left(\frac{1}{220}(11\,x+2)\sqrt{55} - \frac{1}{2}\sqrt{\frac{1}{20}(11\,x+2)^2 + 2}\right)\sqrt{55}$$

After naming $y(x)$ to be the solution obtained in sol, $y(x)$ is graphed with `plot`.

```
> assign(sol):
plot(y(x),x=-5..5);
```

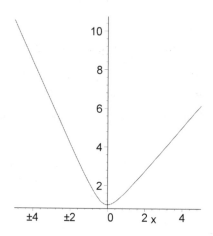

2.5 Linear Equations

In the previous sections, we have seen that calculating explicit or implicit closed-form solutions of most first-order equations may be a formidable task, at best. However, first-order linear equations $a_1(x)\frac{dy}{dx} + a_0(x)y = f(x)$, which we can rewrite in the form

$$\frac{dy}{dx} + p(x)y = q(x),$$

can always be solved, so we discuss their method of solution in this section.

Multiplying the equation $\frac{dy}{dx} + p(x)y = q(x)$ by $e^{\int p(x)dx}$ yields

$$e^{\int p(x)dx}\frac{dy}{dx} + e^{\int p(x)dx}p(x)y = e^{\int p(x)dx}q(x).$$

By the product rule and the fundamental theorem of calculus,

$$\frac{d}{dx}\left(e^{\int p(x)dx}y\right) = e^{\int p(x)dx}\frac{dy}{dx} + e^{\int p(x)dx}p(x)y$$

so we simplify this equation to obtain

$$\frac{d}{dx}\left(e^{\int p(x)dx}y\right) = e^{\int p(x)dx}q(x).$$

Integrating and dividing by $e^{\int p(x)dx}$ yields a general solution of $\frac{dy}{dx} + p(x)y = q(x)$:

$$e^{\int p(x)dx}y = \int e^{\int p(x)dx}q(x)dxy = \frac{\int e^{\int p(x)dx}q(x)dx}{e^{\int p(x)dx}} = e^{-\int p(x)dx}q(x)dx.$$

The term $\mu(x) = e^{\int p(x)dx}$ is called an **integrating factor** for the linear equation $\frac{dy}{dx} + p(x)y = q(x)$ and is useful because $\mu(x)\left(\frac{dy}{dx} + p(x)y\right) = \frac{d}{dx}[\mu(x)y]$ as we saw in the preceding derivation. Therefore, to find a solution to $\frac{dy}{dx} + p(x)y = q(x)$, we solve

$$\frac{d}{dx}[\mu(x)y] = \mu(x)q(x)$$

for y.

As we see with the following commands, `dsolve` and `linearsol`, which are contained in the **DEtools** package, are always able to solve first-order linear differential equations, although the result might contain unevaluated integrals.

> `dsolve(diff(y(x),x)+p(x)*y(x)=q(x),y(x));`

$$y(x) = e^{(-\int \mathrm{p}(x)dx)} \int q(x)e^{(\int p(x)dx)}dx + e^{(-\int p(x)dx)}\_C1$$

> `with(DEtools):`

`linearsol(diff(y(x),x)+p(x)*y(x)=q(x),y(x));`

$$\left\{y(x) = e^{(-\int p(x)dx)} \int q(x)e^{(\int p(x)dx)}dx + e^{(-\int p(x)dx)}\_C_1\right\}$$

EXAMPLE 1: Solve $xdy/dx + y = x\cos x$.

SOLUTION: First, we place the equation in the form used in the preceding derivation. Dividing the equation by x yields

$$\frac{dy}{dx} + \frac{1}{x}y = \cos x,$$

where $p(x) = 1/x$ and $q(x) = \cos x$. Then, an integrating factor is

$$e^{\int dx/x} = e^{\ln|x|} = x, \text{ for } x > 0,$$

and

$$\frac{d}{dx}(xy) = x\frac{dy}{dx} + y = x\cos x$$

so

$$xy = \int x\sin x dx.$$

Using the integration by parts formula, $\int u \, dv = uv - \int v \, du$, with $u = x$ and $dv = \cos x \, dx$, we obtain $du = dx$ and $v = \sin x$, so

$$xy = \int x \cos x \, dx = x \sin x - \int \sin x \, dx = x \sin x + \cos x + C.$$

Therefore, a general solution of the equation $x \, dy/dx + y = x \cos x$ for $x > 0$ is $y = (x \sin x + \cos x + C)/x$. (If we wanted to solve the equation for $x < 0$, then we would let $e^{\int dx/x} = e^{\ln |x|} = -x$ for $x < 0$.)

We see that `dsolve` is also successful in finding a general solution of the equation.

```
> gensol:=dsolve(x*diff(y(x),x)+y(x)=x*cos(x),y(x));
```

$$gensol := y(x) = \frac{\cos(x) + x \sin(x) + \_C1}{x}$$

As we have seen in previous examples, we can graph the solution for various values of the arbitrary constant by generating a set of functions obtained by replacing the arbitrary constant with numbers using `seq` and `subs` and then using `plot` to graph the resulting set of functions.

```
> toplot:={seq(subs(_C1=i,rhs(gensol)),i=-4..4)}:
plot(toplot,x=0..4*Pi,y=-2*Pi..2*Pi);
```

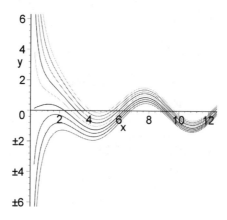

As with other types of equations, we solve initial-value problems by first finding a general solution of the equation and then applying the initial condition to determine the value of the constant.

EXAMPLE 2: Solve the initial-value problem

$$\begin{cases} dy/dx + 5x^4y = x^4 \\ \qquad y(0) = -7 \end{cases}.$$

SOLUTION: As we have seen in many previous examples, dsolve can be used to find a general solution of the equation and the solution to the initial-value problem, as done in gensol and partsol, respectively.

```
> x:='x':y:='y':
gensol:=dsolve(diff(y(x),x)+5*x^4*y(x)=x^4,y(x));
```

$$gensol := y(x) = \frac{1}{5} + e^{(-x^5)}\_C1$$

```
> partsol:=dsolve({diff(y(x),x)+5*x^4*y(x)=x^4,
        y(0)=-7},y(x));
```

$$partsol := y(x) = \frac{1}{5} - \frac{36}{5}e^{(-x^5)}$$

We now graph the solution to the initial-value problem obtained in partsol with plot.

```
> plot(rhs(partsol),x=-1..2);
```

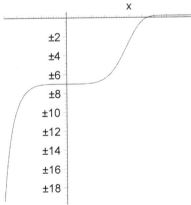

We can also use Maple to carry out the steps necessary to solve first-order linear equations. We begin by identifying the integrating factor $e^{\int 5x^4 dx} = e^{x^5}$, computed as follows with int.

```
> intfac:=exp(int(5*x^4,x));
```

$$intfac := e^{(x^5)}$$

Therefore, the equation can be written as

$$\frac{d}{dx}\left(e^{x^5}y\right) = x^4 e^{x^5}$$

so that integration of both sides of the equation yields

$$e^{x^5}y = \frac{1}{5}e^{x^5} + C.$$

```
> rightside:=int(intfac*x^4,x);
```

$$rightside := \frac{1}{5}e^{(x^5)}$$

Hence, a general solution is $y = \frac{1}{5} + Ce^{-x^5}$. Note that we compute y by using `solve` to solve the equation $e^{x^5}y = \frac{1}{5}e^{x^5} + C$ for y.

```
> step1:=solve(exp(x^5)*y=rightside+c,y);
```

$$step1 := \frac{1}{5}\frac{e^{(x^5)} + 5c}{e^{(x^5)}}$$

We find the unknown constant C by substituting the initial condition $y(0) = -7$ into the general solution and solving for C.

```
> findc:=solve(-7=subs(x=0,step1));
```

$$findc := \frac{-36}{5}$$

Therefore, the solution to the initial-value problem is $y = \frac{1}{5} - \frac{36}{5}e^{-x^5}$.

```
> subs(c=findc,step1);
```

$$\frac{1}{5}\frac{e^{(x^5)} - 36}{e^{(x^5)}}$$

∎

We can use `dsolve` to solve a first-order linear equation even if the coefficient functions are discontinuous.

EXAMPLE 3: If a drug is intoduced into the bloodstream in dosages $D(t)$ and is removed at a rate proportional to the concentration, the concentration $C(t)$ at time t is given by

$$\begin{cases} dC/dt = D(t) - kC \\ \quad C(0) = 0 \end{cases},$$

where $k > 0$ is the constant of proportionality.[*]
Suppose that over a 24-hour period, a drug is introduced into the bloodstream at a rate of $24/t_0$ for exactly t_0 hours and then stopped so that

$$D_{t_0} = \begin{cases} 24/t_0, 0 \leq t \leq t_0 \\ 0, t > t_0 \end{cases}.$$

Calculate and then graph $C(t)$ on the interval $[0, 30]$ if $k = 0.05, 0.10, 0.15$, and 0.20 for $t_0 = 4, 8, 12, 16$, and 20. How does increasing t_0 affect the concentration of the drug in the bloodstream? Increasing k?

SOLUTION: To compute $C(t)$, we must keep in mind that $D_{t_0}(t)$ is a piecewise defined function, which we define using `piecewise`. (Note that we use lowercase letters to avoid any ambiguity with built-in objects such as D.)

```
> c:='c':d:='d':
d:=(t,t0)->piecewise(t>=0 and t<=t0,24/t0,t>t0,0);
```

$$d := (t, t0) \rightarrow \text{piecewise}\left(0 \leq t \text{ and } t \leq t0, 24\frac{1}{t0}, t0 < t, 0\right)$$

For example, entering `d(t,4)` returns $D_4(t) = \begin{cases} 6, 0 \leq t \leq 4 \\ 0, t > 4 \end{cases}$.

```
> simplify(d(t,4));
```

$$\begin{cases} 0 & t < 0 \\ 6 & t \leq 4 \\ 0 & 4 < t \end{cases}$$

We must solve

$$\begin{cases} dC/dt = D_{t_0}(t) - kC \\ \quad C(0) = 0 \end{cases}$$

---

[*]J.D.Murray, *Mathematical Biology*, Springer-Verlag, 1990, pp. 645–649.

for $k = 0.05, 0.10, 0.15$, and 0.20, where $D_{t_0}(t) = D_4(t), D_8(t), \ldots, D_{20}(t)$, which are defined using seq in ds.

```
> t0vals:=[4,8,12,16,20]:
ds:=[seq(simplify(d(t,t0)),t0=t0vals)];
```

$$ds := \left[\left| \left\{ \begin{array}{ll} 0 & t<0 \\ 6 & t \leq 4, \\ 0 & 4<t \end{array} \right. \left\{ \begin{array}{ll} 0 & t<0 \\ 3 & t \leq 8, \\ 0 & 8<t \end{array} \right. \left\{ \begin{array}{ll} 0 & t<0 \\ 2 & t \leq 12, \\ 0 & 12<t \end{array} \right. \left\{ \begin{array}{ll} 0 & t<0 \\ \frac{3}{2} & t \leq 16, \\ 0 & 16<t \end{array} \right. \left\{ \begin{array}{ll} 0 & t<0 \\ \frac{6}{5} & t \leq 20 \\ 0 & 20<t \end{array} \right. \right| \right]$$

Then, for $k = 0.05$ we solve the initial-value probem

$$\begin{cases} dC/dt = D_{t_0}(t) - kC \\ C(0) = 0 \end{cases}$$

for each function $D_{t_0}(t)$ in ds. Note that only the right-hand side of each solution, the explicit formula for the solution, is returned in each case, so the resulting list of functions can easily be graphed with plot next.

```
> k:=0.05:
toplot05:=[seq(rhs(dsolve({diff(c(t),t)=d-k*c(t),c(0)=0},
          c(t))),d=ds)];
```

$$toplot05 := \left[\left| \left\{ \begin{array}{ll} 0 & t \leq 0 \\ 120. - 120.\, e^{(-0.05000000000\, t)} & t \leq 4., \\ 26.5683310\, e^{(-0.05000000000\, t)} & 4.<t \end{array} \right. \right. \right.$$

$$\left\{ \begin{array}{ll} 0 & t \leq 0 \\ -60.\, e^{(-0.05000000000\, t)} + 60. & t<8. \\ undefined & t = 8.\text{'} \\ 29.50948188\, e^{(-0.05000000000\, t)} & 8.<t \end{array} \right.$$

$$\left\{ \begin{array}{ll} 0 & t \leq 0 \\ 40. - 40.\, e^{(-0.05000000000\, t)} & t<12. \\ undefined & t = 12.\text{'} \\ 32.88475200\, e^{(-0.05000000000\, t)} & 12.<t \end{array} \right.$$

$$\left\{ \begin{array}{ll} 0 & t \leq 0 \\ -30.\, e^{(-0.05000000000\, t)} + 30. & t<16. \\ undefined & t = 16.\text{'} \\ 36.76622784\, e^{(-0.05000000000\, t)} & 16.<t \end{array} \right.$$

$$\left. \left\{ \begin{array}{ll} 0 & t \leq 0 \\ 24. - 24.\, e^{(-0.05000000000\, t)} & t \leq 20. \\ 41.23876387\, e^{(-0.05000000000\, t)} & 20.<t \end{array} \right. \right]$$

```
plot(toplot05,t=0..30,c=0..30);
```

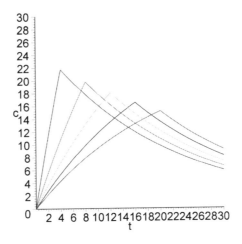

The same steps are repeated for , 0.15, and 0.20.

```
> k:=0.10:
toplot10:=[seq(rhs(dsolve({diff(c(t),t)=d-k*c(t),c(0)=0},
            c(t))),d=ds)]:
plot(toplot10,t=0..30,c=0..30);
```

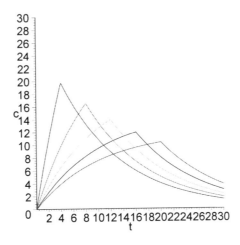

```
> k:=0.15:
toplot15:=[seq(rhs(dsolve({diff(c(t),t)=d-k*c(t),c(0)=0},
            c(t))),d=ds)]:
plot(toplot15,t=0..30,c=0..30);
```

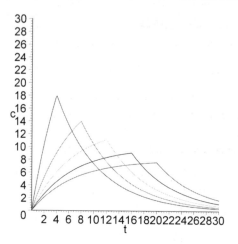

```
k:=0.20:

toplot20:=[seq(rhs(dsolve({diff(c(t),t)=d-k*c(t),c(0)=0},
          c(t))),d=ds)]:

plot(toplot20,t=0..30,c=0..30);
```

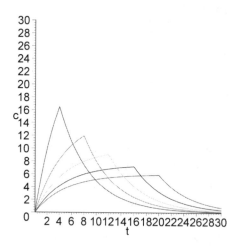

From the graphs, we see that as t_0 is increased, the maximum concentration level decreases and occurs at later times, while increasing k increases the rate at which the drug is removed from the bloodstream.

■

If the integration cannot be carried out, the solution can often be approximated numerically by taking advantage of numerical integration techniques, as illustrated in the following example.

EXAMPLE 4: Graph the solution to the initial-value problem $y' - \sin(2\pi x)y = 1$, $y(0) = 1$ on the interval $[0, 2\pi]$.

SOLUTION: Note that `dsolve` is successful in finding the solution to the initial-value problem even though the result contains unevaluated integrals.

```
partsol:=dsolve({diff(y(x),x)-sin(2*Pi*x)*y(x)=1,
        y(0)=1},y(x));
```

$$partsol := \text{y}(x) = e^{\left(-\frac{1}{2}\frac{\cos(2\pi x)+1}{\pi}\right)} \int_0^x e^{\left(\frac{1}{2}\frac{\cos(2\pi u)+1}{\pi}\right)} du$$

$$+ \frac{e^{\left(-\frac{1}{2}\frac{\cos(2\pi x)+1}{\pi}\right)}}{\cosh\left(\frac{1}{\pi}\right) - \sinh\left(\frac{1}{\pi}\right)}$$

We can evaluate the result for particular numbers. For example, entering

```
> eval(subs(x=1,partsol));
```

$$\text{y}(1) = e^{\left(-\frac{1}{\pi}\right)} \int_0^1 e^{\left(\frac{1}{2}\frac{\cos(2\pi u)+1}{\pi}\right)} du + \frac{e^{\left(-\frac{1}{\pi}\right)}}{\cosh\left(\frac{1}{\pi}\right) - \sinh\left(\frac{1}{\pi}\right)}$$

find the value of the solution to the initial-value problem if $x = 1$. This result is a bit complicated to understand, so we use `evalf` to obtain a numerical approximation.

```
> evalf(subs(x=1,partsol));
```

$$\text{y}(1) = 1.858273585$$

To graph the solution on the interval $[0, 2\pi]$, we use `dsolve` together with the `numeric` option to generate a numerical solution to the initial-value problem

```
numsol:=dsolve({diff(y(x),x)-sin(2*Pi*x)*y(x)=1,y(0)=1},
        y(x),numeric);
```

$$numsol := \mathbf{proc}(rkf45\_x) \dots \ \mathbf{end}$$

We can evaluate the result for particular values of x. For example, entering

> `numsol(1);`

$$[x = 1, y(x) = 1.858273504291522]$$

approximates the value of the solution to the initial-value problem if $x = 1$. Thus, the result means that $y(1) \approx 1.85828$. We can graph results returned by `dsolve` together with the `numeric` option using the `odeplot` command, which is contained in the `plots` package: entering

> `with(plots):`
`odeplot(numsol,[x,y(x)],0..2*Pi);`

graphs the solution to the initial-value problem on the interval $[0, 2\pi]$.

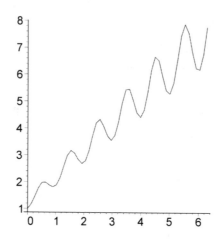

Alternatively, if we had wished only to generate a graph of the solution, we could take advantage of the `DEplot` command, which is contained in the `DEtools` package. For example, entering

> `with(DEtools):`
`DEplot(diff(y(x),x)-sin(2*Pi*x)*y(x)=1,y(x),x=0..2*Pi,`
        `[[y(0)=1],[y(0)=0],[y(0)=3]],`
`y=0..2*Pi,`
            `linecolor=BLACK,color=GRAY,stepsize=0.05);`

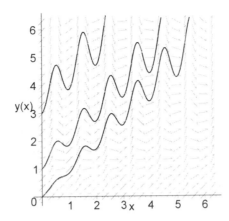

first loads the **DEtools** package and then graphs the solutions to the differential equation that satisfy the initial conditions $y(0) = 1$, $y(0) = 0$, and $y(0) = 3$ together with the direction field for the equation for $0 \leq x \leq 2\pi$ and $0 \leq y \leq 2\pi$. The option `linecolor = BLACK` indicates that the solution curves be displayed in black and the option `color = GRAY` indicates that the direction field be displayed in gray. If the option `arrows = NONE` had been included in the `DEplot` command, the direction field would not have been generated.

■

APPLICATION

Antibiotic Production

When you are injured or sick, your doctor may prescribe antibiotics to prevent or cure infections. In the journal article "Changes in the Protein Profile of *Streptomyces griseus* during a Cycloheximide Fermentation" we see that production of the antibiotic cycloheximide by *Streptomyces* is typical of antibiotic production. During the production of cycloheximide, the mass of *Streptomyces* grows relatively quickly and produces little cycloheximide. After approximately 24 hours, the mass of *Streptomyces* remains relatively constant and cycloheximide accumulates. However, once the concentration of cycloheximide reaches a certain level, extracellular cycloheximide is degraded (**feedback inhibited**). One approach to alleviating this problem to maximize cycloheximide production is to remove extracellular cycloheximide continuously. The rate of growth of *Streptomyces* can be described by the separable equation

$$\frac{dx}{dt} = \mu_{\max}\left(1 - \frac{X}{X_{\max}}\right)X,$$

where X represents the mass concentration in g/L, $\mu_{\max}$ is the maximum specific growth rate, and X_{max} represents the maximum mass concentration. (Note that we convert this equation to a linear equation with the substitution $y = X^{-1}$.)

We now solve the initial-value problem $\begin{cases} sX/dt = \mu_{max}(1 - X/X_{max})X \\ X(0) = 1 \end{cases}$ with dsolve, naming the result sol1.

```
x:='x':
> sol1:=dsolve({diff(x(t),t)=mu[max]*(1-x(t)/xmax)*x(t),
        x(0)=1},x(t));
```

$$sol1 := x(t) = \frac{xmax}{1 + e^{(-\mu_{max}t)}(xmax - 1)}$$

Experimental results have shown that $\mu_{max} = 0.3 \text{ hr}^{-1}$ and $X_{max} = 10 \text{ g/L}$. For these values, we graph $X(t)$ on the interval [0,24]. Then, we use seq and array to determine the mass concentration at the end of 4, 8, 12, 16, 20, and 24 hours.

```
> mu[max]:=0.3:xmax:=10:
plot(x(t),t=0..24);
```

```
> tvals:=seq(4*i,i=1..6):
array([seq([t,rhs(sol1)],t=tvals)]);
```

$$\begin{bmatrix} 4 & 2.694874524 \\ 8 & 5.505208649 \\ 12 & 8.026239369 \\ 16 & 9.310399881 \\ 20 & 9.781780508 \\ 24 & 9.933255753 \end{bmatrix}$$

The rate of accumulation of cycloheximide is the difference between the rate of synthesis and the rate of degradation:

$$\frac{dp}{dt} = R_s - R_d.$$

It is known that $R_d = K_d P$, where $K_d \approx 5 \times 10^{-3} \text{hr}^{-1}$, so $dP/dt = R_s - R_d$ is equivalent to $dP/dt = R_s - K_d P$. Furthermore,

$$R_s = Q_{po} E X (1 + P/K_1)^{-1},$$

where Q_{po} represents the specific enzyme activity with value $Q_{po} \approx 0.6 \text{g CH/g protein} \cdot \text{hr}$ and K_I represents the inhibition constant. E represents the intracellular concentration of an enzyme, which we will assume is constant. For large values of K_I and t, $X(t) \approx 10$ and $(1 + P/K_1)^{-1} \approx 1$. Thus, $R_s \approx 10 Q_{po} E$, so

$$\frac{dP}{dt} = 10 Q_{po} E - K_d P.$$

After defining $K_d \approx 5 \times 10^{-3} \text{hr}^{-1}$ and $Q_{po} \approx 0.6 \text{g CH/g protein} \cdot \text{hr}$, we solve the initial-value problem

$$\begin{cases} dP/dt = 10 Q_{po} E - K_d P \\ p(24) = 0 \end{cases}$$

and then graph $\frac{1}{E} P(t)$ on the interval [24,1000].

```
> p:='p':

k[d]:=5/1000:

Q[po]:=0.6:

sol2:=dsolve({diff(p(t),t)=10*Q[po]*cape-k[d]*p(t),
            p(24)=0},p(t));
```

$$sol2 := p(t) = 1200.\ cape - 1352.996222\ e^{\left(-\frac{1}{200}t\right)} cape$$

```
> toplot:=simplify(rhs(sol2)/cape);
```

$$toplot := 1200. - 1352.996222\ e^{(-0.005000000000\ t)}$$

```
> plot(toplot,t=24..1000);
```

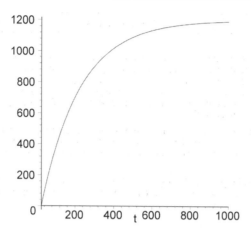

From the graph, we see that the total accumulation of the antiobiotic approaches a limiting value (which in this case is 1200).

Source: Kevin H. Dykstra and Henry Y. Wang, "Changes in the Protein Profile of *Streptomyces Griseus* during a Cycloheximide Fermentation," *Biochemical Engineering V,* Annals of the New York Academy of Sciences, Volume 56, New York Academy of Sciences (1987), pp. 511–522.

2.6 Numerical Approximation of First-Order Equations

Built-In Methods

Numerical approximations of solutions to differential equations can be obtained with dsolve together with the numeric option. This command is particularly useful when working with nonlinear equations, for which dsolve alone is unable to find an explicit solution. This command is entered in the form

$$\text{dsolve}(\{\text{deq, ics}\}, \text{fun, numeric}),$$

where deq is solved for fun. Note that the number of initial conditions in ics must equal the order of the differential equation indicated in deq.

A description of the various options available when using dsolve together with the numeric option can be obtained by entering ?dsolve/numeric.

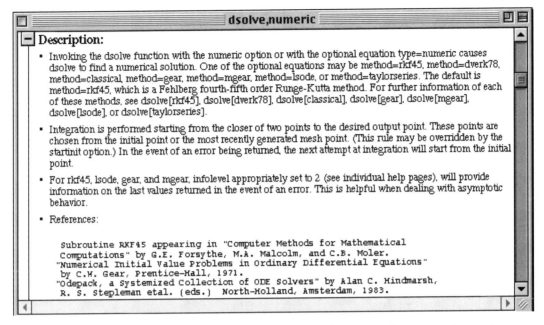

In order to illustrate the command `dsolve` together with the `numeric` option, we consider several nonlinear equations in the following examples.

EXAMPLE 1: Graph the solution to the initial-value problem

$$\begin{cases} dy/dx = \sin(2x - y) \\ \quad y(0) = 0.5 \end{cases},$$

on the interval [0,15]. What is the value of *y(1)?*

SOLUTION: First, we define `eq` to be the equation $dy/dx = \sin(2x - y)$ and then use `dsolve` together with the `numeric` option to approximate the solution of `eq` subject to the initial condition $y(0) = 0.5$, naming the resulting output `sol`. The resulting output is a procedure that represents an approximate function obtained through interpolation.

```
> x:='x':y:='y':
eq:=D(y)(x)=sin(2*x-y(x)):
> sol:=dsolve({eq,y(0)=.5},y(x),numeric);
```

$$sol := \mathbf{proc}(rkf45\_x)\dots\mathbf{end}$$

We can evaluate `sol` for particular values of *x*. For example, entering

```
> sol(1);
```

$$[x = 1, y(x) = 0.8758947797345141]$$

returns an ordered pair corresponding to x and $y(x)$ if $x = 1$. Entering

```
> sol(1)[2];
```

$$y(x) = 0.8758947797345141$$

returns the second part of `sol`. Entering

```
> rhs(sol(1)[2]);
```

$$0.8758947797345141$$

returns the value of the solution, $y(x)$, if $x = 1$.
We then graph the solution by using the command `odeplot`, which is contained in the **plots** package. Generally, we will graph numerical solutions of differential equations obtained with `dsolve` together with the `numeric` option with `odeplot`.

```
> with(plots):
odeplot(sol,[x,y(x)],0..15);
```

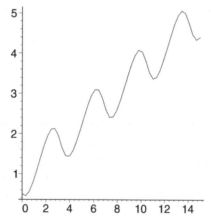

We can also use `DEplot`, which is contained in the **DEtools** package, to graph solutions. However, in this case, a numerical solution that can be evaluated, such as we obtain with `dsolve`, is not generated. For example, entering

```
> with(DEtools):
DEplot(diff(y(x),x)=sin(2*x-y(x)),y(x),
        x=0..15,{[0,1],[0,-1]},
```

```
stepsize=0.05,linecolor=BLACK,color=GRAY);
```

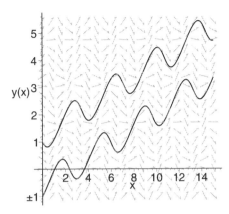

graphs the solutions of the equation satisfying $y(0) = 1$ and $y(0) = -1$ along with the direction field for the equation on the interval [0,15]. The direction field is not displayed if the option `arrows = NONE` is included in the `DEplot` command.

■

We can also use `DEplot` to graph solutions to a differential equation under changing initial conditions.

EXAMPLE 2: Graph the solution of $y' = \sin(xy)$ subject to the initial condition $y(0) = i$ on the interval[0,7] for $i = 0.5, 1.0, 1.5, \ldots, 5.5$, and 6.

SOLUTION: We begin by using `seq` to define the set of ordered pairs $(0, i)$ for $i = 1, 2, \ldots, 12$. *These correspond to the initial conditions $y(0) = 1/2$ for $i = 1, 2, \ldots, 12$.*

```
> inits:={seq([0,i/2],i=1..12)}:
```

Next, we use `DEplot` to graph the solutions to $y' = \sin(x, y)$ for the initial conditions specified in `inits`. The option `arrows = NONE` is included so that the direction fields for the equation are not included in the graph. On the other hand, the option stepsize $= 0.1$ instructs Maple to use a smaller step size, helping ensure that the resulting graphs appear smooth.

```
> with(DEtools):
DEplot(diff(y(x),x)=sin(x*y(x)),y(x),x=0..7,inits,
        arrows=NONE,stepsize=0.1,linecolor=BLACK);
```

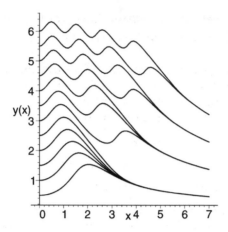

■

APPLICATION

Modeling the Spread of a Disease

Suppose that a disease is spreading among a population of size N. With some diseases, such as chickenpox, once an individual has had the disease, the individual becomes immune to the disease. With other diseases, such as most venereal diseases, once an individual has had the disease and recovers from the disease, the individual does not become immune to the disease; subsequent encounters can lead to recurrences of the infection.

Let $S(t)$ denote the percentage of the population susceptible to a disease at time t, $I(t)$ the percentage of the population infected with the disease, and $R(t)$ the percentage of the population unable to contract the disease. For example, $R(t)$ could represent the percentage of persons who have had a particular disease, recovered, and have subsequently become immune to the disease.

In order to model the spread of various diseases, we begin by making several assumptions and introducing some notation.

1. Susceptible and infected individuals die at a rate proportional to the number of susceptible and infected individuals with proportionality constant μ called the **daily death removal rate**; the number $1/\mu$ is the **average lifetime** or **life expectancy**.
2. The constant λ represents the **daily contact rate**: on average, an infected person will spread the disease to λ people per day.
3. Individuals recover from the disease at a rate proportional to the number infected with the disease with proportionality constant γ. The constant γ is called the **daily recovery removal rate**; the **average period of infectivity** is $1/\gamma$.

4. The **contact number** $\sigma = \lambda/(\gamma + \mu)$ represents the average number of contacts an infected person has with both susceptible and infected persons.

If a person becomes susceptible to a disease after recovering from it (such as gonorrhea, meningitis, and streptococcal sore throat), then the percentage of persons susceptible to becoming infected with the disease, $S(t)$, and the percentage of people in the population infected with the disease, $I(t)$, can be modeled by the system of differential equations

$$\begin{cases} S'(t) = -\lambda IS + \gamma I + \mu = \mu S \\ I'(t) = \lambda IS - \gamma I = \mu I \\ S(0) = S_0, I(0) = I_0, S(t) + I(t) = 1 \end{cases}$$

This model is called an **SIS** (susceptible–infected–susceptible) model because once an individual has recovered from the disease, the individual again becomes susceptible to the disease.

We can write $I'(t) = \lambda IS - \gamma I - \mu I$ as

$$I'(t) = \lambda I(1 - I) - \gamma I - \mu I$$

because $S(t) = 1 - I(t)$ and thus we need to solve the initial-value problem

$$\begin{cases} I'(t) = [\lambda - (\gamma + \mu)]I - \lambda I^2 \\ I(0) = I_0 \end{cases}$$

In the following, we use i to represent I, thus avoiding conflict with the built-in constant $\mathtt{I} = \sqrt{-1}$. After defining eq, we use DSolve to find the solution to the initial-value problem.

> `eq:=diff(i(t),t)+(gamma+mu-lambda)*i(t)=-lambda*i(t)^2;`

$$eq := \left(\frac{\partial}{\partial t} i(t) \right) + (\gamma + \mu - \lambda)i(t) = -\lambda i(t)^2$$

> `sol:=dsolve({eq,i(0)=i0},i(t));`

$$sol := i(t) = (\gamma + \mu - \lambda) / \left(-\lambda + \frac{e^{((\gamma + \mu - \lambda)t)}(\gamma + \mu - \lambda + i0\lambda)\gamma}{i0(\gamma + \mu - \lambda)} \right.$$
$$\left. + \frac{e^{((\gamma + \mu - \lambda)t)}(\gamma + \mu - \lambda + i0\lambda)\mu}{i0(\gamma + \mu - \lambda)} - \frac{e^{((\gamma + \mu - \lambda)t)}(\gamma + \mu - \lambda + i0\lambda)\lambda}{i0(\gamma + \mu - \lambda)} \right)$$

> `simplify(sol);`

$$i(t) = i0(\gamma + \mu - \lambda) / (e^{((\gamma + \mu - \lambda)t)}\mu + e^{((\gamma + \mu - \lambda)t)}\gamma - i0\lambda$$
$$- \lambda e^{((\gamma + \mu - \lambda)t)} + e^{((\gamma + \mu - \lambda)t)}i0\lambda)$$

We can use this result to see how a disease might spread through a population. For example, we compute the solution to the initial-value problem, which is extracted from `sol` with `rhs(sol)`, if $\lambda = 0.50$, and $\mu = .65$. In this case, we see that the contact number is $\sigma = \lambda/(\gamma + \mu) \approx 0.357143$.

```
sigma:=subs({lambda=0.5,gamma=0.75,
        mu=0.65},lambda/(gamma+mu));
toplot:=subs({lambda=0.5,gamma=0.75,
        mu=0.65},rhs(sol));
```

$$\sigma := 0.3571428571$$

$$toplot := 0.90\,\frac{1}{-0.5 + 1.000000000\,\frac{e^{(0.90t)}(0.90+0.5i0)}{i0}}$$

Next, we use `seq` to substitute various initial conditions into `toplot` and graph the resulting list of functions with `plot` $0 \le t \le 5$. Apparently, regardless of the initial percentage of the population infected, under these conditions, the disease is eventually removed from the population. This makes sense because the contact number is less than one.

```
> i0vals:=seq(0.1*k,k=1..9):
plot([seq(subs(i0=k,toplot),k=i0vals)],t=0..5);
```

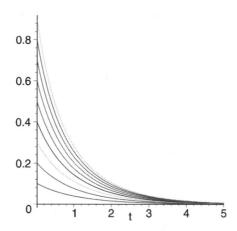

On the other hand, if $\lambda = 1.50$, $\gamma = 0.75$, and $\mu = 0.65$, we see that the contact number is $\sigma = \lambda/(\gamma + \mu) \approx 1.07143$.

```
> sigma:=subs({lambda=1.5,gamma=0.75,
        mu=0.65},lambda/(gamma+mu));
> toplot:=subs({lambda=1.5,gamma=0.75,
        mu=0.65},rhs(sol)):
```

$$\sigma := 1.071428571$$

Proceeding as before, we graph the solution using different initial conditions.

```
> i0vals:=seq(0.1*k,k=1..9):
plot([seq(subs(i0=k,toplot),k=i0vals)],t=0..20);
```

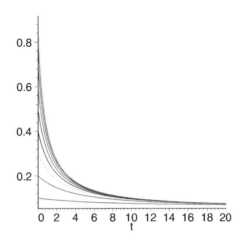

In this case, we see that no matter what percentage of the population is initially infected, a certain percentage of the population is always infected. This makes sense because the contact number is greater than one. In fact, it is a theorem that

$$\lim_{t\to+\infty} I(t) = \begin{cases} 1 - 1/\sigma, & \text{if } \sigma > 1 \\ 0, & \text{if } \sigma \geq 1 \end{cases}.$$

```
> i0vals:=seq(0.01*k,k=1..9):
plot([seq(subs(i0=k,toplot),k=i0vals)],t=0..20);
```

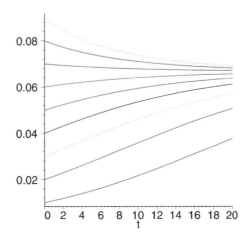

The incidence of some diseases, such as measles, rubella, and gonorrhea, oscillates seasonally. To model these diseases, we may wish to replace the constant contact rate λ by a periodic function $\lambda(t)$. For example, to graph the solution to the SIS model for various initial conditions if (a) $\lambda(t) = 3 - 2.5 \sin 6t$, $\gamma = 2$, and $\mu = 1$ and (b) $\lambda(t) = 3 - 2.5 \sin 6t$, $\gamma = 1$, and $\mu = 1$, we proceed as follows.

For (a), we begin by defining eq.

```
> lambda:=t->2-2.5*sin(6*t):
eq:=diff(i(t),t)=(lambda(t)-3)*i(t)-lambda(t)*i(t)^2;
```

$$eq := \frac{\partial}{\partial t} i(t) = (-1 - 2.5 \sin(6t))i(t) - (2 - 2.5 \sin(6t))i(t)^2$$

We will graph the solutions satisfying the initial conditions $I(0) = i_0$ for $i_0 = 0.1, 0.2, \ldots, 0.9$, so we use seq to define inits to be the set of nine ordered pairs corresponding to the initial conditions. Note that the first coordinate of each pair corresponds to the t-value and the second coordinate corresponds to the value of i for that t-value.

```
> inits:={seq([0,i/10],i=1..9)};
```

Next, we use DEplot to graph eq for $0 \le t \le 10$ subject to the initial conditions specified in inits. The option stepsize = 0.05 reduces the step size in the DEplot calculations, helping ensure that the resulting graphs are smooth. In this case, the direction field for eq is not displayed because we include the option arrows = NONE.

```
> with(DEtools):
DEplot(eq,i(t),0..10,inits,stepsize=0.05,arrows=NONE,
        linecolor=BLACK,thickness=1);
```

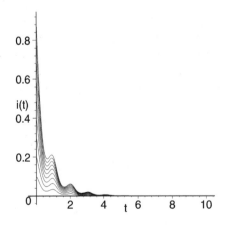

For (b), we proceed in the same manner as in (a).

```
> eq:=diff(i(t),t)=(a(t)-2)*i-a(t)*i^2:

DEplot(eq,i(t),0..10,inits,stepsize=0.05,arrows=NONE,

           linecolor=BLACK,thickness=1);
```

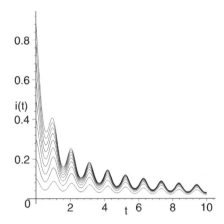

Source: Herbert W. Hethcote, "Three Basic Epidemiological Models," in *Applied Mathematical Ecology,* edited by Simon A. Levin, Thomas G. Hallan, and Louis J. Gross, New York, Springer-Verlag (1989), pp. 119–143.

In other cases, you may wish to implement your own numerical algorithms to approximate solutions of differential equations. We briefly discuss three familiar methods (Euler's method, the improved Euler's method, and the Runge–Kutta method) and illustrate how to implement these algorithms using Mathematica. Details regarding these and other algorithms, including discussions of the error involved in implementing them, can be found in most numerical analysis texts or other references such as the *Handbook of Differential Equations* by Daniel Zwillinger, which is published by Academic Press (San Diego, CA).

Euler's Method

In many cases, we cannot obtain an explicit formula for the solution to an initial-value problem of the form

$$\begin{cases} dy/dx = f(x, y) \\ \quad y(x_0) = y_0 \end{cases}$$

but we can approximate the solution using a numerical method such as **Euler's method,** which is based on tangent line approximations. Let h represent a small change, or **step size,** in the independent variable x. Then, we approximate the value of y at the sequence of x-values, $x_1, x_2, x_3, \ldots,$ where

$$x_1 = x_0 + h$$

$$x_2 = x_1 + h = x_0 + 2h$$

$$x_3 = x_2 + h = x_0 + 3h$$

$$\vdots$$

$$x_n = x_{n-1} + h = x_0 + nh.$$

The slope of the tangent line to the graph of y at these values of x is found with the differential equation $y' = dy/dx = f(x, y)$. For example, at $x = x_0$, the slope of the tangent line is $f(x_0, y(x_0)) = f(x_0, y_0)$. Therefore, the tangent line to the graph of y is

$$y - y_0 = f(x_0, y_0)(x - x_0) \quad \text{or} \quad y = f(x_0, y_0)(x - x_0) + y_0.$$

Using this line to find the value of y at $x = x_1$ (which we call y_1) then yields

$$y_1 = f(x_0, y_0)(x_1 - x_0) + y_0 = hf(x_0, y_0) + y_0.$$

Therefore, we obtain the approximate value of y at $x = x_2$.

Next, we use the point (x_1, y_1) to estimate the value of y when $x = x_2$. Using a similar procedure, we approximate the tangent line at $x = x_1$ with

$$y - y_1 = f(x_1, y_1)(x - x_1) \quad \text{or} \quad y = f(x_1, y_1)(x - x_1) + y_1.$$

Then, at $x = x_2$,

$$y_2 = f(x_1, y_1)(x_2 - x_1) + y_1 = hf(x_1, y_1) + y_1.$$

Continuing with this procedure, we see that at $x = x_n$,

$$y_n = hf(x_{n-1}, y_{n-1}) + y_{n-1}.$$

Using this formula, we obtain a sequence of points of the form (x_n, y_n) $(n = 1, 2, \ldots)$ where y_n is the approximate value of $y(x_n)$.

EXAMPLE 3: Use Euler's method with (a) $h = 0.1$ and (b) $h = 0.05$ to approximate the solution of $y' = xy$, $y(0) = 1$ on $0 \le x \le 1$. Also, determine the exact solution and compare the results.

SOLUTION: Because we will be considering this initial-value problem in subsequent examples, we first determine the exact solution with `dsolve` and graph the result with `plot`, naming the graph p1.

```
> exactsol:=dsolve({diff(y(x),x)=x*y(x),y(0)=1},y(x));
```

$$exactsol := y(x) = e^{\left(\frac{1}{2}x^2\right)}$$

```
> p1:=plot(rhs(exactsol),x=0..1):
```

To implement Euler's method, we note that $f(x, y) = xy$, $x_0 = 0$, and $y_0 = 1$. Then, with $h = 0.1$, we have the formula

$$y_n = hf(x_{n-1}, y_{n-1}) + y_{n-1} = 0.1x_{n-1}y_{n-1} + y_{n-1}.$$

For $x_1 = x_0 + h = 0.1$, we have

$$y_1 = 0.1x_0y_0 + y_0 = 0.1(0)(1) + 1 = 1.$$

Similarly, for $x_2 = x_0 + 2h = 0.2$,

$$y_2 = x.1x_1y_1 + y_1 = 0.1(0.1)(1) + 1 = 1.01.$$

In the following, we define f, h, x, and y to calculate

$$y_n = hf(x_{n-1}, y_{n-1}) + y_n.$$

Note that in defining the recursively defined function y we take advantage of the option remember. This instructs Maple to "remember" the values of y computed, and thus, when computing $y(n)$, Maple need not recompute $y(n - 1)$ if it has previously been computed.

```
> f:=(x,y)->x*y:

h:=0.1:

x:=n->n*h:

y:=proc(n) option remember;

      y(n-1)+h*f(x(n-1),y(n-1))

      end:

y(0):=1:
```

Next, we use seq to calculate the set of ordered pairs (x_n, y_n) for $n = 0, 1, 2, \ldots, 9, 10$, naming the result first, and then array to view first in traditional row-and-column form.

```
> first:=[seq([x(n),y(n)],n=0..10)]:

array(first);
```

$$\begin{vmatrix} 0 & 1 \\ 0.1 & 1 \\ 0.2 & 1.01 \\ 0.3 & 1.0302 \\ 0.4 & 1.061106 \\ 0.5 & 1.10355024 \\ 0.6 & 1.158727752 \\ 0.7 & 1.228251417 \\ 0.8 & 1.314229016 \\ 0.9 & 1.419367337 \\ 1.0 & 1.547110397 \end{vmatrix}$$

To compare these results with the exact solution, we plot the list of points in `first` using `plot` together with the option $\mathtt{style = POINT}$, naming the resulting graph p2. We then display p1 and p2 together with `display`, which is contained in the **plots** package.

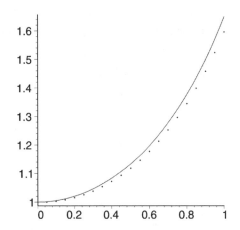

```
> p2:=plot(first,style=POINT,color=BLACK):
with(plots):
display({p1,p2});
```

Then, for $h = 0.05$, we use

$$y_n = hf(x_{n-1}, y_{n-1}) + y_{n-1} = 0.05x_{n-1}y_{n-1} + y_{n-1}$$

to obtain an approximation. In the same manner as in (a), we define f, h, x, and y to calculate

$$y_n = hf(x_{n-1}, y_{n-1}) + y_n.$$

Then, we use `seq` to calculate the set of ordered pairs (x_n, y_n) for $n = 0, 1, 2, \ldots, 19, 20$, naming the result `second`, followed by `array` to view `second` in traditional row-and-column form.

```
> x:='x':y:='y':

h:=0.05:

x:=n->n*h:

y:=proc(n) option remember;

      y(n-1)+h*f(x(n-1),y(n-1))

      end:

y(0):=1:

> second:=[seq([x(n),y(n)],n=0..20)]:

array(second);
```

0	1
0.05	1
0.10	1.0025
0.15	1.00751250
0.20	1.015068844
0.25	1.025219532
0.30	1.038034776
0.35	1.053605298
0.40	1.072043391
0.45	1.093484259
0.50	1.118087655
0.55	1.146039846
0.60	1.177555942
0.65	1.212882620
0.70	1.252301305
0.75	1.296131851
0.80	1.344736795
0.85	1.398526267
0.90	1.457963633
0.95	1.523571997
1.00	1.595941667

We graph the approximation obtained with $h = 0.05$ together with the graph of $y = e^{x^2/2}$. Notice that the approximation is more accurate when h is decreased.

```
> p3:=plot(second,style=POINT,color=BLACK):

display({p1,p3});
```

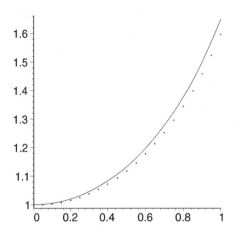

Improved Euler's Method

Euler's method can be improved by using an average slope over each interval. Using the tangent line approximation of the curve through (x_0, y_0), $y = f(x_0, y_0)(x - x_0) + y_0$, we find the approximate value of y at $x = x_1$, which we now call y_1^*. Therefore,

$$y_1^* = hf(x_0, y_0) + y_0$$

Then, with the differential equation $y' = f(x, y)$, we find that the approximate slope of the tangent line at $x = x_1$ is $f(x_1, y_1^*)$. Then, the average of the two slopes, $f(x_0, y_0)$ and $f(x_1, y_1^*)$, is $\frac{1}{2}(f(x_0, y_0) + f(x_1, y_1^*))$, and the equation of the line through (x_0, y_0) with slope $\frac{1}{2}(f(x_0, y_0) + f(x_1, y_1^*))$ is

$$y = \frac{1}{2}(f(x_0, y_0) + f(x_1, y_1^*))(x - x_0) + y_0.$$

Therefore, at $x = x_1$, we find the approximate value of f given by

$$y_1 = \frac{1}{2}(f(x_0, y_0) + f(x_1, y_1^*))(x_1 - x_0) + y_0 = \frac{1}{2}(f(x_0, y_0) + f(x_1, y_1^*))h + y_0.$$

Continuing in this manner, the approximation at each step of the improved Euler method depends on the following two calculations:

$$y_n^* = hf(x_{n-1}, y_{n-1}) + y_{n-1}y_n = \frac{f(x_{n-1}, y_{n-1}) + f(x_n, y_n^*)}{2}h + y_{n-1}.$$

EXAMPLE 4: Use the improved Euler's method to approximate the solution of $y' = xy, y(0) = 1$ on $0 \leq x \leq 1$ for $h = 0.1$. Also, compare the results with the exact solution.

SOLUTION: In this case, $f(x, y) = xy$, $x_0 = 0$, and $y_0 = 1$. Therefore, we use the equations

$$y_n^* hx_{n-1}y_{n-1} + y_{n-1}$$

and

$$y_n = \frac{x_{n-1}y_{n-1} + x_n y_n^*}{2}h + y_{n-1}$$

for $n = 1, 2, \ldots, 10$. For example, if $n = 1$, we have

$$y_1^* = hx_0y_0 + y_0 = (0.1)(0)(1) + 1 = 1$$

and

$$y_1 \frac{x_0y_0 + x_1y_1^*}{2}h + y_0 = \frac{(0)(1) + (0.1)(1)}{2}(0.1) + 1 = 1.005.$$

Then,

$$y_2^* = hx_1y_1 + y_1 = (0.1)(0.1)(1.005) + 1.005 = 1.01505$$

and

$$y_2 \frac{x_1y_1 + x_2y_2^*}{2}h + y_1 = \frac{(0.1)(1.005) + (0.2)(1.01505)}{2}(0.1) + 1.005 = 1.0201755.$$

In the same way as in the previous example, we define f, x, h, and y. We define Y using the option remember so that Maple "remembers" the values of y computed.

```
> x:='x':y:='y':
f:=(x,y)->x*y:
h:=0.1:
x:=n->n*h:
y:=proc(n) option remember;
        y(n-1)+h/2*(f(x(n-1),y(n-1))+
                f(x(n),y(n-1)+h*f(x(n-1),y(n-1))))
        end:
```

```
y(0):=1:
```

We then compute (x_n, y_n) for $n = 0, 1, \ldots, 10$ and name the resulting list of ordered pairs `third`.

```
> third:=[seq([x(n),y(n)],n=0..10)]:
array(third);
```

0	1
0.1	1.005000000
0.2	1.020175500
0.3	1.045985940
0.4	1.083223039
0.5	1.133051299
0.6	1.197068697
0.7	1.277392007
0.8	1.376773105
0.9	1.498755202
1.0	1.647881345

We graph the approximation obtained with $h = 0.10$ together with the graph of $y = e^{x^2/2}$. From the results, we see that the approximation using the improved Euler's method results in a slight improvement over that obtained in Example 3.

```
> p4:=plot(third,style=POINT,color=BLACK):
display({p1,p4});
```

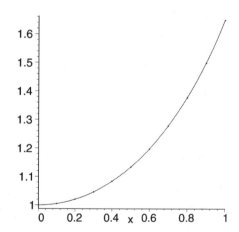

The Runge-Kutta Method

In an attempt to improve on the approximation obtained with Euler's method as well as avoid the analytic differentiation of the function $f(x, y)$ to obtain $y'', y''', \ldots$, the Runge–Kutta method is introduced. Let us begin with the Runge–Kutta method of order two. Suppose that we know the value of y at x_n. We now use the point (x_n, y_n) to approximate the value of y at a nearby value $x = x_n + h$ by assuming that

$$y_{n+1} = y_n + Ak_1 + Bk_2$$

where

$$k_1 = hf(x_n, y_n) \quad \text{and} \quad k_2 = hf(x_n + ah, y_n + bk_1).$$

We can also use the Taylor series expansion of y to obtain another representation of $y_{n+1} = y(x_n + h)$ as follows:

$$y(x_n + h) = y(x_n) + hy'(x_n) + h^2 \frac{y''(x_n)}{2!} + \cdots = y_n + hy'(x_n) + h^2 \frac{y''(x_n)}{2!} + \cdots.$$

Now, because

$$y_{n+1} = y_n + Ak_1 + Bk_2 = y_n + Ahf(x_n, y_n) + Bhf(x_n + ah, y_n + bhf(x_n, y_n)),$$

we wish to determine values of A, B, a, and b such that these two representations of y_{n+1} agree. Notice that if we let $A = 1$ and $B = 0$, then the relationships match up to order h. However, we can choose these parameters more wisely so that agreement occurs up through terms of order h^2. This is accomplished by considering the Taylor expansion of a function F of two variables about (x_0, y_0), which is given by

$$F(x, y) = f(x_0, y_0) + \frac{\partial F}{\partial x}(x_0, y_0)(x - x_0) + \frac{\partial F}{\partial y}(x_0, y_0)(y - y_0) + \cdots.$$

In our case, we have

$$f(x_n + ah, y_n + bhf(x_n, y_n)) = f(x_n, y_n) + ah\frac{\partial f}{\partial x}(x_n, y_n) + bhf(x_n, y_n)\frac{\partial f}{\partial x}(x_n, y_n) + O(h^2).$$

The power series is then substituted into the following expression and simplified to yield:

$$y_{n+1} = y_n + Ahf(x_n, y_n) + Bhf(x_n + ah, y_n + bhf(x_n, y_n))$$

$$= y_n + (A + B)hf(x_n, y_n) + aBh^2 \frac{\partial f}{\partial x}(x_n, y_n) + bBh^2 f(x_n, y_n)\frac{\partial f}{\partial x}(x_n, y_n) + O(h^3).$$

Comparing this expression with the following power series obtained directly from the Taylor series of y,

$$y(x_n + h) = y(x_n) + hf(x_n, y_n) + \frac{1}{2}h^2 \frac{\partial f}{\partial x}(x_n, y_n) + \frac{1}{2}h^2 f(x_n, y_n)\frac{\partial f}{\partial y}(x_n, y_n) + O(h^3)$$

or

$$y_{n+1} = y_n + hf(x_n, y_n) + \frac{1}{2}h^2\frac{\partial f}{\partial x}(x_n, y_n) + \frac{1}{2}h^2 f(x_n, y_n)\frac{\partial f}{\partial y}(x_n, y_n) + O(h^3),$$

we see that A, B, a, and b must satisfy the following system of nonlinear equations:

$$A + B = 1, aA = \frac{1}{2}, \text{ and } bB = \frac{1}{2}.$$

Therefore, choosing $a = b = 1$, the Runge–Kutta method of order two uses the equation:

$$y_{n+1} = y(x_n + h) = y_n + \frac{1}{2}hf(x_n, y_n) + \frac{1}{2}hf(x_n + h, y_n + hf(x_n, y_n)) = y_n + \frac{1}{2}(k_1 + k_2),$$

where $k_1 = hf(x_n, y_n)$ and $k_2 = hf(x_n + h, y_n + k_1)$.

EXAMPLE 5: Use the Runge–Kutta method with $h = 0.1$ to approximate the solution of the initial-value problem $y' = xy$, $y(0) = 1$ on $0 \leq x \leq 1$.

SOLUTION: In this case, $f(x, y) = xy$, $x_0 = 0$, and $y_0 = 1$. Therefore, on each step we use the three equations

$$k_1 = hf(x_n, y_n) = 0.1x_ny_n, k_2 = hf(x_n + h, y_n + k_1) = 0.1(x_n + 0.1)(y_n + k_1),$$

and

$$y_{n+1} = y_n + \frac{1}{2}(k_1 = k_2).$$

For example, if $n = 0$, then

$$k_1 = 0.1x_0y_0 = 0.1(0)(1) = 0, k_2 = 0.1(x_0 + 0.1)(y_0 + k_1) = 0.1(0.1)(1) = 0.01,$$

and

$$y_1 = y_0 + \frac{1}{2}(k_1 + k_2) = 1 + \frac{1}{2}(0.01) = 1.005.$$

Therefore, the Runge–Kutta method of order two approximates that the value of y at $x = 0.1$ is 1.005.

In the same manner as in the previous two examples, we define a function yrk to implement the Runge–Kutta method of order two and use seq to generate a set of approximations for $n = 0, 1, \ldots, 10$.

```
> yrk:='yrk':

f:=(x,y)->x*y:

h:=0.1:

x:=n->n*h:

yrk:=proc(n)
        local k1,k2;
```

```
        option remember;
        k1:=h*f(x(n-1),yrk(n-1));
        k2:=h*f(x(n-1)+h,yrk(n-1)+k1);
        yrk(n-1)+1/2*(k1+k2)
        end:
yrk(0):=1:
> rktable1:=[seq([x(n),yrk(n)],n=0..10)]:
array(rktable1);
```

$$\begin{bmatrix} 0 & 1 \\ 0.1 & 1.005000000 \\ 0.2 & 1.020175500 \\ 0.3 & 1.045985940 \\ 0.4 & 1.083223039 \\ 0.5 & 1.133051299 \\ 0.6 & 1.197068697 \\ 0.7 & 1.277392007 \\ 0.8 & 1.376773105 \\ 0.9 & 1.498755202 \\ 1.0 & 1.647881345 \end{bmatrix}$$

We then use `plot` to graph the set of points determined in `rktable1`. The resulting graphics object, named `p2`, is displayed together with `p1`, which was generated in Example 3, with `display`.

```
> p2:=plot(rktable1,style=POINT,color=BLACK):
display({p1,p2});
```

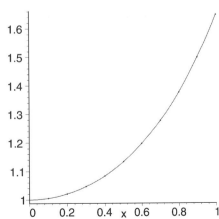

The terms of the power series expansions used in the derivation of the Runge–Kutta method of order two can be made to match up to order four. These computations are rather complicated, so they will not be discussed here. However, after much work, the approximation at each step is found to be made with

$$y_{n+1} = y_n + \frac{h}{6}[k_1 2k_2 + 2k_3 + k_4], \quad n = 0, 1, 2 \ldots$$

where $\quad k_1 = f(x_n, y_n), \qquad k_2 = f\left(x_n + \frac{h}{2}, y_n + \frac{hk_1}{2}\right), \qquad k_3 = f\left(x_n + \frac{h}{2}, y_n + \frac{hk_2}{2}\right) \quad$ and $k_4 = f(x_{n+1}, y_n + hk_3)$.

EXAMPLE 6: Use the fourth-order Runge–Kutta method with $h = 0.1$ to approximate the solution of the problem $y' = xy$, $y(0) = 1$ on $0 \le x \le 1$.

SOLUTION: With $f(x, y) = xy$, $x_0 = -0$, and $y_0 = 1$, the formulas are

$$k_1 = f(x_n, y_n) = x_n y_n,$$

$$k_2 = f\left(x_n + \frac{h}{2}, y_n + \frac{hk_1}{2}\right) = \left(x_n + \frac{0.1}{2}\right)\left(y_n + \frac{0.1k_1}{2}\right),$$

$$k_3 = f\left(x_n + \frac{h}{2}, y_n + \frac{hk_2}{2}\right) = \left(x_n + \frac{0.1}{2}\right)\left(y_n + \frac{0.1k_2}{2}\right), k_4 = f(x_{n+1}, y_n + hk_3) = x_{n+1}(y_n + 0.1k_3),$$

and

$$y_{n+1} = y_n + \frac{h}{6}[k_1 + 2k_2 + 2k_3 + k_4] = y_n + \frac{0.1}{6}[k_1 2k_2 + 2k_3 + k_4].$$

For $\quad n = 0,\quad$ we have $\quad k_1 = x_0 y_0 = (0)(1) = 0, \quad k_2 = \left(x_0 + \frac{0.1}{2}\right)\left(y_0 + \frac{0.1k_1}{2}\right)$
$= (0.05)(1) = 0.05$, $k_3 = \left(x_0 + \frac{0.1}{2}\right)\left(y_0 + \frac{0.1k_2}{2}\right) = (0.05)(1 + 0.0025) = 0.050125$, and
$k_4 = x_1(y_0 + 0.1k_3) = (0.1)(1 + 0.0050125) = 0.10050125$. Therefore,

$$y_1 = y_0 + \frac{0.1}{6}[k_1 + 2k_2 + 2k_3 + k_4] = 1\frac{0.1}{6}[0 + 0.05 + 0.050125 + 0.10050125] = 1.005012521.$$

We list the results for the Runge–Kutta method of order four and compare these results with the exact solution. Notice that this method yields the most accurate approximation of the methods used to this point.

```
> yrk4:='yrk4':

f:=(x,y)->x*y:

h:=0.1:

x:=n->n*h:

yrk4:=proc(n)
        local k1,k2,k3,k4;
```

```
        option remember;
        k1:=f(x(n-1),yrk4(n-1));
        k2:=f(x(n-1)+h/2,yrk4(n-1)+h*k1/2);
        k3:=f(x(n-1)+h/2,yrk4(n-1)+h*k2/2);
        k4:=f(x(n),yrk4(n-1)+h*k3);
        yrk4(n-1)+h/6*(k1+2*k2+2*k3+k4)
        end:
yrk4(0):=1:
> rktable2:=[seq([x(n),yrk4(n)],n=0..10)]:
array(rktable2
```

$$
\begin{bmatrix}
0 & 1 \\
0.1 & 1.005012521 \\
0.2 & 1.020201340 \\
0.3 & 1.046027859 \\
0.4 & 1.083287065 \\
0.5 & 1.133148446 \\
0.6 & 1.197217347 \\
0.7 & 1.277621279 \\
0.8 & 1.377127694 \\
0.9 & 1.499302362 \\
1.0 & 1.648721007
\end{bmatrix}
$$

```
> p3:=plot(rktable2,style=POINT,color=BLACK):
display({p1,p3});
```

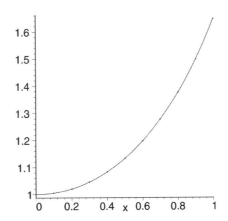

CHAPTER 3

Applications of First-Order Ordinary Differential Equations

When the space shuttle is launched from the Kennedy Space Center, its escape velocity can be determined by solving a first-order ordinary differential equation. The same can be said for finding the flow of electromagnetic forces, the temperature of a cup of coffee, the population of a species, as well as numerous other applications. In this chapter, we show how these problems can be expressed as first-order equations. We will focus our attention on setting up the problems and explaining the meaning of the subsequent solutions because the techniques for solving these problems were discussed in Chapter 2.

3.1 Orthogonal Trajectories

We begin our discussion with a topic that is encountered in the study of electromagnetic fields and heat flow. Before we can give any specific applications, however, we must state the following definition.

Definition
Orthogonal Curves

Two lines, L_1 and L_2, with slopes m_1 and m_2, respectively, are **orthogonal** (or perpendicular) if their slopes satisfy the relationship $m_1 = 1/m_2$. Two curves, C_1 and C_2, are **orthogonal** (or perpendicular) at a point if the respective tangent lines to the curves at that point are perpendicular.

EXAMPLE 1: Use the definition of orthogonality to verify that the curves given by $y' = 1$ and $y = \sqrt{1 - x^2}$ are orthogonal at the point $(\sqrt{2}/2, \sqrt{2}/2)$.

SOLUTION: First note that the point $(\sqrt{2}/2, \sqrt{2}/2)$ lies on the graph of both $y = x$ and $y = \sqrt{1 - x^2}$. The derivatives of the functions are given by $y' = 1$ and $y' = -x/\sqrt{1 - x^2}$, respectively.

```
> y[1]:=x->x:
y[2]:=x->sqrt(1-x^2):
D(y[1])(x);
D(y[2])(x);
```

$$1$$

$$-\frac{x}{\sqrt{1-x^2}}$$

Hence, the slope of the tangent line to $y = x$ at $x = \sqrt{2}/2$ is 1. Substitution of $x = \sqrt{2}/2$ into $y' = -x/\sqrt{1 - x^2}$ yields $-\sqrt{2}/2\sqrt{1 - (\sqrt{2}/2)^2} = -1$ as the slope of the tangent line at $x = \sqrt{2}/2$.

```
> D(y[2])(sqrt(2)/2);
```

$$-1$$

Thus, the curves are orthogonal at the point $(\sqrt{2}/2, \sqrt{2}/2)$ because the slopes of the lines tangent to the graphs of $y = x$ and $y = \sqrt{1 - x^2}$ at the point $(\sqrt{2}/2, \sqrt{2}/2)$ are negative reciprocals. We graph these two curves along with the tangent line to $y = \sqrt{1 - x^2}$ at $(\sqrt{2}/2, \sqrt{2}/2)$ with plot to illustrate that the two are orthogonal. We graph these to curves along with the tangent line to $y = \sqrt{1 - x^2}$ at $(\sqrt{2}/2, \sqrt{2}/2)$.

```
> plot({y[1](x),y[2](x),-x+sqrt(2)},x=-1..1,-.5..1.5);
```

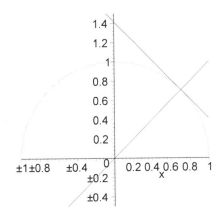

The next step in our discussion of orthogonal curves is to determine the set of orthogonal curves to a given family of curves. Typically, we refer to this set of orthogonal curves as the **family of orthogonal trajectories**. Suppose that a family of curves is defined as $F(x, y) = C$ and that the slope of the tangent line at any point on these curves is $dy/dx = f(x, y)$. Then, the slope of the tangent line on the orthogonal trajectory is $dy/dx = -1/f(x, y)$, so the family of orthogonal trajectories is found by solving the first-order equation $dy/dx = -1/f(x, y)$.

EXAMPLE 2: Determine the family of orthogonal trajectories to the family of curves $y = cx^2$.

SOLUTION: First, we must find the slope of the tangent line at any point on the parabola $y = cx^2$. Differentiating with respect to x results in $dy/dx = 2cx$. However from $y = cx^2$, we have that $c = y/x^2$. Substitution into $dy/dx = 2cx$ then yields $dy/dx = 2cx = 2(y/x^2)x = 2y/x$ on the parabolas. Hence, we must solve $dy/dx = -x/2y$ to determine the orthogonal trajectories. This equation is separable, so we write it as $ydy = -xdx$, and then integrating both sides gives us

$$y^2 + \frac{1}{2}x^2 = k,$$

where k is a constant, which we recognize as a family of ellipses. Note that an equivalent result is obtained with dsolve or with dsolve together with the implicit option.

```
> x:='x':y:='y':
sol:=dsolve(diff(y(x),x)=-x/(2*y(x)),y(x));
```

$$sol := y(x) = \frac{1}{2}\sqrt{-2x^2 + 4\_C1}, y(x) = -\frac{1}{2}\sqrt{-2x^2 + 4\_C1}$$

```
> implicitsol:=dsolve(diff(y(x),x)=-x/(2*y(x)),
          y(x),implicit);
```

$$implicitsol := y(x)^2 + \frac{1}{2}x^2 - \_C1 = 0$$

To graph the family of parabolas $y = cx^2$, the family of ellipses $y^2 + 12x^2 = k$, and the two families of curves together we use the commands `implicitplot` and `display`, both of which are contained in the **plots** package. First, we load the **plots** package.

```
> with(plots):
```

After defining the set of numbers `c_vals`, we use `seq` to define `parabs` to be the set of nine equations consisting of $y = cx^2$ where c has been replaced by each of the numbers in `c_vals`.

```
> c_vals:=seq(-1+i/4,i=0..8):
parabs:={seq(y=c*x^2,c=c_vals)}:
```

We then use `implicitplot` to graph these nine equations on the rectangle $[-3, 3] \times [-3, 3]$, naming the resulting graph `IP_1`. `IP_1` is then displayed with `display`.

```
> IP_1:=implicitplot(parabs,x=-3..3,y=-3..3,
          color=BLACK):
display({IP_1});
```

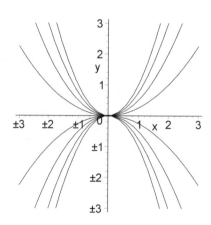

Similarly, we define `ellipses` to be the set of nine equations consisting of $y^2 + 12x^2 = k$ where k is replaced by $1, 2, 3, \ldots, 8$. These eight equations are also graphed on the rectangle $[-3,3] \times [-3,3]$. Then we use `display` to show the eight ellipses and then to show the parabolas and ellipses together.

```
> ellipses:={seq(y^2+x^2/2=k,k=1..8)}:
IP_2:=implicitplot(ellipses,x=-3..3,y=-3..3,color=GRAY):
display({IP_2});
display({IP_1,IP_2});
```

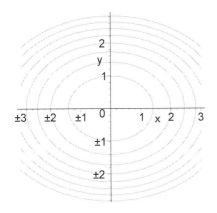

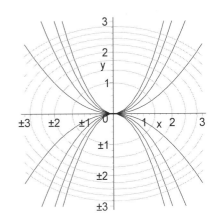

■

EXAMPLE 3: Let $T(x,y)$ represent the temperature at the point (x,y). The curves given by $T(x,y) = c$ (where c is constant) are called `isotherms`. The orthogonal trajectories are curves along which heat will flow. Determine the isotherms if the curves of heat flow are given by $y^2 + 2xy - x^2$.

SOLUTION: We begin by finding the slope of the tangent line at each point on the heat flow curves $y^2 + 2xy - x^2 = c$ using implicit differentiation.
We use `D` to compute the derivative of $y^2 + 2xy - x^2 = c$, naming the resulting output `step_1`.

```
> Eq_1:=y^2+2*x*y-x^2=c:
step_1:=D(Eq_1);
```

$$step\_1 := 2D(y)y + 2D(x)y + 2xD(y) - 2D(x)x = D(c)$$

We then replace each occurrence of `D(x)` in `step_1` by 1 and each occurrence of `D(c)` by 0 (because c represents a constant, $\frac{d}{dx}(c) = 0$) with `subs` and name the resulting output `step_2`. We interpret `step_2` to be equivalent to the equation $2y\frac{dy}{dx} + 2y + 2x\frac{dy}{dx} - 2x = 0$.

```
> step_2:=subs({D(x)=1,D(c)=0},step_1);
```

$$step\_2 := 2D(y)y + 2y + 2xD(y) - 2x = 0$$

We calculate dy/dx by solving `step_2` for `D(y)` with `solve` and name the result `im_deriv`.

```
> im_deriv:=solve(step_2,D(y));
```

$$im\_deriv := -\frac{y - x}{y + x}$$

Thus, $dy/dx = (x - y)/(x + y)$, so the orthogonal trajectories satisfy the differential equation $dy/dx = -(x + y)/(x - y)$.

```
> step_3:=simplify(-1/im_deriv);
```

$$step\_3 := \frac{y + x}{y - x}$$

Writing this equation in differential form as

$$(x + y)dx + (x - y)dy = 0,$$

we see that this equation is exact because $\partial/\partial y(x + y) = 1$ and $\partial/\partial x(x - y) = 1$. Thus, we solve the equation by integrating $x + y$ with respect to x to yield

$$f(x, y) = \frac{1}{2}x^2 + xy + g(y).$$

Differentiating f with respect to y then gives us

$$f_y(x, y) = x + g'(y).$$

Then, because the equation is exact, $x + g'(y) = x - y$. Therefore, $g'(y) = -y$, which implies that $g(y) = -\frac{1}{2}y^2$. This means that the family of orthogonal trajectories (isotherms) is given by $\frac{1}{2}x^2 + xy - \frac{1}{2}y^2 = k$
Note that `exactsol` is able to solve this differential equation.

```
> Diff_Eq:=diff(y(x),x)=subs(y=y(x),step_3);
```

$$Diff\_Eq := \frac{\partial}{\partial x}y(x) = \frac{y(x) + x}{y(x) - x}$$

```
> with(DEtools):

Sol:=exactsol(Diff_Eq,y(x));
```

$$Sol := \{xy(x) + \frac{1}{2}x^2 - \frac{1}{2}y(x)^2 = \_C_1\}$$

To graph $y^2 + 2xy - x^2 = c$ and $12x^2 + xy - 12y^2 = k$ for various values of c and k and see that the curves are orthogonal, we first define Eq_2 to be the equation $12x^2 + xy - 12y^2 = k$.

```
> Eq_2:=subs({y(x)=y,_C[1]=k},Sol[1]);
```

$$Eq\_2 := xy + \frac{1}{2}x^2 - \frac{1}{2}y^2 = k$$

Then, in the same manner as in the previous example, we define set_1 to be the set of seven equations obtained by replacing c in $y^2 + 2xy - x^2 = c$ by $-3, -2, -1, 0, 1, 2,$ and 3 and set_2 to be the set of seven equations obtained by replacing k in $12x^2 + xy - 12y^2 = k$ by $-3, -2, -1, 0, 1, 2,$ and 3. Both sets of equations are graphed with implicitplot and then displayed together with display.

```
> set_1:={seq(subs(c=i,Eq_1),i=-3..3)}:

set_2:={seq(subs(k=i,Eq_2),i=-3..3)}:

with(plots):

IP_1:=implicitplot(set_1,x=-4..4,y=-4..4,color=BLACK):

display({IP_1});

IP_2:=implicitplot(set_2,x=-4..4,y=-4..4,color=GRAY):

display({IP_2});

display({IP_1,IP_2});
```

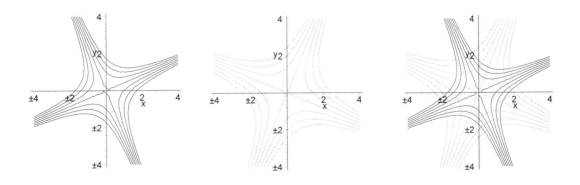

Alternatively, you can display the three graphs side by side with the command
`display(array([IP_1,IP_2,display({IP_1,IP_2})])).`

■

APPLICATION

Oblique Trajectories

If we are given a family of curves that satisfies the differential equation $dy/dx = f(x,y)$ and we want to find a family of curves that intersects this family at a constant angle θ, we must solve the differential equation

$$\frac{dy}{dx} = \frac{f(x,y) \pm \tan\theta}{1 \mp f(x,y)\tan\theta}.$$

For example, to find a family of curves that intersects the family of curves $x^2 + y^2 = c^2$ at an angle of $\pi/6$, we first implicitly differentiate the equation to obtain

$$2x + 2y\frac{dy}{dx} = 0 \Rightarrow \frac{dy}{dx} = -\frac{x}{y} = f(x,y).$$

Because $\tan\theta = \tan\pi/6 = 1/\sqrt{3}$, we solve $dydx = -x/y + 1/\sqrt{3}1 - (-x/y)(1/\sqrt{3}) = -x\sqrt{3} + yy\sqrt{3} + x$, which is a first-order homogeneous equation. With the substitution $x = \nu y$, we obtain the separable equation $1 - \nu\sqrt{3}1 + \nu^2 d\nu = \sqrt{3}ydy$. Integrating yields $\sqrt{3}2\ln(1+\nu^2) + \tan^{-1}\nu = \sqrt{3}\ln|y| + k_1$, so $-\sqrt{3}2\ln(1+x^2y^2) + \tan^{-1}xy = \sqrt{3}\ln|y| + k_1$.

> `soll:=dsolve(diff(y(x),x)=`

`(-sqrt(3)*x+y(x))/(sqrt(3)*y(x)+x),y(x));`

$$sol1 := \ln(x) + \frac{1}{2}\ln\left(\frac{y(x)^2 + x^2}{x^2}\right) + \frac{1}{3}\sqrt{3}\arctan\left(\frac{y(x)}{x}\right) - \_C1 = 0$$

`soll:=subs({ln(x)=log(abs(x)),y(x)=y,_C1=0},lhs(soll));`

$$sol1 := \ln(|x|) + \frac{1}{2}\ln\left(\frac{y^2 + x^2}{x^2}\right) + \frac{1}{3}\sqrt{3}\arctan\left(\frac{y}{x}\right)$$

Similarly, for $dydx = -x/y - 1\sqrt{3}1 + (-x/y)(1/\sqrt{3}) = -x\sqrt{3} - yy\sqrt{3} - x$, we obtain $1 + \nu\sqrt{3}1 + \nu^2 d\nu = -\sqrt{3}ydy$, so the trajectories are

$$\frac{\sqrt{3}}{2} \ln\left(1 + \frac{x^2}{y^2}\right) + \tan^{-1}\frac{x}{y} = -\sqrt{3}\ln|y| + k_2.$$

```
> sol2:=dsolve(diff(y(x),x)=
           (-sqrt(3)*x-y(x))/(sqrt(3)*y(x)-x),y(x));
```

$$sol2 := \ln(x) + \frac{1}{2}\ln\left(\frac{y(x)^2 + x^2}{x^2}\right) - \frac{1}{3}\sqrt{3}\arctan\left(\frac{y(x)}{x}\right) - \_C1 = 0$$

```
> sol2:=subs({ln(x)=log(abs(x)),y(x)=y,_C1=0},lhs(sol2));
```

$$sol2 := \ln(|x|) + \frac{1}{2}\ln\left(\frac{y^2 + x^2}{x^2}\right) - \frac{1}{3}\sqrt{3}\arctan\left(\frac{y}{x}\right)$$

To confirm the result graphically, we graph several members of each family of curves.

```
> with(plots):
cp1:=contourplot(x^2+y^2,x=-10..10,y=-10..10,
           color=GRAY):
display(cp1);
```

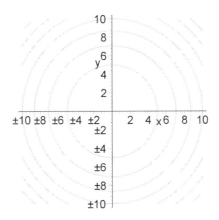

```
> cp2:=contourplot(sol1,x=-10..10,y=-10..10,
           color=BLACK,grid=[70,70]):
cp3:=contourplot(sol2,x=-10..10,y=-10..10,
           color=BLACK,grid=[70,70]):
display(cp2);
display(cp3);
```

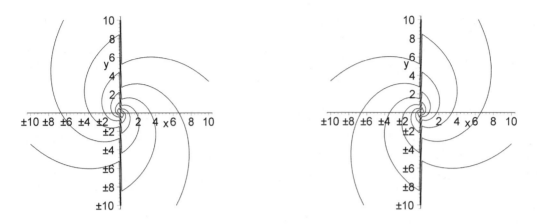

and then show the curves together to see that they intersect at an angle of $\pi/6$.

```
> display({cp1,cp2});

display({cp1,cp3});
```

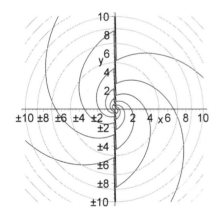 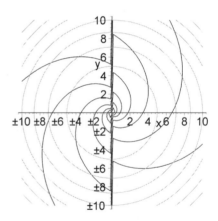

3.2 Population Growth and Decay

Many interesting problems involving population can be solved through the use of first-order differential equations. These include the determination of the number of cells in a bacterial culture, the number of citizens in a country, and the amount of radioactive substance remaining in a fossil. We begin our discussion by solving a population problem.

The Malthus Model

Suppose that the rate at which a population $y(t)$ changes is proportional to the amount present. Mathematically, this statement is represented as the first-order initial-value problem

$$\begin{cases} dy/dt = ky \\ y(0) = y_0 \end{cases},$$

where y_0 is the initial population. If $k>0$, then the population increases (growth), and the population decreases (decay) if $k<0$. Problems of this nature arise in such fields as cell population growth in biology as well as radioactive decay in physics. The model introduced here is known as the **Malthus model** after the work of the English clergyman and economist Thomas R. Malthus.

We solve the Malthus model for all values of k and y_0, which enables us to refer to the solution in other problems without solving the differential equation again. Rewriting $dy/dt = ky$ in the form $dy/y = kdt$, we see that this is a separable differential equation. Integrating and simplifying result in:

$$\int \frac{dy}{y} = \int kdt$$

$$\ln|y| = kt + C_1$$

$$y = Ce^{kt} \left(C = e^{C_1}\right).$$

Notice that because y represents population, $y \geq 0$, and therefore $|y| = y$. To find C, we apply the initial condition, obtaining $y_0 = y(0) = Ce^{k \cdot 0} = C$. Thus, the solution to the initial-value problem is

$$y = y_0 e^{kt}.$$

We obtain the same result with `dsolve`:

```
> dsolve({diff(y(t),t)=k*y(t),y(0)=y0},y(t));
```

$$y(t) = y0e^{(kt)}$$

EXAMPLE 1: Forms of a given element with different numbers of neutrons are called `nuclides`. Some nuclides are not stable. For example, potassium-40 () naturally decays to reach argon-40 (^{40}Ar). This decay that occurs in some nuclides was first observed, but not understood, by Henri Becquerel (1852–1908) in 1896. Marie Curie, however, began studying this decay in 1898, named it **radioactivity**, and discovered the radioactive substances polonium and radium. Marie Curie (1867–1934), along with her husband, Pierre Curie (1859–1906), and

Henri Becquerel, received the Nobel Prize in Physics in 1903 for their work on radioactivity. Marie Curie subsequently received the Nobel Prize in Chemistry in 1910 for discovering polonium and radium.

Given a sample of of sufficient size, after years approximately half of the sample will have decayed to . The **half-life** of a nuclide is the time for half the nuclei in a given sample to decay. We see that the rate of decay of a nuclide is proportional to the amount present because the half-life of a given nuclide is constant and independent of the sample size.

If the half-life of polonium -209 (^{209}Po) is 100 years, determine the percentage of the original amount of (^{209}Po)that remains after 50 years.

SOLUTION: Let y_0 represent the original amount of ^{209}Po that is present. Then the amount present after t years is $y(t) = y_0 e^{kt}$. Because $y(100) = y_0/2$ and $y(100) = y_0 e^{100\,k}$, we solve $y_0 e^{100\,k} = y_0/2$ for e^k:

$$e^{100k} = 1/2$$

$$e^k = 1/2^{1/100}$$

Hence,

$$y(t) = y_0(e^{kt}) = y_0(1/2)^{t/100}.$$

```
> k:=-ln(2)/100;
```

$$k := -\frac{1}{100}\ln(2)$$

```
> y:=t->y0*exp(k*t):
simplify(y(t));
```

$$y0^{2\left(-\frac{1}{100}t\right)}$$

In order to determine the percentage of y_0 that remains, we evaluate

$$y(50) = y_0(1/2)^{50/100} = y_0/\sqrt{2} \approx 0.7071 y_0.$$

```
> y(50);
evalf(y(50));
```

$$\frac{1}{2}y0\sqrt{2} \qquad 0.7071067810\ y0$$

Therefore, approximately 70.71% of the original amount of ^{209}Po remains after 50 years.

∎

In the previous example, we see that we can determine the percentage of y_0 that remains even though we do not know the value of y_0. Hence, instead of letting $y(t)$ represent the amount of the substance present after time t, we can let it represent the fraction (or percentage) of y_0 that remains after time t. In doing this, we use the initial condition $y(0) = 1$ to indicate that 100% of y_0 is present at $t = 0$.

EXAMPLE 2: The wood of an Egyptian sarcophagus (burial case) is found to contain 63% of the carbon-14 found in a present-day sample. What is the age of the sarcophagus?

SOLUTION: The half-life of carbon-14 is 5730 years. Let $y(t)$ be the percentage of carbon-14 in the sample after t years. Then, $y(0) = 1$. Because $y(t) = y_0 e^{kt}, y(5730) = e^{5730k} = 0.5$. Solving for k yields

$$\ln(e^{5730k}) = \ln(0.5)5730k = \ln(0.5)k = \frac{\ln(0.5)}{5730} = \frac{-\ln 2}{5730}.$$

Thus, $y(t) = e^{kt} = e^{-\ln 2/5730\, t} = 2^{-t/5730}$.

```
> k:='k':
k:=solve(exp(5730*k)=1/2,k);
```

$$k := -\frac{1}{5730}\ln(2)$$

```
> y:=t->exp(k*t):
simplify(y(t));
```

$$2^{\left(-\frac{1}{5730}t\right)}$$

In this problem, we must find the value of t for which $y(t) = 0.63$. Solving this equation results in

$$2^{-t/5730} = 0.63 = 63/100$$

$$\ln(2^{-t/5730}) = \ln(63/100)$$

$$\frac{-t}{5730}\ln 2 = \ln(63/100)$$

$$t = \frac{-5730\ln(63/100)}{\ln 2} = \frac{5730(\ln 100 - \ln 63)}{\ln 2} \approx 3819.48.$$

We conclude that the sarcophagus is approximately 3819 years old.

```
> solve(y(t)=.63);
```

$$3819.482006$$

An alternative way to approximate the age of the sarcophagus is first to graph $y(t)$ and the line $y = 0.63$ with `plot`. The age of the sarcophagus is the t-coordinate of the point of intersection of $y(t)$ and $y = 0.63$.

```
> plot({.63,y(t)},t=0..6000);
```

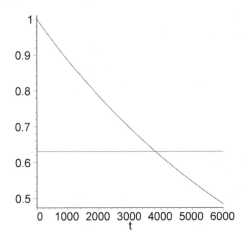

and then use `fsolve` to approximate the solution of $y(t) = 0.63$.

```
> fsolve(y(t)=.63,t,3000..4000);
```

$$3819.482006$$

∎

To observe some of the limitations of the Malthus model, we consider a population problem in which the rate of growth of the population does not depend exclusively on the population present.

EXAMPLE 3: The population of the United States was recorded as 5.3 million in 1800. Use the Malthus model to approximate the population for years after 1800 if

k was experimentally determined to be 0.03. Compare these results with the actual population. Is this a good approximation for years after 1800?

SOLUTION: In this example, $k = 0.03$ and $y_0 = 5.3$ and our model for the population of the United States at time t (where t is the number of years from 1800) is $y(t) = 5.3e^{0.03t}$.

```
> y:='y':k:='k':
peq:=dsolve({diff(y(t),t)=k*y(t),y(0)=y0},y(t));
```

$$peq := y(t) = y0e^{(kt)}$$

```
> pop:=(t,k,y0)->y0*exp(k*t);
```

$$pop := (t, k, y0) \rightarrow y0e^{(tk)}$$

```
> pop(t,0.03,5.3);
```

$$5.3e^{(0.03\ t)}$$

In order to compare this model with the actual population of the United States, census figures for the population of the United States for various years are listed in following table along with the corresponding value of $y(t)$.

Year (t)	Actual Population (in millions)	Value of $y(t) = 5.3e^{0.03t}$	Year (t)	Actual Population (in millions)	Value of $y(t) = 5.3e^{0.03t}$
1800 (0)	5.30	5.30	1870 (70)	38.56	43.28
1810 (10)	7.24	7.15	1880 (80)	50.19	58.42
1820 (20)	9.64	9.66	1890 (90)	62.98	78.86
1830 (30)	12.68	13.04	1900 (100)	76.21	106.45
1840 (40)	17.06	17.60	1910 (110)	92.23	143.70
1850 (50)	23.19	23.75	1920 (120)	106.02	193.97
1860 (60)	31.44	32.06	1930 (130)	123.20	261.83

Although the model appears to approximate the data closely for several years after 1800, the accuracy of the approximation diminishes over time. This is because the population of the United States does not exclusively increase at a rate proportional to the population. Hence, another model that better approximates the population taking other factors into account is needed. The graph of

$y(t) = 5.3e^{0.03t}$ is shown along with the data points to show how the approximation becomes less accurate as t increases.

```
> popplot:=plot(pop(t,0.03,5.3),t=0..100):
pdata:=[[0,5.3],[10,7.2],[20,9.6],[30,12.9],[40,17],
      [50,23.2],[60,31.4],[70,38.6],[80,50.2],
          [90,63],[100,76.2]]:
dataplot:=plot(pdata,style=POINT,color=BLACK):
with(plots):
display({popplot,dataplot});
```

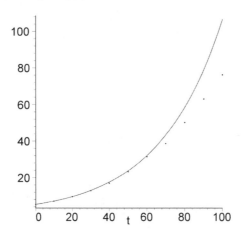

The Logistic Equation

Because the approximation obtained with the Malthus model is less than desirable in the previous example, we see that another model is needed. The **logistic equation** (or **Verhulst equation**) is the equation

$$y'(t) = (r - ay(t))y(t),$$

where r and a are constants, subject to the condition $y(0) = y_0$. This equation was first introduced by the Belgian mathematician Pierre Verhulst to study population growth. The logistic equation differs from the Malthus model in that the term $r - ay(t)$ is not constant. This equation can be written as $dy/dt = (r - ay)y = ry - ay^2$, where the term $-y^2$ represents an inhibitive factor. Under these assumptions, the population is allowed neither to grow out of control nor to grow or decay constantly as it was with the Malthus model.

The logistic equation is separable and, thus, can be solved by separation of variables. Separating variables and using partial fractions to integrate with respect to y, we have

$$\frac{dy}{(r-ay)y} = dt$$

$$\left(\frac{a/r}{r-ay} + \frac{1/r}{y}\right)dy = dt$$

$$\left(\frac{a}{r-ay} + \frac{1}{y}\right)dy = rdt$$

$$-\ln|r-ay| + \ln|y| = rt + c$$

Of course, we would like to solve this expression for y. Using the properties of logarithms yields

$$\ln\left|\frac{y}{r-ay}\right| = rt + c$$

$$\frac{y}{r-ay} = e^{rt+c} = Ke^{rt}(K = e^c)$$

$$y = r\left(\frac{1}{K}e^{-rt} + a\right)^{-1}.$$

Applying the initial condition $y(0) = y_0$ and solving for K, we find that

$$\frac{y_0}{r-ay_0} = K$$

After substituting this value into the general solution and simplifying, the solution can be written as

$$y = \frac{ry_0}{ay_0 + (r-ay_0)e^{-rt}}.$$

Notice that if $r > 0$, $\lim\limits_{t\to\infty} y(t) = r/a$ because $\lim\limits_{t\to\infty} e^{-rt} = 0$. This makes the solution to the logistic equation different from that of the Malthus model in that the solution to the logistic equation approaches a finite nonzero limit as $t\to\infty$ whereas that of the Malthus model approaches either infinity or zero as $t\to\infty$.

We are also able to use `dsolve` to solve this initial-value problem. First, we use `dsolve` to solve the initial-value problem, naming the resulting output `Sol`.

```
> y:='y':
Sol:=dsolve({diff(y(t),t)=(r-a*y(t))*y(t),
        y(0)=y0},y(t));
```

$$Sol := y(t) = \frac{r}{a - \frac{e^{(-rt)}(-r+y0a)}{y0}}$$

Then, we use `assign` to name $y(t)$ the result obtained in `Sol` and `simplify` to simplify $y(t)$.

```
> assign(Sol):
simplify(y(t));
```

$$-\frac{y0r}{-y0a - e^{(-rt)}r + e^{(-rt)}y0a}$$

EXAMPLE 4: Use the logistic equation to approximate the population of the United States using $r = 0.03, a = 0.0001$, and $y_0 = 5.3$. Compare this result with the actual census values. Use the model obtained to predict the population of the United States in the year 2000.

SOLUTION: We substitute the indicated values of r, a, and y_0 into $y = ry_0ay_0 + (r - ay_0)e^{-rt}$ to obtain the approximation of the population of the United States at time t, where t represents the number of years since 1800,

$$y(t) = \frac{0.03 \cdot 5.3}{0.0001 \cdot 5.3 + (0.03 - 0.0001 \cdot 5.3)e^{-.03t}} = \frac{0.159}{0.00053 + 0.02947e^{-0.3t}}.$$

```
> y:='y':
y:=t->0.159/(0.00053+0.02947*exp(-0.03*t)):
```

We compare the approximation of the population of the United States given by $y(t)$ with the actual population obtained from census figures. Note that this model appears to approximate the population more closely over a longer period of time than the Malthus model, which was considered in Example 3, as we can see in the graph.

Year (t)	Actual Population (in millions)	Value of y(t)	Year (t)	Actual Population (in millions)	Value of y(t)
1800 (0)	5.30	5.30	1900 (100)	76.21	79.61
1810 (10)	7.24	7.11	1910 (110)	92.23	98.33
1820 (20)	9.64	9.52	1920 (120)	106.02	119.08
1830 (30)	12.68	12.71	1930 (130)	123.20	141.14
1840 (40)	17.06	16.90	1940 (140)	132.16	163.59
1850 (50)	23.19	22.38	1950 (150)	151.33	185.45
1860 (60)	31.44	29.44	1960 (160)	179.32	205.82
1870 (70)	38.56	38.42	1970 (170)	203.30	224.05
1880 (80)	50.19	49.63	1980 (180)	226.54	239.78
1890 (90)	62.98	63.33	1990 (190)	248.71	252.94

```
> with(plots):
pdata:=[[0,5.3],[10,7.2],[20,9.6],[30,12.9],[40,17],
       [50,23.2],[60,31.4],[70,38.6],[80,50.2],[90,63],
       [100,76.2],[110,92.23],[120,106.02],[130,123.2],
       [140,132.16],[150,151.33],[160,179.32],[70,203.3],
       [180,226.54],[190,248.71]]:
dataplot:=plot(pdata,style=POINT):
plot_y:=plot(y(t),t=0..200):
display({plot_y,dataplot});
```

To predict the population of the United States in the year 200 with this model, we evaluate $y(200)$.

```
> y(200);
```

$$263.6602427$$

Thus, we predict that the population will be approximately 263.66 million in the year 2000. Note that projections of the population of the United States in the year 2000 made by the Bureau of the Census range from 259.57 million to 278.23 million.

■

APPLICATION

Harvesting

If we wish to take a constant harvest rate H (as in hunting, fishing, or disease) into consideration, then we might instead modify the logistic equation and use the equation $dP/dt = rP - aP^2 - H$ to model the population under consideration. Assume that $h \leq r^2 4a$.

```
> gensol:=dsolve (diff (p (t),t)=r*p(t)-a*p(t)^2-h, p(t));
```

$$gensol := \mathrm{p}(t) = \frac{1}{2} \frac{\tan\left(-\frac{1}{2}t\sqrt{4ha - r^2} - \frac{1}{2}\_C1\sqrt{4ha - r^2}\right)\sqrt{4ha - r^2} + r}{a}$$

Suppose that for a certain species it is found that $r = 0.03, a = 0.0001, h = 2.26,$ and $C = 150.1$. We substitute these values into the general solution and graph the result.

```
> toplot:=subs({r=0.03,a=0.0001,h=2.26,_C1=150.1},

        rhs(gensol));
```

$$toplot := 10.00000000 \tan(-0.001000000000\, t - 0.1501000000) + 150.0000000$$

```
> plot(toplot,t=0..1400);
```

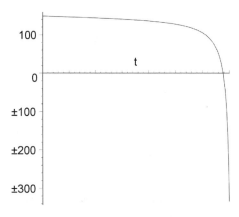

In this case, we see that the species becomes extinct after approximately 1300 years. A more accurate approximation of the time of extinction is obtained with `fsolve`.

```
> fsolve(toplot=0,t,1300..1400);
```

$$1354.128163$$

Source: David A Sanchez, "Populations and Harvesting," *Mathematical Modeling: Classroom Notes in Applied Mathematics*, Murray S. Klamkin, Editor, SIAM (1987), pp. 311–313.

3.3 Newton's Law of Cooling

First-order linear differential equations can be used to solve a variety of problems that involve temperature. For example, a medical examiner can find the time of death in a homicide case, a chemist can determine the time required for a plastic mixture to cool to a hardening temperature, and an engineer can design the cooling and heating system of a manufacturing facility. Although distinct, each of these problems depend on a basic principle, *Newton's law of cooling*, that is used to develop the associated differential equation.

Newton's Law of Cooling

Newton's law of cooling states that the rate at which the temperature $T(t)$ changes in a cooling body is proportional to the difference between the temperature of the body and the constant temperature T_s of the surrounding medium. This situation is represented as the first-order initial-value problem

$$\begin{cases} dT/dt = k(T - T_s) \\ \quad\quad T(0) = T_0 \end{cases},$$

where T_0 is the initial temperature of the body and k is the constant of proportionality. The equation $dT/dt = k(T - T_s)$ is separable and separating variables gives us

$$\frac{dT}{T - T_s} = k\,dt.$$

Hence, $\ln|T - T_s| = kt + C$. Using the properties of the natural logarithm and simplifying yields

$$T = C_1 e^{kt} + T_s,$$

where $C_1 = \pm e^c$. Applying the initial condition implies that $T_0 = C_1 + T_s$, so $C_1 = T_0 - T_s$. Therefore, the solution of the equation is

$$T = (T_0 - T_s)e^{kt} + T_s.$$

We see that an equivalent solution is obtained with `dsolve`, which we name `del` for later use.

```
> del:=dsolve({diff(temp(t),t)=k*(temp(t)-temps),
```

```
temp(0)=temp0},temp(t));
```

$$del := \text{temp}(t) = temps + e^{(kt)}(-temps + temp0)$$

Recall that if $k<0$, $\lim\limits_{t\to\infty} e^{kt} = 0$. Therefore, $\lim\limits_{t\to\infty} T(t) = T_s$, so the temperature of the body approaches that of its surroundings.

EXAMPLE 1: A pie is removed from a 350°F oven and placed to cool in a room with temperature 75°F. In 15 minutes, the pie has a temperature of 150°F. Determine the time required to cool the pie to a temperature of 80°F so that it may be eaten.

SOLUTION: In this example, $T_0 = 350$ and $T_s = 75$. Substituting these values into $T = (T_0 - T_s)e^{kt} + T_s$, we obtain $T(t) = (350 - 75)e^{kt} + 75 = 275e^{kt} + 75$.

```
> step_1:=subs({temp0=350,temps=75},rhs(del));
```

$$step\_1 := 75 + 275e^{(kt)}$$

To solve the problem we must find k. Because we also know that $T(15) = 150$, $T(15) = 275e^{15k} + 75 = 150$. Solving this equation for k gives us

$$275e^{15k} = 75$$

$$e^{15k} = 3/11$$

$$\ln(e^{15k}) = \ln(3/11)$$

$$15k = \ln(3/11)$$

$$k = \ln(3/11)/15 = -\ln(11/3)/15.$$

Thus, $T(t) = 275e^{-t\ln(11/3)/15} + 75 = 275(11/3)^{-t/15} + 75$. (We could have obtained the same solution by solving $T(15) = 275e^{15k} + 75 = 150$ for $e^k = (11/3)^{-1/15}$.)

```
> k:=solve(subs(t=15,step_1)=150);
```

$$k := \frac{1}{15}\ln\left(\frac{3}{11}\right)$$

Thus, $T(t) = 275e^{-t\ln(11/3)/15} + 75 = 275(11/3)^{-t/15} + 75$, as shown next using `simplify` together with the `exp` option.

```
> simplify(step_1,exp);
```

$$75 + 275\left(\frac{3}{11}\right)^{\left(\frac{1}{15}t\right)}$$

To find the value of t for which $T(t) = 80$, we solve the equation $275(113)^{-t/15} + 75 = 80$ for t with `solve`. Thus, the pie will be ready to eat after approximately 46 minutes.

```
> t00:=solve(step_1=80);
evalf(t00);
```

$$t00 := -15\frac{\ln(55)}{\ln\left(\frac{3}{11}\right)} \quad 46.26397676$$

Alternatively, we can graph $T(t) = 275(311)^{t/15} + 75$ with `plot`.

```
> plot({80,step_1},t=0..90);
```

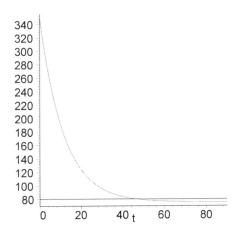

From the graph, we see that the temperature is 80 degrees between $t = 40$ and $t = 50$. To approximate the value of t for which the temperature is 80, we use `fsolve` as shown next.

```
> fsolve(step_1=80,t,40..50);
```

$$46.26397676$$

Thus, the pie will be ready to eat after approximately 46 minutes.
An interesting question associated with cooling problems is to determine whether

the pie reaches room temperature. From the formula, $T(t) = 275(11/3)^{-t/15} + 75$, we see that the component $275(11/3)^{-t/15} > 0$, so $T(t) = 275(11/3)^{-t/15} + 75 > 75$. Therefore, the pie never actually reaches room temperature according to our model. However, we see from the graph that its temperature approaches 75° as t increases.

■

In Example 1, the temperature of the surroundings was assumed to be constant. However, this does not have to be the case. For example, consider the problem of heating and cooling a building. Over the span of a 24-hour day, the outside temperature varies. The problem of determining the temperature inside the building, therefore, becomes more complicated. We assume that the building has no heating or air-conditioning system. Hence, the differential equation that should be solved to find the temperature $u(t)$ at time t inside the building is

$$du/dt = k(C(t) - u(t))$$

where $C(t)$ is a function that describes the outside temperature and $k > 0$ is a constant that depends on the insulation of the building. According to this equation, if $C(t) > u(t)$, then $du/dt > 0$, which implies that u increases. On the other hand, if $C(t) < u(t)$, then $du/dt < 0$, which means that u decreases.

EXAMPLE 2: (a) Suppose that during the month of April in Atlanta, Georgia, the outside temperature in °F is given by $C(t) = 70 - 10\cos(\pi t/12)$, $0 \leq t \leq 24$. (Note: This implies that the average value of $C(t)$ is 70°F.) Determine the temperature in a building that has an initial temperature of 60°F if $k = 1/4$. (b) Compare this with the temperature in June when the outside temperature is $C(t) = 80 - 10\cos(\pi t/12)$ and the initial temperature is 70°F.

SOLUTION: (a) The initial-value problem that we must solve is

$$\begin{cases} du/dt = k[70 - 10\cos(\pi t/12) - u] \\ u(0) = 60 \end{cases}$$

The differential equation can be solved if we write it as $du/dt + ku = k[70 - 10\cos(\pi t/12)]$ and then use an integrating factor. This gives us $d/dt(e^{kt}u) = ke^{kt}[70 - 10\cos(\pi t/12)]$, so we must integrate both sides of the equation. Of course, this is more easily carried out through the use of **dsolve**.

```
> Sol:=dsolve({diff(u(t),t)=1/4*

      (70-10*cos(Pi*t/12)-u(t)),u(0)=60},u(t)):
```

After naming $u(t)$ the solution obtained in `Sol` with `assign`, `simplify` is used to simplify the solution.

```
> assign(Sol):
simplify(u(t));
```

$$-10\frac{-7\pi^2 + e^{\left(-\frac{1}{4}t\right)}\pi^2 + 3\pi\sin\left(\frac{1}{12}\pi t\right) + 9\cos\left(\frac{1}{12}\pi t\right) - 63}{9 + \pi^2}$$

We then use `plot` to graph the solution for $0 \le t \le 24$.

```
> plot(u(t),t=0..24);
```

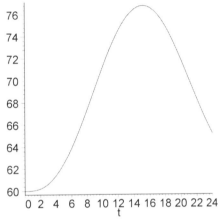

Note that the temperature reaches its maximum (approximately $77°$) near $t = 15.5$ hours, which corresponds to 3:30 p.m. A more accurate estimate is obtained with `fsolve` by setting the first derivative of u equal to zero and solving for t.

```
> fsolve(diff(u(t),t)=0,t,14..16);
```

$$15.15061632$$

(b) This problem is solved in the same manner as the previous case. Be sure to clear the definition of u by entering $u :=' u'$ before entering the `dsolve` command; otherwise, error messages will result because u was assigned a definition in (a).

```
> u:='u':
```

```
Sol:=dsolve({diff(u(t),t)=1/4*
    (80-10*cos(Pi*t/12)-u(t)),u(0)=70},u(t)):
```

Again, we name $u(t)$ the solution obtained in `Sol` and simplify the result.

```
> assign(Sol):
simplify(u(t));
```

$$-10\frac{-8\pi^2 + e^{\left(-\frac{1}{4}t\right)}\pi^2 + 3\pi\sin\left(\frac{1}{12}\pi t\right) + 9\cos\left(\frac{1}{12}\pi t\right) - 72}{9 + \pi^2}$$

The solution is also graphed with `plot`. From the graph, we see that the maximum temperature appears to occur near $t = 15$ hours.

```
> plot(u(t),t=0..24);
```

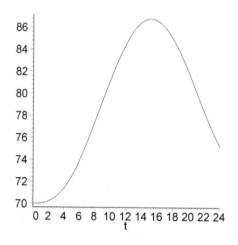

Again, the more accurate value is obtained with `fsolve` by setting the first derivative of u equal to zero and solving for t. This calculation yields approximately 15.15 hours, which is the same as the value in (a).

```
> fsolve(diff(u(t),t)=0,t,14..16);
```

$$15.15061632$$

■

3.4 Free-Falling Bodies

The motion of some objects can be determined through the solution of a first-order equation. We begin by explaining some of the theory that is needed to set up the differential equation that models the situation.

Newton's | The rate at which the momentum of a body changes with respect
Second Law of Motion | to time is equal to the resultant force acting on the body.

Because the body's momentum is defined as the product of its mass and velocity, this statement is modeled as

$$\frac{d}{dt}(mv) = F$$

where m and v represent the body's mass and velocity, respectively, and F is the sum of the forces acting on the body. Because m is constant, differentiation leads to the well-known equation $mdv/dt = F$. If the body is subjected to the force due to gravity, then its velocity is determined by solving the differential equation

$$mdv/dt = mg \text{ or } dv/dt = g,$$

where $g \cong 32\text{ft/s}^2$ (English system) and 9.8m/s^2 (metric system).

This differential equation is applicable only when the resistive force due to the medium (such as air resistance) is ignored. If this offsetting resistance is considered, we must discuss all of the forces acting on the object. Mathematically, we write the equation as

$$m\frac{dv}{dt} = \Sigma \text{ (forces acting on the object)}$$

where the direction of motion is taken to be the positive direction.

Because air resistance acts against the object as it falls and mg acts in the direction of the motion, we state the differential equation in the form

$$mdv/dt = mg + (-F_R) \quad \text{or} \quad mdv/dt = mg - F_R,$$

where F_R represents this resistive force. Note that down is assumed to be the positive direction. The resistive force is typically proportional to the body's velocity (v) or a power of the velocity. Hence, the differential equation is linear or nonlinear based on the resistance of the medium taken into account.

EXAMPLE 1: Determine the velocity and displacement functions of an object with $m = 1$, where $1\text{slug} = 11b - \sec^2\text{ft}$, that is thrown downward with an initial velocity of 2 ft/sec from a height of 1000 feet. Assume that the object is subjected to air resistance that is equivalent to the instantaneous velocity of the object. Also, determine the time at which the object strikes the ground and its velocity when it strikes the ground.

SOLUTION: First, we set up the initial-value problem to determine the velocity of the object. Because the air resistance is equivalent to the instantaneous velocity, we have $F_R = v$. The formula $mdv/dt = mg - F_R$ then gives us $dv/dt = 32 - v$. Of course, we must impose the initial velocity $v(0) = 2$. Therefore, the initial-value problem is

$$\begin{cases} dv/dt = 32 - v \\ \quad v(0) = 2 \end{cases}$$

which can be solved by several methods. We can solve it as a linear first-order equation and use the integrating factor e^t, which results in $d/dt(e^t v) = 32e^t$. Integrating both sides gives us $e^t v = 32e^t + C$, so

$$v = 32e^{-t} + Ce^{-t}.$$

Applying the initial velocity, we have $v(0) = 32e^0 + Ce^0 = 32 + C = 2$. Therefore, the velocity of the object is

$$v = 32 - 30e^{-t}.$$

We obtain the same result with `dsolve`, naming the resulting output `step_1`.

```
> v:='v':s:='s':
step_1:=dsolve({diff(v(t),t)=32-v(t),v(0)=2},v(t));
```

$$step\_1 := v(t) = 32 - 30e^{(-t)}$$

To determine the position, or distance traveled at time t, $s(t)$, we solve the first-order equation

$$ds/dt = v = 32 - 30e^6 - t$$

with initial displacement $s(0) = 0$. Notice that we use the initial displacement as a reference and let s represent the distance traveled from this reference point.

```
> assign(step_1):
step_2:=dsolve({diff(s(t),t)=v(t),s(0)=0},s(t));
```

$$step\_2 := s(t) = 32t + 30e^{(-t)} - 30$$

Thus, the displacement of the object at time t is given by $s = 32t + 30e^{-t} - 30$. Because we are taking $s(0) = 0$ as our starting point, the object strikes the ground when $s(t) = 1000$. Therefore, we must solve $s = 32t + 30e^{-t} - 30 = 1000$. The roots of this equation can be approximated with `fsolve`. We begin by graphing the functions s and the line $s = 1000$ with `plot`.

```
> assign(step_2):

plot({1000,s(t)},t=0..70);
```

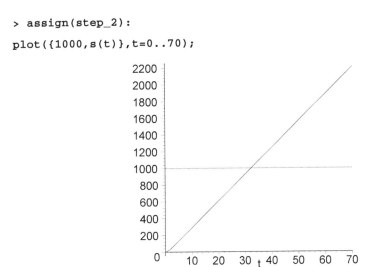

From the graph of this function, we see that $s(t) = 1000$ near $t = 35$. To obtain a better approximation, we use `fsolve`.

```
> t00:=fsolve(s(t)=1000,t,30..40);
```

$$t00 := 32.18750000$$

So, the object strikes the ground after approximately 32.1875 seconds.
The velocity at the point of impact is found to be 32.0 ft/sec by evaluating the derivative, $s'(t) = 32 - 30e^{-t}$, at the time at which the object strikes the ground, $t \approx 32.1875$.

```
> evalf(subs(t=t00,v(t)));
```

$$32.$$

■

EXAMPLE 2: Determine a solution (for the velocity and the displacement) of the differential equation that models the motion of an object of mass m when directed upward with an initial velocity of v_0 from an initial displacement y_0 assuming that the air resistance equals cv (c is constant).

SOLUTION: Because the motion of the object is upward, mg and F_R act against the upward motion of the object; mg and F_R are in the negative direction. Therefore, the differential equation that must be solved in this case is the linear equation $\frac{dv}{dt} = -g - \frac{c}{m}v$. We solve the initial-value problem

$$\begin{cases} \frac{dv}{dt} = -g - \frac{c}{m}v \\ v(0) = v_0 \end{cases}$$

with dsolve naming the resulting output Sol.

```
> v:='v':
Sol:=dsolve({diff(v(t),t)=-g-c/m*v(t),v(0)=v0},v(t));
```

$$Sol := v(t) = -\frac{gm}{c} + \frac{e^{\left(-\frac{ct}{m}\right)}(gm + v0c)}{c}$$

Therefore, the solution to the initial-value problem is

$$v(t) = -\frac{gm}{c} + \frac{cv_0 + gm}{c}e^{-ct/m}.$$

Next, we use Sol to define velocity. This function can be used to investigate numerous situations without solving the differential equation each time.

```
> velocity:=proc(m0,c0,g0,v00,t0)
        subs({m=m0,c=c0,g=g0,v0=v00,t=t0},rhs(Sol))
    end:
velocity(m,c,g,vo,t);
```

$$-\frac{gm}{c} + \frac{e^{\left(-\frac{ct}{m}\right)}(gm + voc)}{c}$$

For example, the velocity function for the case with $m = 1/128$slugs, $c = 1/160$, $g = 32$ft/s^2, and $v_0 = 48$ft/s is $v(t) = 88e^{-4t/5} - 40$.

```
> velocity(1/128,1/160,32,48,t);
```

$$-40 + 88e^{\left(-\frac{4}{5}t\right)}$$

The displacement function $s(t)$ that represents the distance above the ground at time t is determined by integrating the velocity function. This is accomplished with dsolve using the initial position y_0. As in the previous case, the output is named Pos so that the position formula may be extracted from the result for later use.

```
> y:='y':
Pos:=dsolve({diff(y(t),t)=velocity(m,c,g,v0,t),
        y(0)=y0},y(t));
```

$$Pos := y(t) = -\frac{gmt}{c} - \frac{gm^2 e^{\left(-\frac{ct}{m}\right)}}{c^2} - \frac{v0me^{\left(-\frac{ct}{m}\right)}}{c} + \frac{m^2g + v0mc + y0c^2}{c^2}$$

```
> position:=proc(m0,c0,g0,v00,y00,t0)
        subs({m=m0,c=c0,g=g0,v0=v00,y0=y00,t=t0},rhs(Pos))
        end:
position(m,c,g,v0,y0,t);
```

$$-\frac{gmt}{c} - \frac{gm^2 e^{\left(-\frac{ct}{m}\right)}}{c^2} - \frac{v0me^{\left(-\frac{ct}{m}\right)}}{c} + \frac{m^2g + v0mc + y0c^2}{c^2}$$

The displacement and velocity functions are plotted in the following using the parameters $m = 1/128$ slugs, $c = 1/160$, $g = 32\text{ft/s}^2$, and $v_0 = 48$ ft/s as well as $y_0 = 0$.

The time at which the object reaches its maximum height occurs when the derivative of the displacement is equal to zero. From the graph we see that $s'(t) = v(t) = 0$ when $t \approx 1$.

```
> plot({velocity(1/128,1/160,32,48,t),
        position(1/128,1/160,32,48,0,t)},t=0..2);
```

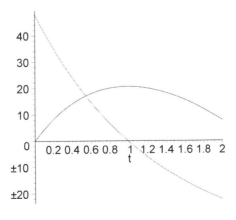

A more accurate approximation $t \approx 0.985572$ is obtained using `solve` together with `evalf`

```
> cn:=solve(diff(position(1/128,1/160,32,48,0,t),t)=0);
```

$$cn := -\frac{5}{4}\ln\left(\frac{5}{11}\right)$$

```
> evalf(cn);
```

$$0.9855717006$$

We now compare the effect that varying the initial velocity and displacement has on the displacement function. Suppose that we use the same values used earlier for m, c, and g. However, we let $v_0 = 48$ in one function and $v_0 = 36$ in the other. We also let $y_0 = 0$ and $y_0 = 6$ in these two functions, respectively.

```
> plot({position(1/128,1/160,32,48,0,t),
        position(1/128,1/160,32,36,6,t)},t=0..2);
```

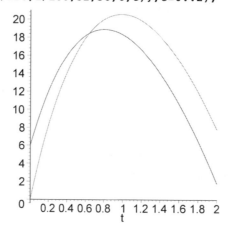

The following plot demonstrates the effect that varying the initial velocity only has on the displacement function. The values of v_0 used are 48, 64, and 80. The darkest curve corresponds to $v_0 = 48$. Notice that as the initial velocity is increased the maximum height attained by the object is increased as well.

```
> plot({position(1/128,1/160,32,48,0,t),
        position(1/128,1/160,32,64,0,t),
        position(1/128,1/160,32,80,0,t)},t=0..2);
```

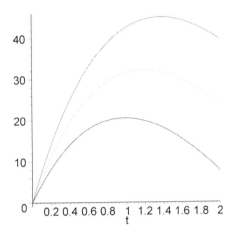

The following graph indicates the effect that varying the initial displacement and holding all other values constant has on the displacement function. We use values of 0, 10, and 20 for y_0. Notice that the value of the initial displacement vertically translates the displacement function.

```
> plot({position(1/128,1/160,32,48,0,t),
        position(1/128,1/160,32,48,10,t),
        position(1/128,1/160,32,48,20,t)},t=0..2);
```

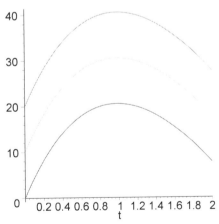

■

We now combine several of the topics discussed in this section to solve the following problem.

EXAMPLE 3: An object of mass $m = 1$ slug is dropped from a height of 50 feet above the surface of a small pond. While the object is in the air, the force due to air resistance is v. However, when the object is in the pond, it is subjected to a buoyancy force equivalent to $6v$. Determine how much time is required for the object to reach a depth of 25 feet in the pond.

SOLUTION: This problem must be broken into two parts: an initial-value problem for the object above the pond and an initial-value problem for the object below the surface of the pond. Using techniques discussed in previous examples, the initial-value problem above the pond's surface is found to be

$$\begin{cases} dv/dt = 32 - v \\ v(0) = 0 \end{cases}.$$

However, to define the initial-value problem to find the velocity of the object beneath the pond's surface, the velocity of the object when it reaches the surface must be known. Hence, the velocity of the object above the surface must be determined by solving the preceding initial-value problem.

```
> s:='s':v:='v':
step_1:=dsolve({diff(v(t),t)=32-v(t),v(0)=0},v(t));
```

$$step\_1 := v(t) = 32 - 32e^{(-t)}$$

In order to find the velocity when the object hits the pond's surface, we must know the time at which the object has traveled 50 feet. Thus, we must find $s(t)$, which is done by integrating the velocity function, obtaining $s(t) = 32e^{-t} + 32t - 32$.

```
> assign(step_1):
step_2:=dsolve({diff(s(t),t)=v(t),s(0)=0},s(t));
```

$$step\_2 := s(t) = 32t + 32e^{(-t)} - 32$$

$s(t) = 32e^{-t} + 32t - 32$ and $s(t) = 50$ are graphed with `plot`. The value of t at which the object has traveled 50 feet appears to be approximately 2.5 seconds.

```
> assign(step_2):
plot({50,s(t)},t=0..5);
```

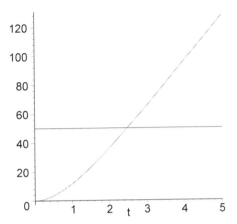

A more accurate value of the time at which the object hits the surface is now found using `fsolve`. In this case, we obtain $t \approx 2.47864$. The velocity at this time is then determined by substitution into the velocity function, resulting in $v(2.47864) \approx 29.3166$. Note that this value is the initial velocity of the object when it hits the surface of the pond.

```
> t1:=fsolve(s(t)=50,t);
```

$$t1 := 2.478643063$$

```
> v1:=evalf(subs(t=t1,v(t)));
```

$$v1 := 29.31657802$$

Thus, the initial-value problem that determines the velocity of the object beneath the surface of the pond is given by $dv/dt = 32 - 6v$, $v(0) = 29.3166$. The solution to this initial-value problem is $v(t) = 163 + 23.9833e^{-t}$:

```
> s:='s':v:='v':
step_3:=dsolve({diff(v(t),t)=32-6*v(t),v(0)=v1},v(t));
```

$$step\_3 := v(t) = \frac{16}{3} + 23.98324469e^{(-6t)}$$

and integrating to obtain the displacement function (the initial displacement is 0), we obtain $s(t) = 3.99722 - 3.99722e^{-6t} + 163t$.

```
> assign(step_3):
step_4:=dsolve({diff(s(t),t)=v(t),s(0)=0},s(t));
```

$$step\_4 := s(t) = -3.997207448e^{(-6.t)} + 5.333333334t + 3.997207448$$

This displacement function is then plotted to determine when the object is 25 feet beneath the surface of the pond. This time appears to be near 4 seconds.

```
> assign(step_4):
plot({25,s(t)},t=0..5);
```

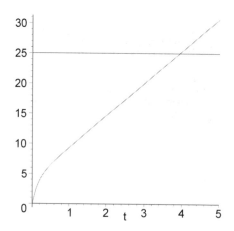

A more accurate approximation of the time at which the object is 25 feet beneath the pond's surface is obtained with `fsolve`. In this case, we obtain . Finally, the time required for the object to reach the pond's surface is added to the time needed for it to travel 25 feet beneath the surface to see that approximately 6.41667 seconds are required for the object to travel from a height of 50 feet above the pond to a depth of 25 feet below the surface.

```
> t2:=fsolve(s(t)=25,t);
```

$$t2 := 3.938023603$$

```
> t1+t2;
```

$$6.416666666$$

■

CHAPTER 4

Higher Order Differential Equations

In Chapters 2 and 3 we saw that first-order differential equations can be used to model a variety of physical situations. However, many physical situations need to be modeled by higher order differential equations. In this chapter, we discuss several methods for solving higher order differential equations.

4.1 Preliminary Definitions and Notation

In the same way as in previous chapters, we can frequently use `dsolve` to generate exact solutions of higher order equations and `dsolve` together with the `numeric` option to generate numerical solutions to higher order initial-value problems.

EXAMPLE 1 (Van der Pol Equation): The **Van der Pol equation**, which arises in the study of nonlinear damping, is the nonlinear second-order equation

$$x'' + \mu(x^2 - 1)x' + x = 0.$$

(a) Find the value of $x(7)$ if $\mu = 1$, $x(0) = 1$, and $x'(0) = 0$. (b) If $x(0) = 1$ and $x'(0) = 0$, graph the solution on the interval $[0,15]$ for $\mu = \frac{1}{32}, \frac{1}{16}, \frac{1}{8}$,

$\frac{1}{4}, \frac{1}{2}, 1, \frac{3}{2}, 2, 3, 5, 7$, and 8. (c) Compare the graphs of these solutions with the graph of the solution to the initial-value problem

$$\begin{cases} x'' + x = 0 \\ x(0) = 1, x'(0) = 0 \end{cases}.$$

SOLUTION: (a) We use `dsolve` together with the `numeric` option to solve the initial-value problem

$$\begin{cases} x'' + \mu(x^2 - 1)x' + x = 0 \\ x(0) = 1, x'(0) = 0 \end{cases}.$$

```
> numsol:=dsolve({diff(x(t),t$2)+
      (x(t)^2-1)*diff(x(t),t)+
         x(t)=0,D(x)(0)=0,x(0)=1},x(t),numeric);
```

$$numsol := \mathbf{proc}(rkf45\_x) \ \dots \ \mathbf{end}$$

Entering `numsol(7)` shows us that the value of $x(7)$ is approximately 1.3093.

```
> numsol(7);
```

$$\left[t = 7, x(t) = 1.309302121508621, \frac{\partial}{\partial t} x(t) = -.9155650438548757 \right]$$

The numerical solution obtained in `numsol` is graphed using the `odeplot` command, which is contained in the `plots` package.

```
> with(plots):
odeplot(numsol,[t,x(t)],0..15);
```

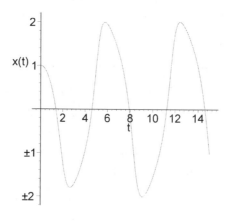

We begin by defining the function vp. Given μ, $vp(\mu)$ graphs the solution to the

initial-value problem $\begin{cases} x'' + \mu(x^2 - 1)x' + x = 0 \\ x(0) = 1, \ x'(0) = 0 \end{cases}$ for $0 \le t \le 15$.

```
> with(DEtools):

vp:=proc(mu)
        DEplot(diff(x(t),t$2)+mu*(x(t)^2-1)*diff(x(t),t)+
            x(t)=0,x(t),t=0..15,[[x(0)=1,D(x)(0)=0]],
                stepsize=0.1,thickness=1,linecolor=BLACK)
    end:
```

For example, entering

```
> vp(6);
```

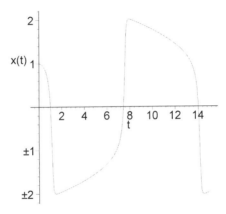

graphs the solution to the initial-value problem $\begin{cases} x'' + 6(x^2 - 1)x' + x = 0 \\ x(0) = 1, \ x'(0) = 0 \end{cases}$ on [0,15]. Thus, entering

```
> A:=array([[vp(1/32),vp(1/16),vp(1/8)],
        [vp(1/4),vp(1/2),vp(1)],
        [vp(3/2),vp(2),vp(3)],[vp(5),vp(7),vp(8)]]):
with(plots):
display(A);
```

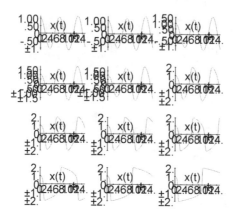

graphs the solution to the initial-value problem on the interval [0,15] for $\mu = \frac{1}{32}, \frac{1}{16}, \frac{1}{8}, \frac{1}{4}, \frac{1}{2}, 1, \frac{3}{2}, 2, 3, 5, 7$, and 8 and then displays the resulting array of graphics objects.

We find the solution to $\begin{cases} x'' + x = 0 \\ x(0) = 1, \ x'(0) = 0 \end{cases}$ with dsolve. The graph of $\cos t$ looks most like the first graph in A, corresponding to $\mu = \frac{1}{32}$.

```
> exactsol:=dsolve({diff(x(t),t$2)+x(t)=0,
            x(0)=1,D(x)(0)=0},x(t));
```

$$exactsol := x(t) = \cos(t)$$

Last, we show the two graphs together to see how similar they are.

```
> p1:=plot(rhs(exactsol),t=0..15,color=GRAY):
with(plots):
display({p1,vp(1/32)});
```

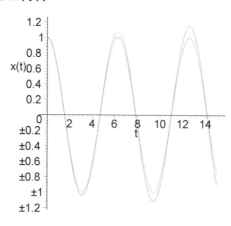

The example illustrates an important difference between linear and nonlinear equations. Exact solutions of linear equations with constant coefficents can often be found. Nonlinear equations can often be approximated by linear equations. Thus, we concentrate our study on linear differential equations.

■

The nth-Order Ordinary Linear Differential Equation

In order to develop the methods needed to solve higher order differential equations, we must state several important definitions and theorems. We begin by introducing the types of higher order equations that we will be solving in this chapter by restating the following definition that was given in Chapter 1.

Definition
*n*th-Order Ordinary
Linear Differential
Equation

An ordinary differential equation of the form

$$a_n(x)y^{(n)}(x) + a_{n-1}(x)y^{(n-1)}(x) + \cdots + a_1(x)y'(x) + a_0(x)y(x) = g(x),$$

where $a_n(x)$ is not the zero function, is called an **nth-order ordinary linear differential equation**. If $g(x)$ is identically the zero function, the equation is said to be **homogeneous**; if $g(x)$ is not the zero function, the equation is said to be **nonhomogeneous**; and if the functions $a_i(x)$, $i = 0, 1, 2, \ldots, n$, are constants, the equation is said to have **constant coefficients**. An *n*th-order equation accompanied by the conditions

$$y(x_0) = y_0, y'(x_0) = y'_0, \ldots, y^{(n-1)}(x_0) = y_0^{(n-1)},$$

where y_0, y_0', $\ldots$, $y_0^{(n-1)}$ are constants, is called an **nth order initial-value problem**.

The following theorem gives sufficient conditions for the existence of a unique solution of the *n*th-order initial-value problem.

Theorem
Existence of a Unique
Solution

> If $a_n(x)$, $a_{n-1}(x)$, $\ldots$, $a_1(x)$, $a_0(x)$ and $g(x)$ are continuous throughout an interval I and $a_n(x) \neq 0$ for all x in the interval I, then for every x_0 in I there is a unique solution to the initial-value problem
>
> $$\begin{cases} a_n(x)y^{(n)}(x) + a_{n-1}(x)y^{(n-1)}(x) + \cdots + a_1(x)y'(x) + a_0(x)y(x) = g(x) \\ \qquad\qquad\qquad y(x_0) = y_0 \\ \qquad\qquad\qquad\quad \vdots \\ \qquad\qquad\quad y^{(n-1)}(x_0) = y_0^{(n)} \end{cases}$$
>
> on I, where y_0, y_0', $\ldots$, $y_0^{(n)}$ represent arbitrary constants.

Now that we have conditions that indicate the existence of solutions, we become familiar with the properties of the functions that form the solution. We will see that solutions to nth-order ordinary linear differential equations require n solutions with the property defined as follows.

Definition
Linearly Dependent
and Linearly
Independent

> Let $f_1(x)$, $f_2(x)$, $f_3(x)$, $\ldots$, $f_{n-1}(x)$, and $f_n(x)$ be a set of n functions and let $S = \{f_1(x), f_2(x), f_3(x), \ldots , f_{n-1}(x), f_n(x)\}$. S is **linearly dependent** on an interval I if there are constants c_1, c_2, $\ldots$, c_n, not all zero, so that
>
> $$c_1 f_1(x) + c_2 f_2(x) + \cdots + c_{n-1}f_{n-1}(x) + c_n f_n(x) = 0$$
>
> for every value of x in the interval I.
> S is **linearly independent** if S is not linearly dependent.

It is a good exercise to use the definition of linear dependence to show that a set of two functions is linearly dependent if and only if the two functions are constant multiples of each other.

Definition
Wronskian

> Let $S = \{f_1(x), f_2(x), f_3(x), \ldots , f_{n-1}(x), f_n(x)\}$ be a set of n functions for which each is differentiable at least $n - 1$ times. The **Wronskian** of S, denoted by
>
> $$W(S) = W(f_1(x), f_2(x), f_3(x), \ldots, f_{n-1}(x), f_n(x)),$$
>
> is the determinant
>
> $$W(S) = \begin{vmatrix} f_1(x) & f_2(x) & \cdots & f_n(x) \\ f_1'(x) & f_2'(x) & \cdots & f_n'(x) \\ \vdots & \vdots & \vdots & \vdots \\ f_1^{(n-1)}(x) & f_2^{(n-1)}(x) & \cdots & f_n^{(n-1)}(x) \end{vmatrix}.$$

EXAMPLE 2: Compute the Wronskian for each of the following sets of functions:
(a) $S = \{\sin x, \cos x\}$ and (b) $S = \{\cos 2x, \sin 2x, \sin x \cos x\}$.

SOLUTION: The 2×2 determinant $\begin{vmatrix} a_{11} & a_{12} \\ a_{21} & a_{22} \end{vmatrix}$ is computed by calculating $a_{11}a_{22} - a_{12}a_{21}$. Thus, for (a) we have

$$W(S) = \begin{vmatrix} \sin x & \cos x \\ \frac{d}{dx}(\sin x) & \frac{d}{dx}(\cos x) \end{vmatrix} = \begin{vmatrix} \sin x & \cos x \\ \cos x & -\sin x \end{vmatrix} = -\sin^2 x - \cos^2 = -1.$$

We can also use Maple to compute the `Wronskian` by taking advantage of the `Wronskian` command contained in the **linalg** package. First, we load the **linalg** package by entering `with(linalg)` and then define S to be the list of functions $S = \{\sin x, \cos x\}$. `Wronskian` is then used to compute the Wronskian matrix, naming the resulting output `ws`. The determinant of `ws` is found with `det`, also contained in the **linalg** package, and named `step_1`. Note that `step_1` is not simplified.

```
> S:=[sin(x),cos(x)]:

ws:=Wronskian(S,x);

step_1:=det(ws);
```

$$ws := \begin{bmatrix} \sin(x) & \cos(x) \\ \cos(x) & -\sin(x). \end{bmatrix}$$

$$step\_1 := -\sin(x)^2 - \cos(x)^2$$

We simplify `step_1` using `simplify`, naming the result `step_2`. In this case, we see that the Wronskian for the set of functions S is -1, agreeing with the result obtained before. You should verify that $\sin x$ and $\cos x$ are linearly independent functions.

```
> simplify(step_1);
```

$$-1$$

For (b), we need to compute the determinant

$$\begin{vmatrix} \cos 2x & \sin 2x & \sin x \cos x \\ \frac{d}{dx}(\cos 2x) & \frac{d}{dx}(\sin 2x) & \frac{d}{dx}(\sin x \cos x) \\ \frac{d^2}{dx^2}(\cos 2x) & \frac{d^2}{dx^2}(\sin 2x) & \frac{d^2}{dx^2}(\sin x \cos x) \end{vmatrix}.$$

The 3×3 determinant $\begin{vmatrix} a_{11} & a_{12} & a_{13} \\ a_{21} & a_{22} & a_{23} \\ a_{31} & a_{32} & a_{33} \end{vmatrix}$ can be computed in several equivalent ways. For example,

$$\begin{vmatrix} a_{11} & a_{12} & a_{13} \\ a_{21} & a_{22} & a_{23} \\ a_{31} & a_{32} & a_{33} \end{vmatrix} = a_{11} \begin{vmatrix} a_{22} & a_{23} \\ a_{32} & a_{33} \end{vmatrix} - a_{12} \begin{vmatrix} a_{21} & a_{23} \\ a_{31} & a_{33} \end{vmatrix} + a_{13} \begin{vmatrix} a_{21} & a_{22} \\ a_{31} & a_{32} \end{vmatrix}.$$

First define S to be the list of functions corresponding to $S = \{\cos 2x, \sin 2x, \sin x \cos x\}$ and then use Wronskian, in the same manner as in (a), to compute the Wronskian matrix for S, naming the resulting output ws.

```
> S:=[cos(2*x),sin(2*x),sin(x)*cos(x)]:

ws:=Wronskian(S,x);
```

$$ws := \begin{bmatrix} \cos(2x) & \sin(2x) & \sin(x)\cos(x) \\ -2\sin(2x) & 2\cos(2x) & \cos(x)^2 - \sin(x)^2 \\ -4\cos(2x) & -4\sin(2x) & -4\sin(x)\cos(x) \end{bmatrix}$$

Similarly, we use det to compute the determinant of ws. In this case, we see that the Wronskian is 0. You should use the identity $\sin 2x = 2\sin x \cos x$ to show that the set of functions $S = \{\cos 2x, \sin 2x, \sin x \cos x\}$ is linearly dependent.

```
> det(ws);
```

$$0$$

∎

In Example 2, we see that in (a) the Wronskian is not 0 but in (b) the Wronskian is 0. Moreover, the set of functions in (a) is linearly independent because $\sin x$ and $\cos x$ are not multiples of each other, whereas the set of functions in (b) is linearly dependent because $\sin 2x = 2 \sin x \cos x$.

In fact, we can use the Wronskian to determine whether a set of functions is linearly dependent or linearly independent.

Theorem

> Let $S = \{f_1(x), f_2(x), f_3(x), \dots, f_{n-1}(x), f_n(x)\}$ be a set of n solutions of
>
> $$a_n(x)y^{(n)}(x) + a_{n-1}(x)y^{(n-1)}(x) + \cdots + a_1(x)y'(x) + a_0(x)y(x) = g(x)$$
>
> on an interval I. S is linearly independent if and only if $W(S) \neq 0$ for at least one value of x in the interval I.

EXAMPLE 3: Use the Wronskian to classify each of the following sets of functions as linearly independent or linearly dependent: (a) $S = \{1 - 2\sin^2 x, \cos 2x\}$ and (b) $S = \{e^x, xe^x, x^2e^x\}$.

SOLUTION: (a) Note that both functions in S are solutions of $y'' + 4y = 0$. Here, we must compute the determinant of the 2×2 matrix

$$\begin{pmatrix} 1 - 2\sin^2 x & \cos 2x \\ \frac{d}{dx}(1 - 2\sin^2 x) & \frac{d}{dx}(\cos 2x) \end{pmatrix}.$$

We use the `Wronskian` and `det` commands, both contained in the **linalg** package, followed by using the `simplify` command to compute the determinant

```
> with(linalg):

S:=[1-2*sin(x)^2,cos(2*x)];

ws:=Wronskian(S,x);

step_1:=det(ws);

step_2:=simplify(step_1);
```

$$S := [1 - 2\sin(x)^2, \cos(2x)]$$

$$ws := \begin{vmatrix} 1 - 2\sin(x)^2 & \cos(2x) \\ -4\sin(x)\cos(x) & -2\sin(2x) \end{vmatrix}$$

$$step\_1 := -2\sin(2x) + 4\sin(2x)\sin(x)^2 + 4\cos(2x)\sin(x)\cos(x)$$

$$step\_2 := 0$$

and see that the result is 0. Therefore, the set of functions $S = \{1 - 2\sin^2 x, \cos 2x\}$ is linearly dependent. This makes sense because these functions are multiples of each other: $\cos 2x = \frac{1}{2}(1 - 2\sin^2 x)$.
(b) Note that all three functions in S are solutions of $y''' - 3y'' + 3y' - y = 0$. Here, we must compute the determinant

$$W(S) = \begin{vmatrix} e^x & xe^x & x^2e^x \\ \frac{d}{dx}(e^x) & \frac{d}{dx}(xe^x) & \frac{d}{dx}(x^2e^x) \\ \frac{d^2}{dx^2}(e^x) & \frac{d^2}{dx^2}(xe^x) & \frac{d^2}{dx^2}(x^2e^x) \end{vmatrix}$$

```
> S:=[exp(x),x*exp(x),x^2*exp(x)]:

ws:=Wronskian(S,x):

det(ws);
```

$$2(e^x)^3$$

We conclude that S is linearly independent because the Wronskian of S is not identically zero.

■

Fundamental Set of Solutions

Obtaining a collection of n linearly independent solutions to an nth-order linear differential equation is of great importance in solving nth-order linear equations. (Note that a **nontrivial solution** is one that is not identically the zero function.)

Definition **Fundamental Set of** **Solutions**	A set $S = \{f_1(x), f_2(x), f_3(x), \ldots, f_{n-1}(x), f_n(x)\}$ of n linearly independent nontrivial solutions of the nth-order linear homogeneous equation $$a_n(x)y^{(n)}(x) + a_{n-1}(x)y^{(n-1)}(x) + \cdots + a_1(x)y'(x) + a_0(x)y(x) = 0$$ is called a **fundamental set of solutions** of the equation.

EXAMPLE 4: Show that $S = \{e^{-5x}, e^{-x}\}$ is a fundamental set of solutions of the equation $y'' + 6y' + 5y = 0$.

SOLUTION: S is linearly independent because

$$W(S) = \begin{vmatrix} e^{-5x} & e^{-x} \\ -5e^{-5x} & -e^{-x} \end{vmatrix} = -e^{-6x} + 5e^{-6x} = 5e^{-6x} \neq 0.$$

Also, we must verify that each function is a solution of the differential equation. Because

$$\frac{d^2}{dx^2}(e^{-5x}) + 6\frac{d}{dx}(e^{-5x}) + 5e^{-5x} = 25e^{-5x} - 30e^{-5x} + 5e^{-5x} = 0 \quad \text{and}$$

$$\frac{d^2}{dx^2}(e^{-x}) + 6\frac{d}{dx}(e^{-x}) + 5e^{-x} = e^{-x} - 6e^{-x} + 5e^{-x} = 0,$$

we conclude that S is a fundamental set of solutions of the equation $y'' + 6y' + 5y = 0$.

Of course, we can perform the same steps with Maple. First, we define S to be the set of functions $S = \{e^{-5x}, e^{-x}\}$. Then, `simplify`, `det`, and `Wronskian` are used to compute the Wronskian of the set of functions S.

```
> with(linalg):
S:=[exp(-5*x),exp(-x)]:
simplify(det(Wronskian(S,x)));
```

$$4e^{(-6x)}$$

Note that the first element of S is extracted from S with $S[1]$; similarly, the second is extracted with $S[2]$.

```
> S[1];
```

$$e^{(-5x)}$$

We verify that each element of S is a solution to the equation with `diff`.

```
> diff(S[1],x$2)+6*diff(S[1],x)+5*S[1];
```

$$0$$

```
> diff(S[2],x$2)+6*diff(S[2],x)+5*S[2];
```

$$0$$

■

We use the fundamental set of solutions to create what is known as a **general solution** of an nth-order linear homogeneous differential equation.

Theorem **Principle of** **Superposition**	If $S = \{f_1(x), f_2(x), f_3(x), \ldots, f_{k-1}(x), f_k(x)\}$ is a set of solutions of the equation $$a_n(x)y^{(n)}(x) + a_{n-1}(x)y^{(n-1)}(x) + \cdots + a_1(x)y'(x) + a_0(x)y(x) = 0$$ and $\{c_1, c_2, \ldots c_{k-1}, c_k\}$ is a set of k constants, then $$f(x) = c_1 f_1(x) + c_2 f_2(x) + \cdots + c_{k-1}f_{k-1}(x) + c_k f_k(x)$$ is also a solution of $$a_n(x)y^{(n)}(x) + a_{n-1}(x)y^{(n-1)}(x) + \cdots + a_1(x)y'(x) + a_0(x)y(x) = 0.$$

$f(x) = c_1 f_1(x) + c_2 f_2(x) + \cdots + c_{k-1} f_{k-1}(x) + c_k f_k(x)$ is called a **linear combination** of functions in the set $S = \{f_1(x), f_2(x), f_3(x), \ldots, f_{k-1}(x), f_k(x)\}$. A consequence of this fact is that the linear combination of the functions in a fundamental set of solutions of an nth-order homogeneous linear differential equation is also a solution of the differential equation, and we call this linear combination a **general solution** of the differential equation.

Corollary

If $S = \{f_1(x), f_2(x), f_3(x), \ldots, f_{n-1}(x), f_n(x)\}$ is a fundamental set of solutions of the nth-order linear homogeneous equation

$$a_n(x)y^{(n)}(x) + a_{n-1}(x)y^{(n-1)}(x) + \cdots + a_1(x)y'(x) + a_0(x)y(x) = 0$$

and $\{c_1, c_2, \ldots c_{n-1}, c_n\}$ is a set of n constants, then

$$f(x) = c_1 f_1(x) + c_2 f_2(x) + \cdots + c_{n-1} f_{n-1}(x) + c_n f_n(x)$$

is also a solution of

$$a_n(x)y^{(n)}(x) + a_{n-1}(x)y^{(n-1)}(x) + \cdots + a_1(x)y'(x) + a_0(x)y(x) = 0.$$

Definition
General Solution

If $S = \{f_1(x), f_2(x), f_3(x), \ldots, f_{n-1}(x), f_n(x)\}$ is a fundamental set of solutions of the nth-order linear homogeneous equation

$$a_n(x)y^{(n)}(x) + a_{n-1}(x)y^{(n-1)}(x) + \cdots + a_1(x)y'(x) + a_0(x)y(x) = 0,$$

then a **general solution** of the equation is

$$f(x) = c_1 f_1(x) + c_2 f_2(x) + \cdots + c_{n-1} f_{n-1}(x) + c_n f_n(x),$$

where $\{c_1, c_2, \ldots c_{n-1}, c_n\}$ is a set of n arbitrary constants.

In other words, if we have a fundamental set of solutions S, then a general solution of the differential equation is formed by taking the **linear combination** of the functions in S.

EXAMPLE 5: Show that $S = \{\cos 2x, \sin 2x\}$ is a fundamental set of solutions of the second-order ordinary linear differential equation with constant coefficients $y'' + 4y = 0$.

SOLUTION: First, we compute the Wronskian

```
> with(linalg):

S:=[cos(2*x),sin(2*x)]:
```

```
simplify(det(Wronskian(S,x)));
```

$$2$$

to show that the functions in S are linearly independent. Next, we verify that both functions are solutions of $y'' + 4y = 0$ by using map to apply the function $y'' + 4y$ to the functions in S. We conclude that both functions in S are solutions of $y'' + 4y = 0$ because the result is a list of two zeros.

```
> map(f->diff(f,x$2)+4*f,S);
```

$$[0,0]$$

By the principle of superposition, $y(x) = c_1 \cos 2x + c_2 \sin 2x$, where c_1 and c_2 are arbitrary constants, is also a solution of the equation. We now graph $y(x)$ for various values of c_1 and c_2. In this case, we use seq and subs to define the set of functions to_plot obtained by replacing c[1] by i and c[2] by j for $i = -1, 0,$ and 1 and $j = -1, 0,$ and 1.

```
> y:=x->c[1]*cos(2*x)+c[2]*sin(2*x):
to_plot:={seq(seq(subs({c[1]=i,c[2]=j},y(x)),
          i=-1..1),j=-1..1)}:
```

The set of functions $S = \{\cos 2x, \sin 2x\}$ and the set of functions to_plot are then each graphed with plot on the interval $[-\pi, 2\pi]$.

```
> plot({cos(2*x),sin(2*x)},x=-Pi..2*Pi);
plot(to_plot,x=-Pi..2*Pi);
```

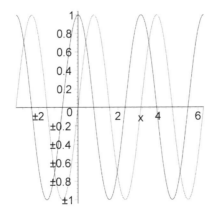
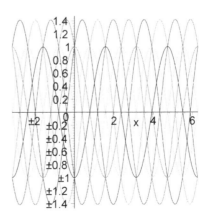

The principle of superposition is a very important property of linear homogeneous equations and is generally not valid for nonlinear equations.

EXAMPLE 6: Is the principle of superposition valid for the nonlinear equation $tx'' - 2xx' = 0$?

SOLUTION: We see that `dsolve` is able to find a general solution of this nonlinear equation.

```
> gensol:=dsolve(t*diff(x(t),t$2)-2*x(t)*diff(x(t),t)=0,
          x(t));
```

$$gensol := x(t) = -\frac{1}{2}\frac{\tan\left(\frac{1}{2}\frac{-\ln(t)+\_C2}{\_C1}\right) + \_C1}{\_C1}$$

$x(t) = -\frac{1}{2}$ is the solution that satisfies $x(1) = -\frac{1}{2}$ and $x'(1) = 0$.

```
> sol1:=dsolve({t*diff(x(t),t$2)-2*x(t)*diff(x(t),t)=0,
          x(1)=-1/2,D(x)(1)=0},x(t));
```

$$sol1 := x(t) = \frac{-1}{2}$$

$x(t) = \frac{1}{2}\left(-1 + \tan\left(\frac{1}{2}\ln t\right)\right)$ is the solution that satisfies $x(1) = -\frac{1}{2}$ and $x'(1) = \frac{1}{4}$.

```
> sol2:=dsolve({t*diff(x(t),t$2)-2*x(t)*diff(x(t),t)=0,
          x(1)=-1/2,D(x)(1)=1/4},x(t));
```

$$sol2 := x(t) = -\frac{1}{2}\tan\left(-\frac{1}{2}\ln(t)\right) - \frac{1}{2}, x(t) = \frac{1}{2}\tan\left(\frac{1}{2}\ln(t)\right) - \frac{1}{2}$$

However, the sum of these two solutions is not a solution to the nonlinear equation because $tf'' - 2ff' \neq 0$; the principle of superposition is *not* valid for this nonlinear equation.

```
> f:=rhs(sol1)+rhs(sol2[1]);
simplify(t*diff(f,t$2)-2*f*diff(f,t));
```

$$f := -1 + \frac{1}{2}\tan\left(\frac{1}{2}\ln(t)\right)$$

$$\frac{1}{4}\frac{1 + \tan\left(\frac{1}{2}\ln(t)\right)^2}{t}$$

■

Existence of a Fundamental Set of Solutions

The following two theorems tell us that under reasonable conditions, the nth-order linear homogeneous equation

$$a_n(x)y^{(n)}(x) + a_{n-1}(x)y^{(n-1)}(x) + \cdots + a_1(x)y'(x) + a_0(x)y(x) = 0$$

has a fundamental set of n solutions.

Theorem

> If $a_i(x)$ is continuous on an open interval I for $i = 0, 1, \ldots, n$, and $a_n(x) \neq 0$ for all x in the interval I, then the nth-order linear homogeneous equation
>
> $$a_n(x)y^{(n)}(x) + a_{n-1}(x)y^{(n-1)}(x) + \cdots + a_1(x)y'(x) + a_0(x)y(x) = 0$$
>
> has a fundamental set of n solutions.

Theorem

> Any set of $n + 1$ solutions of the nth-order linear homogeneous equation
>
> $$a_n(x)y^{(n)}(x) + a_{n-1}(x)y^{(n-1)}(x) + \cdots + a_1(x)y'(x) + a_0(x)y(x) = 0$$
>
> is linearly dependent.

We can summarize the results of these theorems by saying that in order to solve an nth-order linear ordinary differential equation, we must find a set S of n functions that satisfy the differential equation such that $W(S) \neq 0$.

EXAMPLE 7: Show that $y = c_1 e^{-x} + c_2 e^{-2x}$ is a general solution of $y'' + 3y' + 2y = 0$.

SOLUTION: You should verify that $S = \{e^{-x}, e^{-2x}\}$ is a linearly independent set of functions. Differentiating the function $y = c_1 e^{-x} + c_2 e^{-2x}$ yields $y' = -c_1 e^{-x} - 2c_2 e^{-2x}$ and $y'' = c_1 e^{-x} + 4c_2 e^{-2x}$. Substitution then gives us

$$y'' + 3y' + 2y = c_1e^{-x} + 4c_2e^{-2x} + 3(-c_1e^{-x} - 2c_2e^{-2x}) + 2(c_1e^{-x} + c_2e^{-2x})$$
$$= (c_1 - 3c_1 + 2c_1)e^{-x} + (4c_2 - 6c_2 + 2c_2)e^{-2x} = 0.$$

Therefore, $y = c_1e^{-x} + c_2e^{-2x}$ is a general solution of $y'' + 3y' + 2y = 0$. To graph this solution for various values of the constants, we begin by defining $y = c_1e^{-x} + c_2e^{-2x}$.

```
> y:=c->c[1]*exp(-x)+c[2]*exp(-2*x);
```

$$y := c \rightarrow c_1\mathbf{e}^{(-x)} + c_2\mathbf{e}^{(-2x)}$$

We will graph the four solutions obtained by replacing `c[1]` and `c[2]` by 1 and -1, -1 and 1, 2 and 1, and 1 and -2. We define `cvals` to be the set of four ordered pairs $(1, -1)$, $(-1, 1)$, $(2, 1)$, and $(1, -2)$.

```
> cvals:=[[1,-1],[-1,1],[2,1],[1,-2]];
```

$$cvals := [[1, -1], [-1, 1], [2, 1], [1, -2]]$$

We then use `map` to compute `y` for each pair in the set of ordered pairs `cvals`, naming the resulting set of functions `toplot`.

```
> toplot:=map(y,cvals,2);
```

$$toplot := [\mathbf{e}^{(-x)} - \mathbf{e}^{(-2x)}, -\mathbf{e}^{(-x)} + \mathbf{e}^{(-2x)}, 2\mathbf{e}^{(-x)} + \mathbf{e}^{(-2x)},$$
$$\mathbf{e}^{(-x)} - 2\mathbf{e}^{(-2x)}]$$

We then use `plot` to graph the functions in `toplot` on the interval $[-1/2, 3]$.

```
> plot(toplot,x=-1/2..3,y=-1.75..1.75);
```

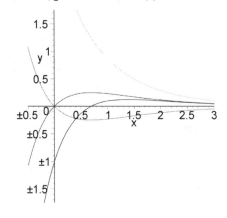

■

EXAMPLE 8: Show that $y = e^{-x}(c_1 \cos 4x + c_2 \sin 4x)$ is a general solution of $y'' + 2y' + 17y = 0$.

SOLUTION: You should verify that $S = \{e^{-x} \cos 4x, e^{-x} \sin 4x\}$ is a linearly independent set of functions. We illustrate how Maple can be used to carry out the steps illustrated in Example 7. After defining y, we use `diff` to compute the first and second derivatives (with respect to x) of y.

```
> y:=c->exp(-x)*(c[1]*cos(4*x)+c[2]*sin(4*x)):
diff(y(c),x$2);
diff(y(c),x);
```

$$\mathbf{e}^{(-x)}(c_1 \cos(4x) + c_2 \sin(4x))$$
$$- 2\mathbf{e}^{(-x)}(-4c_1 \sin(4x) + 4c_2 \cos(4x))$$
$$+ \mathbf{e}^{(-x)}(-16c_1 \cos(4x) - 16c_2 \sin(4x))$$
$$- \mathbf{e}^{(-x)}(c_1 \cos(4x) + c_2 \sin(4x))$$
$$+ \mathbf{e}^{(-x)}(-4c_1 \sin(4x) + 4c_2 \cos(4x))$$

We then compute and simplify $y'' + 2y' + 17y$. Because the result is zero and the set of functions $S = \{e^{-x} \cos 4x, e^{-x} \sin 4x\}$ is linearly independent, $y = e^{-x}(c_1 \cos 4x + c_2 \sin 4x)$ is a general solution of the equation.

```
> simplify(diff(y(c),x$2)+2*diff(y(c),x)+17*y(c));
```

$$0$$

We can graph the solution for various values of the constants in the same manner as in the previous example. We define `cvals` to be the set of ordered pairs consisting of (0,1), (1,0), (2,1), and (1, $-$ 2). Similarly, we use `map` to compute the value of `y` for each ordered pair in `cvals`, naming the resulting set of functions `toplot` and then graphing them on the interval $[-1,2]$ with `plot`.

```
cvals:=[[0,1],[1,0],[2,1],[1,-2]]:
> toplot:=map(y,cvals):
> plot(toplot,x=-1..2);
```

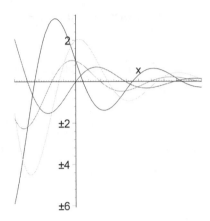

Reduction of Order

In the next section, we learn how to find solutions of homogeneous equations with constant coefficients. In doing this, we will find it necessary to determine a second solution from a known solution. We illustrate this procedure, called **reduction of order**, by considering a second-order equation. In certain situations, we can reduce a second-order equation by making an appropriate substitution to convert the second-order equation to a first-order equation (this reduction in order gives the name to the method). Suppose that we have the equation

$$y'' + p(x)y' + q(x)y = 0,$$

and suppose that $y = f(x)$ is a solution to this equation. Of course we know from our previous discussion that in order to solve the second-order differential equation $y'' + p(x)y' + q(x)y = 0$, we must have two linearly independent solutions. Hence, we must determine a second linearly independent solution. We accomplish this by attempting to find a solution of the form

$$y = v(x)f(x)$$

and solving for $v(x)$. Differentiating with the product rule, we obtain

$$y' = f'v + v'f \quad \text{and} \quad y'' = f''v + 2v'f' + fv''.$$

```
> x:='x':y:='y':f:='f':v:='v':

y:=x->v(x)*f(x):

> diff(y(x),x);

diff(y(x),x$2);
```

$$\left(\frac{\partial}{\partial x}v(x)\right)f(x) + v(x)\left(\frac{\partial}{\partial x}f(x)\right)$$

$$\left(\frac{\partial^2}{\partial x^2}v(x)\right)f(x) + 2\left(\frac{\partial}{\partial x}v(x)\right)\left(\frac{\partial}{\partial x}f(x)\right) + v(x)\left(\frac{\partial^2}{\partial x^2}f(x)\right)$$

Notice that for convenience, we have omitted the argument of these functions. We now substitute y, y', and y'' into the equation $y'' + p(x)y' + q(x)y = 0$. This gives us

$$y'' + p(x)y' + q(x)y = f''v + 2v'f' + fv'' + p(x)(f'v + fv') + q(x)vf$$
$$= y'' + p(x)y' + q(x)y = f''v + 2v'f' + fv'' + p(x)(f'v + fv') + q(x)vf$$
$$= fv'' + (2f' + p(x)f)v'.$$

```
> step1:=collect(diff(y(x),x$2)+p(x)*diff(y(x),x)+q(x)*y(x),
            [v(x),diff(v(x),x),diff(v(x),x$2)]);
```

$$step1 := \left(\left(\frac{\partial^2}{\partial x^2}f(x)\right) + p(x)\left(\frac{\partial}{\partial x}f(x)\right) + q(x)f(x)\right)v(x)$$

$$+ \left(2\left(\frac{\partial}{\partial x}f(x)\right) + p(x)f(x)\right)\left(\frac{\partial}{\partial x}v(x)\right) + \left(\frac{\partial^2}{\partial x^2}v(x)\right)f(x)$$

```
> step2:=subs(diff(f(x),x$2)+p(x)*diff(f(x),x)+q(x)*f(x)=0,
            step1);
```

$$step2 := \left(2\left(\frac{\partial}{\partial x}f(x)\right) + p(x)f(x)\right)\left(\frac{\partial}{\partial x}v(x)\right) + \left(\frac{\partial^2}{\partial x^2}v(x)\right)f(x)$$

Therefore, we have the equation

$$fv'' + (2f' + p(x)f)v' = 0$$

which can be written as a first-order equation by letting $w = v'$. Making this substitution gives us the linear first-order equation

$$fw' + (2f' + p(x)f)w = 0 \quad \text{or} \quad f\frac{dw}{dx} + (2f' + p(x)f)w = 0$$

which is separable, so we obtain the separated equation

$$\frac{dw}{w} = \left(-\frac{2f'}{f} - p\right)dx$$

```
> step3:=subs(diff(v(x),x$2)=diff(w(x),x),
```

```
diff(v(x),x)=w(x),step2);
```

$$step3 := \left(2\left(\frac{\partial}{\partial x}f(x)\right) + p(x)f(x)\right)w(x) + \left(\frac{\partial}{\partial x}w(x)\right)f(x)$$

We can solve this equation by integrating both sides of the equation to yield

$$\ln|w| = \ln\left(1/f^2\right) - \int p(x)dx$$

This means that $w = \frac{1}{f^2}e^{-\int p(x)dx}$,

```
> step4:=dsolve(step3=0,w(x));
```

$$step4 := w(x) = \_C1 e^{\left(-\int 2\frac{\frac{\partial}{\partial x}f(x)}{f(x)} + p(x)dx\right)}$$

```
> step5:=simplify(step4);
```

$$step5 := w(x) = \_C1 e^{\left(-\int \frac{2\left(\frac{\partial}{\partial x}f(x)\right) + p(x)f(x)}{f(x)}dx\right)}$$

so we have the formula $\frac{dv}{dx} = \frac{1}{f^2}e^{-\int p(x)dx}$ or

$$v(x) = \int \frac{e^{-\int p(x)dx}}{[f(x)]^2}dx.$$

```
> step6:=int(rhs(step5),x);
```

$$step6 := \int \_C1 e^{\left(-\int \frac{2\left(\frac{\partial}{\partial x}f(x)\right) + p(x)f(x)}{f(x)}dx\right)}dx$$

Therefore, if we have the solution $f(x)$ of the differential equation $y'' + p(x)y' + q(x)y = 0$, then we can obtain a second linearly independent solution of the form $y = v(x)f(x)$ where

$$v(x) = \int \frac{e^{-\int p(x)dx}}{[f(x)]^2}dx.$$

You can use the command reduceOrder, which is contained in the **DEtools** package, to implement the method of reduction of order.

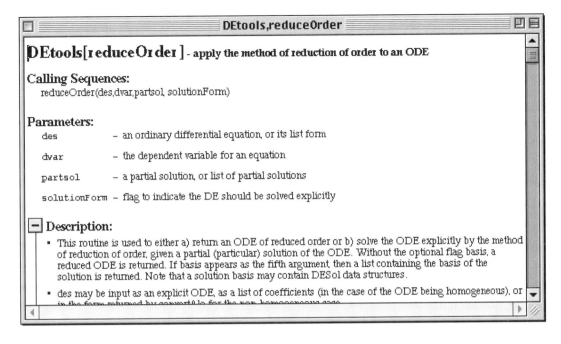

```
> x:='x':
reduceOrder(diff(y(x),x$2)+p(x)*diff(y(x),x)+
          q(x)*y(x)=0,y(x),f(x));
```

$$f(x) \int \frac{e^{(-\int p(x)dx)}}{f(x)^2} dx$$

EXAMPLE 9: Determine a second linearly independent solution to the differential equation $4x^2y'' + 8xy' + y = 0$, $x>0$, if $y = x^{-1/2}$ is a solution.

SOLUTION: In this case, we must divide by $4x^2$ in order to obtain an equation of the form $y'' + p(x)y' + q(x)y = 0$. This gives us the equation $y'' + 2x^{-1}y' + \frac{1}{4}x^{-2}y = 0$. Therefore, $p(x) = 2x^{-1}$, and $f(x) = x^{-1/2}$. Using the formula for v, we obtain

$$v(x) = \int \frac{e^{-\int p(x)dx}}{[f(x)]^2} dx = \int \frac{e^{-\int \frac{2}{x}dx}}{[x^{-1/2}]^2} dx = \int \frac{e^{-2\ln x}}{x^{-1}} dx = \int x^1 dx = \ln x, \qquad x>0.$$

Hence, a second solution is $y = x^{-1/2} \ln x$

Of course, we can take advantage of commands such as `int` to carry out the steps encountered here.

```
> p:=x->2/x:
f:=x->1/sqrt(x):
v:=x->int(exp(-int(p(x),x))/f(x)^2,x):
v(x);
```

$$\ln(x)$$

```
> y:=x->v(x)*f(x):
y(x);
```

$$\frac{\ln(x)}{\sqrt{x}}$$

∎

4.2 Solving Homogeneous Equations with Constant Coefficients

We now turn our attention to solving linear homogeneous equations with constant coefficients. Nonhomogeneous equations are considered in the following sections.

Solutions of any nth-order homogeneous linear differential equations with constant coefficients are determined by the solutions of the **characteristic equation**.

Definition **Characteristic** **Equation**	The equation $$a_n m^n + a_{n-1} m^{n-1} + \cdots + a_1 m + a_0 = 0$$ is called the **characteristic equation** of the nth-order homogeneous linear differential equation with constant coefficients $$a_n y^{(n)}(x) + a_{n-1} y^{(n-1)}(x) + \cdots + a_1 y'(x) + a_0 y(x) = 0.$$

Let us begin our investigation by considering the second-order homogeneous equation with constant coefficients

$$ay'' + by' + cy = 0$$

with characteristic equation

$$am^2 + bm + c = 0.$$

Note that $\bar{m}_1$ is the **complex conjugate** of $m_1 : \bar{m}_1 = \overline{\alpha + i\beta} = \alpha - i\beta.$

Theorem
Solving Second-Order
Equations with
Constant Coefficients

Let $ay'' + by' + cy = 0$ be a homogeneous second-order equation with constant real coefficients and let m_1 and m_2 be the solutions of the equation $am^2 + bm + c = 0$.

1. If $m_1 \neq m_2$ and both m_1 and m_2 are real, $S = \{e^{m_1 x}, e^{m_2 x}\}$ is a fundamental set of solutions to the equation and a general solution of $ay'' + by' + cy = 0$ is

$$y = c_1 e^{m_1 x} + c_2 e^{m_2 x}.$$

2. If $m_1 = m_2$, $S = \{e^{m_1 x}, x e^{m_1 x}\}$ is a fundamental set of solutions to the equation and a general solution of $ay'' + by' + cy = 0$ is

$$y = c_1 e^{m_1 x} + c_2 x e^{m_1 x}.$$

3. If $m_1 = \alpha + i\beta, \beta \neq 0$, and $m_2 = \bar{m}_1 = \alpha - i\beta$, $S = \{e^{\alpha x} \cos \beta x, e^{\alpha x} \sin \beta x\}$ is a fundamental set of solutions to the equation and a general solution of $ay'' + by' + cy = 0$ is

$$y = c_1 e^{\alpha x} \cos \beta x + c_2 e^{\alpha x} \sin \beta x = e^{\alpha x}(c_1 \cos \beta x + c_2 \sin \beta x).$$

Often, you can use `constcoeffsols`, which is contained in the **DEtools** package, to find a fundamental set of solutions to a homogeneous linear equation with constant coefficients.

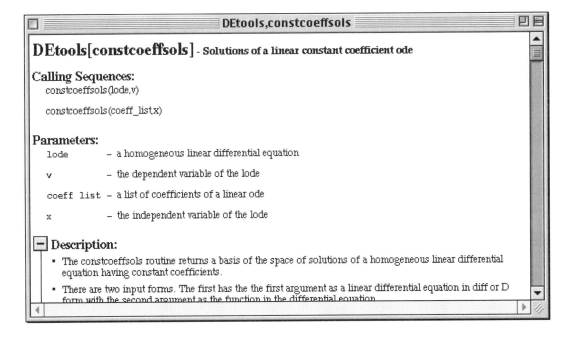

DEtools,constcoeffsols

DEtools[constcoeffsols] - Solutions of a linear constant coefficient ode

Calling Sequences:
 constcoeffsols (lode,v)

 constcoeffsols (coeff_list,x)

Parameters:
 lode   – a homogeneous linear differential equation

 v   – the dependent variable of the lode

 coeff list – a list of coefficients of a linear ode

 x   – the independent variable of the lode

Description:
- The constcoeffsols routine returns a basis of the space of solutions of a homogeneous linear differential equation having constant coefficients.
- There are two input forms. The first has the the first argument as a linear differential equation in diff or D form with the second argument as the function in the differential equation.

The following three examples illustrate each of these situations.

EXAMPLE 1: Solve $y'' + 3y' - 4y = 0$ subject to $y(0) = 1$ and $y'(0) = -1$.

SOLUTION: The characteristic equation of $y'' + 3y' - 4y = 0$ is $m^2 + 3m - 4 = (m + 4)(m - 1) = 0$. A general solution of $y'' + 3y' - 4y = 0$ is

$$y(x) = c_1 e^{-4x} + c_2 e^x$$

because the solutions of the characteristic equation are $m = -4$ and $m = 1$. We see that dsolve finds a general solution as well.

```
> y:='y':x:='x':
diffeqn:=diff(y(x),x$2)+3*diff(y(x),x)-4*y(x)=0:
gensol:=dsolve(diffeqn,y(x));
```

$$gensol := y(x) = \_C1e^x + \_C2e^{(-4x)}$$

Of course, we can use Maple to graph various solutions. We use seq to define toplot to be the list obtained by replacing each occurrence of c_1 in $y(x) = c_1 e^x + c_2 e^{-4x}$ by -1, 0, and 1 and each occurrence of c_2 in y by -1, 0, and 1. Then, we use plot to graph the set of functions toplot on the interval $[-1,1]$.

```
> toplot:=[seq(seq(rhs(gensol),_C1=-1..1),_C2=-1..1)]:
plot(toplot,x=-1..1,y=-10..10);
```

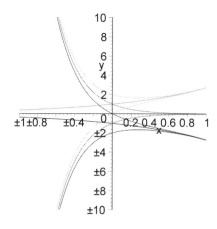

To find a solution of $y'' + 3y' - 4y = 0$ that satisfies $y(0) = 1$ and $y'(0) = -1$, we compute $y'(x) = -4c_1 e^{-4x} + c_2 e^x$. Applying the initial conditions $y(0) = 1$ and $y'(0) = -1$ results in the system of equations

$$\begin{cases} c_1 + c_2 = 1 \\ -4c_1 + c_2 = -1 \end{cases}.$$

Subtracting the second equation from the first results in $5c_1 = 2$ so that $c_1 = 2/5$. Then substituting this value of c_1 into $c_1 + c_2 = 1$ gives us the equation $2/5 + c_2 = 1$ so that $c_2 = 3/5$. This result is confirmed with `solve`:

```
> solve({c1+c2=1,-4*c1+c2=-1});
```

$$\left\{ c1 = \frac{2}{5}, c2 = \frac{3}{5} \right\}$$

Consequently, the desired solution is $y(x) = \frac{2}{5}e^{-4x} + \frac{3}{5}e^x$. `dsolve` is able to solve the initial-value problem

```
> partsol:=dsolve({diffeqn,y(0)=1,D(y)(0)=-1},y(x));
```

$$partsol := y(x) = \frac{3}{5}e^x + \frac{2}{5}e^{(-4x)}$$

which we then graph on the interval $[-1,1]$.

```
> plot(rhs(partsol),x=-1..1);
```

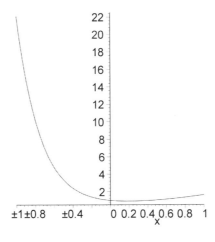

EXAMPLE 2: Solve $y'' + 2y' + y = 0$.

SOLUTION: The characteristic equation of $y'' + 2y' + y = 0$ is $m^2 + 2m + 1 = (m+1)^2 = 0$. A general solution of $y'' + 2y' + y = 0$ is $y(x) = c_1 e^{-x} + c_2 x e^{-x}$ because the solution of the characteristic equation is $m_1 = m_2 = -1$.

We can find a general solution and graph it for various values of the constants in exactly the same manner as in the previous example. Entering

```
> gensol:=dsolve(diff(y(x),x$2)+2*diff(y(x),x)+y(x)=0,
        y(x));
```

$$gensol := y(x) = \_C1 e^{(-x)} + \_C2 e^{(-x)} x$$

finds a general solution of $y'' + 2y' + y = 0$ and names the resulting output gensol. Next, we use seq to define toplot to be the list obtained by replacing each occurrence of _C1 in rhs(gensol) by -1, 0, and 1 and replacing each occurrence of _C2 in rhs(gensol) by -1, 0, and 1.

```
> toplot:=[seq(seq(rhs(gensol),_C1=-1..1),_C2=-1..1)]:
plot(toplot,x=-1..1,y=-5..5);
```

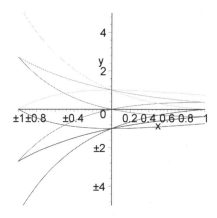

EXAMPLE 3: Solve $y'' + 4y' + 20y = 0$ is subject to $y(0) = 3$ and $y'(0) = -1$.

SOLUTION: The characteristic equation of $y'' + 4y' + 20y = 0$ is $m^2 + 4m + 20 = 0$. The solutions of the characteristic equation are obtained either by using the quadratic formula or by completing the square and are $m = -2 \pm 4i$. A general solution of $y'' + 4y' + 20y = 0$ is

$$y(x) = e^{-2x}(c_1 \cos 4x + c_2 \sin 4x),$$

because the solutions of the characteristic equation are complex conjugates. To find the solution for which $y(0) = 3$ and $y'(0) = -1$, we first calculate

$$y'(x) = 2e^{-2x}(-c_1 \cos 4x + 2c_2 \cos 4x - 2c_1 \sin 4x - c_2 \sin 4x)$$

and then evaluate both $y(0) = C_1$ and $y'(0) = 2(2C_2 - C_1)$, obtaining the system of equations

$$\begin{cases} c_1 = 3 \\ 2(2c_2 - c_1) = -1 \end{cases}$$

Substituting $c_1 = 3$ into the second equation results in $c_2 = 5/4$. Thus, the solution of $y'' + 4y' + 20y = 0$ for which $y(0) = 3$ and $y'(0) = -1$ is $y(x) = e^{-2x}\left(3 \cos 4x + \frac{5}{4} \sin 4x\right)$.

As in Examples 1 and 2, we are able to use `dsolve` to find both a general solution of the equation

```
> gensol:=dsolve(diff(y(x),x$2)+4*diff(y(x),x)+
        20*y(x)=0,y(x));
```

$$gensol := y(x) = \_C1 e^{(-2x)} \cos(4x) + \_C2 e^{(-2x)} \sin(4x)$$

and the solution to the initial-value problem

```
> partsol:=dsolve({diff(y(x),x$2)+4*diff(y(x),x)+
        20*y(x)=0,y(0)=3,D(y)(0)=-1},y(x));
```

$$partsol := y(x) = 3e^{(-2x)} \cos(4x) + \frac{5}{4} e^{(-2x)} \sin(4x)$$

which we then graph on the interval [–1,1] with `plot`.

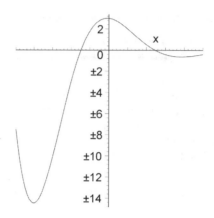

■

As with second-order equations, a general solution of the nth-order homogeneous linear differential equation with constant coefficients is determined by the solutions of its characteristic equation. In order to explain the process of finding a general solution of a higher order equation, we state the following definition.

Definition
Multiplicity

> Suppose that the characteristic equation $a_n m^n + a_{n-1} m^{n-1} + \cdots + a_1 m + a_0 = 0$ can be written in factored form as $(m - m_1)^{k_1}(m - m_2)^{k_2} \cdots (m - m_r)^{k_r} = 0$, where $m_i \neq m_j$ for $i \neq j$ and $k_1 + k_2 + \cdots + k_r = n$. Then the roots of the equation are $m = m_1, m = m_2, \ldots,$ and $m = m_r$, where the roots have **multiplicity** $k_1, k_2, \ldots, k_r$, respectively.

In the same manner as for a second-order homogeneous equation with real constant coefficients, a general solution of an nth-order homogeneous equation with real constant coefficients is determined by the solutions of its characteristic equation. Hence, we state the following rules for finding a general solution of an nth-order equation for the many situations that may be encountered.

1. Let m be a real root of the characteristic equation

$$a_n m^n + a_{n-1} m^{n-1} + \cdots + a_1 m + a_0 = 0$$

of an nth order homogeneous linear differential equation with real constant coefficients. Then, e^{mx} is the solution associated with the root m.

If m is a real root of multiplicity k where $k \geq 2$ of the characteristic equation, then the k solutions associated with m are

$$e^{mx}, xe^{mx}, x^2 e^{mx}, \ldots, x^{k-1} e^{mx}.$$

2. Suppose that m and $\bar{m}$ represent the complex conjugate pair $\alpha \pm \beta i$. Then, the two solutions associated with these two roots are

$$e^{\alpha x} \cos \beta x \quad \text{and} \quad e^{\alpha x} \sin \beta x.$$

If the values $\alpha \pm \beta i$ are each a root of multiplicity k of the characteristic equation, then the other solutions associated with this pair are

$$xe^{\alpha x} \cos \beta x, xe^{\alpha x} \sin \beta x, x^2 e^{\alpha x} \cos \beta x, x^2 e^{\alpha x} \sin \beta x, \ldots, x^{k-1} e^{\alpha x} \cos \beta x, x^{k-1} e^{\alpha x} \sin \beta x.$$

A general solution to the nth-order differential equation is the linear combination of the solutions obtained for all values of m. Note that if $m_1, m_2, \ldots, m_r$ are the roots of the equation of multiplicity $k_1, k_2, \ldots, k_r$ respectively, then $k_1 + k_2 + \cdots + k_r = n$, where n is the order of the differential equation.

We now show how to use these rules to find a general solution. Notice that the key to the process is identifying each root of the characteristic equation and the associated solution(s).

EXAMPLE 4: Find a general solution of each of the following higher order equations: (a) $24y''' + 2y'' - 5y' - y = 0$, (b) $y''' + 3y'' + 3y' + y = 0$, (c) $4y^{(4)} + 12y''' + 49y'' + 42y' + 10y = 0$, and (d) $y^{(4)} + 4y''' + 24y'' + 40y' + 100y = 0$.

SOLUTION: (a)   In   this   case,   the   characteristic   equation   is $24m^3 + 2m^2 - 5m - 1 = 0$. Factoring with `factor`,

```
> factor(24*m^3+2*m^2-5*m-1);
```

$$(2m - 1)(3m + 1)(4m + 1)$$

we have

$$24m^3 + 2m^2 - 5m - 1 = (3m + 1)(2m - 1)(4m + 1) = 0,$$

so the three distinct roots are

$$m_1 = -1/3, m_2 = 1/2, \text{ and } m_3 = -1/4$$

and the corresponding solutions are

$$y = e^{-x/3}, y = e^{x/2}, \text{ and } y = e^{-x/4},$$

respectively. Note that we can obtain the solutions to the characteristic equation directly with `solve`.

```
> solve(24*m^3+2*m^2-5*m-1=0);
```

$$\frac{1}{2'} \frac{-1}{3'} \frac{-1}{4}$$

Roots	Multiplicity	Corresponding Solution
$m = -1/3$	$k = 1$	$y = e^{-x/3}$
$m = 1/2$	$k = 1$	$y = e^{-x/3}$
$m = -1/4$	$k = 1$	$y = e^{-x/4}$

Therefore, a general solution of this differential equation is the linear combination of these functions,

$$y = c_1 e^{-x/3} + c_2 e^{x/2} + c_3 e^{-x/4}.$$

That is, a fundamental set of solutions to the equation is $\{e^{-x/3}, e^{x/2}, e^{-x/4}\}$, as we see with `constcoeffsols`, which is contained in the **DEtools** package.

```
> with(DEtools):
dea:=24*diff(y(x),x$3)+2*diff(y(x),x$2)-
            5*diff(y(x),x)-y(x)=0:
constcoeffsols(dea,y(x));
```

$$\left[e^{\left(-\frac{1}{4}x\right)}, e^{\left(\frac{1}{2}x\right)}, e^{\left(-\frac{1}{3}x\right)} \right]$$

We obtain the same result with `dsolve`.

```
> gensola:=dsolve(dea,y(x));
```

$$gensola := y(x) = \_C1 e^{\left(-\frac{1}{4}x\right)} + \_C2 e^{\left(\frac{1}{2}x\right)} + \_C3 e^{\left(-\frac{1}{3}x\right)}$$

We can graph this general solution for various values of the arbitrary constants in the same way we graph solutions of second-order equations. For example, entering defines `toplot` to be the list of functions obtained by replacing `_C2` in `rhs(gensola)`, which represents the formula for the general solution, by 1; `_C3` by $-1, 0$, and 1; and _C1 by -1 and 1. The list `toplot` is not displayed because a colon is included at the end of the command. The set of functions `toplot` is then graphed with `plot`.

```
> jvals:=[-1,1]:
toplot:={seq(seq(subs({_C1=j,_C2=1,_C3=i},
             rhs(gensola)),i=-1..1),j=jvals)}:
plot(toplot,x=-2..3);
```

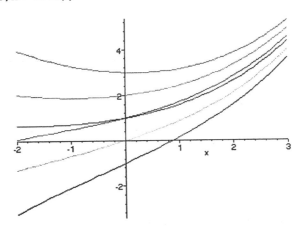

(b) The characteristic equation of $y''' + 3y'' + 3y' + y = 0$ is $m^3 + 3m^2 + 3m + 1 = 0$, and factoring results in

$$m^3 + 3m^2 + 3m + 1 = (m + 1)^3 = 0,$$

so the root is $m = -1$ with multiplicity three. We see that we obtain the same results with `factor` and `solve`.

```
> factor(m^3+3*m^2+3*m+1);
```

$$(m + 1)^3$$

```
> solve(m^3+3*m^2+3*m+1=0);
```

$$-1, -1, -1$$

Therefore, the corresponding solutions are $y = e^{-x}$, $y = xe^{-x}$, and $y = x^2 e^{-x}$, which means that a fundamental set of solutions to the differential equation is $\left\{ e^{-x}, xe^{-x}, x^2 e^{-x} \right\}$.

```
> deb:=diff(y(x),x$3)+3*diff(y(x),x$2)+
         3*diff(y(x),x)+y(x)=0:
constcoeffsols(deb,y(x));
```

$$[e^{(-x)}, e^{(-x)}x, e^{(-x)}x^2]$$

This tells us that a general solution is given by

$$y = c_1 e^{-x} + c_2 x e^{-x} + c_3 x^2 e^{-x}.$$

As in (a), we are also able to use `dsolve` to find a general solution of the equation.

```
> gensolb:=dsolve(deb,y(x));
```

$$gensolb := y(x) = \_C1 e^{(-x)} + \_C2 e^{(-x)} x + \_C3 e^{(-x)} x^2$$

We can graph the general solution for various values of the arbitrary constants in the same manner as in (a).

```
> jvals:=[-1,1]:
toplot:={seq(seq(subs({_C1=i,_C2=j,_C3=1},
             rhs(gensolb)),i=-1..1),j=jvals)}:
plot(toplot,x=-2..3,y=-3..2);
```

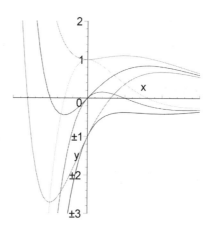

(c) The characteristic equation of $4y^{(4)} + 12y''' + 49y'' + 42y' + 10y = 0$ is $4m^4 + 12m^3 + 49m^2 + 42m + 10 = 0$. We use `factor` to try to factor the polynomial $4m^4 + 12m^3 + 49m^2 + 42m + 10$ but see that Maple does not completely factor the polynomial.

```
> factor(4*m^4+12*m^3+49*m^2+42*m+10);
```

$$(m^2 + 2m + 10)(1 + 2m)^2$$

To factor the polynomial completely, we include the option I in the `factor` command, which instructs Maple to factor the polynomial over the complex field.

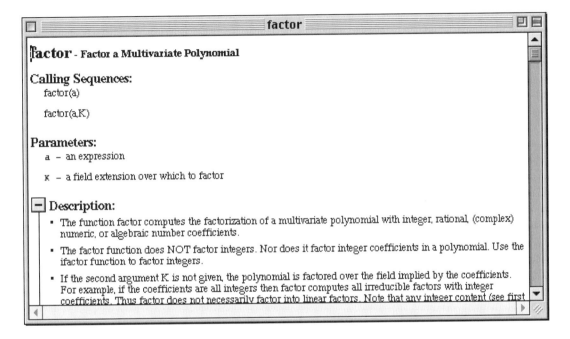

Maple is then able to factor the polynomial completely.

```
> factor(4*m^4+12*m^3+49*m^2+42*m+10,I);
```

$$(m + 1 - 3I)(m + 1 + 3I)(1 + 2m)^2$$

From the results, we see that the solutions of the characteristic equation are $m = -1 \pm 3i$ and $m = -1/2$ with multiplicity 2. As you may suspect, we obtain the same results with `solve`.

```
> solve(4*m^4+12*m^3+49*m^2+42*m+10=0);
```

$$-1 + 3I, -1 - 3I, \frac{-1}{2}, \frac{-1}{2}$$

The corresponding solutions are given by $y = e^{-x}\cos 3x$, $y = e^{-x}\sin 3x$, $y = e^{-x/2}$, and $y = xe^{-x/2}$, which means that a fundamental set of solutions to the equation is $\{e^{-x}\cos 3x, e^{-x}\sin 3x, e^{-x/2}, xe^{-x/2}\}$.

```
> dec:=4*diff(y(x),x$4)+12*diff(y(x),x$3)+
          49*diff(y(x),x$2)+42*diff(y(x),x)+10*y(x)=0:
constcoeffsols(dec,y(x));
```

$$\left[e^{(-x)}\cos(3\ x), e^{(-x)}\sin(3\ x), e^{\left(-\frac{1}{2}x\right)}, e^{\left(-\frac{1}{2}x\right)}x \right]$$

This tells us that a general solution is given by

$$y = e^{-x}(c_1\cos 3x + c_2\sin 3x) + c_3 e^{-x/2} + c_4 x e^{-x/2}.$$

As in (a) and (b), we obtain the same result with `dsolve`. Remember that the formula for the general solution is extracted from `gensolc` with `rhs(gensolc)`.

```
> gensolc:=dsolve(dec,y(x));
```

$$gensolc := y(x) = \_C1e^{\left(-\frac{1}{2}x\right)} + \_C2e^{\left(-\frac{1}{2}x\right)}x + \_C3e^{(-x)}\cos(3\ x)$$
$$+ \_C4e^{(-x)}\sin(3\ x)$$

In this case, we will graph the general solution for $(c_1, c_2, c_3, c_4) = (1, 0, 1, 0)$, $(0, 1, 0, 1)$, $(1, 1, 0, 1)$, $(1, -1, 1, 2)$, $(0, 2, 1, -2)$, and $(1, -2, 1, 2)$. We accomplish this by applying the function

$$\mathtt{f} := \mathtt{c} - > \mathtt{subs}(\{\_C1 = c[1], \_C2 = c[2], \_C3 = c[3], \_C4 = c[4]\}, \mathtt{rhs\ (gensolc)})$$

to the set of ordered quadruples

$$[[1, 0, 1, 0], [0, 1, 0, 1], [1, 1, 0, 1], [1, -1, 1, 2], [0, 2, 1, -2], [1, -2, 1, 2]]$$

with `map`.

```
> cvals:=[[1,0,1,0],[0,1,0,1],[1,1,0,1],[1,-1,1,2],
          [0,2,1,-2],[1,-2,1,2]]:
f:=c->subs({_C1=c[1],_C2=c[2],_C3=c[3],_C4=c[4]},
          rhs(gensolc)):
```

```
toplot:=map(f,cvals);
```

We then graph the set of functions `toplot` on the interval $[-1,2]$ with `plot`.

```
> plot(toplot,x=-1..2);
```

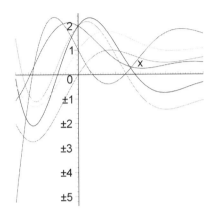

(d) The characteristic equation of $y^{(4)} + 4y''' + 24y'' + 40y' + 100y = 0$ is $m^4 + 4m^3 + 24m^2 + 40m + 100 = 0$, which we can solve by factoring $m^4 + 4m^3 + 24m^2 + 40m + 100$ with `factor` together with the option `I`

```
> factor(m^4+4*m^3+24*m^2+40*m+100,I);
```

$$(m + 1 - 3I)^2 (m + 1 + 3I)^2$$

or using `solve`.

```
> solve(m^4+4*m^3+24*m^2+40*m+100=0,m);
```

$$-1 + 3I, -1 - 3I, -1 + 3I, -1 - 3I$$

Thus, we see that the solutions of the characteristic equation are $m = -1 + 3i$ and $m = -1 - 3i$, each with multiplicity 2, so the corresponding solutions are $y = e^{-x} \cos 3x$, $y = e^{-x} \sin 3x$, $y = xe^{-x} \cos 3x$, and $y = xe^{-x} \sin 3x$, which means that a fundamental set of solutions to the equation is

$$\{e^{-x} \cos 3x, e^{-x} \sin 3x, xe^{-x} \cos 3x, xe^{-x} \sin 3x\}.$$

```
> ded:=diff(y(x),x$4)+4*diff(y(x),x$3)+
        24*diff(y(x),x$2)+40*diff(y(x),x)+100*y(x)=0:
constcoeffsols(ded,y(x));
```

$$[e^{(-x)} \cos(3\,x), e^{(-x)} \cos(3\,x)x, e^{(-x)} \sin(3\,x), e^{(-x)} \sin(3\,x)x]$$

This tells us that a general solution is given by

$$y = e^{-x}(c_1 \cos 3x + c_2 \sin 3x + c_3 x \cos 3x + c_4 x \sin 3x).$$

We obtain the same result using `dsolve`.

```
> gensold:=dsolve(ded,y(x));
```

$gensold := y(x) = \_C1e^{(-x)} \cos(3\,x) + \_C2e^{(-x)} \sin(3\,x)$

$\quad + \_C3e^{(-x)} \sin(3\,x)x + \_C4e^{(-x)} \cos(3\,x)x$

To graph the solution for various values of the constants, we proceed in the same manner as in (c). First, we define a list of ordered quadruples, `cvals`.

```
> cvals:=[[5,0,1,0],[0,1,0,-3],[1,3,0,1],[1,-1,1,2],
            [0,2,1,-2],[1,-2,5,2],[0,-3,0,2],[3,0,0,2],
            [1,1,1,1]]:
```

We then use `map` to replace `_C1` in `rhs(gensold)`, which represents the formula for the solution, by the first part of each quadruple in `cvals`; `_C2` by the second part; `_C3` by the third part; and `_C4` by the fourth part.

```
> toplot:=map(f,cvals):
```

We then use `plot` to graph each function in `toplot` on the interval $[-1/2, 3/2]$.

```
>  plot(toplot,x=-1/2..3/2);
```

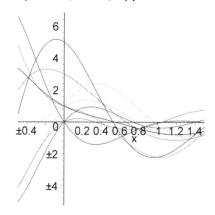

In our previous examples, we have been interested in finding a general solution of a higher order equation. Here, we consider finding a solution to an nth-order initial-value problem

$$\begin{cases} a_n(x)y^{(n)} + a_{n-1}(x)y^{(n-1)} + \cdots + a_1(x)y' + a_0(x)y = 0 \\ y^{(n-1)}(x_0) = y_{n-1}, y^{(n-2)}(x_0) = y_{n-2}, \dots y'(x_0) = y_1, y(x_0) = y_0 \end{cases}.$$

Note that we call the restrictions $\{y^{(n-1)}(x_0) = y_{n-1}, y^{(n-2)}(x_0) = y_{n-2}, \dots, y'(x_0) = y_1, y(x_0) = y_0\}$ **initial conditions** because $y(x)$ and its $(n-1)$ derivatives are evaluated at the same value of x. Because a general solution of an nth-order equation involves n arbitrary constants, the n initial conditions allow us to solve for these constants.

EXAMPLE 5: (a) Find a general solution of $4y''' + 33y' - 37y = 0$. (b) Solve the initial-value problem

$$\begin{cases} 4y''' + 33y' - 37y = 0 \\ y(0) = 0, y'(0) = -1, y''(0) = 3 \end{cases}.$$

SOLUTION: For (a), we use `dsolve` as in the previous examples, naming the result `gensol`.

```
> gensol:=dsolve(4*diff(y(x),x$3)+33*diff(y(x),x)-
        37*y(x)=0,y(x));
```

$$gensol := y(x) = \_C1 e^x + \_C2 e^{\left(-\frac{1}{2}x\right)} \sin(3\,x) + \_C3 e^{\left(-\frac{1}{2}x\right)} \cos(3\,x)$$

We then graph the general solution for various values of the arbitrary constants.

```
> toplot:=[seq(seq(seq(rhs(gensol),_C1=0..1),
        _C2=-1..1),_C3=-1..0)]:
plot(toplot,x=0..2);
```

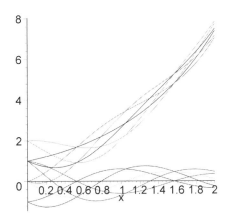

For (b), we again use `dsolve` to solve the initial-value problem.

```
partsol:=dsolve({4*diff(y(x),x$3)+33*diff(y(x),x)-
        37*y(x)=0,y(0)=0,D(y)(0)=-1,(D@@2)(y)(0)=3},
      y(x));
```

$$partsol := y(x) = \frac{8}{45}e^x - \frac{19}{45}e^{\left(-\frac{1}{2}x\right)}\sin(3\,x) - \frac{8}{45}e^{\left(-\frac{1}{2}x\right)}\cos(3\,x)$$

We then use `plot` to graph the solution on the interval [0,2].

```
> plot(rhs(partsol),x=0..2,y=-0.5..1.5);
```

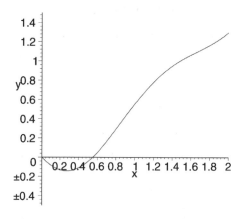

■

EXAMPLE 6: If a is a positive constant, find conditions on the constant b so that $y(x)$ satisfies

$$\begin{cases} y''' + 0.344425y'' + 12.4454y' - 4.50057y = 0 \\ \qquad y(0), = 0, y'(0) = a, y''(0) = b \end{cases}$$

and $\lim_{x\to\infty} y(x) = 0$. (b) For this function, find and classify the first critical point on the interval $[0, \infty)$.

SOLUTION: We use `solve` to find (accurate approximations of) the solutions of the characteristic equation $m^3 + 0.344425m^2 + 12.4454m - 4.50047 = 0$.

```
> solve(m^3+0.344425*m^2+12.4454*m-4.50047=0);
```

$$-.3494908783 - 3.545572567\ I, -.3494908783 + 3.545572567\ I, .3545567566$$

Then, a general solution of the equation is

$$y = c_1 e^{0.354557x} + e^{-0.349491x}(c_2 \cos 3.54557x + c_3 \sin 3.54557x).$$

```
> y:=x->c[1]*exp(.3545567566*x)+
      exp(-.3494908783*x)*(c[2]*cos(3.545572567*x)+
          c[3]*sin(-3.545572567*x)):
```

We now apply the initial conditions and solve for c_1, c_2, and c_3.

```
> sys:={y(0)=0,D(y)(0)=a,(D@@2)(y)(0)=b};
```

$sys := \{.1257104937\ c_1 - 12.44894096\ c_2 + 2.478290541\ c_3 = b,\ c_1 + c_2 = 0,$

$\quad .3545567566\ c_1 - .3494908783\ c_2 - 3.545572567\ c_3 = a\}$

```
> cvals:=solve(sys,{c[1],c[2],c[3]});
```

$cvals := \{c_3 = -.2714197214\ a + .0159663695\ b,$

$\quad c_1 = .05349308732\ a + .07653001930\ b, c_2 = -.05349308732\ a - .07653001930\ b\}$

We obtain the solution to the initial-value problem by substituting these values back into the general solution.

```
> sol:=subs(cvals,y(x));
```

$sol := (.05349308732\ a + .07653001930\ b)\mathbf{e}^{(.3545567566\ x)} +$

$\quad \mathbf{e}^{(-.3494908783\ x)}((-.05349308732\ a - .07653001930\ b) \cos(3.545572567\ x)$

$\quad - (-.271419124\ a + .01519663695\ b) \sin(3.545572567\ x))$

These results indicate that $\lim_{x \to \infty} y(x) = 0$ if $0.0534931a + 0.07653b = 0$, which leads to $b = -0.698982a$.

```
> bval:=solve(.5349308732e-1*a+.7653001930e-1*b=0,b);
```

$$bval := -.6989817566\ a$$

Substituting back into the solution to the initial-value problem yields $y = 0.282042ae^{-0.349491x} \sin 3.54557x$.

```
> sol:=subs(b=bval,sol);
```

$$sol := .2820418934e^{(-.3494908783\,x)}a\sin(3.545572567\,x)$$

To find and classify the first critical point of $y = 0.282042ae^{-0.349491x} \sin 3.54557x$, we compute y'

```
> dsol:=diff(sol,x);
```

$dsol := -.09857106904e^{(-.3494908783\,x)}a\sin(3.545572567\,x)$

$\qquad + 1.000000000e^{(-.3494908783\,x)}a\cos(3.545572567\,x)$

and graph y'/a to locate the first zero of y'.

```
> plot(dsol/a,x=0..1);
```

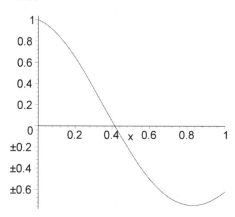

From the graph, we see that the first zero occurs near 0.4 and with `fsolve` we obtain the critical number $x = 0.415319$.

```
> critval:=fsolve(simplify(dsol/a=0),x=0.3..0.5);
```

$$critval := .4153187190$$

At this critical number, we use `subs` to find that $y(0.415319) = 0.242759a$, so by the first derivative test $(0.415319, 0.242759a)$ is a local maximum.

```
> evalf(subs(x=critval,sol));
```

$$.2427594016a$$

To see that is the *absolute maximum*, we graph y for various values of a with `plot`.

```
> avals:=seq(0.5+0.5*i,i=1..9):

toplot:=[seq(subs(a=i,sol),i=avals)];

plot(toplot,x=0..6);
```

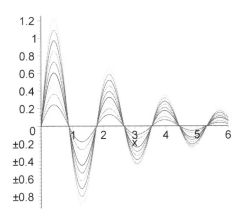

APPLICATION

Testing for Diabetes

Diabetes mellitus affects approximately 12 million Americans; approximately one-half of these people are unaware that they have diabetes. Diabetes is a serious disease: it is the leading cause of blindness in adults, is the leading cause of renal failure, and is responsible for approximately one-half of all nontraumatic amputations in the United States. In addition, people with diabetes have an increased rate of coronary artery disease and strokes. People at risk for developing diabetes include those who are obese; those suffering from excessive thirst, hunger, urination, and weight loss; women who have given birth to a baby with weight greater than 9 pounds; those with a family history of diabetes; and those who are older than 40 years of age. People with diabetes cannot metabolize glucose because their pancreas produces an inadequate or ineffective supply of insulin. Subsequently, glucose levels rise. The body attempts to remove the excess glucose through the kidneys: the glucose acts as a diuretic, resulting in increased water consumption. Because some cells require energy, which is not being provided by glucose, fat and protein are broken down and ketone levels rise.

Although there is no cure for diabetes at this time, many cases can be effectively managed by a balanced diet and insulin therapy in addition to maintaining an optimal weight. Diabetes can be diagnosed by several tests. In the **fasting blood sugar** test, a patient fasts for at least 4 hours, and then the glucose level is measured. In a fasting state, the glucose level in normal adults ranges from 70 to 110 milligrams per milliliter. An adult in a fasting state with consistent readings over 150 milligrams probably has diabetes. However, people with mild cases of diabetes might have fasting state glucose levels within the normal range because individuals vary greatly. In these cases, a highly accurate test that is frequently used to diagnose mild diabetes is the **glucose tolerance test** (GTT), which was developed by Drs. Rosevear and Molnar of the Mayo Clinic and Drs. Ackerman and Gatewood of the University of Minnesota. During the GTT, blood and urine samples are taken from a patient in a fasting state to measure the glucose, G_0; hormone, H_0; and glycosuria levels, respectively. We assume that these values are equilibrium values. The patient is then given 100 grams of glucose. Blood and urine samples are then taken at intervals of 1, 2, 3, and 4 hours. In a person without diabetes, glucose levels return to normal after 2 hours; in diabetics, the blood sugar levels either take longer to return or never return to normal levels. Let G denote the cumulative level of glucose in the blood, $g = G - G_0$, and H be the cumulative level of hormones that affect insulin production (e.g., glucagon, epinephrine, cortisone, and thyroxine), $h = H - H_0$. Notice that g and h represent the fluctuations of the cumulative levels of glucose and hormones from their equilibrium values. The relationship between the rate of change of glucose in the blood and the rate of change of the cumulative levels of the hormones in the blood that affect insulin production is

$$\begin{cases} g' = f_1(g, h) + J(t) \\ h' = f_2(g, h) \end{cases},$$

where $J(t)$ represents the **external** rate at which the blood glucose concentration is being increased. If we assume that f_1 and f_2 are linear functions, then this system of equations becomes

$$\begin{cases} g' = -ag - bh + J(t) \\ h' = -ch + dg \end{cases},$$

where a, b, c, and d represent positive numbers. We define these equations in eq1 and eq2.

```
> eq1:=diff(g(t),t)=-a*g(t)-b*h(t)+J(t);

eq2:=diff(h(t),t)=-c*h(t)+d*g(t);
```

$$eq1 := \frac{\partial}{\partial t} g(t) = -a\, g(t) - b\, h(t) + J(t)$$

$$eq2 := \frac{\partial}{\partial t} h(t) = -c\, h(t) + d\, g(t)$$

Next, we solve the first equation for $h(t)$

```
> step_1:=solve(eq1,h(t));
```

$$step\_1 := -\frac{\left(\frac{\partial}{\partial t}g(t)\right) + a\ g(t) - J(t)}{b}$$

and differentiate this result with respect to t to obtain $h'(t)$.

> `step_2:=diff(step_1,t);`

$$step\_2 := -\frac{\left(\frac{\partial^2}{\partial t^2}g(t)\right) + a\left(\frac{\partial}{\partial t}g(t)\right) - \left(\frac{\partial}{\partial t}J(t)\right)}{b}$$

Substituting these results into the second equation yields the second-order equation

$$\frac{1}{b}(-g'' - ag' + J') = -\frac{c}{b}(-g' - ag + J) + dgg'' + (a+c)g' + (ac+bd)g = J' + cJ.$$

> `step_3:=subs({h(t)=step_1,diff(h(t),t)=step_2},eq2);`

$$step\_3 := -\frac{\left(\frac{\partial^2}{\partial t^2}g(t)\right) + a\left(\frac{\partial}{\partial t}g(t)\right) - \left(\frac{\partial}{\partial t}J(t)\right)}{b} = \frac{c\left(\left(\frac{\partial}{\partial t}g(t)\right) + a\ g(t) - J(t)\right)}{b} + d\ g(t)$$

> `step_4:=expand(b*step_3);`

$$step\_4 := -\left(\frac{\partial^2}{\partial t^2}g(t)\right) - a\left(\frac{\partial}{\partial t}g(t)\right) + \left(\frac{\partial}{\partial t}J(t)\right) = c\left(\frac{\partial}{\partial t}g(t)\right) + c\ a\ g(t) - c\ Jt + b\ d\ g(t)$$

For $t > 0$ we have that $J(t) = 0$ and $J'(t) = 0$ because the glucose solution is consumed at $t = 0$, so for $t > 0$ we can rewrite the equation as

$$g'' + (a+c)g' + (ac+bd)g = 0.$$

> `step_5:=subs({diff(J(t),t)=0,J(t)=0},step_4);`

$$step\_5 := -\left(\frac{\partial^2}{\partial t^2}g(t)\right) - a\left(\frac{\partial}{\partial t}g(t)\right) = c\left(\frac{\partial}{\partial t}g(t)\right) + c\ a\ g(t) + b\ d\ g(t)$$

We now use `dsolve` to solve this second-order equation.

> `sol:=dsolve(step_5,g(t));`

$$sol := g(t) = \_C1\,e^{\left(-\frac{1}{2}(a+c-\sqrt{a^2-2\,c\,a+c^2-4\,d\,b})t\right)}$$
$$+ \_C2\,e^{\left(-\frac{1}{2}(a+c+\sqrt{a^2-2\,c\,a+c^2-4\,d\,b})t\right)}$$

It might be reasonable to assume that glucose levels fluctuate in a periodic fashion so that the solutions to the equation involve periodic functions. In order to have periodic functions in the solution (such as sine and cosine), we must have that $(a + c)^2 - 4(ac + bd) < 0$. We now replace $(a + c)^2 - 4(ac + bd)$ with $-4\omega^2$ and $a + c$ with 2α using subs.

```
> step_6:=subs({a+c=2*alpha,
              a^2-2*c*a+c^2-4*d*b=-4*omega^2},rhs(sol));
```

$step\_6 :=$

$$\_C1e^{\left(-\frac{1}{2}(a+c-\sqrt{-4\omega^2})t\right)} + \_C2e^{\left(-\frac{1}{2}(a+c+\sqrt{-4\omega^2})t\right)}$$

We then simplify the result with simplify together with the symbolic option.

```
> step_7:=simplify(step_6,symbolic);
```

$$step\_7 := \_C1e^{\left(-\frac{1}{2}(a+c-2I\omega)t\right)} + \_C2e^{\left(-\frac{1}{2}(a+c+2I\omega)t\right)}$$

To write this complex solution in terms of trigonometric functions, use convert together with the trig and then the expsincos options to rewrite step_7 in terms of trigonometric functions.

```
> step_8:=convert(step_7,trig);
```

$step\_8 :=$

$$\_C1\left(\cosh\left(\frac{1}{2}(a+c)t\right) - \sinh\left(\frac{1}{2}(a+c)t\right)\right)(\cos(\omega t) + I\sin(\omega t))$$

$$+ \_C2\left(\cosh\left(\frac{1}{2}(a+c)t\right) - \sinh\left(\frac{1}{2}(a+c)t\right)\right)(\cos(\omega t) - I\sin(\omega t))$$

```
> step_9:=convert(step_8,expsincos);
```

$$step\_9 := \frac{\_C1(\cos(\omega t) + I\sin(\omega t))}{e^{\left(\frac{1}{2}(a+c)t\right)}} + \frac{\_C2(\cos(\omega t) - I\sin(\omega t))}{e^{\left(\frac{1}{2}(a+c)t\right)}}$$

We want to choose the constants _C1 and _C2 so that the result is a real-valued function. We begin by using collect to collect together the terms involving $\cos \omega t$ and $\sin \omega t$.

```
> step_10:=collect(step_9,{cos(omega*t),sin(omega*t)});
```

$$step\_10 := \left(\frac{I\_C1}{\mathbf{e}^{\left(\frac{1}{2}(a+c)t\right)}} - \frac{I\_C2}{\mathbf{e}^{\left(\frac{1}{2}(a+c)t\right)}} \right) \sin(\omega t)$$

$$+ \left(\frac{\_C1}{\mathbf{e}^{\left(\frac{1}{2}(a+c)t\right)}} + \frac{\_C2}{\mathbf{e}^{\left(\frac{1}{2}(a+c)t\right)}} \right) \cos(\omega t)$$

If possible, we would like to choose _C1 and _C2 so that _C1 + _C2 can be replaced by an arbitrary real constant c1 and - I _C1 + I _C2 can be replaced by an arbitrary real constant c2. To see that this is possible, we solve this system of equations for _C1 and _C2 with solve.

> `toapply:=solve({_C1+_C2=c[1],-I*_C1+I*_C2=c[2]},{_C1,_C2});`

$$toapply := \left\{ \_C2 = -\frac{1}{2}I(Ic_1 + c_2), \_C1 = \frac{1}{2}c_1 + \frac{1}{2}Ic_2 \right\}$$

Replacing _C1 and _C2 by the values obtained in toapply yields our model.

> `model:=simplify(subs(toapply,step_10));`

$$model := (-c_2 \sin(\omega t) + c_1 \cos(\omega t))\mathbf{e}^{\left(-\frac{1}{2}(a+c)t\right)}$$

Thus, $g(t) = e^{-\alpha t}(C_1 \cos \omega t + c_2 \sin \omega t)$ and $G(t) = G_0 + e^{-\alpha t}(c_1 \cos \omega t + c_2 \sin \omega t)$.

Research has shown that laboratory results of $2\pi/\omega > 4$ indicate a mild case of diabetes.

Sources: D. N. Burghess and M. S. Borrie, *Modeling with Differential Equations*, Ellis Horwood Limited, pp. 113–116. Joyce M. Black and Esther Matassarin-Jacobs, *Luckman and Sorensen's Medical–Surgical Nursing: A Psychophysiologic Approach*, fourth edition, W. B. Saunders Company (1993), pp. 1775–1808.

4.3 Introduction to Solving Nonhomogeneous Equations with Constant Coefficients

In the previous section, we learned how to solve nth-order linear homogeneous equations with real constant coefficients. These techniques are also useful in solving some nonhomogeneous equations of the form

$$a_n y^{(n)}(x) + a_{n-1} y^{(n-1)}(x) + \cdots + a_1 y'(x) + a_0 y(x) = g(x).$$

Before describing how to obtain solutions of some nonhomogeneous equations, we need to describe what is meant by a *general solution of a nonhomogeneous equation.*

Definition **Particular Solution**	A **particular solution,** $y_p(x)$, of the nonhomogeneous differential equation $a_n y^{(n)}(x) + a_{n-1} y^{(n-1)}(x) + \cdots + a_1 y'(x) + a_0 y(x) = g(x)$ is a specific function that contains no arbitrary constants and satisfies the differential equation.

EXAMPLE 1: Verify that $y_p(x) = -\frac{3}{2} \sin x$ is a particular solution of $y'' - 2y' + y = 3 \cos x$.

SOLUTION: After defining $y_p(x) = -\frac{3}{2} \sin x$,

```
> y[p]:=x->-3*sin(x)/2;
```

$$y_p := x \rightarrow -\frac{3}{2} \sin(x)$$

we compute and simplify $y_p'' - 2y_p' + y_p$

```
> diff(y[p](x),x$2)-2*diff(y[p](x),x)+y[p](x);
```

$$3 \cos(x)$$

and see that the result is 3 cos x.

■

If $y_p(x)$ is a particular solution of the nonhomogeneous equation

$$a_n y^{(n)}(x) + a_{n-1} y^{(n-1)}(x) + \cdots + a_1 y'(x) + a_0 y(x) = f(x)$$

and

$$y_h(x) = c_1 f_1(x) + c_2 f_2(x) + \cdots + c_n f_n(x)$$

is a general solution of the corresponding homogeneous equation

$$a_n y^{(n)}(x) + a_{n-1} y^{(n-1)}(x) + \cdots + a_1 y'(x) + a_0 y(x) = 0,$$

where $\{f_1(x), f_2(x), \ldots, f_n(x)\}$ is a fundamental set of solutions for the corresponding homogeneous equation, then every solution, $y(x)$, of the nonhomogeneous equation can be written in the form

$$y(x) = y_h(x) + y_p(x),$$

for some choice of $c_1, c_2, \ldots, c_n$.

Definition **General Solution of a** **Nonhomogeneous** **Equation**	**A general solution to the nonhomogeneous equation** $$a_n y^{(n)}(x) + a_{n-1} y^{(n-1)}(x) + \cdots + a_1 y'(x) + a_0 y(x) = f(x)$$ is $$y(x) = y_h(x) + y_p(x)$$ where $y_h(x)$ is a general solution of the corresponding homogeneous equation $$a_n y^{(n)}(x) + a_{n-1} y^{(n-1)}(x) + \cdots + a_1 y'(x) + a_0 y(x) = 0,$$ and $y_p(x)$ is a particular solution of the nonhomogeneous equation.

EXAMPLE 2: Let $y_p(x) = -\frac{1}{5} e^{-2x} \cos x + \frac{2}{5} e^{-2x} \sin x$ and $y_h(x) = e^{-3x}(c_1 \cos 2x + c_2 \sin 2x)$. Show that $y(x) = y_h(x) + y_p(x)$ is a general solution of $y'' + 6y' + 13y = 2 e^{-2x} \sin x$.

SOLUTION: We first show that $y_p(x) = -\frac{1}{5} e^{-2x} \cos x + \frac{2}{5} e^{-2x} \sin x$ is a particular solution of $y'' + 6y' + 13y = 2 e^{-2x} \sin x$. After defining $y_p(x) = -\frac{1}{5} e^{-2x} \cos x + \frac{2}{5} e^{-2x} \sin x$, we calculate $y_p'' + 6y_p' + 13y_p$

```
> y[p]:=x->-1/5*exp(-2*x)*cos(x)+2/5*exp(-2*x)*sin(x):
```

```
> diff(y[p](x),x$2)+6*diff(y[p](x),x)+13*y[p](x);
```

$$2\mathbf{e}^{(-2x)} \sin(x)$$

and see that the result is $2 e^{-2x} \sin x$. We see that $y_h(x) = e^{-3x}(c_1 \cos 2x + c_2 \sin 2x)$ is a general solution of the corresponding homogeneous equation with `dsolve`.

```
> dsolve(diff(Y(x),x$2)+6*diff(Y(x),x)+13*Y(x)=0,Y(x));
```

$$Y(x) = \_C1e^{(-3x)}\sin(2x) + \_C2e^{(-3x)}\cos(2x)$$

We now graph the general solution for various values of the arbitrary constant. First we define $y_h(x) = e^{-3x}(c_1\cos 2x + c_2\sin 2x)$ and $y(x) = y_h(x) + y_p(x)$.

```
> y[h]:=x->_C1*exp(-3*x)*sin(2*x)+_C2*exp(-3*x)*cos(2*x):

Y:=x->y[h](x)+y[p](x):
```

Then, we use `seq` to create a list of functions obtained by replacing c_1 in $y(x)$ by -1, 0, and 1 and c_2 by -1, 0, and 1. The resulting list of functions `toplot` is graphed with `plot`.

```
> toplot:=[seq(seq(Y(x),_C1=-1..1),_C2=-1..1)]:

> plot(toplot,x=0..2);
```

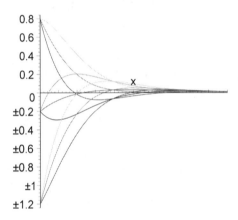

Techniques for solving nonhomogeneous equations with constant coefficients are discussed in the next two sections. In addition, you can often use `dsolve` to find a general solution of a nonhomogeneous equation.

EXAMPLE 3: (a) Find a general solution of $y'' - 2y' + y = 2e^{-2x}\sin x$. What is a particular solution of the equation? (b) Solve the initial value problem
$\begin{cases} y'' + y = \cos\omega x \\ by(0) = y'(0) = 0 \end{cases}$. Graph the solution for various values of $\omega = 1$.

SOLUTION: (a) We use `dsolve` to find a general solution of the equation

```
> y:='y':
```

```
gensol:=dsolve(diff(y(x),x$2)-2*diff(y(x),x)+

        y(x)=2*exp(-2*x)*sin(x),y(x));
```

$$gensol := y(x) = \frac{3}{25}e^{(-2x)}\cos(x) + \frac{4}{25}e^{(-2x)}\sin(x) + \_C1e^x + \_C2e^x x$$

The results indicate that a general solution of the corresponding homogeneous equation is $y_h = c_1 e^x + c_2 x e^x$ and a particular solution of the nonhomogeneous equation is $y_p = \frac{3}{25}e^{-2x}\cos x + \frac{4}{25}e^{-2x}\sin x$.

(b) We first use `dsolve` to solve the initial-value problem. Note that the result is not valid if $\omega = 1$.

```
> partsol:=dsolve({diff(y(x),x$2)+y(x)=cos(omega*x),

        y(0)=0,D(y)(0)=0},y(x));
```

$$partsol := y(x) = \left(\frac{1}{2}\frac{\sin((-1+\omega)x)}{-1+\omega} + \frac{1}{2}\frac{\sin((1+\omega)x)}{1+\omega}\right)\sin(x)$$
$$+ \left(\frac{1}{2}\frac{\cos((1+\omega)x)}{1+\omega} - \frac{1}{2}\frac{\cos((-1+\omega)x)}{-1+\omega}\right)\cos(x) + \frac{\cos(x)}{-1+\omega^2}$$

```
> step2:=combine(partsol,trig);
```

```
> step2:=combine(partsol,trig);
```

$$step2 := y(x) = \frac{-\cos(\omega x) + \cos(x)}{-1+\omega^2}$$

In fact, when we graph this solution for various values of ω, we are certain to avoid $\omega = 1$.

```
> omegavals:=[seq(2/9*i,i=0..8)]:
```

```
> somegraphs:=[seq(plot(subs(omega=i,rhs(step2)),x=0..12*Pi,

        color=BLACK),i=omegavals)]:
```

From the graphs, we see that the solution to the initial-value problem is bounded and periodic if $\omega \neq 1$.

```
> with(plots):
```

```
anarray:=display(somegraphs,insequence=true):
```

```
display(anarray);
```

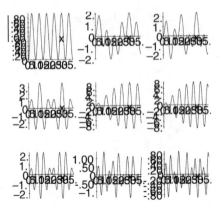

We consider $\omega = 1$ separately.

```
> partsol:=dsolve({diff(y(x),x$2)+y(x)=cos(x),
            y(0)=0,D(y)(0)=0},y(x));
```

$$partsol := y(x) = \left(\frac{1}{2}\sin(x)\cos(x) + \frac{1}{2}x\right)\sin(x) - \frac{1}{2}\sin(x)^2\cos(x)$$

```
> step2:=combine(partsol,trig);
```

$$step2 := y(x) = \frac{1}{2}\sin(x)x$$

```
> plot(rhs(step2),x=0..12*Pi);
```

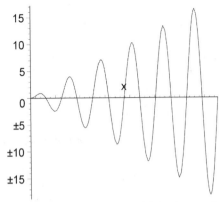

We see that if $\omega = 1$ the solution is unbounded and not periodic.

∎

4.4 Nonhomogeneous Equations with Constant Coefficients: The Method of Undetermined Coefficients

Consider the nonhomogeneous linear nth-order differential equation with constant coefficients

$$a_n y^{(n)}(x) + a_{n-1} y^{(n-1)}(x) + \cdots + a_1 y'(x) + a_0 y(x) = f(x).$$

We know that a general solution of this differential equation is given by

$$y(x) = y_h(x) + y_p(x)$$

where $y_h(x)$ is a solution of the corresponding homogeneous equation

$$a_n y^{(n)}(x) + a_{n-1} y^{(n-1)}(x) + \cdots + a_1 y'(x) + a_0 y(x) = 0,$$

and $y_p(x)$ is a particular solution involving no arbitrary constants of the nonhomogeneous equation

$$a_n y^{(n)}(x) + a_{n-1} y^{(n-1)}(x) + \cdots + a_1 y'(x) + a_0 y(x) = f(x).$$

If $f(x)$ is a linear combination of the functions 1, x, x^2, ..., e^{kx}, $x e^{kx}$, $x^2 e^{kx}$, ..., $e^{\alpha x} \cos \beta x, x e^{\alpha x} \cos \beta x, x^2 e^{\alpha x} \cos \beta x, \ldots e^{\alpha x} \sin \beta x, x e^{\alpha x} \sin \beta x, x^2 e^{\alpha x} \sin \beta x, \ldots$, the method of undetermined coefficients can be used to determine a particular solution of the equation.

Outline of the Method of Undetermined Coefficients

1. Solve the corresponding homogeneous equation for $y_h(x)$.
2. Determine the form of a particular solution $y_p(x)$. (See *Determining the Form of $y_p(x)$* next.)
3. Determine the unknown coefficients in $y_p(x)$ by substituting $y_p(x)$ into the nonhomogeneous equation and equating the coefficients of like terms.
4. Form a general solution with $y(x) = y_h(x) + y_p(x)$.

Determining the Form of $y_p(x)$ (Step 2):

Suppose that $f(x) = b_1 f_1(x) + b_2 f_2(x) + \ldots + b_j f_j(x)$ where $b_1, b_2, \ldots, b_j$ are constants and each $f_i(x)$, $i = 1, 2, \ldots, j$, is a function of the form x^m, $x^m e^{kx}$, $x^m e^{\alpha x} \cos \beta x$, or $x^m e^{\alpha x} \sin \beta x$.

(a) If $f_i(x) = x^m$, the associated set of functions is

$$S = \left\{ x^m, x^{m-1}, \ldots x^2, x, 1 \right\}.$$

(b) If $f_i(x) = x^m e^{kx}$, the associated set of functions is

$$S = \left\{ x^m e^{kx}, x^{m-1} e^{kx}, \ldots, x^2 e^{kx}, x e^{kx}, e^{kx} \right\}.$$

(c) If $f_i(x) = x^m e^{\alpha x} \cos \beta x$ or $f_i(x) = x^m e^{\alpha x} \sin \beta x$, the associated set of functions is

$$S = \left\{ x^m e^{\alpha x} \cos \beta x, x^{m-1} e^{\alpha x} \cos \beta x, \ldots x^2 e^{\alpha x} \cos \beta x, e^{\alpha x} \cos \beta x, e^{\alpha x} \cos \beta x, \right.$$
$$\left. x^m e^{\alpha x} \sin \beta x, x^{m-1} e^{\alpha x} \sin \beta x, \ldots x^2 e^{\alpha x} \sin \beta x, x e^{\alpha x} \sin \beta x, e^{\alpha x} \sin \beta x \right\}.$$

For each function $f_i(x)$ in $f(x)$, determine the associated set of functions S. If any of the functions in S appears in the general solution to the corresponding homogeneous equation, $y_h(x)$, multiply each function in S by x^r to obtain a new set S', where r is the smallest positive integer so that each function in S' is not a function in $y_h(x)$. A particular solution is obtained by taking the linear combination of all functions in the associated sets where repeated functions should appear only once in the particular solution.

EXAMPLE 1: Solve the nonhomogeneous equations (a) $y'' + 5y' + 6y = 2e^x$ and (b) $y'' + 5y' + 6y = 3e^{-2x}$.

SOLUTION: (a) The corresponding homogeneous equation $y'' + 5y' + 6y = 0$ has general solution $y_h(x) = c_1 e^{-2x} + c_2 e^{-3x}$.

```
> Hom_Sol:=dsolve(diff(y(x),x$2)+5*diff(y(x),x)+6*y(x)=0,

        y(x));
```

$$Hom\_Sol := y(x) = \_C1e^{(-2x)} + \_C2e^{(-3x)}$$

Next, we determine the form of $y_p(x)$. We choose

$$S = \{e^x\}.$$

because $f(x) = 2e^x$. Notice that e^x is not a solution to the homogeneous equation, so we take $y_p(x)$ to be the linear combination of the functions in S. Therefore,

$$y_p(x) = Ae^x.$$

```
> yp:=x->A*exp(x);
```

$$yp := x \rightarrow Ae^x$$

Substituting this solution into $y'' + 5y' + 6y = 2e^x$, we have

$$Ae^x + 5Ae^x + 6Ae^x = 12Ae^x = 2e^x.$$

```
> LHSeqn:=diff(yp(x),x$2)+5*diff(yp(x),x)+6*yp(x);
```

$$LHeqn := 12Ae^x$$

Equating the coefficients of e^x with `match` then gives us $A = 1/6$.

```
> match(2*exp(x)=12*A*exp(x),x,'Val'):
Val;
```

$$\left\{A = \frac{1}{6}\right\}$$

Hence, a particular solution is $y_p(x) = \frac{1}{6}e^x$,

```
> assign(Val):
yp(x);
```

$$\frac{1}{6}\mathbf{e}^x$$

and a general solution of $y'' + 5y' + 6y = 2\,e^x$ is

$$y(x) = y_h(x) + y_p(x) = c_1 e^{-2x} + c_2 e^{-3x} + \frac{1}{6}e^x.$$

The same result is obtained with `dsolve` as shown next.

```
> Gen_Sol:=dsolve(diff(y(x),x$2)+5*diff(y(x),x)+
        6*y(x)=2*exp(x),y(x));
```

$$Gen\_Sol := y(x) = \frac{1}{6}\mathbf{e}^x + \_C1\mathbf{e}^{(-2x)} + \_C2\mathbf{e}^{(-3x)}$$

We then graph the general solution for various values of the arbitrary constants in the same way as in other examples.

```
> to_plot:=[seq(seq(rhs(Gen_Sol),
        _C1=-1..1),_C2=-1..1)]:
plot(to_plot,x=-3..5,y=-3..5);
```

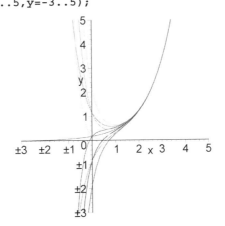

(b) In this case, we see that $f(x) = 3e^{-2x}$, so the associated set is $S = \{e^{-2x}\}$. However, because e^{-2x} is a solution to the corresponding homogeneous equation, we must multiply this function by x^r so that it is no longer a solution. We multiply the element of S by x to obtain $S' = \{xe^{-2x}\}$. Because xe^{-2x} is not a solution of $y'' + 5y' + 6y = 0$, we search for a particular solution of the form $y_p(x) = Axe^{-2x}$. Differentiating $y_p(x)$ twice

```
> A:='A':yp:='yp':

yp:=x->A*x*exp(-2*x):

> Diff(yp(x),x)=diff(yp(x),x);
```

$$\frac{\partial}{\partial x} Axe^{(-2x)} = Ae^{(-2x)} - 2Axe^{(-2x)}$$

```
> Diff(yp(x),x$2)=diff(yp(x),x$2);
```

$$\frac{\partial^2}{\partial x^2} Axe^{(-2x)} = -4Ae^{(-2x)} + 4Axe^{(-2x)}$$

and substituting into the equation yields

$$y'' + 5y' + 6y = -4Ae^{-2x} + 4Axe^{-2x} + 5(Ae^{-2x} - 2Axe^{-2x}) + 6Ae^{-2x} = Ae^{-2x} = 3e^{-2x}.$$

```
> eqn:=diff(yp(x),x$2)+5*diff(yp(x),x)+6*yp(x)=3*exp(-2*x);
```

$$eqn := Ae^{(-2x)} = 3e^{(-2x)}$$

Thus, $A = 3$,

```
> aval:=solve(eqn,A);
```

$$aval := 3$$

so $y_p(x) = 3xe^{-2x}$ and a general solution of $y'' + 5y' + 6y = 3e^{-2x}$ is

$$y(x) = y_h(x) + y_p(x) = c_1e^{-2x} + c_2e^{-3x} + 3xe^{-2x}.$$

As in (a), we can use `dsolve` to obtain equivalent results. For example, entering

```
> y:='y':

part_sol:=dsolve({diff(y(x),x$2)+5*diff(y(x),x)+
    6*y(x)=3*exp(-2*x),y(0)=a,D(y)(0)=b},y(x));
```

$$part\_sol := y(x) = 3xe^{(-2x)} - 3e^{(-2x)} + (3a+b)e^{(-2x)} + (3-2a-b)e^{(-3x)}$$

solves the equation subject to the initial conditions $y(0) = a$ and $y'(0) = b$ and names the resulting output `part_sol`. Thus, entering

```
> to_plot:=[seq(seq(rhs(part_sol),
        a=0..2),b=-1..1)]:

plot(to_plot,x=-1/2..3/2);
```

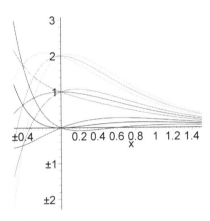

defines `toplot` to be the set consisting of nine functions corresponding to the solutions of $y'' + 5y' + 6y = 3e^{-2x}$ that satisfy the initial conditions $y(0) = a$ and $y'(0) = b$ for $a = -1$, 0, and 1 and $b = -1$, 0, and 1 and then graphs the set of functions `toplot` on the interval $[-1/2, 3/2]$.

■

In order to solve an initial-value problem, first determine the general solution and then use the initial conditions to solve for the unknown constants in the general solution.

EXAMPLE 2: Solve $y''' + 4y'' + 14y' + 20y = 10e^{-2x} - e^{-x}\cos 3x$ subject to the initial conditions $y(0) = 5$, $y'(0) = 0$, and $y''(0) = -1/2$.

SOLUTION: First, we solve the corresponding homogeneous equation $y''' + 4y'' + 14y' + 20y = 0$.

```
> y:='y':
Hom_sol:=dsolve(diff(y(x),x$3)+4*diff(y(x),x$2)+
```

```
14*diff(y(x),x)+20*y(x)=0,y(x));
```

$$Hom\_sol := y(x) =$$

$$\_C1e^{(-2x)} + \_C2e^{(-x)}\cos(3x) + \_C3e^{(-x)}\sin(3x)$$

In this case, $f(x) = 10\,e^{-2x} - e^{-x}\cos 3x$. The set of functions associated with $10\,e^{-2x}$ is $S_1 = \{e^{-2x}\}$, and the set of functions associated with $-e^{-x}\cos 3x$ is $S_2 = \{e^{-x}\cos 3x,\ e^{-x}\sin 3x\}$. However, we note that both e^{-2x} and $e^{-x}\cos 3x$ appear in the solution of the corresponding homogeneous equation. Because $x\,e^{-2x}$ does not appear in the solution of $y''' + 4y'' + 14y' + 20y = 0$, we take $S_1' = \{xe^{-2x}\}$. Similarly, we take $S_2' = \{xe^{-x}\cos 3x, x\,e^{-x}\sin 3x\}$. Thus, we must find numbers A, B, and C so that $y_p(x) = A\,x\,e^{-2x} + x\,e^{-x}(B\cos 3x + C\sin 3x)$ is a particular solution of the nonhomogeneous equation. After defining yp,

```
> A:='A':B:='B':C:='C':

yp:=x->A*x*exp(-2*x)+x*exp(-x)*(B*cos(3*x)+C*sin(3*x)):
```

We compute and simplify $y_p''' + 4y_p'' + 14y_p' + 20y_p$ and set the result equal to $10\,e^{-2x} - e^{-x}\cos 3x$.

```
> eqn:=simplify(diff(yp(x),x$3)+4*diff(yp(x),x$2)+
        14*diff(yp(x),x)+20*yp(x))=10*exp(-2*x)-
        exp(-x)*cos(3*x);
```

$$eqn := 10Ae^{(-2x)} - 18e^{(-x)}B\cos(3x)$$

$$- 18e^{(-x)}C\sin(3x) - 6e^{(-x)}B\sin(3x)$$

$$+ 6e^{(-x)}C\cos(3x) = 10e^{(-2x)} - e^{(-x)}\cos(3x)$$

This equation is true for all values of x. In particular, substituting $x = 0$, $x = \pi/6$ and $x = \pi/3$ into this equation leads to three equations

```
> eqn1:=eval(subs(x=0,eqn));
```

$$eqn1 := 10A - 18B + 6C = 9$$

```
> eqn2:=eval(subs(x=Pi/6,eqn));
```

$$eqn2 := 10Ae^{\left(-\frac{1}{3}\pi\right)} - 18e^{\left(-\frac{1}{6}\pi\right)}C - 6e^{\left(-\frac{1}{6}\pi\right)}B =$$

$$10e^{\left(-\frac{1}{3}\pi\right)}$$

> eqn3:=eval(subs(x=Pi/3,eqn));

$$eqn3 := 10Ae^{\left(-\frac{2}{3}\pi\right)} + 18e^{\left(-\frac{1}{3}\pi\right)}B - 6e^{\left(-\frac{1}{3}\pi\right)}C =$$
$$10e^{\left(-\frac{2}{3}\pi\right)} + e^{\left(-\frac{1}{3}\pi\right)}$$

that we then solve to determine the values of A, B, and C.

> consts:=solve({eqn1,eqn2,eqn3});

$$consts := \left\{ B = \frac{1}{20}, A = 1, C = \frac{-1}{60} \right\}$$

A particular solution of the nonhomogeneous equation is determined by substituting these values into yp.

> Part_sol:=subs(consts,yp(x));

$$Part\_sol := xe^{(-2x)} + xe^{(-x)}\left(\frac{1}{20}\cos(3x) - \frac{1}{60}\sin(3x)\right)$$

A general solution is given by the sum of the particular solution and the general solution of the corresponding homogeneous equation.

> y:=rhs(Hom_sol)+Part_sol;

$$y := \_C1e^{(-2x)} + \_C2e^{(-x)}\cos(3x) + \_C3e(-x)\sin(3x)$$
$$+ xe^{(-2x)} + xe^{(-x)}\left(\frac{1}{20}\cos(3x) - \frac{1}{60}\sin(3x)\right)$$

To find the values of c_1, c_2, and c_3 so that $y(x)$ satisfies the conditions $y(0) = 5$, $y'(0) = 0$, and $y''(0) = -1/2$, we solve the equations $y(0) = 5$, $y'(0) = 0$, and $y''(0) = -1/2$

eq1:=eval(subs(x=0,y))=5;

$$eq1 := \_C1 + \_C2 = 5$$

> eq2:=eval(subs(x=0,diff(y,x)))=0;

$$eq2 := -2\_C1 - \_C2 + \frac{21}{20} + 3\_C_3 = 0$$

```
> eq3:=eval(subs(x=0,diff(y,x$2)))=-1/2;
```

$$eq3 : 4\_C1 - 8\_C2 - \frac{21}{5} - 6\_C3 = \frac{-1}{2}$$

for c_1, c_2, and c_3 with `solve`.

```
> cvals:=solve({eq1,eq2,eq3});
```

$$cvals := \left\{ \_C2 = \frac{-4}{25}, \_C1 = \frac{129}{25}, \_C3 = \frac{911}{300} \right\}$$

Then, we substitute these values into $y(x)$ to obtain the solution to the initial-value problem.

```
> assign(cvals):
y;
```

$$\frac{129}{25}e^{(-2x)} - \frac{4}{25}e^{(-x)}\cos(3x) + \frac{911}{300}e^{(-x)}\sin(3x)$$

$$+ xe^{(-2x)} + xe^{(-x)}\left(\frac{1}{20}\cos(3x) - \frac{1}{60}\sin(3x)\right)$$

We graph this solution on the interval [0,2].

```
> plot(y,x=0..2);
```

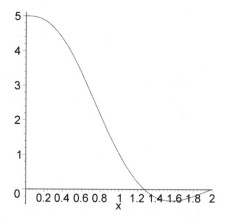

Equivalent results are obtained with `dsolve`. For example, entering

```
> y:='y':
```

```
Gen_Sol:=dsolve(diff(y(x),x$3)+4*diff(y(x),x$2)+
        14*diff(y(x),x)+20*y(x)=
    10*exp(-2*x)-exp(-x)*cos(3*x),y(x));
```

$$Gen\_Sol := y(x) = xe^{(-2x)} - \frac{1}{20}e^{(-x)}\cos(3x)\cos(x)^2$$

$$+ \frac{1}{20}e^{(-x)}\cos(3x)x - \frac{4}{15}e^{(-x)}\sin(3x)\cos(x)^6$$

$$+ \frac{2}{5}e^{(-x)}\sin(3x)\cos(x)^4 - \frac{3}{20}e^{(-x)}\sin(3x)\cos(x)^2 + \frac{1}{10}e^{(-2x)}$$

$$- \frac{1}{60}e^{(-x)}\sin(3x)x - \frac{4}{45}e^{(-x)}\cos(3x)\cos(x)^6$$

$$+ \frac{2}{15}e^{(-x)}\cos(3x)\cos(x)^4 + \_C4e^{(-2x)} - \frac{1}{400}e^{(-x)}\cos(3x)$$

$$+ \frac{4}{15}e^{(-x)}\cos(3x)\cos(x)^5\sin(x) - \frac{144}{205}\cos(x)^8\sin(x)e^{(-x)}$$

$$+ \frac{252}{205}\cos(x)^6\sin(x)e^{(-x)} - \frac{16}{205}\cos(x)^9e^{(-x)}$$

$$+ \frac{4}{45}e^{(-x)}\sin(3x)\cos(x)^3\sin(x) - \frac{1}{60}e^{(-x)}\sin(3x)\cos(x)\sin(x)$$

$$- \frac{27}{205}\cos(x)^5e^{(-x)} - \frac{4}{45}e^{(-x)}\sin(3x)\cos(x)^5\sin(x)$$

$$+ \frac{2}{5}e^{(-2x)}\sin(3x)\cos(x)^2\sin(x) + \frac{36}{205}\cos(x)^7e^{(-x)}$$

$$+ \frac{27}{4100}\cos(x)^3e^{(-x)} + \frac{1}{20}e^{(-x)}\cos(3x)\cos(x)\sin(x)$$

$$+ \frac{1}{3280}\cos(9x)e^{(-x)} + \frac{9}{3280}\sin(9x)e^{(-x)}$$

$$- \frac{4}{15}e^{(-x)}\cos(3x)\cos(x)^3\sin(x) - \frac{3}{400}e^{(-x)}\sin(3x)$$

$$+ \frac{1}{3}e^{(-2x)}\cos(3x)\sin(3x) + \frac{1}{10}e^{(-2x)}\sin(3x)\cos(x)$$

$$- \frac{2}{15}e^{(-2x)}\sin(3x)\cos(x)^3 - \frac{1}{10}e^{(-2x)}\sin(3x)\sin(x)$$

$$+ \frac{3}{10}e^{(-2x)}\cos(3x)\sin(x) - \frac{3}{10}e^{(-2x)}\cos(3x)\cos(x)$$

$$+ \frac{2}{5}3^{(-2x)}\cos(3x)\cos(x)^3 + \frac{81}{4100}e^{(-x)}\cos(x) + \frac{81}{4100}e^{(-x)}\sin(x)$$

$$+ \frac{81}{4100}\cos(x)^2\sin(x)e^{(-x)} - \frac{27}{41}\cos(x)^4\sin(x)e^{(-x)}$$

$$- \frac{6}{5}e^{(-2x)}\cos(3x)\cos(x)^2\sin(x) + \_C5e^{(-x)}\cos(3x)$$

$$+ \_C6e^{(-x)}\sin(3x)$$

```
> combine(Gen_Sol,trig);
```

$$y(x) = xe^{(-2x)} + \frac{1}{20}e^{(-x)}\cos(3x)x + \frac{1}{5}e^{(-2x)} - \frac{1}{60}e^{(-x)}\sin(3x)x$$

$$+ \_C4e^{(-2x)} - \frac{7}{450}e^{(-x)}\cos(3x) - \frac{3}{100}e^{(-x)}\sin(3x)$$

$$+ \_C5e^{(-x)}\cos(3x) + \_C6e^{(-x)}\sin(3x)$$

finds a general solution of the equation. In contrast, entering

```
> y:='y':
Init_Sol:=dsolve({diff(y(x),x$3)+4*diff(y(x),x$2)+
        14*diff(y(x),x)+20*y(x)=10*exp(-2*x)-exp(-x)*cos(3*x),
        y(0)=y0,D(y)(0)=y1,(D@@2)(y)(0)=y2},y(x));
step2:=combine(Init_Sol,trig);
```

$$step2 := y(x) = xe^{(-2x)} + \frac{1}{20}e^{(-x)}\cos(3x)x + \frac{21}{100}e^{(-2x)}$$

$$- \frac{1}{60}e^{(-x)}\sin(3x)x - \frac{21}{100}e^{(-x)}\cos(3x) + \frac{1}{10}e^{(-2x)}y2$$

$$+ \frac{1}{5}e^{(-2x)}y1 + e^{(-2x)}y0 - \frac{7}{25}e^{(-x)}\sin(3x) - \frac{1}{10}e^{(-x)}\cos(3x)y2$$

$$- \frac{1}{5}e^{(-x)}\cos(3x)y1 + \frac{1}{30}e^{(-x)}\sin(3x)y2 + \frac{2}{5}e^{(-x)}\sin(3x)y1$$

$$+ \frac{2}{3}e^{(-x)}\sin(3x)y0$$

solves the equation $y''' + 4y'' + 14y' + 20y = 10e^{-2x} - e^{-x}\cos 3x$ subject to the initial conditions $y(0) = y_0$, $y'(0) = y_1$, and $y''(0) = y_2$. (You can obtain the solution to the initial-value problem by replacing y_0, y_1, and y_2 by 5, 0, and $-1/2$, respectively.) Thus, entering

```
> y0_vals:={-1,2}:y1_vals:={-3,0,3}:y2_vals:={-2,2}:
to_plot:={seq(seq(seq(rhs(step2),
        y0=y0_vals),y1=y1_vals),y2=y2_vals)}:
plot(to_plot,x=-1/2..3/2);
```

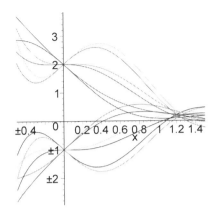

defines `to_plot` to be the set consisting of 12 functions corresponding to the solutions of the nonhomogeneous equation that satisfy the initial conditions $y_0 = -1$ and 2; $y_1 = -3$, 0, and 3; and $y_2 = -2$ and 2 and then graphs the set of functions `to_plot` on the interval $[-1/2, 3/2]$.

■

EXAMPLE 3: Show that the boundary value problem $\begin{cases} 4y'' + 4y' + 37y = \cos 3x \\ y(0) = y(\pi) \end{cases}$ has infinitely many solutions.

SOLUTION: First, we find a general solution of the corresponding homogeneous equation.

```
> y:='y':
homsol:=dsolve(4*diff(y(x),x$2)+4*diff(y(x),x)+
        37*y(x)=0,y(x));
```

$$homsol := y(x) = \_C4e^{\left(-\frac{1}{2}x\right)}\cos(3x) + \_C5e^{\left(-\frac{1}{2}x\right)}\sin(3x)$$

Using the method of undetermined coefficients, we find a particular solution to the nonhomogeneous equation of the form $y_p = A\cos 3x + B\sin 3x$. Substitution into the nonhomogeneous equation yields

```
> A:='A':B:='B':
yp:=x->A*cos(3*x)+B*sin(3*x):
step1:=simplify(4*diff(yp(x),x$2)+4*diff(yp(x),x)+
        37*yp(x))=cos(3*x);
```

$$step1 := A\cos(3x) + B\sin(3x) - 12A\sin(3x) + 12B\cos(3x) = \cos(3x)$$

This equation is true for all values of x. In particular, substituting $x = 0$ and $x = \pi/6$ yields two equations

```
> eq1:=eval(subs(x=0,step1));
```

$$eq1 := A + 12B = 1$$

```
> eq2:=eval(subs(x=Pi/6,step1));
```

$$eq2 := B - 12A = 0$$

which we then solve for A and B

```
> vals:=solve({eq1,eq2});
```

$$vals := \left\{ B = \frac{12}{145}, A = \frac{1}{145} \right\}$$

to see that $A = 1/145$ and $B = 12/145$.

```
> assign(vals):
yp(x);
```

$$\frac{1}{145}\cos(3x) + \frac{12}{145}\sin(3x)$$

```
> y:='y':
y:=rhs(homsol)+yp(x);
```

$$y := \_C4 e^{\left(-\frac{1}{2}x\right)}\cos(3x) + \_C5 e^{\left(-\frac{1}{2}x\right)}\sin(3x) + \frac{1}{145}\cos(3x) + \frac{12}{145}\sin(3x)$$

Applying the boundary conditions indicates that $\frac{1}{145} + c_4 = -\frac{1}{145} - e^{-\pi/2}c_4$

```
> eval(subs(x=0,y));
```

$$\_C4 + \frac{1}{145}$$

```
> eval(subs(x=Pi,y));
```

$$-\_C4\mathbf{e}^{\left(-\frac{1}{2}\pi\right)} - \frac{1}{145}$$

```
> cval:=solve(eval(subs(x=0,y))=eval(subs(x=Pi,y)));
```

$$cval := -\frac{2}{145}\frac{1}{1 + \mathbf{e}^{\left(-\frac{1}{2}\pi\right)}}$$

```
> sol:=subs(_C4=cval,y);
```

$$sol :=$$

$$-\frac{2}{145}\frac{\mathbf{e}^{\left(-\frac{1}{2}x\right)}\cos(3x)}{1 + \mathbf{e}^{\left(-\frac{1}{2}\pi\right)}} + \_C5\mathbf{e}^{\left(-\frac{1}{2}x\right)}\sin(3x) + \frac{1}{145}\cos(3x) + \frac{12}{145}\sin(3x)$$

Several solutions are then graphed with `plot`.

```
> toplot:={seq(sol,_C5=-5..5)}:
plot(toplot,x=0..2*Pi);
```

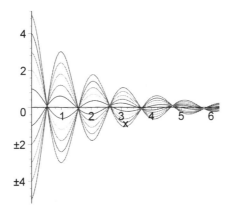

◼

EXAMPLE 4: Graph the solution to the initial-value problem $\begin{cases} x'' + 4x = \sin \omega t \\ x(0) = 1, \ x'(0) = 0 \end{cases}$ for various values of ω, including $\omega = 2$.

SOLUTION: First, we find a general solution of the corresponding homogeneous equation.

```
> x:='x':
homsol:=dsolve(diff(x(t),t$2)+4*x(t)=0,x(t));
```

$$homsol := \mathrm{x}(t) = \_C4 \cos(2\ t) + \_C5 \sin(2\ t)$$

If $\omega \neq 2$, we can find a particular solution to the nonhomogeneous equation of the form $x_p = A \cos \omega t + B \sin \omega t$. We substitute this function into the nonhomogeneous equation and simplify the result.

```
> A:='A':B:='B':
xp:=t->A*cos(omega*t)+B*sin(omega*t):
step1:=simplify(diff(xp(t),t$2)+4*xp(t))=sin(omega*t);
```

$$step1 :=$$
$$-A\ \cos(\omega\ t)\ \omega^2 - B\ \sin(\omega\ t)\ \omega^2 + 4\ A\cos(\omega\ t) + 4\ B\ \sin(\omega\ t) = \sin(\omega\ t)$$

This equation is true for all values of t. In particular, substituting $t = 0$ and $t = \frac{\pi}{2\omega}$ yields two equations

```
> eqn1:=eval(subs(t=0,step1));
```

$$eqn1 := -A\ \omega^2 + 4\ A = 0$$

```
> eqn2:=eval(subs(t=Pi/(2*omega),step1));
```

$$eqn2 := -B\ \omega^2 + 4\ B = 1$$

which we then solve to determine A and B.

```
> vals:=solve({eqn1,eqn2},{A,B});
```

$$vals := \left\{ A = 0, \ B = -\frac{1}{\omega^2 - 4} \right\}$$

We then form a particular solution to the nonhomogeneous equation x_p and a general solution to the nonhomogeneous equation $x = x_h + x_p$.

```
> assign(vals):

x:=rhs(homsol)+xp(t);
```

$$x := \_C4 \cos(2\ t) + \_C5 \sin(2\ t) - \frac{\sin(\omega\ t)}{\omega^2 - 4}$$

The solution to the initial-value problem is found by applying the initial conditions

```
> cvals:=solve({eval(subs(t=0,x))=1,
          eval(subs(t=0,diff(x,t)))=0},{_C4,_C5});
```

$$cvals := \left\{ \_C5 = \frac{1}{2} \frac{\omega}{\omega^2 - 4}, \_C4 = 1 \right\}$$

and substituting back into the general solution.

```
> sola:=subs(cvals,x);
```

$$sola := \cos(2\ t) + \frac{1}{2} \frac{\omega \sin(2\ t)}{\omega^2 - 4} - \frac{\sin(\omega\ t)}{\omega^2 - 4}$$

If $\omega = 2$, we can find a particular solution to the nonhomogeneous equation of the form $x_p = t(A \cos 2t + B \sin 2t)$. We proceed in the same manner as before.

```
> A:='A':B:='B':

xp:=t->t*(A*cos(2*t)+B*sin(2*t)):

step1:=simplify(diff(xp(t),t$2)+4*xp(t))=sin(2*t);
```

$$step1 := -4\ A \sin(2\ t) + 4\ B \cos(2\ t) = \sin(2\ t)$$

```
> eqn1:=eval(subs(t=0,step1));
```

$$eqn1 := 4\ B = 0$$

```
> eqn2:=eval(subs(t=Pi/12,step1));
```

$$eqn2 := -2\ A + 2\ B\ \sqrt{3} = \frac{1}{2}$$

```
> vals:=solve({eqn1,eqn2});
```

$$vals := \left\{ B = 0, A = \frac{-1}{4} \right\}$$

```
> assign(vals):
```

```
> x:=rhs(homsol)+xp(t);
```

$$x := \_C4\cos(2\ t) + \_C5\sin(2\ t) - \frac{1}{4}\ t\ \cos(2\ t)$$

```
> cvals:=solve({eval(subs(t=0,x))=1,
           eval(subs(t=0,diff(x,t)))=0},{_C4,_C5});
```

$$cvals := \left\{ \_C5 = \frac{1}{8},\ \_C4 = 1 \right\}$$

```
> solb:=subs(cvals,x);
```

$$solb := \cos(2\ t) + \frac{1}{8}\sin(2\ t) - \frac{1}{4}\ t\ \cos(2\ t)$$

We use these results to graph the solution for various values of ω.

```
> omegavals:=[seq(.9+.25*i,i=0..8)]:
somegraphs:=[seq(plot(subs(omega=i,sola),
       t=0..18*Pi,x=-1..1,
            color=BLACK),i=omegavals)]:
> with(plots):
anarray:=display(somegraphs,insequence=true):
display(anarray);
```

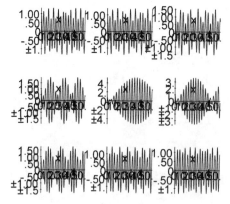

```
> plot(solb,t=0..18*Pi);
```

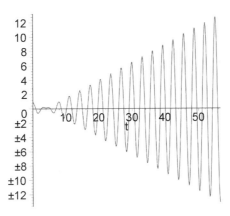

The graphs indicate that if $\omega \neq 2$ the solution to the initial-value problem is bounded and periodic; if $\omega = 2$ the solution is unbounded. We investigate this type of behavior further in Chapter 5.

We see that `dsolve` is able to solve the initial-value problem as well. Note that the result returned is valid for $\omega \neq 2$.

```
> x:='x':t:='t':
solc:=dsolve({diff(x(t),t$2)+4*x(t)=
        sin(omega*t),x(0)=1,D(x)(0)=0},x(t));
```

$$solc := x(t) = \left(-\frac{1}{4}\frac{\sin((\omega - 2)\,t)}{\omega - 2} + \frac{1}{4}\frac{\sin((\omega + 2)\,t)}{\omega + 2}\right)\cos(2\,t)$$

$$+ \left(-\frac{1}{4}\frac{\cos((\omega + 2)\,t)}{\omega + 2} - \frac{1}{4}\frac{\cos((\omega - 2)\,t)}{\omega - 2}\right)\sin(2\,t) + \cos(2\,t) + \frac{1}{2}\frac{\omega\,\sin(2\,t)}{\omega^2 - 4}$$

We solve the equation for the case $\omega = 2$ separately.

```
> sold:=dsolve({diff(x(t),t$2)+4*x(t)=
        sin(2*t),x(0)=1,D(x)(0)=0},x(t));
```

$$sold := x(t) = \left(\frac{1}{8}\cos(2\,t)\,\sin(2\,t) - \frac{1}{4}t\right)\cos(2\,t)$$

$$-\frac{1}{8}\cos(2\,t)^2\,\sin(2\,t) + \cos(2\,t) + \frac{1}{8}\sin(2\,t)$$

```
> combine(sold,trig);
```

$$x(t) = \cos(2\ t) + \frac{1}{8}\sin(2\ t) - \frac{1}{4}\ t\ \cos(2\ t)$$

■

4.5 Nonhomogeneous Equations with Constant Coefficients: Variation of Parameters

Second-Order Equations

We know that in order to solve the second-order homogeneous linear differential equation

$$a_2(x)y'' + a_1(x)y' + a_0(x)y = 0,$$

we need two linearly independent solutions $y_1(x)$ and $y_2(x)$. A general solution is then given by $y(x) = c_1y_1(x) + c_2y_2(x)$.

Using the method of variation of parameters to construct a particular solution of the nonhomogeneous equation

$$a_2(x)y'' + a_1(x)y' + a_0(x)y = g(x),$$

we first divide by $a_2(x)$ to rewrite it in the form $y'' + p(x)y' + q(x)y = f(x)$, assume that a particular solution has a form similar to the general solution by "varying" the parameters c_1 and c_2, and let

$$y_p(x) = u_1(x)y_1(x) + u_2(x)y_2(x)$$

where $y_1(x)$ and $y_2(x)$ are linearly independent solutions of the corresponding homogeneous equation.

```
> y:='y':y1:='y1':y2:='y2':u1:='u1':u2:='u2':
     p:='p':q:='q':f:='f':yp:='yp':
  yp:=x->y1(x)*u1(x)+y2(x)*u2(x);
```

$$yp := x \rightarrow y1(x)u1(x) + y2(x)u2(x)$$

We need two equations in order to determine the two unknown functions $u_1(x)$ and $u_2(x)$. We obtain these equations by substituting $y_p(x) = u_1(x)y_1(x) + u_2(x)y_2(x)$ into the nonhomogeneous differential equation $y'' + p(x)y' + q(x)y = f(x)$. Differentiating $y_p(x)$, we obtain

$$y_p'(x) = u_1(x)y_1'(x) + u_1'(x)y_1(x) + u_2(x)y_2'(x) + u_2'(x)y_2(x)$$

```
> step1:=diff(yp(x),x);
```

$$step1 := \left(\frac{\partial}{\partial x}y1(x)\right)u1(x) + y1(x)\left(\frac{\partial}{\partial x}u1(x)\right) + \left(\frac{\partial}{\partial x}y2(x)\right)u2(x) + y2(x)\left(\frac{\partial}{\partial x}u2(x)\right)$$

which can be simplified to

$$y_p(x) = u_1(x)y_1'(x) + u_2(x)y_2'(x)$$

with the assumption that $u_1'(x)y_1(x) + u_2'(x)y_2(x) = 0$. (Observe that we use subs to replace the terms $u_1'(x)y_1(x)$ and $u_2'(x)y_2(x)$ with 0.)

```
> step2:=subs({y1(x)*diff(u1(x),x)=0,
            y2(x)*diff(u2(x),x)=0},step1);
```

$$step2 := \left(\frac{\partial}{\partial x}y1(x)\right)u1(x) + \left(\frac{\partial}{\partial x}y2(x)\right)u2(x)$$

This becomes our first equation for $u_1(x)$ and $u_2(x)$. The second derivative is $y_p''(x) = u_1(x)y_1''(x) + u_1'(x)y_1'(x) + u_2(x)y_2''(x) + u_2'(x)y_2'(x)$.

```
> second:=diff(step2,x);
```

$$second := \left(\frac{\partial^2}{\partial x^2}y1(x)\right)u1(x) + \left(\frac{\partial}{\partial x}y1(x)\right)\left(\frac{\partial}{\partial x}u1(x)\right) + \left(\frac{\partial^2}{\partial x^2}y2(x)\right)u2(x)$$

$$+ \left(\frac{\partial}{\partial x}y2(x)\right)\left(\frac{\partial}{\partial x}u2(x)\right)$$

Substitution into $y'' + p(x)y' + q(x)y = f(x)$ then yields

$$\begin{aligned} y'' + p(x)y' + q(x)y &= u_1(x)[y_1''(x) + p(x)y_1'(x) + q(x)y_1(x)] \\ &\quad + u_2(x)[y_2''(x) + p(x)y_2'(x) + q(x)y_2(x)] \\ &\quad + u_1'(x)y_1'(x) + u_2'(x)y_2'(x) \\ &= u_1'(x)y_1'(x) + u_2'(x)y_2'(x) = f(x) \end{aligned}$$

because $y_1(x)$ and $y_2(x)$ are solutions of the corresponding homogeneous equation.

```
> step3:=expand(second+p(x)*step2+q(x)*yp(x));
```

$$step3 := \left(\frac{\partial^2}{\partial x^2}y1(x)\right)u1(x) + \left(\frac{\partial}{\partial x}y1(x)\right)\left(\frac{\partial}{\partial x}u1(x)\right) + \left(\frac{\partial^2}{\partial x^2}y2(x)\right)u2(x)$$

$$+ \left(\frac{\partial}{\partial x}y2(x)\right)\left(\frac{\partial}{\partial x}u2(x)\right) + p(x)\left(\frac{\partial}{\partial x}y1(x)\right)u1(x) + p(x)\left(\frac{\partial}{\partial x}y2(x)\right)u2(x)$$

$$+ q(x)y1(x)u1(x) + q(x)y2(x)u2(x)$$

```
> step4:=collect(step3,{u1(x),u2(x)});
```

$$step4 := \left(p(x)\left(\frac{\partial}{\partial x}y2(x)\right) + \left(\frac{\partial^2}{\partial x^2}y2(x)\right) + q(x)y2(x)\right)u2(x)$$

$$+ \left(\left(\frac{\partial^2}{\partial x^2}y1(x)\right) + q(x)y1(x) + p(x)\left(\frac{\partial}{\partial x}y1(x)\right)\right)u1(x)$$

$$+ \left(\frac{\partial}{\partial x}y1(x)\right)\left(\frac{\partial}{\partial x}u1(x)\right) + \left(\frac{\partial}{\partial x}y2(x)\right)\left(\frac{\partial}{\partial x}u2(x)\right)$$

```
> step5:=subs({diff(y2(x),x$2)+p(x)*diff(y2(x),x)+
           q(x)*y2(x)=0,
       diff(y1(x),x$2)+p(x)*diff(y1(x),x)+q(x)*y1(x)=0},
           step4);
```

$$step5 := \left(\frac{\partial}{\partial x}y1(x)\right)\left(\frac{\partial}{\partial x}u1(x)\right) + \left(\frac{\partial}{\partial x}y2(x)\right)\left(\frac{\partial}{\partial x}u2(x)\right)$$

Therefore, our second equation for determining $u_1(x)$ and $u_2(x)$ is $u_1'(x)y_1'(x) + u_2'(x)y_2'(x) = f(x)$. Hence, we have the system

$$\begin{cases} u_1'(x)y_1(x) + u_2'(x)y_2(x) = 0 \\ u_1'(x)y_1'(x) + u_2'(x)y_2'(x) = f(x) \end{cases}$$

which is written in matrix form as

$$\begin{pmatrix} y_1(x) & y_2(x) \\ y_1'(x) & y_2'(x) \end{pmatrix} \begin{pmatrix} u_1'(x) \\ u_2'(x) \end{pmatrix} = \begin{pmatrix} 0 \\ f(x) \end{pmatrix}.$$

In linear algebra, we learn that this system has a unique solution if and only if

$$\begin{vmatrix} y_1(x) & y_2(x) \\ y_1'(x) & y_2'(x) \end{vmatrix} \neq 0.$$

Notice that this determinant is the Wronskian, $W(S)$, of the set $S = \{y_1(x), \ y_2(x)\}$. We stated in Section 4.1 that $W(S) \neq 0$ if the functions $y_1(x)$ and $y_2(x)$ in the set S are linearly independent.

Because $S = \{y_1(x),\ y_2(x)\}$ represents a fundamental set of solutions of the corresponding homogeneous equation, $W(S) \neq 0$. Hence, this system has a unique solution, which can be found with Cramer's rule to be

$$u_1'(x) = \frac{\begin{vmatrix} 0 & y_2(x) \\ f(x) & y_2'(x) \end{vmatrix}}{W(S)} = \frac{-y_2(x)f(x)}{W(S)} \quad \text{and} \quad u_2'(x) = \frac{\begin{vmatrix} y_1(x) & 0 \\ y_1'(x) & f(x) \end{vmatrix}}{W(S)} = \frac{y_1(x)f(x)}{W(S)}$$

where $S = \{y_1(x),\ y_2(x)\}$ is a fundamental set of solutions of the corresponding homogeneous equation.

```
> uprimes:=solve({step5=f(x),
        diff(u1(x),x)*y1(x)+diff(u2(x),x)*y2(x)=0},
            {diff(u1(x),x),diff(u2(x),x)});
```

$$uprimes := \left\{ \frac{\partial}{\partial x}u2(x) = \frac{y1(x)f(x)}{-\left(\frac{\partial}{\partial x}y1(x)\right)y2(x) + \left(\frac{\partial}{\partial x}y2(x)\right)y1(x)}, \right.$$

$$\left. \frac{\partial}{\partial x}u1(x) = \frac{f(x)y2(x)}{\left(\frac{\partial}{\partial x}y1(x)\right)y2(x) - \left(\frac{\partial}{\partial x}y2(x)\right)y1(x)} \right\}$$

Integrating, we then obtain

$$u_1(x) = \int \frac{-y_2(x)f(x)}{W(S)}\,dx \quad \text{and} \quad u_2(x) = \int \frac{y_1(x)f(x)}{W(S)}\,dx,$$

so a general solution of the nonhomogeneous equation $y'' + p(x)y' + q(x)y = f(x)$ is

$$y(x) = y_h(x) + y_p(x) = c_1 y_1(x) + c_2 y_2(x) + u_1(x)y_1(x) + u_2(x)y_2(x).$$

Of course, the problems that were solved in the preceding section by the method of undetermined coefficients can be solved by variation of parameters as well.

Summary of Variation of Parameters for Second-Order Equations

Given the second-order equation $a_2(x)y'' + a_1(x)y' + a_0(x)y = g(x)$.

1. Divide by $a_2(x)$ to rewrite the equation in the form $y'' + p(x)y' + q(x)y = f(x)$.
2. Find a general solution $y_h = c_1 y_1 + c_2 y_2$ of the corresponding homogeneous equation $y'' + p(x)y' + q(x)y = 0$.

3. Let $W = \begin{vmatrix} y_1 & y_2 \\ y_1' & y_2' \end{vmatrix}$.

4. Let $u_1' = \frac{-y_2 f(x)}{W}$ and $u_2' = \frac{y_1 f(x)}{W}$.

5. Integrate to obtain u_1 and u_2.
6. A particular solution of $a_2(x)y'' + a_1(x)y' + a_0(x)y = g(x)$ is given by $y_p = u_1 y_1 + u_2 y_2$.
7. A general solution of $a_2(x)y'' + a_1(x)y' + a_0(x)y = g(x)$ is given by $y = y_h + y_p$.

Generally, if S is a fundamental set of solutions of $a_2(x)y'' + a_1(x)y' + a_0(x)y = 0$, the command

$$\texttt{varparam (S, g(x), x)}$$

uses variation of parameters to find a general solution of $a_2(x)y'' + a_1(x)y' + a_0(x)y = g(x)$. Note that `varparam` is contained in the **DEtools** package.

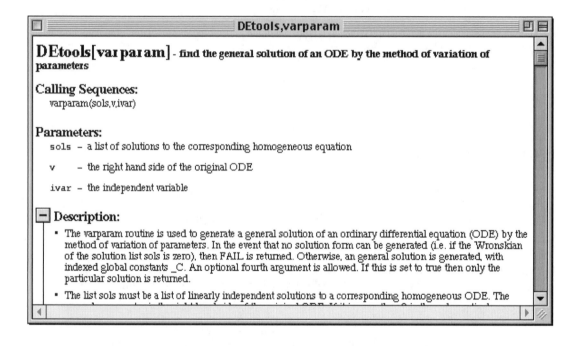

DEtools,varparam

DEtools[varparam] - find the general solution of an ODE by the method of variation of parameters

Calling Sequences:
varparam(sols,v,ivar)

Parameters:
sols – a list of solutions to the corresponding homogeneous equation

v   – the right hand side of the original ODE

ivar – the independent variable

Description:

- The varparam routine is used to generate a general solution of an ordinary differential equation (ODE) by the method of variation of parameters. In the event that no solution form can be generated (i.e. if the Wronskian of the solution list sols is zero), then FAIL is returned. Otherwise, an general solution is generated, with indexed global constants _C. An optional fourth argument is allowed. If this is set to true then only the particular solution is returned.

- The list sols must be a list of linearly independent solutions to a corresponding homogeneous ODE. The

---

EXAMPLE 1: Solve $y'' + y/4 = \sec(x/2) + \csc(x/2), 0 < x < \pi$.

SOLUTION: A general solution of the corresponding homogeneous equation is
$y_h(x) = c_1 \cos(x/2) + c_2 \sin(x/2)$.

```
> y:='y':

Hom_Sol:=dsolve(diff(y(x),x$2)+y(x)/4=0,y(x));
```

$$Hom\_Sol := y(x) = \_C1 \sin\left(\frac{1}{2}x\right) + \_C2 \cos\left(\frac{1}{2}x\right)$$

Hence, $S = \{y_1(x), y_2(x)\} = \{\cos(x/2), \sin(x/2)\}$ and

$$W(S) = \begin{vmatrix} \cos(x/2) & \sin(x/2) \\ -\frac{1}{2}\sin(x/2) & \frac{1}{2}\cos(x/2) \end{vmatrix} = \frac{1}{2}\cos^2(x/2) + \frac{1}{2}\sin^2(x/2) = \frac{1}{2},$$

which we confirm using `constcoeffsols`, which is contained in the **DEtools** package, and `wronskian`, which is contained in the **linalg** package.

```
> with(DEtools):
> S:=constcoeffsols(diff(y(x),x$2)+y(x)/4=0,y(x));
```

$$S := \left[\cos\left(\frac{1}{2}x\right) \quad \sin\left(\frac{1}{2}x\right)\right]$$

```
> with(linalg):
ws:=simplify(det(wronskian(S,x)));
```

$$ws := \frac{1}{2}$$

We therefore have the following calculations:

$$u_1(x) = \int \frac{-\sin(x/2)(\sec(x/2) + \csc(x/2))}{1/2}\,dx = -2\int\left(\frac{\sin(x/2)}{\cos(x/2)} + 1\right)dx = -2x + 4\ln|\cos(x/2)|$$

and

$$u_2(x) = \int \frac{\cos(x/2)(\sec(x/2) + \csc(x/2))}{1/2} = 2\int\left(1 + \frac{\cos(x/2)}{\sin(x/2)} +\right)dx = 2x + 4\ln|\sin(x/2)|.$$

(Notice that `S[2]` returns the second element of $S = \{y_1(x), y_2(x)\}$ $= \{\cos(x/2), \sin(x/2)\}$ and `S[1]` returns the first.)

```
> f:=x->sec(x/2)+csc(x/2):
u1:=x->-int(f(x)*S[2]/ws,x);
```

$$u1 := x \rightarrow -\int \frac{f(x)S_2}{ws} - dx$$

```
> u1(x);
```

$$4\ln\left(\cos\left(\frac{1}{2}x\right)\right) - 2x$$

```
> u2:=x->int(f(x)*S[1]/ws,x):
u2(x);
```

$$2x + 4\ln\left(\sin\left(\frac{1}{2}x\right)\right)$$

Then by variation of parameters a particular solution of the nonhomogeneous equation is

$$y_p(x) = \cos(x/2)[-2x + 4\ln|\cos(x/2)|] + \sin(x/2)[2x + 4\ln|\sin(x/2)|].$$

```
> yp:=x->S[1]*u1(x)+S[2]*u2(x):
yp(x);
```

$$\cos\left(\frac{1}{2}x\right)\left(4\ln\left(\cos\left(\frac{1}{2}x\right)\right) - 2x\right) + \sin\left(\frac{1}{2}x\right)\left(2x + 4\ln\left(\sin\left(\frac{1}{2}x\right)\right)\right)$$

and

$$y(x) = c_1\cos(x/2) + c_2\sin(x/2) + y_p(x)$$
$$= c_1\cos(x/2) + c_2\sin(x/2) + \cos(x/2)[-2x + 4\ln|\cos(x/2)|] + \sin(x/2)[2x + 4\ln|\sin(x/2)|]$$

is a general solution.

```
> Gen_Sol:=rhs(Hom_Sol)+yp(x);
```

$$Gen\_Sol := \_C1\sin\left(\frac{1}{2}x\right) + \_C2\cos\left(\frac{1}{2}x\right) + \cos\left(\frac{1}{2}x\right)\left(4\ln\left(\cos\left(\frac{1}{2}x\right)\right) - 2x\right)$$
$$+ \sin\left(\frac{1}{2}x\right)\left(2x + 4\ln\left(\sin\left(\frac{1}{2}x\right)\right)\right)$$

Note that $\cos(x/2) > 0$ and $\sin(x/2) > 0$ for $0 < x < \pi$, so the absolute value signs can be eliminated.

As we have seen in previous examples, we are able to graph the general solution obtained for various values of the arbitrary constants with `plot`.

```
> toplot:={seq(seq(Gen_Sol,_C1=-1..1),_C2=-1..1)}:
plot(toplot,x=0..Pi);
```

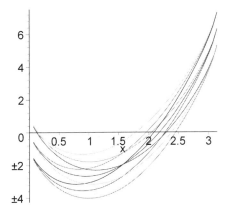

In addition, we see that both `dsolve` and `varparam` are successful in finding a general solution of the equation.

```
> dsolve(diff(y(x),x$2)+y(x)/4=sec(x/2)+csc(x/2),y(x));
```

$$y(x) = \_C1 \sin\left(\frac{1}{2}x\right) + \_C2 \cos\left(\frac{1}{2}x\right) + \cos\left(\frac{1}{2}x\right)\left(4\ln\left(\cos\left(\frac{1}{2}x\right)\right) - 2x\right)$$
$$+ \sin\left(\frac{1}{2}x\right)\left(2x + 4\ln\left(\sin\left(\frac{1}{2}x\right)\right)\right)$$

```
> varparam(S,sec(x/2)+csc(x/2),x);
```

$$\_C_1 \cos\left(\frac{1}{2}x\right) + \_C_2 \sin\left(\frac{1}{2}x\right) + \cos\left(\frac{1}{2}x\right)\left(4\ \ln\left(\cos\left(\frac{1}{2}x\right)\right) - 2x\right)$$
$$+ \sin\left(\frac{1}{2}x\right)\left(2x + 4\ \ln\left(\sin\left(\frac{1}{2}x\right)\right)\right)$$

■

EXAMPLE 2: Find a general solution of $y'' + 4y' + 13y = x\cos^2 3x$.

SOLUTION: We see that `dsolve` is able to find a general solution of the equation.

```
> y:='y':
gensol:=dsolve(diff(y(x),x$2)+4*diff(y(x),x)+13*y(x)=
```

```
x*cos(3*x)^2,y(x));
```

$$gensol := y(x) = \frac{3}{340}\cos(3x)\cos(9x)x - \frac{3}{7225}\cos(3x)\cos(9x)$$

$$- \frac{1}{510}\cos(3x)\sin(9x)x - \frac{77}{86700}\cos(3x)\sin(9x)$$

$$- \frac{1}{26}x\cos(3x)^2 + \frac{2}{169}\cos 3x^2 + \frac{1}{39}\cos(3x)\sin(3x)x$$

$$+ \frac{5}{1014}\cos(3x)\sin(3x) + \frac{1}{510}\sin(3x)\cos(9x)x$$

$$+ \frac{77}{86700}\sin(3x)\cos(9x) + \frac{3}{340}\sin(3x)\sin(9x)x$$

$$- \frac{3}{7225}\sin(3x)\sin(9x) + \frac{3}{52}x - \frac{3}{169} + \_C1e^{(-2x)}\cos(3x)$$

$$+ \_C2e^{(-2x)}\sin(3x)$$

```
> combine(gensol,trig);
```

$$y(x) = \frac{1}{26}x + \frac{1926}{1221025}\sin(6x) + \frac{12}{1105}x\sin(6x)$$

$$+ \_C2e^{(-2x)}\sin(3x) - \frac{23}{2210}x\cos(6x) + \frac{6718}{1221025}\cos(6x)$$

$$- \frac{2}{169} + \_C1e^{(-2x)}\cos(3x)$$

Alternatively, we can use Maple to help us implement the variation of parameters procedure. The characteristic equation of the corresponding homogeneous equation is $m^2 + 4m + 13 = 0$, which is solved with `solve`.

```
> solve(m^2+4*m+13=0);
```

$$-2 + 3I, -2 - 3I$$

The result means that a general solution of $y'' + 4y' + 13y = 0$ is $y_h(x) = e^{-2x}(c_1\cos 3x + c_2\sin 3x)$ and a fundamental set of solutions for the corresponding homogeneous equation is $S = \{y_1(x), y_2(x)\}$ $= \{e^{-2x}\cos 3x, e^{-2x}\sin 3x\}$. These results are confirmed with `constcoeffsols`.

```
> y:='y':

with(DEtools):

S:=constcoeffsols(diff(y(x),x$2)+4*diff(y(x),x)+
```

```
13*y(x)=0,y(x));
```

$$S := [e^{(-2x)} \cos(3x), e^{(-2x)} \sin(3x)]$$

Next, we define $f(x)$ and then compute and simplify the Wronskian of the set of functions S, naming the result ws.

```
> f:=x->x*cos(3*x)^2:
with(linalg):
ws:=simplify(det(wronskian(S,x)));
```

$$ws := 3e^{(-4x)}$$

To calculate $u_1(x) = \int \frac{-y_2(x)f(x)}{3e^{-4x}} \, dx$ and $u_2(x) = \int \frac{y_1(x)f(x)}{3e^{-4x}} \, dx$, we use int.

```
> u1prime:=simplify(-S[2]*f(x)/ws);
```

$$u1prime := -\frac{1}{3} \sin(3x) x \cos(3x)^2 e^{(2x)}$$

```
> u1:=x->int(u1prime,x):
u1(x);
```

$$-\frac{1}{12}\left(-\frac{9}{85}x + \frac{36}{7225}\right) e^{(2x)} \cos(9x)$$

$$-\frac{1}{12}\left(\frac{2}{85}x + \frac{77}{7225}\right) e^{(2x)} \sin(9x)$$

$$-\frac{1}{12}\left(-\frac{3}{13}x + \frac{12}{169}\right) e^{(2x)} \cos(3x)$$

$$-\frac{1}{12}\left(\frac{2}{13}x + \frac{5}{169}\right) e^{(2x)} \sin(3x)$$

```
> u2prime:=simplify(S[1]*f(x)/ws);
```

$$u2prime := \frac{1}{3} \cos(3x)^3 x e^{(2x)}$$

```
> u2:=x->int(u2prime,x):
u2(x);
```

$$\frac{1}{12}\left(\frac{2}{85}x+\frac{77}{7225}\right)e^{(2x)}\cos(9x)$$

$$-\frac{1}{12}\left(-\frac{9}{85}x+\frac{36}{7225}\right)e^{(2x)}\sin(9x)$$

$$+\frac{1}{4}\left(\frac{2}{13}x+\frac{5}{169}\right)e^{(2x)}\cos(3x)$$

$$-\frac{1}{4}\left(-\frac{3}{13}x+\frac{12}{169}\right)e^{(2x)}\sin(3x)$$

Then, a particular solution of the nonhomogeneous equation is given by $y_p(x) = u_1(x)y_1(x) + u_2(x)y_2(x)$,

```
> yp:=x->S[1]*u1(x)+S[2]*u2(x):
combine(yp(x),trig);
```

$$-\frac{23}{2210}e^{(-2x)}e^{(2x)}x\cos(6x)$$

$$+\frac{6718}{1221025}e^{(-2x)}e^{(2x)}\cos(6x)$$

$$+\frac{12}{1105}e^{(-2x)}e^{(2x)}x\sin(6x)+\frac{1926}{1221025}e^{(-2x)}e^{(2x)}\sin(6x)$$

$$+\frac{1}{26}e^{(-2x)}e^{(2x)}x-\frac{2}{169}e^{(-2x)}e^{(2x)}$$

and a general solution of the nonhomogeneous equation is given by $y(x) = y_h(x) + y_p(x)$.

```
> yh:=x->c[1]*S[1]+c[2]*S[2]:
y:=combine(yh(x)+yp(x),trig);
```

$$y := c_1 e^{(-2x)}\cos(3x) + c_2 e^{(-2x)}\sin(3x)$$

$$-\frac{23}{2210}e^{(-2x)}e^{(2x)}x\cos(6x)$$

$$+\frac{6718}{1221025}e^{(-2x)}e^{(2x)}\cos(6x)$$

$$+\frac{12}{115}e^{(-2x)}e^{(2x)}x\sin(6x)+\frac{1926}{1221025}e^{(-2x)}e^{(2x)}\sin(6x)$$

$$+\frac{1}{26}e^{(-2x)}e^{(2x)}x-\frac{2}{169}e^{(-2x)}e^{(2x)}$$

We then graph the general solution obtained for various values of the arbitrary constants. Note that the option `view=[-1..1,-20..20]` instructs Maple that the numbers displayed on the x-axis (the horizontal axis) correspond to the interval $[-1,1]$ and the numbers displayed on the y-axis (the vertical axis) correspond to the interval $[-20,20]$.

```
> c1vals:=seq(-5+5*i,i=0..2):

c2vals:=seq(-4+2*i,i=0..4):

toplot:={seq(seq(y,c[1]=c1vals),c[2]=c2vals)}:

plot(toplot,x=-1..1,view=[-1..1,-20..20]);
```

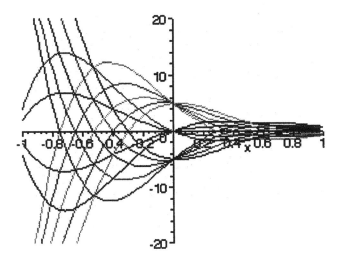

■

Higher Order Nonhomogeneous Equations

Nonhomogeneous higher order linear equations can be solved through variation of parameters as well. In general, if we are given the nonhomogeneous equation

$$y^{(n)}(x) + a_{n-1}(x)y^{(n-1)}(x) + \cdots + a_1(x)y'(x) + a_0(x)y(x) = f(x)$$

and a fundamental set of solutions $y_1(x)$, $y_2(x)$, $\ldots$, $y_n(x)$ of the associated homogeneous equation

$$y^{(n)}(x) + a_{n-1}(x)y^{(n-1)}(x) + \cdots + a_1(x)y'(x) + a_0(x)y(x) = 0,$$

we can extend the method for second order equations to find $u_1(x)$, $u_2(x)$, $\ldots$, $u_n(x)$ such that

$$y_p(x) = u_1(x)y_1(x) + u_2(x)y_2(x) + \cdots + u_n(x)y_n(x) = \sum_{i=1}^{n} u_i(x)y_i(x)$$

is a particular solution of the nonhomogeneous equation.

If

$$u_1'(x)y_1(x) + u_2'(x)y_2(x) + \cdots + u_n'(x)y_n(x) = \sum_{i=1}^{n} u_i'(x)y_i(x) = 0,$$

then

$$y_p^{(m)}(x) = u_1(x)y_1^{(m)}(x) + u_2(x)y_2^{(m)}(x) + \cdots + u_n(x)y_n^{(m)}(x) = \sum_{i=1}^{n} u_i(x)y_i^{(m)}(x)$$

for $m = 0, 1, 2, \ldots , n-1$ and if

$$u_1'(x)y_1^{(m-1)}(x) + u_2'(x)y_2^{(m-1)}(x) + \cdots + u_n'(x)y_n^{(m-1)}(x) = \sum_{i=1}^{n} u_i'(x)y_i^{(m-1)}(x) = 0$$

for $m = 0, 1, 2, \ldots , n-1$, then

$$y_p^{(n)}(x) = \sum_{i=1}^{n} u_i(x)y_i^{(n)}(x) + \sum_{i=1}^{n} u_i'(x)y_i^{(n-1)}(x).$$

Therefore, we obtain the system of n equations

$$\begin{cases} \sum_{i=1}^{n} y_i(x)u_i'(x) = 0 \\ \sum_{i=1}^{n} y_i'(x)u_i'(x) = 0 \\ \quad\vdots \\ \sum_{i=1}^{n} y_i^{(n-1)}(x)u_i'(x) = f(x) \end{cases}$$

which can be solved for $u_1'(x)$, $u_2'(x)$, $\ldots$, $u_n'(x)$ using Cramer's rule.

Let $W_m(y_1(x), y_2(x), \ldots , y_n(x))$ denote the determinant of the matrix obtained by replacing the mth column of

$$\begin{pmatrix} y_1(x) & y_2(x) & \cdots & y_n(x) \\ y_1'(x) & y_2'(x) & \cdots & y_n'(x) \\ \vdots & \vdots & \vdots & \vdots \\ y_1^{(n-1)}(x) & y_2^{(n-1)}(x) & \cdots & y_n^{(n-1)}(x) \end{pmatrix}$$

by the column $\begin{pmatrix} 0 \\ 0 \\ \vdots \\ 0 \\ f(x) \end{pmatrix}$. Then, by Cramer's rule,

$$u_i'(x) = \frac{f(x)W_i(y_1(x), y_2(x), \dots, y_n(x))}{W(y_1(x), y_2(x), \dots, y_n(x))}$$

and

$$u_i(x) = \int \frac{f(x)W_i(y_1(x), y_2(x), \dots, y_n(x))}{W(y_1(x), y_2(x), \dots, y_n(x))} \, dx,$$

for $i = 1, 2, \dots, n$, so

$$y_p(x) = \sum_{i=1}^{n} u_i(x)y_i(x) = \sum_{i=1}^{n} y_i(x) \int \frac{f(x)W_i(y_1(x), y_2(x), \dots, y_n(x))}{W(y_1(x), y_2(x), \dots, y_n(x))} \, dx$$

is a particular solution of the nonhomogeneous equation.

A general solution of the nonhomogeneous equation is given by $y(x) = y_h(x) + y_p(x)$, where $y_h(x)$ is a general solution of the corresponding homogeneous equation.

EXAMPLE 3: Solve $y''' + y' = \tan x$, $-\pi < x < \pi/2$. Find the solution that satisfies the initial conditions $y(0) = 0$, $y'(0) = 1$, and $y''(0) = 0$.

SOLUTION: We use Maple to assist us in implementing the Method of Variation of Parameters. We first find that a general solution of the corresponding homogeneous equation is $y_h(x) = c_1 + c_2 \cos x + c_3 \sin x$

```
> homsol:=dsolve(diff(y(x),x$3)+diff(y(x),x)=0,y(x));
```

$$homsol := \mathrm{y}(x) = \_C1 + \_C2 \cos(x) + \_C3 \sin(x)$$

and a fundamental set of solutions is $S = \{\cos x, \sin x, 1\}$.

```
> with(DEtools):
S:=constcoeffsols(diff(y(x),x$3)+diff(y(x),x)=0,y(x));
```

$$S := [\cos(x), \sin(x), 1]$$

```
> with(linalg):
ws:=wronskian(S,x);
```

$$ws := \begin{vmatrix} \cos(x) & \sin(x) & 1 \\ -\sin(x) & \cos(x) & 0 \\ -\cos(x) & -\sin(x) & 0 \end{vmatrix}$$

Therefore, we must solve the system

$$\begin{pmatrix} \cos x & \sin x & 1 \\ -\sin x & \cos x & 0 \\ -\cos x & -\sin x & 0 \end{pmatrix} \begin{pmatrix} u_i'(x) \\ u_2'(x) \\ u_3'(x) \end{pmatrix} = \begin{pmatrix} 0 \\ 0 \\ \tan x \end{pmatrix}.$$

We take advantage of the `linsolve` command, which is contained in the **linalg** package.

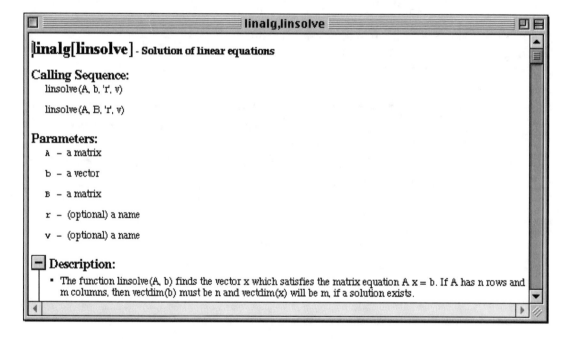

linalg,linsolve

linalg[linsolve] - Solution of linear equations

Calling Sequence:
  linsolve (A, b, 'r', v)

  linsolve (A, B, 'r', v)

Parameters:
  A – a matrix

  b – a vector

  B – a matrix

  r – (optional) a name

  v – (optional) a name

Description:
  ▪ The function linsolve (A, b) finds the vector x which satisfies the matrix equation A x = b. If A has n rows and m columns, then vectdim(b) must be n and vectdim(x) will be m, if a solution exists.

to solve this system for $u_1'(x)$, $u_2'(x)$, and $u_3'(x)$.

```
> b := vector([0,0,tan(x)]):

uprimes:=simplify(linsolve(ws,b));
```

$$uprimes := \begin{vmatrix} -\sin(x), \dfrac{-1 + \cos(x)^2}{\cos(x)}, \tan(x) \end{vmatrix}$$

Next, we use `int` together with `map` to determine $u_1(x)$, $u_2(x)$, and $u_3(x)$.

```
> us:=map(int,uprimes,x);
```

$$us := [\cos(x), -\ln(\sec(x) + \tan(x)) + \sin(x), -\ln(\cos(x))]$$

We then use `dotprod`, which is contained in the **linalg** package, to compute

$$y_p(x) = u_1(x)y_1(x) + u_2(x)y_2(x) + u_3(x)y_3(x) = \langle u_1(x), u_2(x), u_3(x) \rangle \cdot \langle y_1(x), y_2(x), y_3(x) \rangle,$$

a particular solution of the nonhomogeneous equation

```
> yp:=dotprod(us,S,orthogonal);
```

$$yp := \cos(x)^2 + (-\ln(\sec(x) + \tan(x)) + \sin(x))\sin(x) - \ln(\cos(x))$$

and then a general solution.

```
> gensol:=rhs(homsol)+yp;
```

$$gensol := \_C1 + \_C2\cos(x) + \_C3\sin(x) + \cos(x)^2$$
$$+ (-\ln(\sec(x) + \tan(x)) + \sin(x))\sin(x) - \ln(\cos(x))$$

As we have seen with previous examples, we can graph the solution for various values of the constants. Entering

```
> toplot:={seq(seq(seq(gensol,_C1=-1..1),_C2=0..1),
            _C3=-1..0)}:
plot(toplot,x=-Pi/2..Pi/2);
```

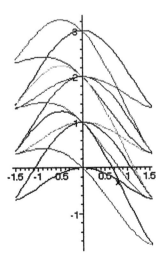

defines `toplot` be a set of 12 functions and then graphs the functions in `toplot` on the interval $(-\pi/2, \pi/2)$.

In this case, we see that `dsolve` can find a general solution of the equation

```
> gensol:=dsolve(diff(y(x),x$3)+diff(y(x),x)=tan(x),y(x));
```

$$gensol := y(x) = \frac{1}{2} - \ln(\cos(x)) - \sin(x)\ln\left(\frac{1 + \sin(x)}{\cos(x)}\right) + \_C1$$
$$+ \_C2\cos(x) + \_C3\sin(x)$$

```
> partsol:=dsolve({diff(y(x),x$3)+diff(y(x),x)=tan(x),
          y(0)=0,D(y)(0)=1,(D@@2)(y)(0)=0},y(x));
```

$partsol := $

$$y(x) = 1 - \ln(\cos(x)) - \sin(x)\ln\left(\frac{1 + \sin(x)}{\cos(x)}\right) - \cos(x) + \sin(x)$$

as well as find the solution that satisfies the initial conditions $y(0) = 0$, $y'(0) = 1$, and $y''(0) = 0$, which we then graph with `plot`.

```
> plot(rhs(partsol),x=-Pi/2..Pi/2);
```

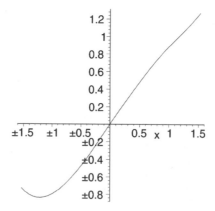

We are often able to use `dsolve` to solve equations directly, which is especially useful when constructing a solution by traditional methods is tedious, time consuming, or both.

EXAMPLE 4: Solve $y'' - 4y = x^{-3} e^{-4x}$. Find the solution that satisfies $y(1) = 0$ and $y'(1) = 1$.

SOLUTION: Entering

```
> gensol:=dsolve(diff(y(x),x$2)-4*y(x)=exp(-4*x)/x^3,y(x)):
> gensol:=simplify(convert(gensol,expsincos));
```

$$gensol := y(x) = \frac{1}{2}e^{(-4x)}(\_C1xe^{(2x)} - \text{Chi}(2x)xe^{(2x)}$$

$$- \_C2xe^{(2x)}x\text{Shi}(2x)xe^{(2x)} + 1 - 9\text{Shi}(6x)xe^{(6x)}$$

$$+ 9\text{Chi}(6x)xe^{6x} + \_C2xe^{(6x)} + \_C1xe^{(6x)})/x$$

finds a general solution of the nonhomogeneous equation and names the result gensol. The solution is very long and so is not displayed because a colon is included at the end of the dsolve command. Instead, we use convert together with the expsincos option and simplify to convert gensol to an expression involving exponentials and simplify the result. Note that the functions Shi and Chi represent the **hyperbolic sine integral** and **hyperbolic cosine integral**, respectively:

$$Shi(x) = \int_0^x \frac{\sinh t}{t}\,dt \quad \text{and} \quad Chi(x) = \gamma + \ln x + \int_0^x \frac{\cosh t - 1}{t}\,dt.$$

Similarly, entering

```
> partsol:=dsolve({diff(y(x),x$2)-4*y(x)=exp(-4*x)/x^3,
                    y(1)=0,D(y)(1)=1},y(x));
```

$$partsol := y(x) = \left(\frac{1}{8}\frac{\cosh(2x)}{x^2} + \frac{1}{4}\frac{\sinh(2x)}{x} - \frac{1}{2}\text{Chi}(2x)\right.$$

$$- \frac{1}{8}\frac{\cosh(6x)}{x^2} - \frac{3}{4}\frac{\sinh(6x)}{x} + \frac{9}{2}\text{Chi}(6x) - \frac{1}{8}\frac{\sinh(2x)}{x^2}$$

$$- \frac{1}{4}\frac{\cosh(2x)}{x} + \frac{1}{2}\text{Shi}(2x) + \frac{1}{8}\frac{\sinh(6x)}{x^2} + \frac{3}{4}\frac{\cosh(6x)}{x}$$

$$\left. - \frac{9}{2}\text{Shi}(6x)\right)\cosh(2x) + \left(-\frac{1}{8}\frac{\cosh(2x)}{x^2} - \frac{1}{4}\frac{\sinh(2x)}{x}\right.$$

$$+ \frac{1}{2}\text{Chi}(2x) - \frac{1}{8}\frac{\cosh(6x)}{x^2} - \frac{3}{4}\frac{\sinh(6x)}{x} + \frac{9}{2}\text{Chi}(6x)$$

$$+ \frac{1}{8}\frac{\sinh(2x)}{x^2} + \frac{1}{4}\frac{\cosh(2x)}{x} - \frac{1}{2}\text{Shi}(2x) + \frac{1}{8}\frac{\sinh(6x)}{x^2}$$

$$\left. + \frac{3}{4}\frac{\cosh(6x)}{x} - \frac{9}{2}\text{Shi}(6x)\right)\sinh(2x) + \frac{1}{8}$$

$$\left(\left(3(e^2)^2 - 4(e^2)^3 \mathrm{Ei}(1,2) - 5 + 36\mathrm{Ei}(1,6)(e^2)^3 - 2(e^2)^4\right)\right.$$

$$\cosh(2x))/(e^2)^3 + \frac{1}{8}$$

$$\left(\left(2(e^2)^4 + (e^2)^3 \mathrm{Ei}(1,2) + 36\mathrm{Ei}(1,6)(e^2)^3 + (e^2)^2 - 5\right)\right.$$

$$\sinh(2x))/(e^2)^3$$

solves the initial-value problem. Note that the Ei function appearing in the result represents the **exponential integral function**.

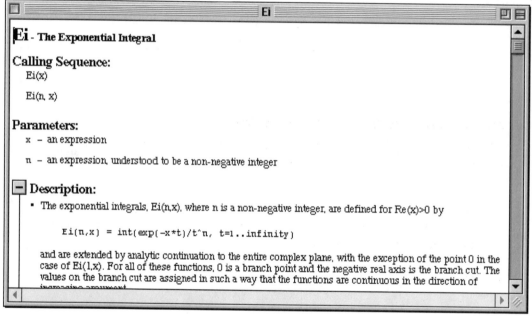

We graph this function with `plot`.

```
> plot(rhs(partsol),x=0..4,y=-1..3);
```

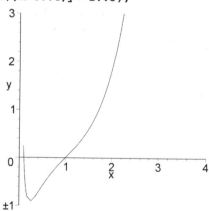

4.6. Cauchy–Euler Equations

Generally, solving an arbitrary differential equation is a formidable if not impossible task, particularly in the case that the coefficients are not constants. However, we are able to solve certain equations with variable coefficients using techniques similar to those discussed previously.

**Definition
Cauchy–Euler
Equation**

A **Cauchy–Euler** differential equation is an equation of the form

$$a_n x^n y^{(n)} + a_{n-1} x^{n-1} y^{(n-1)} + \cdots + a_1 x y' + a_0 y = f(x)$$

where $a_0, a_1, a_2, \ldots, a_n$ are constants.

Second-Order Cauchy–Euler Equations

We begin our study by considering the second-order homogeneous Cauchy–Euler equation

$$ax^2 y'' + bxy' + cy = 0.$$

Notice that because the coefficient of y'' is zero if $x = 0$, we must restrict our domain to either $x > 0$ or $x < 0$ in order to ensure that the theory of second-order equations stated in Section 4.1 holds.

Let $y = x^m$ for some constant m. Substitution of $y = x^m$ with derivatives $y' = mx^{m-1}$ and $y' = m(m-1)x^{m-2}$ yields

$$ax^2 y'' + bxy' + cy = ax^2 m(m-1)x^{m-2} + bxmx^{m-1} + cx^m = x^m(am(m-1) + bm + c) = 0.$$

Then, $y = x^m$ is a solution of $ax^2 y'' + bxy' + cy = 0$ if m satisfies

$$am(m-1) + bm + c = 0,$$

which is called the **auxiliary** (or **characteristic**) **equation** associated with the Cauchy–Euler equation of order two. The solutions of the auxiliary equation completely determine the general solution of the homogeneous Cauchy–Euler equation of order two.

Let m_1 and m_2 denote two solutions of the auxiliary (or characteristic) equation. Notice that the roots of the equation

$$am(m-1) + bm + c = am^2 + (b-a)m + c = 0$$

are

$$m_{1,2} = \frac{-(b-a) \pm \sqrt{(b-a)^2 - 4ac}}{2a}$$

Hence, we can obtain two real roots, one repeated real root, or a complex conjugate pair, depending on the values of a, b, and c. We state a general solution that corresponds to the different types of roots.

1. If $m_1 \neq m_2$ are real, a general solution is

$$y = c_1 x^{m_1} + c_2 x^{m_2}.$$

2. If m_1 and m_2 are real and $m_1 = m_2$, a general solution is

$$y = c_1 x^{m_1} + c_2 x^{m_1} \ln x.$$

3. If $m_1 = \overline{m_2} = \alpha + i\beta$, $\beta \neq 0$, a general solution is

$$y = x^{\alpha}[c_1 \cos(\beta \ln x) + c_2 \sin(\beta \ln x)].$$

EXAMPLE 1: Solve $3x^2 y'' - 2xy' + 2y = 0$, $x > 0$.

SOLUTION: If $y = x^m$, $y' = m x^{m-1}$ and $y'' = m(m-1)x^{m-2}$, substitution into the differential equation yields

$$3x^2 y'' - 2xy' + 2y = 3x^2 m(m-1)x^{m-2} - 2xmx^{m-1} + 2x^m = x^m(3m(m-1) - 2m + 2) = 0.$$

Hence, the auxiliary equation is

$$3m(m-1) - 2m + 2 = 3m^2 - 5m + 2 = (3m - 2)(m - 1) = 0$$

with roots $m_1 = 2/3$ and $m_2 = 1$. Therefore, a general solution is $y = c_1 x^{2/3} + c_2 x$. We obtain the same results with `dsolve`. Entering

```
> y:='y':
gensol:=dsolve(3*x^2*diff(y(x),x$2)-
       2*x*diff(y(x),x)+2*y(x)=0,y(x));
```

$$gensol := y(x) = \_C1\,x + \_C2\,x^{\left(\frac{2}{3}\right)}$$

finds a general solution of the equation, naming the result `gensol`, and then entering

```
> cvals:=seq(-2+2*i,i=0..2):
```

```
toplot:={seq(seq(rhs(gensol),_C1=cvals),_C2=cvals)}:
plot(toplot,x=0..12,y=-6..6);
```

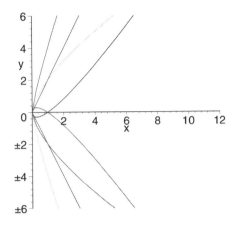

defines `toplot` to be the list of functions obtained by replacing `_C1` in `rhs(gensol)` by $-2, 0,$ and 2 and `_C2` in `rhs(gensol)` by $-2, 0,$ and 2 and graphs the set of functions `toplot` on the interval $[0,12]$.

■

EXAMPLE 2: Solve $x^2 y'' - x y' + y = 0,\ x > 0$.

SOLUTION: In this case, the auxiliary equation is

$$m(m-1) - m + 1 = m^2 - 2m + 1 = (m-1)^2 = 0$$

with root $m_1 = m_2 = 1$ of multiplicity two. Hence, a general solution is $y = c_1 x + c_2 x \ln x$. As in the previous example, we see that we obtain the same results with `dsolve`.

```
> y:='y':
gensol:=dsolve(x^2*diff(y(x),x$2)-
        x*diff(y(x),x)+y(x)=0,y(x));
```

$$gensol := y(x) = \_C1\,x + \_C2\,x\ln(x)$$

```
> toplot:={seq(seq(rhs(gensol),_C1=-1..1),_C2=-1..1)}:
plot(toplot,x=0..10,y=-5..5);
```

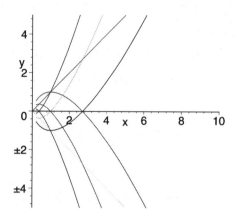

■

EXAMPLE 3: Solve $x^2 y'' - 5xy' + 10y = 0$, $x > 0$.

SOLUTION: The auxiliary equation is given by

$$m(m - 1) - 5m + 10 = m^2 - 6m + 10 = 0$$

with complex conjugate roots $m = \frac{1}{2}(6 \pm \sqrt{36 - 40}) = 3 \pm i$. Thus, a general solution is

$$y = x^3[c_1 \cos(\ln x) + c_2 \sin(\ln x)].$$

Again, we see that we obtain equivalent results with dsolve. First, we find a general solution of the equation, naming the resulting output gensol.

```
> y:='y':
gensol:=dsolve(
      x^2*diff(y(x),x$2)-5*x*diff(y(x),x)+10*y(x)=0,y(x));
```

$$gensol := y(x) = \_C1x^3 \, \sin(\ln(x)) + \_C2x^3 \, \cos(\ln(x))$$

Now, we name $y(x)$ to be the general solution obtained in gensol.

```
> assign(gensol);
```

To find the values of _C1 and _C2 so that the solution satisfies the conditions $y(1) = a$ and $y'(1) = b$, we use `solve` and name the resulting list `cvals`.

```
> cvals:=eval(solve({subs(x=1,y(x))=a,
           subs(x=1,diff(y(x),x))=b},{_C1,_C2}));
```

$$cvals := \{\_C2 = -3\,a + b, \_C1 = a\}$$

The solution to the initial-value problem

$$\begin{cases} x^2 y'' - 5xy' + 10y = 0 \\ \quad y(1) = a, y'(1) = b \end{cases}$$

is obtained by replacing _C1 and _C2 in $y(x)$ by the values found in `cvals`.

```
> sol:=subs(cvals,y(x));
```

$$sol := a\,x^3\,\cos(\ln(x)) + (-3\,a + b)\,x^3\,\sin(\ln(x))$$

This solution is then graphed for various initial conditions.

```
> vals:=seq(-2+2*i,i=0..2):
toplot:={seq(seq(sol,a=vals),b=vals)}:
plot(toplot,x=0..30);
```

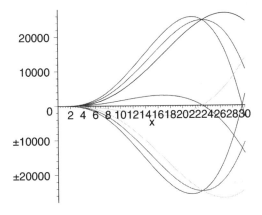

Higher Order Cauchy–Euler Equations

The auxiliary equation of higher order Cauchy–Euler equations is defined in the same way and solutions of higher order homogeneous Cauchy–Euler equations are determined in the same manner as solutions of higher order homogeneous differential equations with constant coefficients.

In the case of higher order Cauchy–Euler equations, note that if a real root r of the auxiliary equation is repeated m times, m linearly independent solutions that correspond to r are x^r, $x^r \ln x$, $x^r (\ln x)^2$, ..., $x^r (\ln x)^{m-1}$; solutions corresponding to repeated complex roots are generated similarly.

EXAMPLE 4: Solve $2x^3 y''' - 4x^2 y'' - 20xy' = 0$, $x > 0$.

SOLUTION: In this case, if we assume that $y = x^m$ for $x > 0$, we have the derivatives $y' = mx^{m-1}$, $y'' = m(m-1)x^{m-2}$, and $y''' = m(m-1)(m-2)x^{m-3}$. Substitution into the differential equation and simplification then yield $\left(2m^3 - 10m^2 - 12m\right)x^m = 0$.

```
> x:='x':y:='y':

eq:=2*x^3*diff(y(x),x$3)-4*x^2*diff(y(x),x$2)-
            20*x*diff(y(x),x)=0:

> y:=x->x^m:

> eq;
```

$$2x^3 \left(\frac{x^m m^3}{x^3} - 3\frac{x^m m^2}{x^3} + 2\frac{x^m m}{x^3} \right)$$

$$- 4x^2 \left(\frac{x^m m^2}{x^2} - \frac{x^m m}{x^3} \right) - 20 x^m m = 0$$

```
> factor(eq);
```

$$2 x^m m(m+1)(m-6) = 0$$

We must solve

$$\left(2m^3 - 10m^2 - 12m\right) = m(2m+2)(m-6) = 0$$

for m because $x^m \neq 0$.

```
> mvals:=solve(factor(eq),m);
```

$$mvals := 0, -1, 6$$

We see that the solutions are $m_1 = 0$, $m_2 = -1$, and $m_3 = 6$, so a general solution is $y(x) = c_1 + c_2 x^{-1} + c_3 x^6$. As in the previous examples, we see that we obtain the same results with `dsolve`.

```
> y:='y':
gensol:=dsolve(eq,y(x));
```

$$gensol := y(x) = \_C1 + \frac{\_C2}{x} + \_\_C3\,x^6$$

We graph this solution for various values of the arbitrary constants in the same way as we graph solutions of other equations.

```
> toplot:={seq(seq(subs(_C1=0,rhs(gensol)),
            _C2=-1..1),_C3=-1..1)}:

plot(toplot,x=0..2,y=-10..10);
```

◼

EXAMPLE 5: Solve the initial-value problem

$$\begin{cases} x^4 y^{(4)} + 4x^3 y''' + 11x^2 y'' - 9xy' + 9y = 0, x > 0 \\ y(1) = 1, y'(1) = -9, y''(1) = 27, y'''(1) = 1 \end{cases}$$

SOLUTION: Substitution of $y = x^m, x > 0$, into the differential equation results in

$$x^4 y^{(4)} + 4x^3 y''' + 11x^2 y'' - 9xy' + 9y = 0$$

and simplification leads to the equation

$$\left(m^4 - m^3 + 8m^2 - 9m - 9\right)x^m = 0.$$

```
> y:='y':
eq:=x^4*diff(y(x),x$4)+4*x^3*diff(y(x),x$3)+
        11*x^2*diff(y(x),x$2)-9*x*diff(y(x),x)+
            9*y(x)=0:
> step_1:=factor(eval(subs(y(x)=x^m,eq)));
```

$$step\_1 := x^m(m^2 + 9)(m - 1)^2 = 0$$

We solve

$$(m^4 - m^3 + 8m^2 - 9m - 9) = (m - 1)^2 = 0$$

for m because $x^m \neq 0$.

```
> step_2:=solve(step_1,m);
```

$$step\_2 := 3I, -3I, 1, 1$$

Hence, $m = \pm 3i$, and $m = 1$ is a root of multiplicity two, so a general solution is

$$y(x) = c_1 \cos(3 \ln x) + c_2 \sin(3 \ln x) + c_3 x + c_4 x \ln x$$

with first, second, and third derivatives computed as follows.

```
> y:=x->c[1]*cos(3*ln(x))+c[2]*sin(3*ln(x))+
     c[3]*x+c[4]*x*ln(x):
> d1:=simplify(diff(y(x),x));
d2:=simplify(diff(y(x),x$2));
d3:=simplify(diff(y(x),x$3));
```

$d1 :=$

$\quad (-3\, c_1 \sin(3\ \ln(x)) + 3\, c_2 \cos(3\ \ln(x)) + c_3\, x + c_4\, x\ln(x) + c_4\, x)/x$

$\quad d2 := (-9\, c_1 \cos(3\ \ln(x)) + 3\, c_1 \sin(3\ \ln(x)) - 9\, c_2 \sin(3\ \ln(x))$

$\qquad - 3\, c_2 \cos(3\ \ln(x)) + c_4\, x)/x^2$

$\quad d3 := -(-21\, c_1 \sin(3\ln(x)) - 27\, c_1 \cos(3\ln(x))$

$\qquad + 21\, c_2 \cos(3\ \ln(x)) - 27\, c_2\ \sin(3\ln(x)) + c_4\, x)/x^3$

Substitution of the initial conditions then yields the system of equations

$$\{c_1 + c_3 = 1, c_3 + c_4 + 3c_2 = -9, c_4 - 9c_1 - 3c_2 = 27, -c_4 + 27c_1 - 21c_2 = 1\}$$

```
> cvals:=solve({y(1)=1,D(y)(1)=-9,
          (D@@2)(y)(1)=27,(D@@3)(y)(1)=1});
```

$$cvals := \left\{ c_3 = \frac{17}{5}, c_1 = \frac{-12}{5}, c_4 = \frac{-7}{2}, c_2 = \frac{-89}{30} \right\}$$

which has the solution $\{c_1 = -12/5, c_2 = -89/30, c_3 = 17/5, c_4 = -7/2\}$.

Therefore, the solution to the initial-value problem is $y(x) = -\frac{12}{5}\cos(3 \ln x) - \frac{89}{30}\sin(3 \ln x) + \frac{17}{5}x - \frac{7}{2}x \ln x$.

```
> sol:=subs(cvals,y(x));
```

$$sol := -\frac{12}{5}\cos(3\ln(x)) - \frac{89}{30}\sin(3\ln(x)) + \frac{17}{5}x - \frac{7}{2}x\ln(x)$$

We graph this solution with `plot`. From the graph produced, we see that it *might appear* to be the case that $\lim_{x \to 0^+} y(x)$ exists.

```
> plot(sol,x=0..1);
```

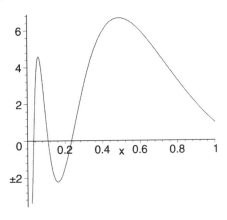

However, when we graph the solution on "small" intervals close to the origin, we see that $\lim_{x \to 0^+} y(x)$ does not exist.

```
> plot(sol,x=0..0.1);

plot(sol,x=0..0.001);

plot(sol,x=0..0.00001);
```

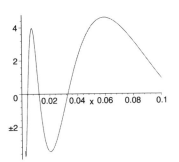
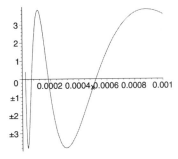
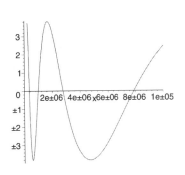

As expected, we see that `dsolve` can be used to solve the initial-value problem directly.

```
> y:='y':
dsolve({Eq,y(1)=1,D(y)(1)=-9,(D@@2)(y)(1)=27,
        (D@@3)(y)(1)=1},y(x));
```

$$y(x) = -\frac{12}{5}\cos(3\ln(x)) - \frac{89}{30}\sin(3\ln(x)) + \frac{17}{5}x - \frac{7}{2}x\ln(x)$$

Variation of Parameters

Of course, Cauchy–Euler equations can be nonhomogeneous, in which case the method of variation of parameters can be used to solve the problem. Note that the command `eulersols`, which is contained in the **DEtools** package, can be used to find a fundamental set of solutions to the corresponding homogeneous equation of a nonhomogeneous Cauchy–Euler equation.

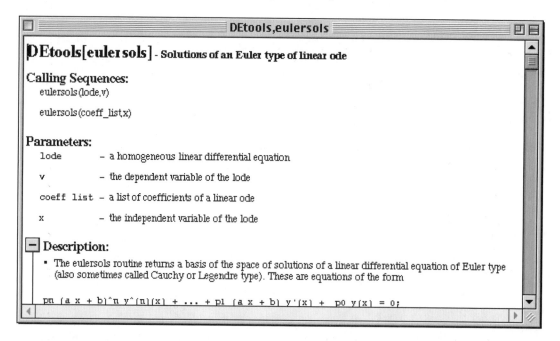

EXAMPLE 6: Solve $x^2 y'' - xy' + 5y = x,\ x>0$.

SOLUTION: We first note that `dsolve` can be used to find a general solution of the equation directly.

```
> x:='x':y:='y':
gensol:=dsolve(x^2*diff(y(x),x$2)-
              x*diff(y(x),x)+5*y(x)=x,y(x));
```

$$gensol := y(x) = \frac{1}{4}x + \_C1\ x\cos(2\ln(x)) + \_C2\ x\sin(2\ln(x))$$

Alternatively, we can use Maple to help us implement variation of parameters. We begin by finding a general solution to the corresponding homogeneous equation $x^2 y'' - x y' + 5y = 0$.

```
> homsol:=dsolve(x^2*diff(y(x),x$2)-x*diff(y(x),x)+
       5*y(x)=0,y(x));
```

$$homsol := y(x) = \_C1\ x\cos(2\ln(x)) + \_C2\ x\ \sin(2\ln(x))$$

We see that a general solution of the corresponding homogeneous equation is $y_h(x) = x\left(c_1 \cos\left(2\ln x\right) + c_2 \sin\left(2\ln x\right)\right)$. A fundamental set of solutions for the homogeneous equation is $S = \{x \cos\left(2\ln x\right), x \sin\left(2\ln x\right)\}$, which is confirmed using `eulersols`.

```
> S:=eulersols(x^2*diff(y(x),x$2)-
         x*diff(y(x),x)+5*y(x)=0,y(x));
```

$$S := [x\ \cos(2\ \ln(x)), x\ \sin(2\ \ln(x))]$$

The Wronskian is $W(S) = 2x$.

```
> with(linalg):
ws:=simplify(det(wronskian(S,x)));
```

$$ws := 2x$$

To implement variation of parameters, we rewrite the equation in standard form, $y'' - \frac{1}{x}y' + \frac{5}{x^2}y = \frac{1}{x}$, and identify $f(x) = 1/x$. We then use `int` to compute $u_1(x) = \int \frac{-f(x)\,y_2(x)}{2x}\,dx$ and $u_2(x) = \int \frac{f(x)\,y_1(x)}{2x}\,dx$.

```
> f:=x->1/x:
```

```
> u1prime:=-S[2]*f(x)/ws:
```

```
u2prime:=S[1]*f(x)/ws:
> u1:=x->int(u1prime,x):
u1(x);
```

$$\frac{1}{4}\cos(2\,\ln(x))$$

```
> u2:=x->int(u2prime,x):
u2(x);
```

$$\frac{1}{4}\sin(2\ln(x))$$

A particular solution of the nonhomogeneous equation is given by $y_p(x) = y_1(x)u_1(x) + y_2(x)u_2(x)$

```
> yp:=x->S[1]*u1(x)+S[2]*u2(x):
simplify(yp(x));
```

$$\frac{1}{4}x$$

and a general solution is given by $y(x) = y_h(x) + y_p(x)$.

```
> yh:=x->c[1]*S[1]+c[2]*S[2]:
yh(x);
y:=x->yh(x)+yp(x):
y(x);
```

$$c_1 x\cos(2\ln(x)) + c_2 x\sin(2\ln(x))$$

$$c_1 x\cos(2\ln(x)) + c_2 x\sin(2\ln(x)) + \frac{1}{4}x$$

As in previous examples, we graph this general solution for various values of the arbitrary constants.

```
> toplot:={seq(seq(y(x),c[1]=-1..1),c[2]=-1..1)}:
plot(toplot,x=0..2);
```

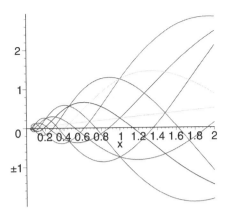

4.7 Series Solutions of Ordinary Differential Equations

Many users will find numerical solutions obtained using dsolve together with the numeric option adequate for most purposes. In some situations, however, a series solution may be useful, and in those cases it can be obtained using dsolve together with the 'type = series' option.

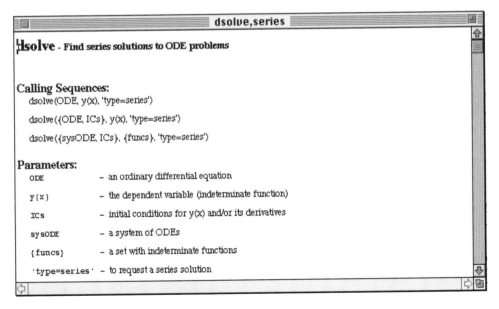

Series Solutions about Ordinary Points

Consider the equation $a_2(x)\,y'' + a_1(x)\,y' + a_0(x)y = 0$ and let $p(x) = a_1(x)/a_2(x)$ and $q(x) = a_0(x)/a_2(x)$. Then, $a_2(x)\,y'' + a_1(x)\,y' + a_0(x)y = 0$ is equivalent to $y'' + p(x)\,y' + q(x)y = 0$, which is called the **standard form** of the equation. A number x_0 is an **ordinary point** of the differential equation means that both $p(x)$ and $q(x)$ are analytic at x_0. If x_0 is not an ordinary point, x_0 is called a **singular point**.

If x_0 is an ordinary point of the differential equation $y'' + p(x)\,y' + q(x)y = 0$, we can write $p(x) = \sum\limits_{n=0}^{\infty} b_n(x - x_0)^n$, where $b_n = \frac{p^{(n)}(x_0)}{n}$, and $q(x) = \sum\limits_{n=0}^{\infty} c_n(x - x_0)^n$, where $c_n = \frac{q^{(n)}(x_0)}{n}$. Substitution into the equation $y'' + p(x)\,y' + q(x)y = 0$ results in

$$y'' + y' \sum_{n=0}^{\infty} b_n(x - x_0)^n + y \sum_{n=0}^{\infty} c_n(x - x_0)^n = 0.$$

If we assume that y is analytic at x_0, we can write $y(x) = \sum\limits_{n=0}^{\infty} a_n(x - x_0)^n$. Because a power series can be differentiated term by term, we can compute the first and second derivatives of y and substitute back into the equation to calculate the coefficients a_n. Thus, we obtain a power series solution of the equation.

EXAMPLE 1: Let k and m be fixed integers. **Legendre's equation** is the equation

$$(1 - x^2)\frac{d^2y}{dx^2} - 2x\frac{dy}{dx} + \left[k(k+1) - \frac{m^2}{1 - x^2}\right]y = 0.$$

If $m = 0$, Legendre's equation becomes $(1 - x^2)\frac{d^2y}{dx^2} - 2x\frac{dy}{dx} + k(k+1)y = 0$. Find a general solution to Legendre's equation if $m = 0$.

SOLUTION: In standard form, the equation is $y'' - \frac{2x}{1 - x^2}y' + \frac{k(k+1)}{1 - x^2} = 0$. There is a solution to the equation of the form $y = \sum\limits_{n=0}^{\infty} a_n x^n$ because $x = 0$ is an ordinary point. This solution will converge at least on the interval $(-1, 1)$ because the closest singular points to $x = 0$ are $x = \pm 1$. Substituting of this function and its derivatives $y' = \sum\limits_{n=0}^{\infty} (n+1)a_{n+1}x^n$ and $y'' = \sum\limits_{n=0}^{\infty} (n+1)(n+2)a_{n+2}x^n$ into the differential equation and simplifying the results yield

$$[2a_2 + k(k+1)a_0]x^0 + [-2a_1 + k(k+1)a_1 + 6a_3]x$$

$$+ \sum_{n=4}^{\infty} n(n-1)a_n x^{n-2} - \sum_{n=2}^{\infty} n(n-1)a_n x^n - \sum_{n=2}^{\infty} 2na_n x^n + \sum_{n=2}^{\infty} k(k+1)a_n x^n = 0.$$

After substituting $n + 2$ for each occurrence of n in the first series and simplifying, we have

$$[2a_2 + k(k+1)a_0]x^0 + [-2a_1 + k(k+1)a_1 + 6a_3]x$$

$$+ \sum_{n=2}^{\infty} \{(n+2)(n+1)a_{n+2} + [-n(n-1) - 2n + k(k+1)]a_n\}x^n = 0.$$

Equating the coefficients to zero, we find a_2, a_3, and a_{n+2} with `solve`.

```
solve(2*a[2]+k*(k+1)*a[0]=0,a[2]);
```

$$-\frac{1}{2}k(k+1)a_0$$

```
> solve(-2*a[1]+k*(k+1)*a[1]+6*a[3]=0,a[3]);
```

$$\frac{1}{3}a_1 - \frac{1}{6}k^2 a_1 - \frac{1}{6}ka_1$$

```
> genform:=solve((n+1)*(n+1)*a[n+2]+
        (-n*(n-1)-2*n+k*(k+1))*a[n]=0,a[n+2]);
```

$$genform := -\frac{(-n^2 - n + k^2 + k)a_n}{(n+1)^2}$$

We obtain a formula for a_n by replacing each occurrence of n in a_{n+2} by $n - 2$

```
> subs(n=n-2,genform);
```

$$-\frac{(-(n-2)^2 - n + 2 + k^2 + k)a_n - 2}{(n-1)^2}$$

Notice how we use the `remember` option in the procedure that defines the recursively defined function a.

```
> a:= proc(n) option remember;
        -(-(n-2)^2-n+2+k^2+k)*a(n-2)/((n-1)^2) end:
```

Using this formula, we find several coefficients with `seq`.

```
> a(1):=a[1]:
a(0):=a[0]:
array([seq([i,a(i)],i=2..10
```

$[2, -(k^2 + k)a_0]$

$\left[3, -\frac{1}{4}(-2 + k^2 + k)a_1\right]$

$\left[4, \frac{1}{9}(-6 + k^2 + k)(k^2 + k)a_0\right]$

$\left[5, \frac{1}{64}(-12 + k^2 + k)(-2 + k^2 + k)a_1\right]$

$\left[6, -\frac{1}{225}(-20 + k^2 + k)(-6 + k^2 + k)(k^2 + k)a_0\right]$

$\left[7, -\frac{1}{2304}(-30 + k^2 + k)(-12 + k^2 + k)(-2 + k^2 + k)a_1\right]$

$\left[8, \frac{1}{11025}(-42 + k^2 + k)(-20 + k^2 + k)(-6 + k^2 + k)(k^2 + k)a_0\right]$

$\left[9, \frac{1}{147456}\right.$

$(-56 + k^2 + k)(-30 + k^2 + k)(-12 + k^2 + k)(-2 + k^2 + k)a_1]$

$\left[10, -\frac{1}{893025}(-72 + k^2 + k)(-42 + k^2 + k)(-20 + k^2 + k)\right.$

$(-6 + k^2 + k)(k^2 + k)a_0]$

Hence, we have the two linearly independent solutions

$$y_1(x) = a_0\left(1 - \frac{k(k+1)}{2!}x^2 + \frac{(2-k)(3+k)k(k+1)}{4!}x^4 - \frac{(4-k)(5+k)(2-k)(3+k)k(k+1)}{6!}x^6 + \cdots\right)$$

and

$$y_2(x) = a_1\left(x - \frac{(k-1)(k+2)}{6}x^3 + \frac{(3-k)(4+k)(k-1)(k+2)}{5!}x^5 \right.$$
$$\left. - \frac{(5-k)(6+k)(3-k)(4+k)(k-1)(k+2)}{7!}x^7 + \cdots\right),$$

so a general solution is

$$y = a_0\left(1 - \frac{k(k+1)}{2!}x^2 + \frac{(2-k)(3+k)k(k+1)}{4!}x^4 - \frac{(4-k)(5+k)(2-k)(3+k)k(k+1)}{6!}x^6 + \cdots\right)$$
$$+ a_1\left(x - \frac{(k-1)(k+2)}{6}x^3 + \frac{(3-k)(4+k)(k-1)(k+2)}{5!}x^5 \right.$$
$$\left. - \frac{(5-k)(6+k)(3-k)(4+k)(k-1)(k+2)}{7!}x^7 + \cdots\right).$$

We see that `dsolve` is able to find a general solution of the equation

```
> sola:=dsolve((1-x^2)*diff(y(x),x$2)-
         2*x*diff(y(x),x)+k*(k+1)*y(x)=0,y(x));
```

$$sola := y(x) = \_C1 \, \text{LegendreP}(k, \, x) + \_C2 \, \text{LegendreQ}(k, \, x)$$

as well as a series solution when we include the option `'type=series'`.

```
> solb:=dsolve((1-x^2)*diff(y(x),x$2)-
         2*x*diff(y(x),x)+k*(k+1)*y(x)=0,y(x),
         'type=series');
```

$$solb := y(x) = y(0) + D(y)(0) \, x + \left(-\frac{1}{2}k^2 \, y(0) - \frac{1}{2}k \, y(0) \right) x^2$$

$$+ \left(-\frac{1}{6} k \, D(y)(0) - \frac{1}{6} k^2 \, D(y)(0) + \frac{1}{3} D(y)(0) \right) x^3$$

$$+ \left(-\frac{5}{24} k^2 \, y(0) + \frac{1}{24} k^4 \, y(0) + \frac{1}{12} k^3 \, y(0) - \frac{1}{4} k \, y(0) \right) x^4.$$

$$+ \left(\frac{1}{60} k^3 \, D(y)(0) + \frac{1}{120} k^4 \, D(y)(0) - \frac{13}{120} k^2 \, D(y)(0) - \frac{7}{60} k \, D(y)(0) \right.$$

$$\left. + \frac{1}{5} D(y)(0) \right) x^5 + O(x^6)$$

In the first result, `LegendreP` and `LegendreQ` represent the Legendre functions of the first and second kind, respectively, and are linearly independent solutions to the equation.

■

An interesting observation from the general solution to Legendre's equation is that the series solutions terminate for integer values of k. If k is an even integer, the first series terminates, whereas if k is an odd integer the second series terminates. Therefore, polynomial solutions are found for integer values of k.

We compute these polynomial solutions for several values of k using a Maple spreadsheet inserted into the Maple notebook. Spreadsheets can be inserted into a Maple notebook by going to **Insert** followed by **Spreadsheet**.

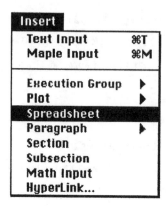

Once a spreadsheet has been inserted, input into cells can be entered manually. In addition, references can be made to other cells. For example, in the following spreadsheet, we manually entered n, 1, 2, 3, and 4 in the first five cells in column A.

	A	B	C
1	n	$(1-x^2)\left(\frac{\partial^2}{\partial x^2}y(x)\right) - 2x\left(\frac{\partial}{\partial x}y(x)\right) + n(n+1)y(x) = 0$	$y(x) = \_C1\,\text{LegendreP}(n,x) + \_C2\,\text{LegendreQ}(n,x)$
2	1	$(1-x^2)\left(\frac{\partial^2}{\partial x^2}y(x)\right) - 2x\left(\frac{\partial}{\partial x}y(x)\right) + 2y(x) = 0$	$y(x) = \_C1\,x + \_C2\left(\frac{1}{2}x\ln\left(-\frac{x+1}{-1+x}\right) - 1\right)$
3	2	$(1-x^2)\left(\frac{\partial^2}{\partial x^2}y(x)\right) - 2x\left(\frac{\partial}{\partial x}y(x)\right) + 6y(x) = 0$	$y(x) = \_C1\,(3x^2-1)$ $+ \_C2\left(\frac{3}{4}\ln\left(-\frac{x+1}{-1+x}\right)x^2 - \frac{1}{4}\ln\left(-\frac{x+1}{-1+x}\right) - \frac{3}{2}x\right)$
4	3	$(1-x^2)\left(\frac{\partial^2}{\partial x^2}y(x)\right) - 2x\left(\frac{\partial}{\partial x}y(x)\right) + 12y(x) = 0$	$y(x) = \_C1\,(5x^3-3x)$ $+ \_C2\left(\frac{5}{4}\ln\left(-\frac{x+1}{-1+x}\right)x^3 - \frac{3}{4}x\ln\left(-\frac{x+1}{-1+x}\right) - \frac{5}{2}x^2 + \frac{2}{3}\right)$
5	4	$(1-x^2)\left(\frac{\partial^2}{\partial x^2}y(x)\right) - 2x\left(\frac{\partial}{\partial x}y(x)\right) + 20y(x) = 0$	$y(x) = \_C1\,(35x^4-30x^2+3) + \_C2\Big($ $\frac{35}{16}\ln\left(-\frac{x+1}{-1+x}\right)x^4 - \frac{15}{8}\ln\left(-\frac{x+1}{-1+x}\right)x^2$ $+ \frac{3}{16}\ln\left(-\frac{x+1}{-1+x}\right) - \frac{35}{8}x^3 + \frac{55}{24}x\Big)$

The contents of the cell in the row 1 and column B are

$$(1 - x^2)^*\text{diff}(y(x),'\$'(x,2)) - 2^*x^*\text{diff}(y(x),x) + {\sim}A1^*({\sim}A1+1)^*y(x) = 0$$

Note that ${\sim}\texttt{A1}$ is a relative reference to the first entry in the row. Similarly, the contents of the cell in the first row and column C are

$$\text{dsolve}({\sim}B1, y(x))$$

and $\sim$ B1 is a relative reference to the second entry in the row.

When we select the cells in rows 1 through 5 and columns B and C and then go to **Spreadsheet** and select **Fill** followed by Down, the formulas are pasted into each cell and the differential equation is solved.

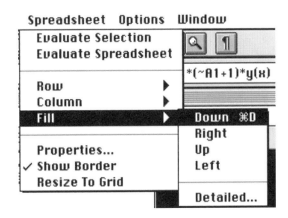

Thus, column A represents the value of n, column B the equation $(1 - x^2)\frac{d^2y}{dx^2} - 2x\frac{dy}{dx} + n(n+1)y = 0$, and column C its solution.

Because these polynomials are useful and are encountered in numerous applications, we have a special notation for them: $P_n(x)$ is called the **Legendre polynomial of degree n** and represents an nth-degree polynomial solution to Legendre's equation. The Maple command P(n,x), which is contained in the orthopoly package, returns $P_n(x)$. After loading the orthopoly package, we use seq together with P to list the first few Legendre polynomials.

```
> with(orthopoly):

toplot:=seq(P(n,x),n=0..5);
```

toplot :=

$$1, x, \frac{3}{2}x^2 - \frac{1}{2}, \frac{5}{2}x^3 - \frac{3}{2}x, \frac{35}{8}x^4 - \frac{15}{4}x^2 + \frac{3}{8}, \frac{63}{8}x^5 - \frac{35}{4}x^3 + \frac{15}{8}x$$

We graph these polynomials for $-2 \le x \le 2$.

```
> plot({toplot},x=-2..2,y=-2..2);
```

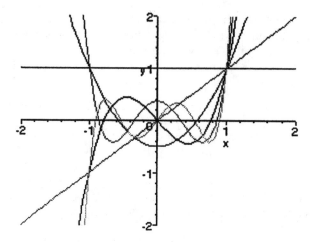

Another interesting observation is about the Legendre polynomials is that they satisfy the relationship $\int_{-1}^{1} P_m(x)P_n(x)dx = 0$, $m \neq n$, called an **orthogonality condition**, which we verify with int for $m, n = 0, 1, \ldots, 6$.

```
array([seq([seq(int(P(n,x)*P(m,x),
          x=-1..1),n=0..6)],m=0..6)]);
```

$$\begin{vmatrix} 2 & 0 & 0 & 0 & 0 & 0 & 0 \\ 0 & \frac{2}{3} & 0 & 0 & 0 & 0 & 0 \\ 0 & 0 & \frac{2}{5} & 0 & 0 & 0 & 0 \\ 0 & 0 & 0 & \frac{2}{7} & 0 & 0 & 0 \\ 0 & 0 & 0 & 0 & \frac{2}{9} & 0 & 0 \\ 0 & 0 & 0 & 0 & 0 & \frac{2}{11} & 0 \\ 0 & 0 & 0 & 0 & 0 & 0 & \frac{2}{13} \end{vmatrix}$$

Note that the entries down the diagonal of this result correspond to the value of $\int_{-1}^{1} [P_n(x)]^2 dx$ for $n = 0, 1, \ldots, 6$ and indicate that $\int_{-1}^{1} [P_n(x)]^2 dx = \frac{2}{2n+1}$.

Regular and Irregular Singular Points and the Method of Frobenius

Let x_0 be singular point of $y'' + p(x)y' + q(x)y = 0$. x_0 is a **regular singular point** of the equation if both $(x - x_0)p(x)$ and $(x - x_0)^2 q(x)$ are analytic at $x = x_0$. If x_0 is not a regular singular point, then x_0 is called an **irregular singular point** of the equation.

Theorem (**Method of Frobenius**)	Let x_0 be a regular singular point of $y'' + p(x)y' + q(x)y = 0$. Then this differential equation has at least one solution of the form $$y = \sum_{n=0}^{\infty} a_n(x - x_0)^{n+r},$$ where r is a constant that must be determined. This solution is convergent at least on some interval $|x - x_0| < r$, $R > 0$.

Suppose that $x = 0$ is a regular singular point of the differential equation $y'' + p(x)y' + q(x)y = 0$. Then the functions $x\,p(x)$ and $x^2 q(x)$ are analytic, which means that both of these functions have a power series in x with a positive radius of convergence. Hence,

$$xp(x) = p_0 + p_1 x + p_2 x^2 + \cdots \quad \text{and} \quad x^2 q(x) = q_0 + q_1 x + q_2 x^2 + \cdots.$$

Therefore,

$$p(x) = \frac{p_0}{x} + p_1 + p_2 x + p_3 x^2 + p_4 x^3 + \cdots \quad \text{and} \quad q(x) = \frac{q_0}{x^2} + \frac{q_1}{x} + q_2 + q_3 x + q_4 x^2 + q_5 x^3 + \cdots.$$

Substituting these series into the differential equation $y'' + p(x)y' + q(x)y = 0$ and multiplying through by the first term in the power series for $p(x)$ and $q(x)$, we see that the lowest term in the series involves x^{n+r-2}:

$$\left(\sum_{n=0}^{\infty} a_n(n + r)(n + r - 1)x^{n+r-2} \right) + \left(\sum_{n=0}^{\infty} a_n p_0(n + r)x^{n+r-2} \right)$$

$$+ \left(p_1 + p_2 x + p_3 x^2 + p_4 x^3 + \cdots \right)\left(\sum_{n=0}^{\infty} a_n(n + r)x^{n+r-1} \right) + \left(\sum_{n=0}^{\infty} a_n q_0 x^{n+r-2} \right)$$

$$+ \left(\frac{q_1}{x} + q_2 + q_3 x + q_4 x^2 + q_5 x^3 + \cdots \right)\left(\sum_{n=0}^{\infty} a_n x^{n+r} \right) = 0.$$

Then, with $n = 0$, we find that the coefficient of x^{r-2} is

$$-ra_0 + r^2 a_0 + ra_0 p_0 + a_0 q_0 = a_0 \left(r^2 + (p_0 - 1)r + q_0 \right) = a_0(r(r - 1) + p_0 r + q_0).$$

Thus, for any equation of the form $y'' + p(x)y' + q(x)y = 0$ with regular singular point $x = 0$, we have the **indicial equation**

$$r(r - 1) + p_0 r + q_0 = 0.$$

The command `indicialeq`, which is contained in the `DEtools` package, can be used to find the indicial equation.

The values of r that satisfy this equation are called the **exponents** or **indicial roots** and are

$$r_1 = \frac{1}{2}\left(1 - p_0 + \sqrt{1 - 2p_0 + p_0^2 - 4q_0} \right) \quad \text{and} \quad r_2 = \frac{1}{2}\left(1 - p_0 - \sqrt{1 - 2p_0 + p_0^2 - 4q_0} \right).$$

Note that $r_1 \geq r_2$ and $r_1 - r_2 = \sqrt{1 - 2p_0 + p_0^2 - 4q_0}$.
Several situations can arise when finding the roots of the indicial equation.

1. If $r_1 \neq r_2$ and $r_1 - r_2 = \sqrt{1 - 2p_0 + p_0^2 - 4q_0}$ is not an integer, then there are two linearly independent solutions of the equation of the form $y_1(x) = x^{r_1} \sum_{n=0}^{\infty} a_n x^n$ and $y_2(x) = x^{r_2} \sum_{n=0}^{\infty} b_n x^n$.

2. If $r_1 \neq r_2$ and $r_1 - r_2 = \sqrt{1 - 2p_0 + p_0^2 - 4q_0}$ is an integer, then there are two linearly independent solutions of the equation of the form $y_1(x) = x^{r_1} \sum_{n=0}^{\infty} a_n x^n$ and $y_2(x) = c y_1(x) \ln x + x^{r_2} \sum_{n=0}^{\infty} b_n x^n$.

3. If $r_1 - r_2 = \sqrt{1 - 2p_0 + p_0^2 - 4q_0} = 0$, then there are two linearly independent solutions of the problem of the form $y_1(x) = x^{r_1} \sum_{n=0}^{\infty} a_n x^n$ and $y_2(x) = y_1(x) \ln x + x^{r_1} \sum_{n=0}^{\infty} b_n x^n$.

In any case, if $y_1(x)$ is a solution of the equation, a second linearly independent solution is given by

$$y_2(x) = y_1(x) \int \frac{e^{-\int p(x)dx}}{[y_1(x)]^2} \, dx$$

which can be obtained through reduction of order.

EXAMPLE 2: Bessel's equation (of order μ) is the equation

$$x^2 y'' + xy' + (x^2 - \mu^2)y = 0$$

where $\mu \geq 0$ is a constant. Solve Bessel's equation.

SOLUTION: To use a power series method to solve Bessel's equation, we first write the equation in standard form as

$$y'' + \frac{1}{x} y' + \frac{x^2 - \mu^2}{x^2} y = 0.$$

so $x = 0$ is a regular singular point. Using the method of Frobenius, we assume that there is a solution of the form $y = \sum_{n=0}^{\infty} a_n x^{n+r}$. We determine the value(s) of r with the indicial equation. Because $xp(x) = x \cdot 1/x = 1$ and $x^2 q(x) = x^2 \cdot (x^2 - \mu^2)/x^2 = x^2 - \mu^2$, $p_0 = 1$ and $q_0 = -\mu^2$. Hence, the indicial equation is

$$r(r - 1) + p_0 r + q_0 = r(r - 1) + r - \mu^2 = r^2 - \mu^2 = 0$$

with roots $r_1 = \mu$ and $r_2 = -\mu$, which we confirm using `indicialeq`.

```
> eq:=x^2*diff(y(x),x$2)+x*diff(y(x),x)+
            (x^2-mu^2)*y(x)=0:
with(DEtools):
indicialeq(eq,x,0,y(x));
```

$$x^2 - \mu^2 = 0$$

Therefore, we assume that $y = \sum\limits_{n=0}^{\infty} a_n x^{n+\mu}$ with derivatives $y' = \sum\limits_{n=0}^{\infty} (n+\mu)a_n x^{n+\mu-1}$ and

$y'' = \sum\limits_{n=0}^{\infty} (n+\mu)(n+\mu-1)a_n x^{n+\mu-2}$. Substituting into Bessel's equation and simplifying the result yield

$$\left[\mu(\mu-1)+\mu-\mu^2\right]a_0 x^\mu + \left[(1+\mu)\mu + (1+\mu) - \mu^2\right]a_1 x^{\mu+1}$$
$$+ \sum\limits_{n=2}^{\infty} \left\{[(n+\mu)(n+\mu-1)+(n+\mu)-\mu^2]a_n + a_{n-2}\right\} x^{n+\mu} = 0.$$

Notice that the coefficient of $a_0 x^\mu$ is zero. After simplifying the other coefficients and equating them to zero, we have $(1+2\mu)a_1 = 0$ and $\left[(n+\mu)(n+\mu-1)+(n+\mu)-\mu^2\right]a_n + a_{n-2} = 0$, which we solve for a_n.

```
> a:='a':
solve(((n+mu)*(n+mu-1)+
            (n+mu)-mu^2)*a[n]+a[n-2]=0,a[n]);
```

$$-\frac{a_n-2}{n(n+2\mu)}$$

From the first equation, $a_1 = 0$. Therefore, from $a_n = -\frac{a_{n-2}}{n(n+2\mu)}$, $n \geq 2$, $a_n = 0$ for all odd n. We use the formula for a_n to calculate several of the coefficients that correspond to even indices.

```
> a:= proc(n) option remember;
            -a(n-2)/(n*(n+2*mu)) end:
a(0):=a[0]:
> nvals:=seq(2*i,i=1..5):
array([seq([n,a(n)],n=nvals)]);
```

$$\begin{bmatrix} 2, & -\frac{1}{2}\frac{a_0}{2+2\mu} \\ 4, & \frac{1}{8}\frac{a_0}{(2+2\mu)(4+2\mu)} \\ 6, & -\frac{1}{48}\frac{a_0}{(2+2\mu)(4+2\mu)(6+2\mu)} \\ 8, & \frac{1}{384}\frac{a_0}{(2+2\mu)(4+2\mu)(6+2\mu)(8+2\mu)} \\ 10, & -\frac{1}{3840}\frac{a_0}{(2+2\mu)(4+2\mu)(6+2\mu)(8+2\mu)(10+2\mu)} \end{bmatrix}$$

A general formula for these coefficients is given by $a_{2n} = \frac{(-1)^n a_0}{2^{2n}(1+\mu)(2+\mu)(3+\mu)\cdots(n+\mu)}$, $n \geq 2$. Our solution can then be written as

$$y = \sum_{n=0}^{\infty} a_{2n}x^{2n+\mu} = \sum_{n=0}^{\infty} \frac{(-1)^n x^{2n+\mu}}{2^{2n}(1+\mu)(2+\mu)(3+\mu)\cdots(n+\mu)} = \sum_{n=0}^{\infty} \frac{(1)^n 2^\mu}{(1+\mu)(2+\mu)(3+\mu)\cdots(n+\mu)}\left(\frac{x}{2}\right)^{2n+\mu}.$$

If μ is an integer, then by using the gamma function $\Gamma(x)$, we can write this solution as

$$y = \sum_{n=0}^{\infty} \frac{(-1)^n}{n!\Gamma(1+\mu+n)}\left(\frac{x}{2}\right)^{2n+\mu}.$$

This function, denoted $J_\mu(x)$, is called the **Bessel function of the first kind of order** μ. The command `BesselJ(mu,x)` returns $J_\mu(x)$.

We use `BesselJ` to graph $J_\mu(x)$ for $\mu = 0$, 1, 2, 3, and 4. Notice that these functions have numerous zeros. We will need to know these values in subsequent sections.

```
> plot({seq(BesselJ(mu,x),mu=0..4)},x=0..10);
```

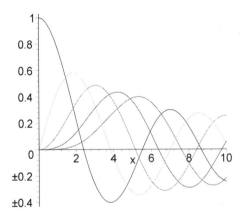

For the other root $r_2 = -\mu$, a similar derivation yields a second solution

$$y = \sum_{n=0}^{\infty} \frac{(-1)^n}{n!\Gamma(1-\mu+n)}\left(\frac{x}{2}\right)^{2n-\mu},$$

which is the **Bessel function of the first kind of order** $-\mu$ and is denoted $J_{-\mu}(x)$. Now, we must determine whether the functions $J_{\mu}(x)$ and $J_{-\mu}(x)$ are linearly independent. Notice that if $\mu = 0$, then these two functions are the same. If $\mu > 0$, then $r_1 - r_2 = \mu - (-\mu) = 2\mu$. If 2μ is not an integer, then by the method of Frobenius, the two solutions $J_{\mu}(x)$ and $J_{-\mu}(x)$ are linearly independent. Also, we can show that if 2μ is an odd integer, then $J_{\mu}(x)$ and $J_{-\mu}(x)$ are linearly independent.

If μ is not an integer, we define the **Bessel function of the second kind of order** μ by the linear combination of the functions $J_{\mu}(x)$ and $J_{-\mu}(x)$. This function, denoted by $Y_{\mu}(x)$, is given by

$$Y_{\mu}(x) = \frac{\cos \mu \pi J_{\mu}(x) - J_{-\mu}(x)}{\sin \mu \pi}.$$

The command `BesselY(m,x)` returns $Y_{\mu}(x)$. We can show that $J_{\mu}(x)$ and $Y_{\mu}(x)$ are linearly independent, so a general solution of Bessel's equation of order μ can be represented by $y = c_1 J_{\mu}(x) + c_2 Y_{\mu}(x)$. In fact, when we use Maple to solve the differential equation, it returns the solution $y = c_1 J_{\mu}(x) + c_2 Y_{\mu}(x)$.

```
> dsolve(eq,y(x));
```

$$y(x) = \_C1 \mathrm{BesselJ}(\mu, x) + \_C2 \mathrm{BesselY}(\mu, x)$$

We use `BesselY` to graph the functions $Y_{\mu}(x)$ for $\mu = 0$, 1, 2, 3, and 4. Notice that $\lim_{x \to 0^+} Y_{\mu}(x) = -\infty$. This property will be important in several applications in later chapters.

```
> plot({seq(BesselY(mu,x),mu=0..4)},x=0..10,y=-9..1);
```

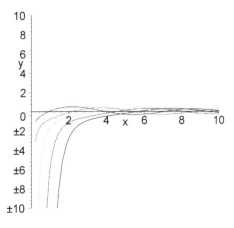

Alternatively, a series solution can be found using `dsolve` together with the
option `'type = series'`.

```
> dsolve(eq,y(x),'type=series');
```

$y(x) =$

$$\_C1x^{(-\mu)}\left(1+\frac{1}{4\mu-4}x^2+\frac{1}{(8\mu-16)(4\mu-4)}x^4+O(x^6)\right)$$

$$+\_C2x^\mu\left(1+\frac{1}{-4\mu-4}x^2+\frac{1}{(-8\mu-16)(-4\mu-4)}x^4+O(x^6)\right)$$

■

CHAPTER 5

Applications of Higher Order Differential Equations

In Chapter 4, we discussed several techniques for solving higher order differential equations. In this chapter, we illustrate how some of these methods can be used to solve initial-value problems that model physical situations.

5.1 Simple Harmonic Motion

Suppose that a mass is attached to an elastic spring that is suspended from a rigid support such as a ceiling. The mass causes the spring to stretch to a distance s from its natural position. The position at which it comes to rest is called the equilibrium position. According to Hooke's law, the spring exerts a restoring force in the upward direction that is proportional to the distance s that the spring is stretched. Mathematically, Hooke's law is stated as

Hooke's Law

$$F = ks$$

where $k > 0$ is the constant of proportionality or **spring constant** and s is the distance that the spring is stretched.

A spring has natural length b. When a mass is attached to the spring, it is stretched s units past its natural length to the equilibrium position $x = 0$. When the system is put into motion, the displacement from $x = 0$ at time t is given by $x(t)$.

By Newton's second law of motion, $F = ma = md^2x/dt^2$, where m represents mass and a represents acceleration. If we assume that there are no other forces acting on the mass, then we determine the differential equation that models this situation in the following way:

$$m\frac{d^2x}{dt^2} = \sum(\text{forces acting on the system}) = -k(s+x) + mg = -ks - kx + mg.$$

At equilibrium $ks = mg$, so after simplification, we obtain the differential equation

$$m\frac{d^2x}{dt^2} = -kx \quad \text{or} \quad m\frac{d^2x}{dt^2} + kx = 0.$$

The two initial conditions that are used with this problem are the initial displacement $x(0) = \alpha$ and initial velocity $\frac{dx}{dt}(0) = \beta$. Hence, the function $x(t)$ that describes the displacement of the mass with respect to the equilibrium position is found by solving the initial-value problem

$$\begin{cases} mx'' + kx = 0 \\ x(0) = \alpha, x'(0) = \beta. \end{cases}$$

The solution $x(t)$ to this problem represents the displacement of the mass at time t. Based on an assumption made in deriving the differential equation (the positive direction is down), positive values of $x(t)$ indicate that the mass is beneath the equilibrium position and negative values of $x(t)$ indicate that the mass is above the equilibrium position.

EXAMPLE 1: A mass weighing 60 lb stretches a spring 6 inches. Determine the function $x(t)$ that describes the displacement of the mass if the mass is released from rest 12 inches below the equilibrium position.

SOLUTION: First, the spring constant k must be determined from the given information. By Hooke's law, $F = ks$, so we have $60 = k \cdot 0.5$. Therefore, $k = 120$ lb/ft. Next, the mass m must be determined using $F = mg$. In this case, $60 = m \cdot 32$, so $m = 15/8$ slugs. Because $k/m = 64$ and 12 inches is equivalent to 1 foot, the initial-value problem that needs to be solved is

$$\begin{cases} \frac{d^2x}{dt^2} + 64x = 0 \\ x(0) = 1, \frac{dx}{dt}(0) = 0 \end{cases}.$$

This problem is now solved with `dsolve`, and the resulting output is named `de1`.

```
> x:='x':
de1:=dsolve({diff(x(t),t$2)+64*x(t)=0,
      x(0)=1,D(x)(0)=0},x(t));
```

$$del := x(t) = \cos(8t)$$

We then graph the solution on the interval $[0, \pi/2]$ with `plot`.

```
> assign(de1):
plot(x(t),t=0..Pi/2);
```

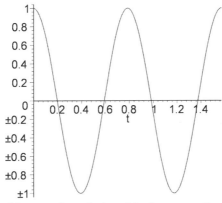

In order to understand better the relationship between the formula just obtained and the motion of the mass on the spring, an alternative approach is taken next. We begin by assigning n and `eps` the values 15 and 0.1.

```
> n:=15:
eps:=0.1:
```

Next, we define the procedure `spring`. Given `t0`, the procedure `spring` first declares the variables `xt0` and `pts` local to the procedure `spring`; then defines `xt0` to be the value of *x(t)* if `t = t0`, `pts` to be the list of points corresponding to (0,xt0),

$$\left(-\text{eps, xt0} + \tfrac{1-\text{xt0}}{n}\right), \left(\text{eps, xt0} + 2\tfrac{1-\text{xt0}}{n}\right), \dots, \left(\text{eps}(-1^{n-1}, \text{xt0} + (n-1)\tfrac{1-\text{xt0}}{n}\right),$$

(0,1); and then uses `plot` to display the list of points `pts`. Note that in the resulting plot, the points are connected with line segments so the result *looks* like a spring.

```
> spring:=proc(t0)
        local xt0,pts;
        xt0:=evalf(subs(t=t0,x(t)));
        pts:=[[0,xt0],
                seq([eps*(-1)^m,xt0+m*(1-xt0)/n],m=1..n-1),
                    [0,1]];
        plot(pts,view=[-1..1,-1.2..1.2],
                xtickmarks=2,ytickmarks=2);
        end:
```

Next, we load the **plots** package, define k_vals to be the list of numbers $k\frac{4}{49}$ for $k = 0, 1, 2, \cdots, 49,$ and then use seq to generate the list of graphs spring(k) for the values of k in k_vals, naming the resulting list of graphics objects to_animate.

```
> with(plots):

k_vals:=seq(k*4/49,k=0..49):

to_animate:=[seq(spring(k),k=k_vals)]:
```

The 50 graphs in to_animate are then animated using the display function, which is contained in the **plots** package, together with the option insequence = true. One frame from the resulting animation is shown in the following screen shot.

```
> display(to_animate,insequence=true);
```

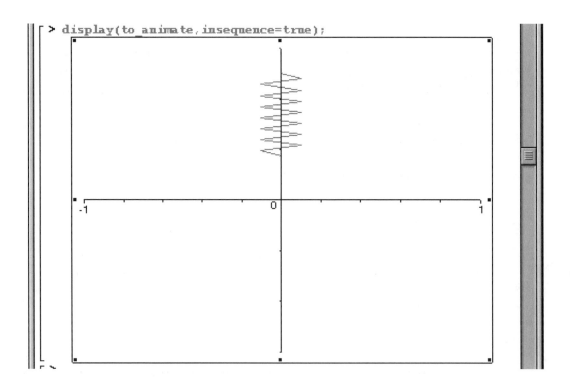

Alternatively, entering

```
> k_vals:=seq(k*4/15,k=0..15):

an_array:=display([seq(spring(k),k=k_vals)],

            insequence=true):
```

```
display(an_array);
```

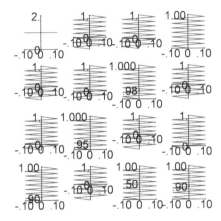

displays the list `an_array` as an array.

Notice that the position function $x(t) = \cos 8t$ indicates that the spring–mass system never comes to rest once it is set into motion. The solution is periodic, so the mass moves vertically, retracing its motion. Hence, motion of this type is called **simple harmonic motion**.

■

EXAMPLE 2: An object with mass $m = 1$ slug is attached to a spring with spring constant $k = 4$. (a) Determine the displacement function of the object if $x(0) = \alpha$ and $x'(0) = 0$. Plot the solution for $\alpha = 1, 4, -2$. How does varying the value of α affect the solution? Does it change the values of t at which the mass passes through the equilibrium position? (b) Determine the displacement function of the object if $x(0) = 0$ and $x'(0) = \beta$. Plot the solution for $\beta = 1, 4, -2$. How does varying the value of β affect the solution? Does it change the values of t at which the mass passes through the equilibrium position?

SOLUTION: For (a), the initial-value problem we need to solve is

$$\begin{cases} x'' + 4x = 0 \\ x(0) = \alpha, x'(0) = 0 \end{cases},$$

for $\alpha = 1, 4, -2$.

We now determine the solution to each of the three problems with `dsolve`. For example, entering

```
> x:='x':
de2:=dsolve({diff(x(t),t$2)+4*x(t)=0,
        x(0)=1,D(x)(0)=0},x(t));
```

$$de2 := x(t) = \cos(2t)$$

solves the initial-value problem if $\alpha = 1$ and names the result `de2`. Note that the formula for the solution can be extracted from `de2` with `rhs` as shown next.

```
> rhs(de2);
```

$$\cos(2t)$$

Similarly, entering

```
> de3:=dsolve({diff(x(t),t$2)+4*x(t)=0,
        x(0)=4,D(x)(0)=0},x(t));
de4:= dsolve({diff(x(t),t$2)+4*x(t)=0,
        x(0)=-2,D(x)(0)=0},x(t));
```

$$de3 := x(t) = 4\cos(2t)$$
$$de4 := x(t) = -2\cos(2t)$$

solves

$$\begin{cases} x'' + 4x = 0 \\ x(0) = 4, x'(0) = 0 \end{cases} \quad \text{and} \quad \begin{cases} x'' + 4x = 0 \\ x(0) = -2, x'(0) = 0 \end{cases},$$

naming the results `de3` and `de4`, respectively. We graph the solutions on the interval $[0,\pi]$ with `plot`. Note how we use `map` to extract the formula for each solution from `de2`, `de3`, and `de4`.

```
> plot(map(rhs,{de2,de3,de4}),t=0..Pi);
```

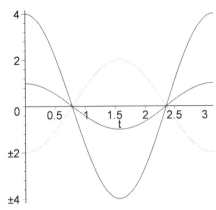

We see that the initial position affects only the amplitude of the function (and direction in the case of the negative initial position). The mass passes through the equilibrium position $(x = 0)$ at the same time in all three cases.

For (b), we need to solve the initial-value problem

$$\begin{cases} x'' + 4x = 0 \\ x(0) = 0, x'(0) = \beta \end{cases}$$

for $\beta = 1, 4, -2$.

In this case, we define a procedure d that, given β, returns the solution to the initial-value problem.

```
> d:=proc(beta)
      dsolve({diff(x(t),t$2)+4*x(t)=0,
            x(0)=0,D(x)(0)=beta},x(t))
  end:
```

We then use map to apply d to the list of numbers [1,4,-2] and name the resulting output sols.

```
> sols:=map(d,[1,4,-2]);
```

$$sols := \left[x(t) = \frac{1}{2}\sin(2t), x(t) = 2\sin(2t), x(t) = -\sin(2t) \right]$$

The explicit formulas for the solutions are obtained by mapping rhs to the list sol.

```
> toplot:=map(rhs,sols);
```

$$toplot := \left[\frac{1}{2}\sin(2t), 2\sin(2t), -\sin(2t) \right]$$

All three solutions are then graphed together on [0,2π] with `plot`.

```
> plot(toplot,t=0..2*Pi);
```

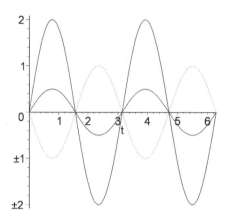

Notice that varying the initial velocity affects the amplitude (and direction in the case of the negative initial velocity) of each function. The mass passes through the equilibrium position ($x = 0$) at the same time in all three cases.

■

5.2 Damped Motion

Because the differential equation derived in Section 5.1 disregarded all retarding forces acting on the motion of the mass, a more realistic model that takes these forces into account is needed. Studies in mechanics reveal that the resistive force due to damping is a function of the velocity of the motion. Therefore for $c > 0$, functions such as

$$F_R = cdx/dt, F_R = c(dx/dt)^3, \text{ and } F_R = c\text{sgn}(dx/dt), \text{ where sgn } (dx/dt) = \begin{cases} 1, dx/dt > 0 \\ 0, dx/dt = 0 \\ -1, dx/dt < 0 \end{cases}$$

can be used to represent the damping force. We follow procedures similar to those used in Section 5.1 to model simple harmonic motion in order to determine a differential equation that models the spring–mass system including damping. Assuming that $F_R = cdx/dt$, we have after summing the forces acting on the spring–mass system

$$m\frac{d^2x}{dt^2} = -c\frac{dx}{dt} - kx \quad \text{or} \quad m\frac{d^2x}{dt^2} + c\frac{dx}{dt} + kx = 0.$$

Thus, the displacement function is found by solving the initial-value problem

$$\begin{cases} m\frac{d^2x}{dt^2} + c\frac{dx}{dt} + kx = 0 \\ x(0) = \alpha, \frac{dx}{dt}(0) = \beta. \end{cases}$$

From our experience with second-order ordinary differential equations with constant coefficients in Chapter 4, the solutions to initial-value problems of this type greatly depend on the values of m, k, and c.

Suppose we assume that solutions of the differential equation have the form $x(t) = e^{rt}$. (Note that m is not used in the exponent as it was in Chapter 4 to avoid confusion with the mass m. Otherwise, this calculation is identical to those followed in Chapter 4.) Because $dx/dt = re^{rt}$ and $d^2x/dt^2 = r^2e^{rt}$, we have by substitution into the differential equation $mr^2e^{rt} + cre^{rt} + ke^{rt} = 0$, so $e^{rt}(mr^2 + cr + k) = 0$.

The solutions to the characteristic equation are

$$r = \frac{-c \pm \sqrt{c^2 - 4mk}}{2a}.$$

Hence, the solution depends on the value of the quantity $c^2 - 4mk$. In fact, problems of this type are characterized by the value of $c^2 - 4mk$ as follows.

Case 1: $c^2 - 4mk > 0$

> This situation is said to be **overdamped** because the damping coefficient c is large in comparison with the spring constant k.

Case 2: $c^2 - 4mk = 0$

> This situation is described as **critically damped** because the resulting motion is oscillatory with a slight decrease in the damping coefficient c.

Case 3: $c^2 - 4mk < 0$

> This situation is called **underdamped** because the damping coefficient c is small in comparison with the spring constant k.

EXAMPLE 1: An 8 lb weight is attached to a spring of length 4 feet. At equilibrium, the spring has length 6 feet. Determine the displacement function if $F_R = 2dx/dt$ and (a) the mass is released from its equilibrium position with a downward initial velocity of 1 ft/sec; (b) the mass released 6 inches above the equilibrium with an initial velocity of 5 ft/sec in the upward direction.

SOLUTION: Notice that $s = 6 - 4 = 2$ and that $F = 4$. Hence, we find the spring constant with $8 = k \cdot 2$, so $k = 4$. Also, the mass of the object is $m = 8/32 = 1/4$ slug. Therefore, the differential equation that models this spring-mass system is

$$\frac{1}{4}\frac{d^2x}{dt^2} + 2\frac{dx}{dt} + 4x = 0 \quad \text{or} \quad \frac{d^2x}{dt^2} + 8\frac{dx}{dt} + 16x = 0.$$

(a) In this case, the initial conditions are $x(0) = 0$ and $\frac{dx}{dt}(0) = 1$. We solve the initial-value problem with dsolve and see that the solution is $x(t) = te^{-4t}$.

```
> x:='x':
eq:=diff(x(t),t$2)+8*diff(x(t),t)+16*x(t)=0:
> de:=dsolve({eq,x(0)=0,D(x)(0)=1},x(t));
```

$$de := x(t) = e^{(-4t)}t$$

The graph of the solution is generated with plot. Notice that it is always positive, so the mass is always below the equilibrium position and approaches zero (the equilibrium position) as t approaches infinity. Because of the resistive forces due to damping, the mass is not allowed to pass through its equilibrium position.

```
> assign(de):
plot(x(t),t=0..3);
```

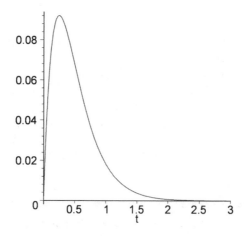

(b) In this case, $x(0) = -1/2$ and $\frac{dx}{dt}(0) = 5$. First, we use `dsolve` to solve the equation

```
> x:='x':
DE:=dsolve({eq,x(0)=-0.5,D(x)(0)=5},x(t));
```

$$DE := x(t) = -0.5000000000\,e^{(-4t)} + 3.\,e^{(-4t)}t$$

and then graph the solution on the interval [0,2].

```
> assign(DE):
plot(x(t),t=0..2);
```

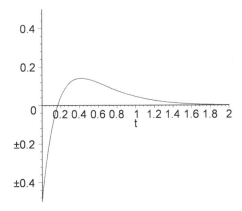

This graph indicates the importance of the initial conditions for the resulting motion. In this case, the displacement is negative (above the equilibrium position) initially, but the positive initial velocity causes the function to become positive (below the equilibrium position) before approaching zero.

■

EXAMPLE 2: A 32 lb weight stretches a spring 8 feet. If the resistive force due to damping is $F_R = 5dx/dt$, then determine the displacement function if the mass is released from 1 foot below the equilibrium position with (a) an upward velocity of 1 ft/sec; (b) an upward velocity of 6 ft/sec.

SOLUTION: Because $F = 32$ lb, the spring constant is found with $32 = k \cdot 8$, so $k = 4$ lb/ft. Also, $m = 32/32 = 1$ slug. Therefore, the differential equation that models this situation is $x'' + 5x' + 4x = 0$. The initial position is $x(0) = 1$ and the initial velocity in (a) is $x'(0) = -1$. Using dsolve,

```
> x:='x':
eq:=diff(x(t),t$2)+5*diff(x(t),t)+4*x(t)=0:
> de:=dsolve({eq,x(0)=1,D(x)(0)=-1},x(t));
```

$$de := x(t) = e^{(-t)}$$

we see that the solution of the initial-value problem is $x(t) = e^{-t}$, which is graphed with plot.

```
> assign(de):
plot(x(t),t=0..5);
```

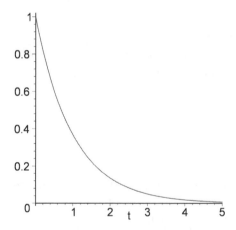

Notice that the solution is always positive and, due to the damping, approaches zero as t approaches infinity. Therefore, the mass is always below its equilibrium position. For (b) we use the initial velocity $x'(0) = -6$ and use dsolve to see that the solution of the initial-value problem is $x(t) = \frac{5}{3}e^{4t} - \frac{2}{3}e^{-t}$.

```
> x:='x':
DEq:=dsolve({eq,x(0)=1,D(x)(0)=-6},x(t));
```

$$de := x(t) = \frac{5}{3}e^{(-4t)} - \frac{2}{3}e^{(-t)}$$

```
> assign(de):

plot(x(t),t=0..5);
```

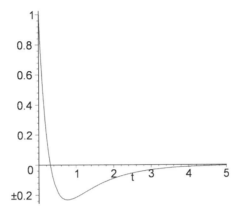

These results indicate the importance of the initial conditions for the resulting motion. In this case, the displacement is positive (below its equilibrium) initially, but the larger negative initial velocity causes the function to become negative (above its equilibrium) before approaching zero. Therefore, we see that the initial velocity in part (b) causes the mass to pass through its equilibrium position.

■

EXAMPLE 3: A 16 lb weight stretches a spring 2 feet. Determine the displacement function if the resistive force due to damping is $F_R = \frac{1}{2} dx/dt$ and the mass is released from the equilibrium position with a downward velocity of 1 ft/sec.

SOLUTION: Because $F = 16$ lb, the spring constant is determined with $16 = k \cdot 2$. Hence, $k = 8$ lb/ft. Also, $m = 16/32 = 1/2$ slug. Therefore, the differential equation is $\frac{1}{2}s'' + \frac{1}{2}x' + 8x = 0$ or $x'' + x' + 16x = 0$. The initial position is $x(0) = 0$ and the initial velocity is $x'(0) = 1$. Thus, we must solve the initial-value problem

$$\begin{cases} \frac{d^2x}{dt^2} + \frac{dx}{dt} + 16x = 0 \\ x(0) = 0, \frac{dx}{dt}(0) = 1 \end{cases},$$

which is now solved with dsolve.

```
> x:='x':
sol:=dsolve({diff(x(t),t$2)+diff(x(t),t)+16*x(t)=0,
      x(0)=0,D(x)(0)=1},x(t));
```

$$sol := x(t) = \frac{2}{21}\sqrt{7}e^{(-\frac{1}{2}t)}\sin\left(\frac{3}{2}\sqrt{7}t\right)$$

Solutions of this type have several interesting properties. First, the trigonometric component of the solution causes the motion to oscillate. Also, the exponential portion forces the solution to approach zero as t approaches infinity. These qualities are illustrated in the following graph.

```
> assign(sol):
Plot_1:=plot(x(t),t=0..2*Pi):
with(plots):
display({Plot_1});
```

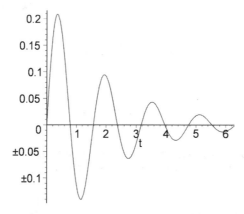

Physically, the displacement of the mass in this case oscillates about the equilibrium position and eventually comes to rest in the equilibrium position. Of course, with our model the displacement function $x(t) \to 0$ as $t \to \infty$, but there is no number T such that $x(t) = 0$ for $t > T$ as we might expect from the physical situation. Hence, our model only approximates the behavior of the mass. Notice also that the solution is bounded above and below by the exponential term of the solution $e^{-t/2}$ and its reflection through the horizontal axis, $-e^{-t/2}$. This is illustrated with the simultaneous display of these functions.

```
> Plot_2:=plot({2/21*7^(1/2)*exp(-1/2*t),
      -2/21*7^(1/2)*exp(-1/2*t)},t=0..2*Pi):
display({Plot_1,Plot_2});
```

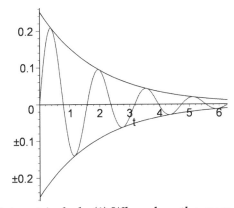

Other questions of interest include (1) When does the mass first pass through its equilibrium point? (2) What is the maximum displacement of the spring?

The time at which the mass passes through $x = 0$ can be determined in several ways. The solution equals zero at the time that $\sin(\frac{3}{2}\sqrt{7}t)$ first equals zero after $t = 0$, which occurs when $\frac{3}{2}\sqrt{7}t = \pi$. We use `solve` to solve this equation for t and then use `evalf` to approximate the time. (Note that `%` refers to the most recent output.)

```
> solve(3/2*sqrt(7)*t=Pi,t);
```

$$\frac{2}{21}\pi\sqrt{7}$$

```
> evalf(%);
```

$$0.7916069413$$

Similarly, the maximum displacement of the spring is found by finding the first value of t for which the derivative of the solution is equal to zero, as done here with `solve`.

```
cp:=solve(diff(x(t),t)=0);
```

$$cp := \frac{2}{21}\arctan(3\sqrt{7})\sqrt{7}$$

An approximation of the result given is obtained with `evalf`.

```
> cp1:=evalf(cp);
```

$$cp1 := 0.3642238255$$

The maximum displacement is then given by evaluating the solution for the value of t obtained with `evalf`.

```
> evalf(subs(t=cp1,x(t)));
```

$$0.2083770137$$

Another interesting characteristic of solutions to undamped problems is the time between successive maxima and minima of the solution, called the **quasiperiod**. This quantity is found by first determining the time at which the second maximum occurs with `fsolve`. Then, the difference between these values of t is taken to obtain the value 1.58321.

```
> cp2:=fsolve(diff(x(t),t)=0,t=1.5..2.0);
```

$$cp2 := 1.947437708$$

```
> cp2-cp1;
```

$$1.583213883$$

To investigate the solution further, an animation can be created with the `spring` command defined in Example 1 in Section 5.1. Here, we redefine `spring`

```
> spring:='spring':
n:=15:eps:=0.1:
spring:=proc(t0)
        local m,xt0,pts;
        xt0:=evalf(subs(t=t0,x(t)));
        pts:=[[0,xt0],
              seq([eps*(-1)^m,xt0+m*(0.25-xt0)/n],m=1..n-1),
                  [0,0.25]];
        plot(pts,view=[-1..1,-0.25..0.25],
              xtickmarks=2,ytickmarks=2);
        end:
```

Next, we load the **plots** package, define k_vals to be the list of numbers $k\frac{4}{49}$ for $k = 0, 1, 2, \ldots, 49$, and then use `seq` to generate the list of graphs `spring(k)` for the values of k in k_vals, naming the resulting list of graphics objects to_animate. The 50 graphs in to_animate are then animated using the `display` function, which is contained in the **plots** package, together with the option insequence = true. One frame from the resulting animation is shown in the following screen shot.

```
> with(plots):
```

```
k_vals:=seq(k*6/59,k=0..59):
```

```
to_animate:=[seq(spring(k),k=k_vals)]:
```

```
> display(to_animate,insequence=true);
```

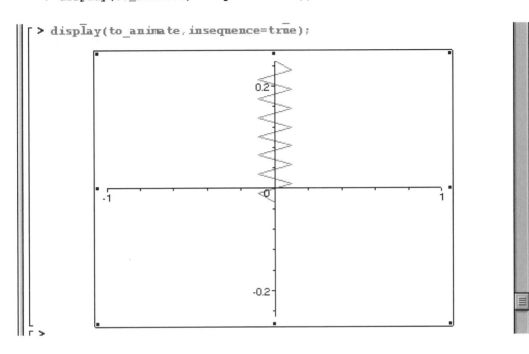

Alternatively, entering

```
> k_vals:=seq(k*4/15,k=0..15):
```

```
an_array:=display([seq(spring(k),k=k_vals)],
                  insequence=true):
```

```
display(an_array);
```

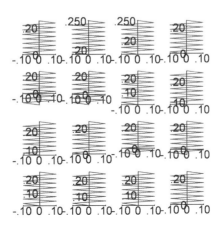

displays the list `an_array` as an array.

■

EXAMPLE 4: Suppose that we have the initial-value problem

$$\begin{cases} x'' + cx' + 6x = 0 \\ x(0) = 0, x'(0) = 1 \end{cases},$$

where $c = 2\sqrt{6}, 4\sqrt{6}$, and $\sqrt{6}$. Determine how the value of c affects the solution of the initial-value problem.

SOLUTION: We begin by defining the function d. Given c, d(c) solves the initial-value problem

$$\begin{cases} x'' + cx' + 6x = 0 \\ x(0) = 0, x'(0) = 1 \end{cases}.$$

(Be sure to use (lowercase) d instead of (uppercase) D to avoid conflict with the built-in function D.)

```
> x:='x':
d:=proc(c)
      dsolve({diff(x(t),t$2)+c*diff(x(t),t)+6*x(t)=0,
          x(0)=0,D(x)(0)=1},x(t))
   end:
```

We then use `map` and `d` to find the solution of the initial-value problem for each value of c, naming the resulting list `somesols`.

```
> somesols:=map(d,[2*sqrt(6),4*sqrt(6),sqrt(6)]);
```

$$somesols := \left[x(t) = te^{(-\sqrt{6}t)} \right.$$

$$x(t) = \frac{1}{72}\sqrt{72}\left(e^{\left(\left(-2\sqrt{6}+\frac{1}{2}\sqrt{72}\right)t\right)} e^{\left(\left(-2\sqrt{6}-\frac{1}{2}\sqrt{72}\right)t\right)} \right),$$

$$\left. x(t) = -\frac{1}{18}\sqrt{-18}\left(e^{\left(\left(-\frac{1}{2}\sqrt{6}+\frac{1}{2}\sqrt{-18}\right)t\right)} - e^{\left(\left(-\frac{1}{2}\sqrt{6}-\frac{1}{2}\sqrt{-18}\right)t\right)} \right) \right]$$

Note that each case results in a different classification: $c = 2\sqrt{6}$, critically damped; $c = 4\sqrt{6}$, overdamped; and $c = \sqrt{6}$, underdamped.

All three solutions are graphed together on the interval [0,4]. Note that because the values of c vary more widely than those considered in the previous example, the behavior of the solutions differs more obviously as well.

```
> plot(map(rhs,somesols),t=0..4);
```

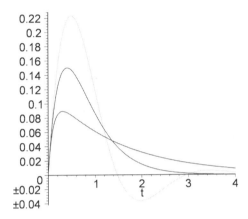

EXAMPLE 5: Consider the system

$$\begin{cases} \frac{d^2x}{dt^2} + f(t)\frac{dx}{dt} + \frac{5}{4}x = 0, t>0 \\ x(0) = 0, \frac{dx}{dt}(0) = 1 \end{cases} \quad , \quad \text{where } f(t) = \begin{cases} 1, 0 \le t < \pi \\ 3, \pi \le t < 2\pi \end{cases} , f(t) = f(t-2\pi),$$

in which damping oscillates periodically: the rate at which energy is taken away from the system fluctuates periodically. Find the displacement for $x(t)$ for $0 \le t \le 4\pi$.

SOLUTION: We solve the initial-value problem

$$\begin{cases} \frac{d^2X}{dt^2} + \frac{dx}{dt} + \frac{5}{4}x = 0 \\ x(0) = 0, \frac{dx}{dt}(0) = 1 \end{cases}$$

for $0 \le t < \pi$ and name the result `sol`. Similarly, we solve

$$\begin{cases} \frac{d^2x}{dt^2} + 3\frac{dx}{dt} + \frac{5}{4}x = 0 \\ x(0) = a, \frac{dx}{dt}(0) = b \end{cases}$$

where a and b are the value of `sol` and the derivative of `sol`, respectively, if $t = \pi$ for $\pi \le t < 2\pi$ and name the result `sol2`.

```
> sol:=dsolve({diff(x(t),t$2)+diff(x(t),t)+5/4*x(t)=0,
        x(0)=0,D(x)(0)=1},x(t));
```

$$sol := x(t) = e^{\left(-\frac{1}{2}t\right)} \sin(t)$$

```
> a:=eval(subs(t=Pi,rhs(sol)));
```

$$a := 0$$

```
> b:=eval(subs(t=Pi,diff(rhs(sol),t)));
```

$$b := -e^{\left(-\frac{1}{2}\pi\right)}$$

```
> sol2:=dsolve({diff(x(t),t$2)+3*diff(x(t),t)+5/4*x(t)=0,
        x(Pi)=a,D(x)(Pi)=b},x(t));
```

$$sol2 := x(t) = -\frac{1}{2}\frac{e^{\left(-\frac{1}{2}\pi\right)}e^{\left(-\frac{1}{2}t\right)}}{e^{(2\pi)}e^{\left(-\frac{5}{2}\pi\right)}} + \frac{1}{2}\frac{e^{\left(-\frac{1}{2}\pi\right)}e^{\left(-\frac{5}{2}t\right)}}{e^{\left(-\frac{5}{2}\pi\right)}}$$

In a similar way, we find the solution for $2\pi \le t < 3\pi$ in `sol3` and the solution for $3\pi < t < 4\pi$ in `sol4`.

```
> a:=eval(subs(t=2*Pi,rhs(sol2)));
```

$$a := \frac{1}{2}\frac{e^{\left(-\frac{1}{2}\pi\right)}e^{(-\pi)}}{e^{(2\pi)}e^{\left(-\frac{5}{2}\pi\right)}} + \frac{1}{2}\frac{e^{\left(-\frac{1}{2}\pi\right)}e^{(-5\pi)}}{e^{\left(-\frac{5}{2}\pi\right)}}$$

```
> b:=eval(subs(t=2*Pi,diff(rhs(sol2),t)));
```

$$b := \frac{1}{4}\frac{e^{\left(-\frac{1}{2}\pi\right)}e^{(-\pi)}}{e^{(2\pi)}e^{\left(-\frac{5}{2}\pi\right)}} - \frac{5}{4}\frac{e^{\left(-\frac{1}{2}\pi\right)}e^{(-5\pi)}}{e^{\left(-\frac{5}{2}\pi\right)}}$$

```
> sol3:=dsolve({diff(x(t),t$2)+diff(x(t),t)+5/4*x(t)=0,
        x(2*Pi)=a,D(x)(2*Pi)=b},x(t)):
sol3a:=convert(sol3,expsincos):
```

```
sol3b:=simplify(sol3a);
```

$$sol3b := x(t) = -\frac{1}{2}e^{\left(-2\pi-\frac{1}{2}t\right)}\left(2\sin(t)+\cos(t)e^{(2\pi)}-\cos(t)\right)$$

```
> a:=eval(subs(t=2*Pi,rhs(sol3))):
b:=eval(subs(t=2*Pi,diff(rhs(sol3),t))):
sol4:=dsolve({diff(x(t),t$2)+3*diff(x(t),t)+5/4*x(t)=0,
        x(2*Pi)=a,D(x)(2*Pi)=b},x(t));
```

$$sol4 := x(t) = -\frac{1}{2}e^{\left(-\frac{1}{2}t\right)}+\frac{1}{2}\left(e^{\left(\frac{1}{2}\pi\right)}\right)^4 e^{\left(-\frac{5}{2}t\right)}$$

We see the damped motion of the system by graphing the pieces of the solution individually and then together with `display`.

```
> p1:=plot(rhs(sol),t=0..Pi):
p2:=plot(rhs(sol2),t=Pi..2*Pi):
p3:=plot(rhs(sol3b),t=2*Pi..3*Pi):
p4:=plot(rhs(sol4),t=3*Pi..4*Pi):
with(plots):
display({p1,p2,p3,p4});
```

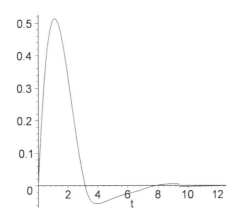

5.3 Forced Motion

In some cases, the motion of the spring is influenced by an external driving force, $f(t)$. Mathematically, this force is included in the differential equation that models the situation as follows:

$$m\frac{d^2x}{dt^2} = -kx - c\frac{dx}{dt} + f(t).$$

The resulting initial-value problem is

$$\begin{cases} m\frac{d^2x}{dt^2} + c\frac{dx}{dt} + kx = f(t) \\ x(0) = \alpha, \frac{dx}{dt}(0) = \beta \end{cases}.$$

Therefore, differential equations modeling forced motion are nonhomogeneous and require the method of undetermined coefficients or variation of parameters for solution. We first consider forced motion that is undamped.

EXAMPLE 1: An object of mass $m = 1$ slug is attached to a spring with spring constant $k = 4$. Assuming there is no damping and that the object is released from rest in the equilibrium position, determine the position function of the object if it is subjected to an external force of (a) $f(t) = 0$, (b) $f(t) = 1$, (c) $f(t) = cos, t$, and (d) $f(t) = sin, t$.

SOLUTION: First, we note that we must solve the initial-value problem

$$\begin{cases} \frac{d^2x}{dt^2} + 4x = f(t) \\ x(0) = 0, \frac{dx}{dt}(0) = 0 \end{cases}$$

for each of the forcing functions in (a), (b), (c), and (d). Because we will be solving this initial-value problem for various forcing functions, we begin by defining the function fm. Given a function $f = f(t)$, fm(f) returns the formula for the solution to the initial-value problem

$$\begin{cases} \frac{d^2x}{dt^2} + 4x = f(t) \\ x(0) = 0, \frac{dx}{dt}(0) = 0 \end{cases}.$$

```
> x:='x':

fm:=proc(f)

        dsolve({diff(x(t),t$2)+4*x(t)=f,x(0)=0,D(x)(0)=0},x(t))

        end:
```

Next, we define fs to be the forcing functions in (a)–(d).

> `fs:=[0,1,cos(t),sin(t)]:`

We then use `map` to apply `fm` to `fs` and name the resulting list of functions `somesols`.

> `somesols:=map(fm,fs):`

`somesols:=combine(somesols,trig);`

$$somesols := \left[x(t) = 0, x(t) = \frac{1}{4} - \frac{1}{4}\cos(2t), \right.$$

$$\left. x(t) = \frac{1}{3}\cos(t) - \frac{1}{3}\cos(2t), x(t) = \frac{1}{3}\sin(t) - \frac{1}{6}\sin(2t) \right]$$

From the result, we see that for (a) the solution is $x(t) = 0$. Physically, this solution indicates that because there is no forcing function, no initial displacement from the equilibrium position, and no initial velocity, the object does not move from the equilibrium position.
These three functions are then graphed on the interval $[0, 2\pi]$ with `plot`.

> `plot(map(rhs,somesols),t=0..2*Pi);`

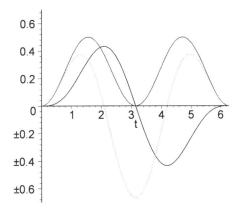

From the graph, we see that for (b) the object never moves above the equilibrium position. (Negative values of x indicate that the mass is *above* the equilibrium position; positive values indicate that the mass is *below* the equilibrium position.) This makes sense because $0 \le \cos 2t \le 1$: $x(t) = \frac{1}{4}(1 - \cos 2t) \ge 0$ for all t. For (c), we see that the mass passes through the equilibrium position twice (near $t = 2$ and $t = 4$) over the period. For (d), we again see that the resulting motion is periodic, although different from that observed in (c).

∎

When we studied nonhomogeneous equations, we considered equations in which the nonhomogeneous function was a solution of the corresponding homogeneous equation. This situation is modeled by the initial-value problem

$$\begin{cases} \frac{d^2x}{dt^2} + \omega^2 x = F_1 \cos \omega t + F_2 \sin \omega t + G(t) \\ x(0) = \alpha, \frac{dx}{dt}(0) = \beta. \end{cases}$$

where F_1 and F_2 are constants and G is any function of t. (Note that one of the constants F_1 and F_2 can equal zero and G can be identically the zero function.) In this case, we say that ω is the **natural frequency of the system** because the homogeneous solution is $x_h(t) = c_1 \cos \omega t + c_2 \sin \omega t$. We consider a problem of this type in the following example.

EXAMPLE 2: Investigate the effect that the forcing functions (a) $f(t) = \cos 2t$ and (b) $f(t) = \sin 2t$ have on the solution of the initial-value problem

$$\begin{cases} x'' + 4x = f(t) \\ x(0) = 0, x'(0) = 0 \end{cases}.$$

SOLUTION: We take advantage of the function fm defined in Example 1. In the same manner as in Example 1, we use map to apply fm to each of the forcing functions in (a) and (b).

```
> moresols:=map(fm,[cos(2*t),sin(2*t)]):

moresols:=combine(moresols,trig);
```

$$moresols := \left[x(t) = \frac{1}{4}\sin(2t)t, x(t) = \frac{1}{8}\sin(2t) - \frac{1}{4}\cos(2t)t \right]$$

From the result, we see that the nonperiodic function $t \sin 2t$ appears in the result for (a) and the nonperiodic function $t \cos 2t$ appears in the result for (b). In each case, we see that the amplitude increases without bound as t increases, as illustrated in the following graph. This indicates that the spring–mass system will encounter a serious problem in that the mass will eventually hit its support (such as a ceiling or beam) or its lower boundary (such as the ground or floor).

```
> plot(map(rhs,moresols),t=0..3*Pi);
```

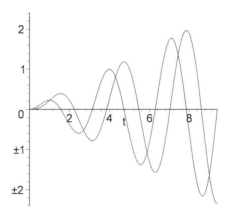

The phenomenon illustrated in the previous example is called **resonance** and can be extended to other situations such as vibrations in an aircraft wing, skyscraper, glass, or bridge. Some of the sources of excitation that lead to the vibration of these structures include unbalanced rotating devices, vortex shedding, strong winds, rough surfaces, and moving vehicles. Therefore, the engineer has to overcome many problems when structures and machines are subjected to forced vibrations.

EXAMPLE 3: How does slightly changing the value of the argument of the forcing function $f(t) = \cos 2t$ change the solution of the initial-value problem given in Example 2? Use the functions (a) $f(t) = \cos 1.9t$ and (b) $f(t) = \cos 2.1t$ with the initial-value problem.

SOLUTION: In the same manner as in Example 2, we use `map` to apply `fm` to each of the forcing functions in (a) and (b).

```
> moresols:=map(fm,[cos(1.9*t),sin(2.1*t)]):
moresols:=combine(moresols,trig);
```

$$moresols := [$$
$$x(t) = 2.564102564 \cos(1.900000000\,t) - 2.564102564 \cos(2t),$$
$$x(t) = -2.439024390 \sin(2.100000000\,t) + 2.560975610 \sin(2t)]$$

The result shows that each solution is periodic and bounded. These solutions are then graphed to reveal the behavior of the curves. If the solutions are plotted over only a small interval, however, resonance *seems* to be present. (Compare the following graph with the graph generated in Example 2.)

```
> plot(map(rhs,moresols),t=0..3*Pi);
```

However, the functions obtained with fm clearly indicate that there is no resonance. This is further indicated with the second graph.

```
> plot(rhs(moresols[1]),t=0..40*Pi);

plot(rhs(moresols[2]),t=0..40*Pi);
```

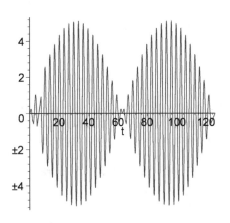 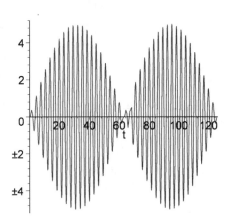

■

Let us investigate in detail initial-value problems of the form

$$\begin{cases} \frac{d^2x}{dt^2} + \omega^2 x = F \cos \beta t, \omega \neq \beta \\ x(0) = 0, \frac{dx}{dt}(0) = 0 \end{cases}.$$

A general solution of the corresponding homogeneous equation is $x_h(t) = c_1 \cos \omega t + c_2 \sin \omega t$. Using the method of undetermined coefficients, we assume that there is a particular solution to the nonhomogeneous equation of the form $x_p(t) = A \cos \beta t + B \sin \beta t$.

```
> xp:=t->a*cos(beta*t)+b*sin(beta*t):diff(xp(t),t);
```

$$-a \sin(\beta t)\beta + b \cos(\beta t)\beta$$

Next, we calculate the corresponding derivatives of this solution

```
> diff(xp(t),t$2);
```

$$-a\cos(\beta t)\beta^2 - b\sin(\beta t)\beta^2$$

and substitute into the nonhomogeneous equation $\frac{d^2x}{dt^2} + \omega^2 x = F\cos\beta t$.

```
> step1:=simplify(diff(xp(t),t$2)+

       omega^2*diff(xp(t),t)=f*cos(beta*t));
```

$step1 :=$

$$- a\cos(\beta t)\beta^2 - b\sin(\beta t)\beta^2 - \omega^2 a\sin(\beta t)\beta + \omega^2 b\cos(\beta t)\beta$$

$$= f\cos(\beta t)$$

This equation is true for all values of t. In particular, substituting $t = 0$ and $t = \frac{\pi}{2\beta}$ yields two equations

```
> eq1:=eval(subs(t=0,step1));
```

$$eq1 := -a\beta^2 + \omega^2 b\beta = f$$

```
> eq2:=eval(subs(t=Pi/(2*beta),step1));
```

$$eq2 := -b\beta^2 - \omega^2 a\beta = 0$$

that we then solve for a and b.

```
> vals:=solve({eq1,eq2},{a,b});
```

$$vals := \left\{ a = -\frac{f}{\beta^2 + \omega^4}, \, b = \frac{\omega^2 f}{(\beta^2 + \omega^4)\beta} \right\}$$

Thus, $A = F/(\omega^2 - \beta^2)$ and $B = 0$ and a general solution of the nonhomogeneous equation is

$$x(t) = c_1\cos\omega t + c_2\sin\omega t + \frac{F}{\omega^2 - \beta^2}\cos\beta t$$

Application of the initial conditions yields the solution

$$x(t) = \frac{F}{\omega^2 - \beta^2}(\cos\beta t - \cos\omega t) = \frac{F}{\beta^2 - \omega^2}(\cos\omega t - \cos\beta t).$$

We can use `dsolve` and `combine` together with the `trig` option to solve the initial-value problem as well.

```
> gensol:=dsolve({diff(x(t),t$2)+
              omega^2*x(t)=f*cos(beta*t),x(0)=0,D(x)(0)=0},x(t)):
combine(gensol,trig);
```

$$x(t) = \frac{-f\cos(\beta t) + f\cos(\omega t)}{-\omega^2 + \beta^2}$$

Using the trigonometric identity $\frac{1}{2}[\cos(A - B) - \cos(A + B)] = \sin A \sin B$, we have

$$x(t) = \frac{2F}{\omega^2 - \beta^2} \sin\frac{(\omega + \beta)t}{2} \sin\frac{(\omega - \beta)t}{2}.$$

These solutions are of interest because of what they indicate about the motion of the spring under consideration. Notice that the solution can be represented as

$$x(t) = A(t)\sin\frac{(\omega + \beta)t}{2}, \quad \text{where } A(t) = \frac{2F}{\omega^2 - \beta^2}\sin\frac{(\omega - \beta)t}{2}.$$

Therefore, if the quantity $\omega - \beta$ is small, $\omega + \beta$ is relatively large in comparison. Hence, the function $\sin\frac{(\omega+\beta)t}{2}$ oscillates quite frequently because it has period $\frac{\pi}{\omega+\beta}$. Meanwhile, the function $\sin\frac{(\omega-\beta)t}{2}$ oscillates relatively slowly because it has period $\frac{\pi}{|\omega-\beta|}$, so the functions $\pm\frac{2F}{\omega^2-\beta^2}\sin\frac{(\omega-\beta)t}{2}$ form an **envelope** for the solution.

EXAMPLE 4: Solve the initial-value problem

$$\begin{cases} \frac{d^2x}{dt^2} + 4x = f(t) \\ x(0) = 0, \frac{dx}{dt}(0) = 0 \end{cases}$$

with (a) $f(t) = \cos 3t$ and (b) $f(t) = \cos 5t$.

SOLUTION: Again, we use `fm`, defined in Example 1, to solve the initial-value problem in each case.

```
> fs:=[cos(3*t),cos(5*t)]:
somesols:=map(fm,fs):
somesols:=combine(somesols,trig);
```

somesols :=

$$\left[x(t) = -\frac{1}{5}\cos(3\ t) + \frac{1}{5}\cos(2\ t), x(t) = -\frac{1}{21}\cos(5\ t) + \frac{1}{21}\cos(2\ t) \right]$$

The solution for (a) is graphed and named `p1` for later use.

```
> p1:=plot(rhs(somesols[1]),t=0..6*Pi,color=BLACK):
with(plots):
display(p1);
```

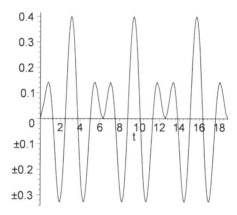

Using the formula obtained earlier for the functions that "envelope" the solution, we have $x(t) = \frac{2}{5}\sin{(t/2)}$ and $x(t) = -\frac{2}{5}\sin{(t/2)}$. These functions are graphed in `p2` and displayed with `p1` with `display`.

```
> p2:=plot([2/5*sin(t/2),-2/5*sin(t/2)],t=0..6*Pi,
        color=GRAY):
display({p1,p2});
```

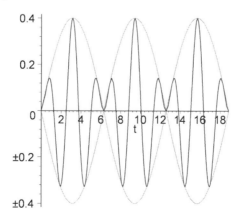

For (b), the graph of the solution with the envelope functions $x(t) = \frac{2}{21} \sin (3t/2)$ and $x(t) = -\frac{2}{21} \sin (3t/2)$ is as follows.

```
> plot([rhs(somesols[2]),2/21*sin(3*t/2),
        -2/21*sin(3*t/2)],t=0..4*Pi,
          color=[BLACK,GRAY,GRAY]);
```

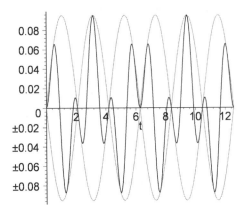

Oscillations such as those illustrated in the previous example are called **beats** because of the periodic variation of amplitude. This phenomenon is commonly encountered when two musicians (especially bad ones) try to tune their instruments simultaneously or when two tuning forks with almost equivalent frequencies are played at the same time.

EXAMPLE 5: Investigate the effect that the forcing function $f(t) = e^{-t} \cos 2t$ has on the initial-value problem

$$\begin{cases} \frac{d^2x}{dt^2} + 4x = f(t) \\ x(0) = 0, \frac{dx}{dt}(0) = 0. \end{cases}$$

SOLUTION: The initial-value problem is solved with fm, which was defined in Example 1, and the result is graphed with plot.

```
> del1:=fm(exp(-t)*cos(2*t));

del1:=combine(del1,trig);
```

$$de11 := x(t) = \frac{1}{34} (\sin(2\ t)e^{(-t)}e^t + 2e^{(-t)}e^t \cos(2\ t)$$

$$- 9\sin(2\ t) + 9\sin(2\ t)e^t - 2\cos(2\ t)^t)/e^t$$

```
> plot(rhs(de11),t=0..4*Pi);
```

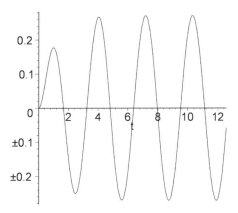

Notice that the effect of terms involving the exponential function diminishes as t increases. In this case, the forcing function $f(t) = e^{-t} \cos 2t$ approaches zero as t increases. Hence, over time, the solution of the nonhomogeneous equation approaches that of the homogeneous equation.

■

We now consider spring problems that involve forces due to damping as well as external forces. In particular, consider the following initial-value problem:

$$\begin{cases} m\frac{d^2x}{dt^2} + c\frac{dx}{dt} + kx = \rho \cos \lambda t \\ x(0) = \alpha, \frac{dx}{dt}(0) = \beta \end{cases}.$$

Problems of this nature have solutions of the form $x(t) = h(t) + s(t)$, where $\lim_{t \to \infty} h(t) = 0$ and $s(t) = c_1 \cos \lambda t + c_2 \sin \lambda t$.

The function $h(t)$ is called the **transient** solution and $s(t)$ is known as the **steady-state** solution. Therefore, as t approaches infinity, the solution $x(t)$ approaches the steady-state solution. Note that the steady-state solution corresponds to the particular solution obtained through the method of undetermined coefficients or variation of parameters.

EXAMPLE 6: Solve the initial-value problem

$$\begin{cases} \frac{d^2x}{dt^2} + 4\frac{dx}{dt} + 13x = \cos t \\ x(0) = 0, \frac{dx}{dt}(0) = 1. \end{cases}$$

that models the motion of an object of mass $m = 1$ attached to a spring with spring constant $k = 13$ that is subjected to a resistive force of $F_R = 4dx/dt$ and an external force of $f(t) = \cos t$. Identify the transient and steady-state solutions.

SOLUTION: First, dsolve is used to obtain the solution of this nonhomogeneous problem. (The method of undetermined coefficients could be used to find this solution as well.)

```
> deq:=dsolve({diff(x(t),t$2)+4*diff(x(t),t)+13*x(t)=cos(t),
      x(0)=0,D(x)(0)=1},x(t)):
deq:=combine(deq,trig);
```

$$deq := x(t) = \frac{1}{40}\sin(t) + \frac{3}{40}\cos(t) + \frac{11}{40}e^{(-2t)}\sin(3t)$$
$$- \frac{3}{40}e^{(-2t)}\cos(3t)$$

The solution is then graphed over the interval $[0,5\pi]$ in plot1 to illustrate the behavior of this solution.

```
> plot1:=plot(rhs(deq),t=0..5*Pi,color=BLACK):
with(plots):
display(plot1);
```

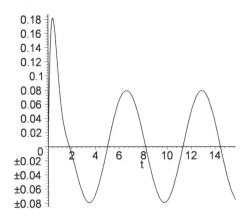

The transient solution is $e^{-2t}\left(-\frac{3}{40}\cos 3t + \frac{11}{40}\sin 3t\right)$ and the steady-state solution is $\frac{3}{40}\cos t + \frac{1}{40}\sin t$. We graph the steady-state solution over the same interval so that it can be compared with `plot1` and then we show the two graphs together with `display`.

```
> ss:=t->1/40*(3*cos(t)+sin(t)):

ssplot:=plot(ss(t),t=0..5*Pi,color=GRAY):

display({plot1,ssplot});
```

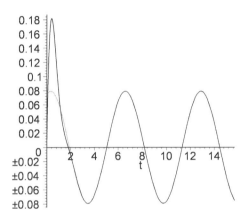

Notice that the two curves appear identical for $t > 2.5$. The reason for this is shown in the subsequent plot of the transient solution, which becomes quite small near $t = 2.5$.

```
> plot(1/40*exp(-2*t)*(-3/40*cos(3*t)+
            11/40*sin(3*t)),t=0..Pi);
```

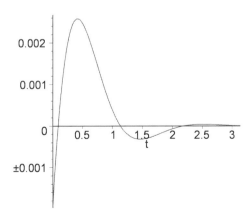

Notice also that the steady-state solution corresponds to a particular solution to the nonhomogeneous differential equation as verified here with `simplify`.

```
> simplify(diff(ss(t),t$2)+4*diff(ss(t),t)+13*ss(t));
```

$$\cos(t)$$

■

Instead of solving initial-value problems that model the motion of damped and undamped systems as functions of time only, we can consider problems that involve an arbitrary parameter. In doing this, we can obtain a new understanding of the phenomena of resonance and beats.

EXAMPLE 7: Solve (a)

$$\begin{cases} \frac{d^2x}{dt^2} + 4\frac{dx}{dt} + 13x = \cos \omega t \\ x(0) = 0, \ \frac{dx}{dt}(0) = 0 \end{cases} ;$$

(b)

$$\begin{cases} \frac{d^2x}{dt^2} + 4x = \cos \omega t \\ x(0) = 0, \ \frac{dx}{dt}(0) = 0 \end{cases}$$

Plot the solution for various values of ω near the natural frequency of the system.

SOLUTION: (a) We solve the initial-value problem and simplify the result with `combine` for arbitrary ω in `sol` and then define the solution to be `u(t,omega)`.

```
> sol:=dsolve({diff(x(t),t$2)+4*diff(x(t),t)+
        13*x(t)=cos(omega*t),x(0)=0,D(x)(0)=0},x(t)):
sol:=combine(sol,trig);
```

$$sol := x(t) = (39\cos(\omega t) + 12\omega\sin(\omega t) - 3\omega^2\cos(\omega t)$$

$$- 39e^{(-2\ t)}\cos(3\ t) + 3e^{(-2\ t)}\cos(3\ t)\omega^2$$

$$- 2e^{(-2t)}\sin(3\ t)\omega^2 - 26e^{(-2\ t)}\sin(3\ t))/$$

$$(3\omega^4 - 30\omega^2 + 507)$$

```
> u:=(t0,omega0)->eval(subs({t=t0,omega=omega0},rhs(sol))):
```

We can graph the solution for various values of ω and animate the results using the `animate` command, which is contained in the **plots** package.

```
> with(plots):
```

```
?animate
```

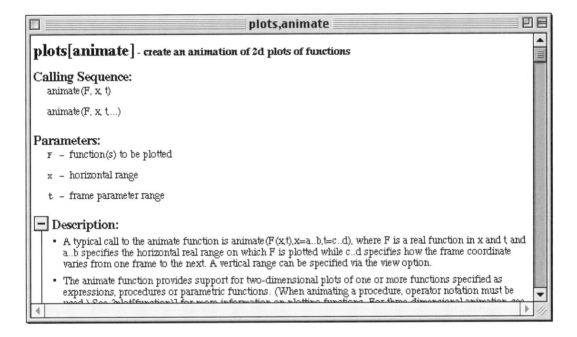

In the following command, we use `animate` to create a 40-frame animation of u for 40 equally spaced values of ω between 0 and 6. We show a screen shot from the resulting animation.

```
> animate(u(t,omega),t=0..10,omega=0..6,frames=40,
view=[0..10,-0.1..0.1],color=BLACK);
```

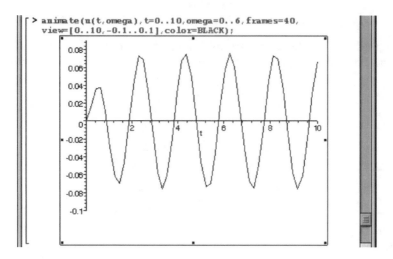

We can also observe how the motion approaches and then moves away from resonance using a graphics array. In the following, we use `animate` to create a 12-frame animation of u for 12 equally spaced values of ω between 0 and 6. The result is displayed with `display`.

```
> toshow:=animate(u(t,omega),t=0..10,omega=0..6,frames=12,

         view=[0..10,-0.1..0.1],color=BLACK):

display(toshow);
```

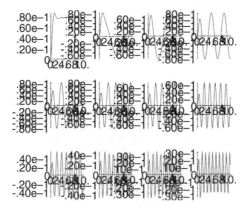

On the other hand, we can graph the three-dimensional surface u(t, omega) to see how the motion depends on the value of ω.

```
> plot3d(u(t,omega),t=0..10,omega=0..6,

         grid=[40,40],axes=BOXED);
```

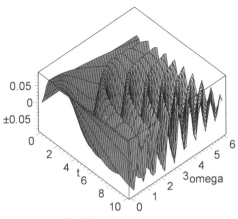

(b) In a similar way, we solve $\begin{cases} \frac{d^2x}{dt^2} + 4x = \cos \omega t \\ x(0) = 0, \ \frac{dx}{dt}(0) = 0 \end{cases}$ for arbitrary ω in `sol`.

```
> sol:=dsolve({diff(x(t),t$2)+4*x(t)=cos(omega*t),
         x(0)=0,D(x)(0)=0},x(t)):
sol:=combine(sol,trig);
```

$$sol := x(t) = \frac{-\cos(\omega t) + \cos(2\,t)}{-4 + \omega^2}$$

```
> u:=(t0,omega0)->eval(subs({t=t0,omega=omega0},rhs(sol))):
```

We animate the solution for $0 \le \omega \le 4$ using 55 equally spaced values of ω between 0 and 4 to observe how the solution behaves as ω approaches the natural frequency of the system, 2. We show a screen shot from the resulting animation.

```
> animate(u(t,omega),t=0..20,omega=0..4,frames=55);
```

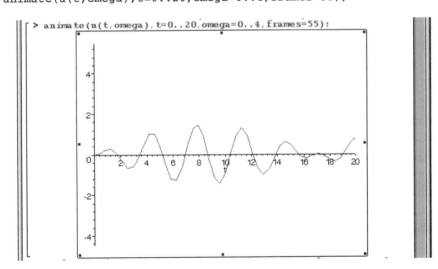

In addition, we can use a graphics array to observe the behavior of the function.

```
> toshow:=animate(u(t,omega),t=0..20,omega=0..4,
           frames=12,color=BLACK):
display(toshow);
```

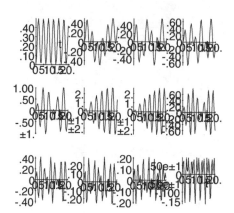

We can see this behavior in the three-dimensional graph of `u(t,omega)` as well.

```
> plot3d(u(t,omega),t=0..10,omega=0..3,
           grid=[40,40],axes=BOXED);
```

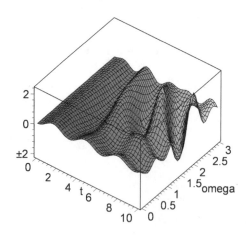

■

5.4 Other Applications

L–R–C *Circuits*

Second-order nonhomogeneous linear ordinary differential equations arise in the study of electrical circuits after the application of *Kirchhoff's law.* Suppose that $I(t)$ is the current in the L–R–C series electrical circuit where L, R, and C represent the inductance, resistance, and capacitance of the circuit, respectively.

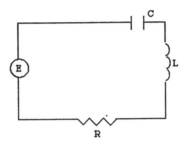

The voltage drops across the circuit elements shown in the following table have been obtained from experimental data where Q is the charge of the capacitor and $dQ/dt = I$.

Circuit Element	Voltage Drop
Inductor	$L \frac{dI}{dt}$
Resistor	RI
Capacitor	$\frac{1}{C} Q$

Our goal is to model this physical situation with an initial-value problem so that we can determine the current and charge in the circuit. For convenience, the terminology used in this section is summarized in the following table.

Electrical Quantities	Units
Inductance (L)	Henrys (H)
Resistance (R)	Ohms (Ω)
Capacitance (C)	Farads (F)
Charge (Q)	Coulombs (C)
Current (I)	Amperes (A)

The physical principle needed to derive the differential equation that models the L–R–C series circuit is stated as follows.

Kirchhoff's law

> The sum of the voltage drops across the circuit elements is equivalent to the voltage $E(t)$ impressed on the circuit.

Applying Kirchhoff's law, therefore, yields the differential equation $L\frac{dI}{dt} + RI + \frac{1}{C}Q = E(t)$. Using the fact that $dQ/dt = I$, we also have $d^2Q/dt^2 = dI/dt$. Therefore, the equation becomes $L\frac{d^2Q}{dt^2} + R\frac{dQ}{dt} + \frac{1}{C}Q = E(t)$, which can be solved by the method of undetermined coefficients or the method of variation of parameters. Hence, if the initial charge and current are $Q(0) = Q_0$ and $I(0) = \frac{dQ}{dt}(0) = I_0$, we must solve the initial-value problem

$$\begin{cases} L\frac{d^2Q}{dt^2} + R\frac{dQ}{dt} + \frac{1}{C}Q = E(t) \\ Q(0) = Q_0, I(0) = \frac{dQ}{dt}(0) = I_0 \end{cases}$$

for the charge $Q(t)$. This solution can then be differentiated to find the current $I(t)$.

EXAMPLE 1: Consider the L–R–C circuit with $L = 1$ henry, $R = 40$ ohms, $C = 4000$ farads, and $E(t) = 24$ volts. Determine the current in this circuit if there is zero initial current and zero initial charge.

SOLUTION: Using the indicated values, the initial-value problem that we must solve is

$$\begin{cases} \frac{d^2Q}{dt^2} + 40\frac{dQ}{dt} + 4000Q = 24 \\ Q(0) = 0, I(0) = \frac{dQ}{dt}(0) = 0. \end{cases}$$

`dsolve` is used to obtain the solution to the nonhomogeneous problem in `cir1`.

```
> cir1:=dsolve({
      diff(q(t),t$2)+40*diff(q(t),t)+4000*q(t)=24,
      q(0)=0,D(q)(0)=0},q(t));
```

$$cir1 := q(t) = \frac{3}{500} - \frac{3}{500}e^{(-20t)}\cos(60t) - \frac{1}{500}e^{(-20t)}\sin(60t)$$

These results indicate that in time the charge approaches the constant value of $3/500$, which is known as the **steady-state charge**. Also, due to the exponential term, the current approaches zero as t increases. This limit is indicated by the graph of $Q(t)$ as well.

```
> assign(cir1):
plot(q(t),t=0..0.35);
```

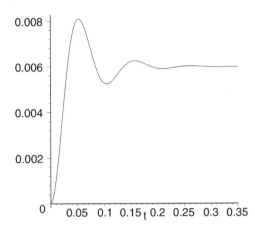

The current $I(t)$ is obtained by differentiating the solution $Q(t)$.

```
> dq:=diff(q(t),t);
```

$$dq := \frac{2}{5} e^{(-20t)} \sin(60t)$$

This function is graphed as follows.

```
> plot(dq,t=0..0.35);
```

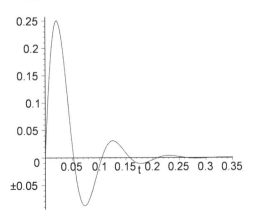

■

Deflection of a Beam

An important mechanical model involves the deflection of a long beam that is supported at one or both ends as shown in the following figure.

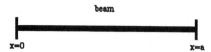

Assuming that in its undeflected form the beam is horizontal, then the deflection of the beam can be expressed as a function of x.

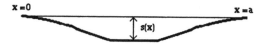

Suppose that the shape of the beam when it is deflected is given by the graph of the function $y(x) = -s(x)$ where x is the distance from one end of the beam and s the measurement of the vertical deflection from the equilibrium position. The boundary-value problem that models this situation is derived as follows.

Let $m(x)$ equal the turning moment of the force relative to the point x and $w(x)$ represent the weight distribution of the beam. These two functions are related by the equation

$$\frac{d^2 m}{dx^2} = w(x).$$

Also, the turning moment is proportional to the curvature of the beam. Hence,

$$m(x) = \frac{EI}{\left(\sqrt{1 + (ds/dx)^2}\right)^3} \frac{d^2 s}{dx^2}$$

where E and I are constants related to the composition of the beam and the shape and size of a cross section of the beam, respectively. Notice that this equation is, unfortunately, nonlinear. However, this difficulty is overcome with an approximation. For small values of s, the denominator of the right-hand side of the equation can be approximated by the constant 1. Therefore, the equation is simplified to

$$m(x) = EI \frac{d^2 s}{dx^2}$$

This equation is linear and can be differentiated twice to obtain

$$\frac{d^2 m}{dx^2} = EI \frac{d^4 s}{dx^4}.$$

This equation can then be used with the preceding equation relating $m(x)$ and $w(x)$ to obtain the single fourth-order linear nonhomogeneous differential equation

$$EI\frac{d^4s}{dx^4} = w(x).$$

Boundary conditions for this problem may vary. In most cases, two conditions are given for each end of the beam. Some of these conditions are specified in pairs. For example, at $x = a$ these include $s(a) = 0$, $\frac{ds}{dx}(a) = 0$ (fixed end); $\frac{d^2s}{dx^2}(a) = 0$, $\frac{d^3s}{dx^3}(a) = 0$ (free end); $s(a) = 0$, $\frac{d^2s}{dx^2}(a) = 0$ (simple support); and $\frac{ds}{dx}(a) = 0$, $\frac{d^3s}{dx^3}(a) = 0$ (sliding clamped end).

The following example investigates the effects that a constant weight distribution function $w(x)$ has on the solution to these boundary-value problems.

EXAMPLE 2: Solve the beam equation over the interval $0 \le x \le 1$ if $E = I = 1$, $w(x) = 48$, and the following boundary conditions are used: $s(0) = 0$, $\frac{ds}{dx}(0) = 0$ (fixed end at $x = 0$); and

(a) $s(1) = 0$, $\frac{d^2s}{dx^2}(1) = 0$ (simple support at $x = 1$);

(b) $\frac{d^2s}{dx^2}(1) = 0$, $\frac{d^3s}{dx^3}(1) = 0$ (free end at $x = 1$);

(c) $\frac{ds}{dx}(1) = 0$, $\frac{d^3s}{dx^3}(1) = 0$ (sliding clamped end at $x = 1$); and

(d) $s(1) = 0$, $\frac{ds}{dx}(1) = 0$ (fixed end at $x = 1$).

SOLUTION: `dsolve` is used to obtain the solution to this nonhomogeneous problem. In `de1`, the solution that depends on E, I, and w is given.

```
> de1:=dsolve({e*i*diff(s(x),x$4)=w,
      s(0)=0,D(s)(0)=0,s(1)=0,(D@@2)(s)(1)=0},s(x));
```

$$de1 := s(x) = \frac{1}{24}\frac{wx^4}{ei} - \frac{5}{48}\frac{wx^3}{ei} + \frac{1}{16}\frac{wx^2}{ei}$$

We can visualize the shape of the beam by graphing $y = -s(x)$. Thus, we define `toplot1` to be the negative of the solution obtained in `de1`.

```
> toplot1:=-subs({e=1,i=1,w=48},rhs(de1));
```

$$toplot1 := -2x^4 + 5x^3 - 3x^2$$

Similar steps are followed to determine the solution to each of the other three boundary-value problems. The corresponding functions to be graphed are named `toplot2`, `toplot3`, and `toplot4`.

```
> de2:=dsolve({e*i*diff(s(x),x$4)=w,
      s(0)=0,D(s)(0)=0,(D@@3)(s)(1)=0,
```

```
(D@@2)(s)(1)=0},s(x));
```

$$de2 := s(x) = \frac{1}{24}\frac{wx^4}{ei} - \frac{1}{6}\frac{wx^3}{ei} + \frac{1}{4}\frac{wx^2}{ei}$$

```
> toplot2:=-subs({e=1,i=1,w=48},rhs(de2));
```

$$toplot2 := -2x^4 + 8x^3 - 12x^2$$

```
> de3:=dsolve({e*i*diff(s(x),x$4)=w,s(0)=0,D(s)(0)=0,
      (D@@3)(s)(1)=0,(D)(s)(1)=0},s(x));
```

$$de3 := s(x) = \frac{1}{24}\frac{wx^4}{ei} - \frac{1}{6}\frac{wx^3}{ei} + \frac{1}{6}\frac{wx^2}{ei}$$

```
> toplot3:=-subs({e=1,i=1,w=48},rhs(de3));
```

$$toplot3 := -2x^4 + 8x^3 - 8x^2$$

```
> de4:=dsolve({e*i*diff(s(x),x$4)=w,s(0)=0,D(s)(0)=0,
      s(1)=0,(D)(s)(1)=0},s(x));
```

$$de4 := s(x) = \frac{1}{24}\frac{wx^4}{ei} - \frac{1}{12}\frac{wx^3}{ei} + \frac{1}{24}\frac{wx^2}{ei}$$

```
> toplot4:=-subs({e=1,i=1,w=48},rhs(de4));
```

$$toplot4 := -2x^4 + 4x^3 - 2x^2$$

In order to compare the effects that the varying boundary conditions have on the resulting solutions, all four functions are graphed together with plot on the interval [0,1].

```
> plot({toplot1,toplot2,toplot3,toplot4},x=0..1);
```

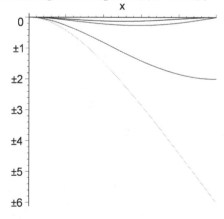

Soft Springs

In the case of a **soft spring,** the spring force weakens with compression or extension. For springs of this type, we model the physical system with the nonlinear equation

$$\begin{cases} m\frac{d^2x}{dt^2} + c\frac{dx}{dt} + kx - jx^3 = f(t) \\ x(0) = \alpha, \frac{dx}{dt}(0) = \beta \end{cases}$$

where j is a positive constant.

EXAMPLE 3: Approximate the solution to

$$\begin{cases} \frac{d^2x}{dt^2} + 0.2\frac{dx}{dt} + 10x - 0.2x^3 = -9.8 \\ x(0) = \alpha, \frac{dx}{dt}(0) = \beta \end{cases}$$

for various values of α and β in the initial conditions.

SOLUTION: After stating this nonlinear differential equation in eq, we load the **DEtools** package.

```
> eq:=diff(x(t),t$2)+0.2*diff(x(t),t)+
          10*x(t)-0.2*x(t)^3=-9.8:
> with(DEtools);
```

We then define several initial conditions in vals1. Notice that these conditions correspond to fixing the initial velocity but varying the initial displacement. We will graph the solution that satisfies the conditions for each ordered pair in vals1.

```
> vals1:=[seq([x(0)=-1+0.5*i,D(x)(0)=0],i=0..4)];
```

$$vals1 := [[x(0) = -1., D(x)(0) = 0], [x(0) = -0.5, D(x)(0) = 0],$$
$$[x(0) = 0, D(x)(0) = 0], [x(0) = 0.5, D(x)(0) = 0],$$
$$[x(0) = 1.0, D(x)(0) = 0]]$$

The graphs of the solutions that satisfy the initial conditions specified in vals1 are generated with DEplot, which is contained in the **DEtools** package. We notice that $x(t) \to -1$ as $t \to \infty$.(The spring does not converge to its equilibrium position.)

```
> DEplot(eq,x(t),t=0..15,vals1,thickness=1,
          stepsize=0.05,linecolor=BLACK);
```

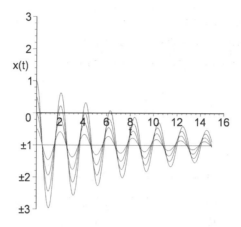

Similarly, in `vals2`, we fix the initial displacement but vary the initial velocity. We graph these approximate solutions with `DEplot` as well.

```
> vals2:=[seq([x(0)=0,D(x)(0)=-1+0.5*i],i=0..4)]:
```

```
> DEplot(eq,x(t),t=0..15,vals2,thickness=1,

        stepsize=0.05,linecolor=BLACK);
```

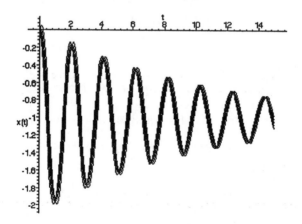

In each of the two previous sets of initial conditions, we see that $x(t)$ approaches a limit as $t \to \infty$. However, this is not always the case. If we consider larger values of the initial displacement as defined in `vals3`, we find that solutions are unbounded. (The spring becomes weak.) Because of this, we must use a smaller interval for t in the `DEplot` command. Otherwise, we do not obtain meaningful results.

```
> vals3:=[[x(0)=-10,D(x)(0)=0],[x(0)=-9,D(x)(0)=0],

        [x(0)=-8,D(x)(0)=0],[x(0)=8,D(x)(0)=0],
```

$$[x(0)=9,D(x)(0)=0],[x(0)=10,D(x)(0)=0]]:$$

```
DEplot(eq,x(t),t=0..0.4,vals3,thickness=1,
       stepsize=0.01,linecolor=BLACK);
```

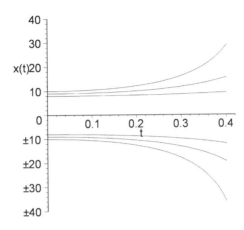

Hard Springs

In the case of a **hard spring,** the spring force strengthens with compression or extension. For springs of this type, we model the physical system with

$$\begin{cases} m\frac{d^2x}{dt^2} + c\frac{dx}{dt} + kx + jx^3 = f(t) \\ x(0) = \alpha, \frac{dx}{dt}(0) = \beta \end{cases},$$

where j is a positive constant.

EXAMPLE 4: Approximate the solution to

$$\begin{cases} \frac{d^2x}{dt^2} + 0.3x + 0.04x^3 = 0 \\ x(0) = \alpha, \frac{dx}{dt}(0) = \beta \end{cases}$$

for various values of α and β in the initial conditions.

SOLUTION: First, we define the undamped nonlinear differential equation in eq.

```
> eq:=diff(x(t),t$2)+0.3*x(t)+0.04*x(t)^3=0:
```

We then proceed in the same way as in Example 3 using DEplot. We graph the approximate solution using the initial conditions defined in vals4, which fix the initial velocity but vary the initial displacement, $x(0)$, with DEplot. Notice that solutions with larger amplitudes have smaller periods as expected with a hard spring.

```
> with(DEtools):
vals4:=[seq([x(0)=i,D(x)(0)=0],i=1..5)]:
DEplot(eq,x(t),t=0..15,vals4,thickness=1,
          stepsize=0.01,linecolor=BLACK);
```

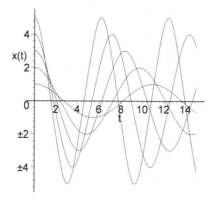

In a similar manner, we fix the intial displacement but vary the initial velocity. Again, we see that when the amplitude is large, the spring strengthens so that the period of the motion is decreased.

```
> vals5:=[seq([x(0)=0,D(x)(0)=i],i=1..5)]:
> DEplot(eq,x(t),t=0..15,vals5,thickness=1,
          stepsize=0.01,linecolor=BLACK);
```

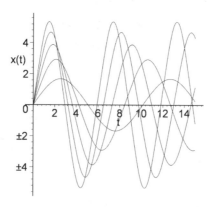

Aging Springs

In the case of an **aging spring,** the spring constant weakens with time. For springs of this type, we model the physical system with

$$\begin{cases} m\frac{d^2x}{dt^2} + c\frac{dx}{dt} + k(t)x = f(t) \\ x(0) = \alpha, \frac{dx}{dt}(0) = \beta \end{cases},$$

where $k(t) \to 0$ as $t \to \infty$.

EXAMPLE 5: Approximate the solution to

$$\begin{cases} \frac{d^2x}{dt^2} + 4e^{-t/4}x = 0 \\ x(0) = \alpha, \frac{dx}{dt}(0) = \beta \end{cases}$$

for various values of α and β in the initial conditions.

SOLUTION: First, we define the differential equation in `eq`.

```
> eq:=diff(x(t),t$2)+4*exp(-t/4)*x(t)=0:
```

Notice that `dsolve` is able to find a general solution of the equation

```
> dsolve(eq,x(t));
```

$$x(t) =$$

$$\_C1 \text{ BesselY}\left(0, 16\sqrt{e^{\left(-\frac{1}{4}t\right)}}\right) + \_C2 \text{ BesselJ}\left(0, 16\sqrt{e^{\left(-\frac{1}{4}t\right)}}\right)$$

as well as solve the initial-value problem

$$\begin{cases} \frac{d^2x}{dt^2} + 4e^{-t/4}x = 0 \\ x(0) = \alpha, \frac{dx}{dt}(0) = \beta \end{cases},$$

although the result is given in terms of the Bessel functions, `BesselJ` and `BesselY`.

```
> partsol:=dsolve({eq,x(0)=alpha,D(x)(0)=beta},x(t));
```

partsol := x(*t*)

$$= -\frac{1}{2}\left(\text{BesselJ}(0, 16)\beta - 2\alpha\ \text{BesselJ}(1, 16)) \text{BesselY}\left(0, 16\sqrt{e^{\left(-\frac{1}{4}t\right)}}\right)\right)$$

/(-BesselY(1, 16) BesselJ(0, 16) + BesselJ(1, 16) BesselY(0, 16))+

$$\frac{1}{2}\left((-2\ \text{BesselY}(1, 16)\alpha + \beta\ \text{BesselY}(0, 16))\right.$$

$$\left.\text{BesselJ}\left(0.16\sqrt{e^{\left(-\frac{1}{4}t\right)}}\right)\right)/$$

(-BesselY(1, 16) BesselJ(0, 16) + BesselJ(1, 16) BesselY(0, 16))

We then use the results obtained in `partsol` to graph the solution for various initial conditions with `plot`. Notice that the period of the oscillations increases over time due to the diminishing value of the spring constant.

```
> toplot1:=[seq(subs({alpha=i,beta=0},rhs(partsol)),i=1..5)]:
plot(toplot1,t=0..30,color=BLACK);
```

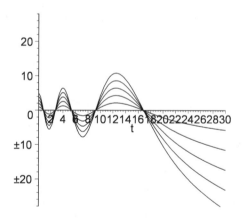

However, when we choose a longer time interval, as we do in the following graph, we see that eventually the motion is not oscillatory.

```
> plot(toplot1,t=0..100,color=BLACK);
```

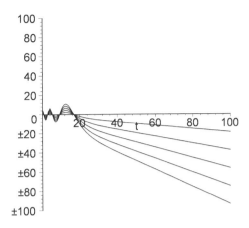

Bodé Plots

Consider the differential equation $\frac{d^2x}{dt^2} + 2c\frac{dx}{dt} + k^2x = F_0 \sin \omega t$ where c and $k7$ are positive constants such that $c < k$. Therefore, the system is underdamped. To find a particular solution, we can consider the complex exponential form of the forcing function, $F_0 e^{i\omega t}$, which has imaginary part $F_0 \sin \omega t$. Assuming a solution of the form $z_p(t) = Ae^{i\omega t}$, substitution into the differential equation yields $A[-\omega^2 + 2ic\omega + k^2] = F_0 e^{i\omega t}$. Because $k^2 - \omega^2 + 2ic\omega = 0$ only when $k = \omega$ and $c = 0$, we find that $A = \frac{F_0}{k^2 - \omega^2 + 2ic\omega}$ or $A = \frac{F_0}{k^2 - \omega^2 + 2ic\omega} \cdot \frac{k^2 - \omega^2 - 2ic\omega}{k^2 - \omega^2 - 2ic\omega} = \frac{k^2 - \omega^2 - 2ic\omega}{(k^2 - \omega^2)^2 + 4c^2\omega^2} F_0 = H(i\omega)F_0$. Therefore, a particular solution is $z_p(t) = H(i\omega)F_0 e^{i\omega t}$. Now, we can write $H(i\omega)$ in polar form as $H(i\omega) = M(\omega)e^{i\phi(\omega)}$ where $M(\omega) = \frac{1}{\sqrt{(k^2-\omega^2)^2 + 4c^2\omega^2}}$ and $\phi(\omega) = \cot^{-1}\left(\frac{\omega^2 - k^2}{2c\omega}\right)(-\pi \leq \phi \leq 0)$. A particular solution can then be written as $z_p(t) = M(\omega)F_0 e^{i\omega t}e^{i\phi(\omega)} = M(\omega)F_0 e^{i(\omega t + \phi(\omega))}$ with imaginary part $M(\omega)F_0 \sin(\omega t + \phi(\omega))$, so we take the particular solution to be $x_p(t) = M(\omega)F_0 \sin(\omega t + \phi(\omega))$. Comparing the forcing function to x_p, we see that the two functions have the same form but with different amplitudes and phase shifts. The ratio of the amplitude of the particular solution (or steady state), $M(\omega)F_0$, to that of the forcing function, F_0, is $M(\omega)$ and is called the **gain**. Also, x_p is shifted in time by $\frac{|\phi(\omega)|}{\omega}$ radians to the right, so $\phi(\omega)$ is called the **phase shift**. When we graph the gain and the phase shift against ω (using a $\log_{10}$ on the ω axis) we obtain the **Bodé plots**. Engineers refer to the value of $20\log_{10} M(\omega)$ as the gain in **decibels**.

EXAMPLE 6: Solve the initial-value problem

$$\begin{cases} \frac{d^2x}{dt^2} + 2\frac{dx}{dt} + 4x = \sin 2t \\ x(0) = \frac{1}{2}, \frac{dx}{dt}(0) = 1 \end{cases}$$

(a) Graph the solution simultaneously with the forcing function $f(t) = \sin 2t$. Approximate $M(2)$ and $\phi(2)$ using this graph. (b) Graph the corresponding Bodé plots. Compare the values of $M(2)$ and $\phi(2)$ with those obtained in (a).

SOLUTION: First, we define the nonhomogeneous differential equation in eq. Next, we solve the initial-value problem in sol.

```
> eq:=diff(x(t),t$2)+2*diff(x(t),t)+4*x(t)=sin(2*t):
```

```
> sol:=dsolve({eq,x(0)=1/2,D(x)(0)=1},x(t)):
```

```
sol:=combine(sol,trig);
```

$$sol := x(t) = \frac{3}{4}e^{(-t)}\cos(\sqrt{3}t) + \frac{7}{12}e^{(-t)}\sin(\sqrt{3}t)\sqrt{3} - \frac{1}{4}\cos(2t)$$

We extract the solution with rhs(sol) and graph it simultaneously with $f(t) = \sin 2t$ using a lighter level of gray for the graph of $f(t) = \sin 2t$.

```
> plot([rhs(sol),sin(2*t)],t=0..10,color=[BLACK,GRAY]);
```

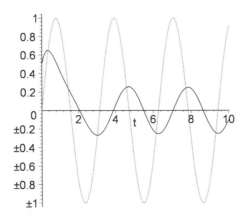

Clicking inside the graphics cell and clicking the mouse, we use the cursor to see that a minimum value of the forcing function occurs near 5.49 and a minimum value of the solution occurs near 6.26.

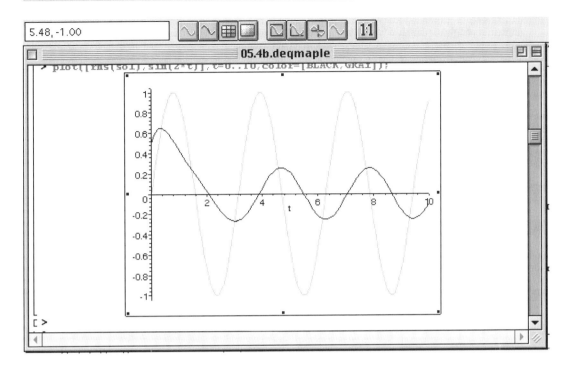

Therefore, the solution is shifted approximately $6.26 - 5.49 = 0.77$ units to the right. Returning to the solution containing $\omega t + \phi(\omega) = 2\left(t + \frac{1}{2}\phi(2)\right)$, we see that $\frac{1}{2}\phi(2) \approx -0.77$, so $\phi(2) \approx -1.54$. Using a similar technique, we approximate the amplitude of the steady-state solution (after it dies down) to be 0.255. Therefore, from the graph, $M(2) \approx 0.255$.

```
> 6.26-5.49;
```

$$0.77$$

(b) In the equation $\frac{d^2x}{dt^2} + 2\frac{dx}{dt} + 4x = \sin 2t$, $2c = 2$ and $k^2 = 4$. Therefore, $c = 1$ and $k = 2$. We define the gain function based on these constants in m(w).

```
> k:=2:
```

```
> c:=1:
```

```
> m:=w->1/sqrt((k^2-w^2)^2+4*c^2*w^2);
```

$$m := w \to \frac{1}{\sqrt{\left(k^2 - w^2\right)^2 + 4c^2 w^2}}$$

Because the graph of $M(\omega)$ is a log–log graph, we load the **plots** package to take advantage of the `loglogplot` command.

```
> with(plots);
```

```
> ?loglogplot
```

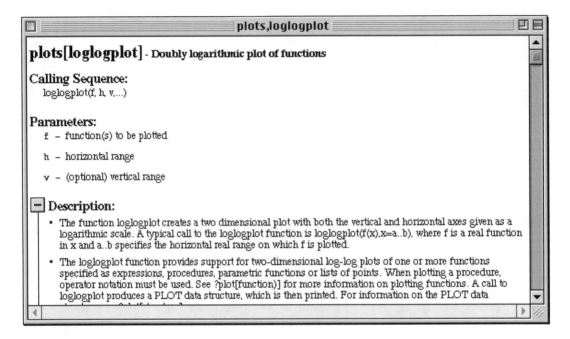

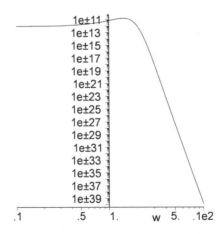

We graph `m(w)^20` because engineers are interested in $20\log_{10} M(\omega) = \log_{10} M(\omega)^{20}$. In (a), we obtained $M(2) \approx 0.255$. With the formula for $M(\omega)$, we find that $M(2) = 0.25$.

```
> loglogplot(m(w)^20,w=0.1..10);
```

Using `arccot`, we are able to graph $\phi(\omega)$. We define $\phi(\omega)$ so that it is returned as an angle between $-180°$ and $0°$. In (a), we found $\phi(2) \approx -1.54$ (radians). Here, we see that $\phi(2) = -90°$. However, $-\frac{\pi}{2} \approx -1.57$, so the approximations of $M(2)$ and $\phi(2)$ obtained in (a) are accurate.

```
> phi:=w->180*arccot((w^2-k^2)/(2*c*w))/Pi-180:

plot(phi(w),w=0.1..10);
```

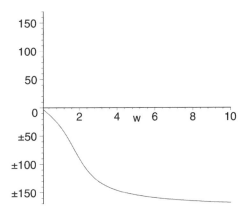

```
> evalf(phi(2));
```

$$-90$$

■

5.5 The Pendulum Problem

Suppose that a mass m is attached to the end of a rod of length L, the weight of which is negligible.

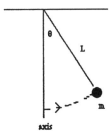

We want to determine the equation that describes the motion of the mass in terms of the displacement $\theta(t)$, which is measured counterclockwise in radians from the axis shown in the illustration. This is possible if we are given an initial displacement and an initial velocity of the mass. A force diagram for this situation is shown as follows.

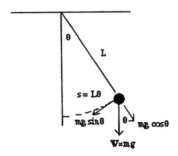

Notice that the forces are determined with trigonometry using the diagram. Here, $\cos \theta = mg/x$ and $\sin \theta = mg/y$, so we obtain the forces $x = mg \cos \theta$ and $y = mg \sin \theta$, indicated as follows.

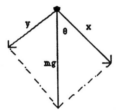

Because the momentum of the mass is given by $m\,ds/dt$, the rate of change of the momentum is

$$\frac{d}{dt}\left(m\frac{ds}{dt}\right) = m\frac{d^2s}{dt^2}$$

(where s represents the length of the arc formed by the motion of the mass). Then, because the force $mg \sin \theta$ acts in the opposite direction to the motion of the mass, we have the equation

$$m\frac{d^2s}{dt^2} = -mg \sin \theta.$$

Using the relationship from geometry between the length of the arc, the length of the rod, and the angle θ, $s = L\theta$, we have the relationship

$$\frac{d^2s}{dt^2} = \frac{d^2}{dt^2}(L\theta) = L\frac{d^2\theta}{dt^2}.$$

Hence, the displacement $\theta(t)$ satisfies $mL\frac{d^2\theta}{dt^2} = -mg \sin \theta$ or

$$mL\frac{d^2\theta}{dt^2} + mg \sin \theta = 0,$$

which is a nonlinear equation. However, because we are concerned only with small displacements, we note from the Maclaurin series for sin θ, $\sin \theta = \theta - \frac{\theta^3}{3!} + \frac{\theta^5}{5!} - \cdots$, that for small values of θ, $\sin \theta \approx \theta$. Therefore, we obtain the linear equation $mL\frac{d^2\theta}{dt^2} + mg\theta = 0$ or

$$\frac{d^2\theta}{dt^2} + \frac{g}{L}\theta = 0,$$

which approximates the original problem. If the initial displacement is given by $\theta(0) = \theta_0$ and the initial velocity is given by $\frac{d\theta}{dt}(0) = v_0$, then we have the initial-value problem

$$\begin{cases} \frac{d^2\theta}{dt^2} + \frac{g}{L}\theta = 0 \\ \theta(0) = \theta_0, \frac{d\theta}{dt}(0) = v_0 \end{cases}$$

to find the displacement function $\theta(t)$.

Suppose that $\omega^2 = g/L$ so that the differential equation becomes $\frac{d^2\theta}{dt^2} + \omega^2\theta = 0$. Therefore, functions of the form

$$\theta(t) = c_1 \cos \omega t + c_2 \sin \omega t,$$

where $\omega = \sqrt{g/L}$ satisfy the equation $\frac{d^2\theta}{dt^2} + \frac{g}{L}\theta = 0$. When we use the conditions $\theta(0) = \theta_0$ and $\frac{d\theta}{dt}(0) = v_0$, we find that the function

$$\theta(t) = \theta_0 \cos \omega t + \frac{v_0}{\omega} \sin \omega t$$

satisfies the equation as well as the initial displacement and velocity conditions. We can write this function solely in terms of a cosine function that includes a phase shift with

$$\theta(t) = \sqrt{\theta_0{}^2 + v_0^2/\omega^2} \cos(\omega t - \phi),$$

where $\phi = \cos^{-1}\left(\frac{\theta_0}{\sqrt{\theta_0^2 + v_0^2/\omega^2}}\right)$ and $\omega = \sqrt{g/L}$.

Note that the approximate period of $\theta(t)$ is $T = 2\pi/\omega = 2\pi\sqrt{L/g}$.

EXAMPLE 1: Determine the displacement of a pendulum of length $L = 32$ feet if $\theta(0) = 0$ and $\frac{d\theta}{dt}(0) = 1/2$ using both the linear and nonlinear models. What is the period? If the pendulum is part of a clock that ticks once for each time the pendulum makes a complete swing, how many ticks does the clock make in 1 minute?

SOLUTION: The linear initial-value problem that models this situation is

$$\begin{cases} \frac{d^2\theta}{dt^2} + \theta = 0 \\ \theta(0) = 0, \frac{d\theta}{dt}(0) = 1/2 \end{cases}$$

because $g/L = 32/32 = 1$.

We use `dsolve` to find a general solution of the equation

```
> gensol:=dsolve(diff(x(t),t$2)+x(t)=0,x(t));
```

$$gensol := x(t) = \_C1 \sin(t) + \_C2 \cos(t)$$

and the solution to the initial-value problem

$$\begin{cases} \frac{d^2\theta}{dt^2} + \theta = 0 \\ \theta(0) = a, \frac{d\theta}{dt}(0) = b \end{cases}.$$

```
> eq:=dsolve({diff(x(t),t$2)+x(t)=0,
        x(0)=a,D(x)(0)=b},x(t));
```

$$eq := x(t) = b \sin(t) + a \cos(t)$$

In this case, we have that $a = 0$ and $b = 1/2$, so substituting these values into `rhs(eq)` results in the solution to the initial-value problem.

```
> pen:=subs({a=0,b=1/2},rhs(eq));
```

$$pen := \frac{1}{2} \sin(t)$$

The period of this function is

$$T = 2\pi\sqrt{L/g} = 2\pi\sqrt{32 \text{ ft } /32 \text{ ft } /s^2} = 2\pi \sec.$$

Therefore, the number of ticks made by the clock per minute is calculated with the conversion $\frac{1 \text{ rev}}{2\pi \sec} \times \frac{1 \text{ tick}}{1 \text{ rev}} \times \frac{60 \sec}{1 \text{ min}} \approx 9.55$ ticks/min. Hence, the clock makes approximately 9.55 ticks in 1 minute.

To solve the nonlinear equation, we use `dsolve` together with the `numeric` option to generate a numerical solution to the initial-value problem.

```
> numsol:=dsolve({diff(x(t),t$2)+sin(x(t))=0,
        x(0)=0,D(x)(0)=1/2},x(t),numeric);
```

$$numsol := \mathbf{Proc}(rkf45\_x) \ \ldots \ \mathbf{end}$$

We then graph the approximate solution obtained in `pen` (with `plot`) together with the numerical solution obtained in `numsol` (with `odeplot`, which is contained in the **plots** package) on the interval [0,20] and display the graphs together with `display`.

```
> with(plots):
plot1:=plot(pen,t=0..20,color=BLACK):
plot2:=odeplot(numsol,[t,x(t)],0..20,color=GRAY):
display({plot1,plot2});
```

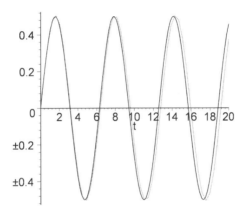

The graphs indicate that the error between the two functions increases as t increases, which is confirmed by graphing the absolute value of the difference of the two functions.

```
> plot(abs(pen-'rhs(numsol(t)[2])'),'t'=0..20,color=BLACK);
```

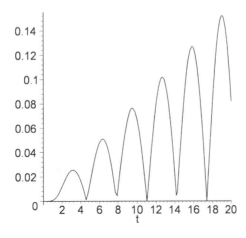

(Note that rhs(numsol(t)[2]) returns the value of $x(t)$ for a given value of t, as illustrated as follows with $t = 3$.)

> **numsol(3);**

$$\left[t = 3, x(t) = 0.09564929921788888, \frac{\partial}{\partial t} x(t) = -0.4907730638487397 \right]$$

> **rhs(numsol(3)[2]);**

$$0.09564929921788888$$

∎

Suppose that the pendulum undergoes a damping force that is proportional to the instantaneous velocity. Hence, the force due to damping is given as

$$F_R = b \frac{d\theta}{dt}.$$

Incorporating this force into the sum of the forces acting on the pendulum, we have the nonlinear equation $L \frac{d^2\theta}{dt^2} + b \frac{d\theta}{dt} + g \sin \theta = 0$. Again, using the approximation $\sin \theta \approx \theta$ for small values of θ, we obtain the linear equation $L \frac{d^2\theta}{dt^2} + b \frac{d\theta}{dt} + g\theta = 0$ that approximates the situation. Thus, we solve the initial-value problem

$$\begin{cases} L \frac{d^2\theta}{dt^2} + b \frac{d\theta}{dt} + g\theta = 0 \\ \theta(0) = \theta_0, \frac{d\theta}{dt}(0) = v_0. \end{cases}$$

to find the displacement function $\theta(t)$.

We investigate properties of solutions of this problem in the following example.

EXAMPLE 2: A pendulum of length $L = 8/5$ ft is subjected to the resistive force $F_R = 32/5 \frac{d\theta}{dt}$ due to damping. Determine the displacement function if $\theta(0) = 1$ and $\frac{d\theta}{dt}(0) = 2$.

SOLUTION: The initial-value problem that models this situation is

$$\begin{cases} \frac{8}{5} \frac{d^2\theta}{dt^2} + \frac{32}{5} \frac{d\theta}{dt} + 32\theta = 0 \\ \theta(0) = 1, \frac{d\theta}{dt}(0) = 2. \end{cases}$$

Simplifying the differential equation, we obtain $\frac{d^2\theta}{dt^2} + 4 \frac{d\theta}{dt} + 20\theta = 0$, and then using dsolve, we find the solution to the initial-value problem

```
> sol:=dsolve({diff(theta(t),t$2)+4*diff(theta(t),t)+
        20*theta(t)=0,
              theta(0)=1,D(theta)(0)=2},theta(t));
```

$$sol := \theta(t) = e^{(-2t)}\cos(4t) + e^{(-2t)}\sin(4t)$$

which is then graphed with `plot`.

```
> assign(sol):
plot(theta(t),t=0..2);
```

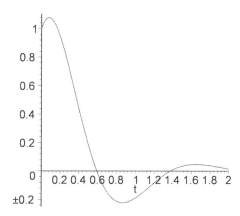

Notice that the damping causes the displacement of the pendulum to decrease over time.

To see the pendulum move, we define the procedure `pendulum`. Given `t0`, `pendulum` declares the variables `pt1` and `xt0` local to the procedure `pendulum`, defines `xt0` to be the value of $\theta(t)$ for $t = $ `t0`, defines `pt1` to be the point $\left(\frac{8}{5}\cos\left(\frac{3\pi}{2} + xt0\right), \frac{8}{5}\sin\left(\frac{3\pi}{2} + xt0\right)\right)$, and then displays the line segment connecting $(0,0)$ and `pt1` on the rectangle $[-2,2] \times [-2,0]$. The result *looks* like the pendulum at time $t = $ `t0`.

```
> pendulum:=proc(t0)
        local pt1,xt0;
        xt0:=evalf(subs(t=t0,theta(t)));
        pt1:=[8/5*cos(3*Pi/2+xt0),8/5*sin(3*Pi/2+xt0)];
        plot([[0,0],pt1],xtickmarks=2,ytickmarks=2,
              view=[-2..2,-2..0]);
        end:
```

To watch the pendulum move for $0 \le t \le 2$, we define k_vals to be 50 equally spaced numbers between 0 and 2 and then define to_animate to be the list of graphs pendulum(k) for k in k_vals. The resulting list of graphics is animated using the display function, which is contained in the **plots** package, together with the option insequence = true. We show one frame from the resulting animation in the following screen shot.

```
> k_vals:=seq(k*2/49,k=0..49):

to_animate:=[seq(pendulum(k),k=k_vals)]:

with(plots):

display(to_animate,insequence=true);
```

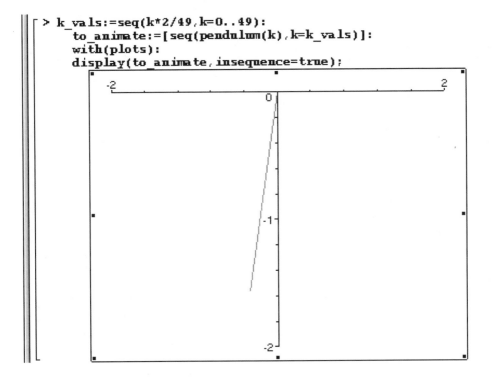

On the other hand, entering

```
> k_vals:=seq(k*2/11,k=0..11):

to_animate:=[seq(pendulum(k),k=k_vals)]:

with(plots):

to_show:=display(to_animate,insequence=true):

display(to_show);
```

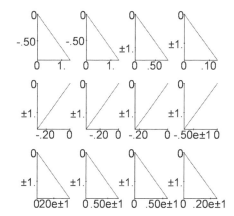

defines k_vals to be 12 equally spaced numbers between 0 and 2 and then defines to_animate to be the list of graphs pendulum(k) for *k* in k_vals. The resulting list of graphics is animated using the display function together with the option insequence = true but not displayed because a colon is included at the end of the display command. Instead, we use display again to display the result as a graphics array.

Notice that from our approximate solution, the displacement of the pendulum becomes very close to zero near *t* = 2, which was our observation from the graph of $\theta(t) = e^{-2t}(\cos 4t + 2 \sin 4t)$.

■

Our last example investigates the properties of the nonlinear differential equation.

EXAMPLE 3: Graph the solution to the initial-value problem

$$\begin{cases} \frac{d^2\theta}{dt^2} + 0.5\frac{d\theta}{dt} + \sin\theta = 0 \\ \theta(0) = \theta_0, \frac{d\theta}{dt}(0) = v_0 \end{cases}$$

subject to the following initial conditions.

θ_0	v_0	θ_0	v_0	θ_0	v_0	θ_0	v_0
− 1	0	− 0.5	0	0.5	0	1	0
0	− 2	0	− 1	0	1	0	2
1	1	1	− 1	− 1	1	− 1	− 1
1	2	1	3	− 1	4	− 1	5
− 1	2	− 1	3	1	− 4	1	− 5

SOLUTION: We begin by defining eq to be $\frac{d^2\theta}{dt^2} + 0.5\frac{d\theta}{dt} + \sin\theta = 0$.

```
> theta:='theta':
eq:=diff(theta(t),t$2)+0.5*diff(theta(t),t)+sin(theta(t))=0;
```

$$eq := \left(\frac{\partial^2}{\partial t^2}\theta(t)\right) + 0.5\left(\frac{\partial}{\partial t}\theta(t)\right) + \sin(\theta(t)) = 0$$

To avoid retyping the same commands, we define the procedure s. Given an ordered pair pair, s returns a numerical solution of eq that satisfies the initial conditions $\theta(0)$ equals the first coordinate of pair and $\theta'(0)$ equals the second coordinate of pair.

```
> s:=proc(pair)
      dsolve({Eq,theta(0)=pair[1],D(theta)(0)=pair[2]},
             theta(t),numeric)
      end:
```

For example, next we define t1 to be the list of ordered pairs corresponding to the initial conditions

θ_0	v_0	θ_0	v_0	θ_0	v_0	θ_0	v_0
-1	0	-0.5	0	0.5	0	1	0z

and then use map to apply s to each ordered pair in t1, naming the resulting list of processes to_graph.

```
> t1:=[[?1,0],[?0.5,0],[.5,0],[1,0]]:
to_graph:=map(s,t1);
```

$$to\_graph := [\mathbf{proc}(rkf45\_x) \ \dots \ \mathbf{end}, \mathbf{proc}(rkf45\_x) \ \dots \ \mathbf{end}, \mathbf{proc}(rkf45\_x) \ \dots \ \mathbf{end},$$

$$\mathbf{proc}(rkf45\_x) \ \dots \ \mathbf{end}]$$

Each of the processes in to_graph is then graphed using the odeplot command, which is contained in the **plots** package, on the interval [0,15] and all four graphs contained in the list `to_show` are displayed together using `display`, which is also contained in the **plots** package.

```
> with(plots):

to_show:=map(odeplot,to_graph,[t,theta(t)],0..15,

          color=BLACK):

display(to_show);
```

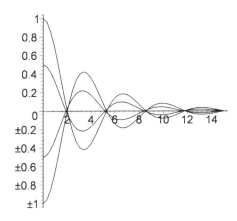

Similarly, entering

```
> t2:=[[0,?2],[0,?1],[0,1],[0,2]]:

to_graph:=map(s,t2):

to_show:=map(odeplot,to_graph,[t,theta(t)],0..15,

          color=BLACK):

display(to_show);
```

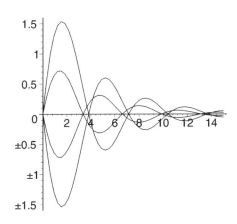

defines t2 to be the list of ordered pairs corresponding to the initial conditions

θ_0	v_0	θ_0	v_0	θ_0	v_0	θ_0	v_0
0	−2	0	−1	0	1	0	2

to_graph to be the resulting list of processes obtained by applying s to t2, and to_show to be the list of graphs of each of the processes in to_graph, and displays the list of graphs to_show.
The solutions that satisfy the remaining initial conditions are graphed in the same manner. Thus, entering

```
> t3:=[[1,1],[1,?1],[?1,1],[?1,?1]]:

to_graph:=map(s,t3):

to_show:=map(odeplot,to_graph,[t,theta(t)],0..15,

        color=BLACK):

display(to_show);
```

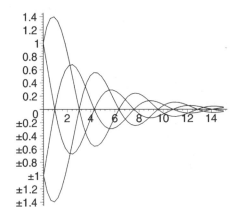

displays the graphs of the solutions that satisfy the initial conditions

θ_0	v_0	θ_0	v_0	θ_0	v_0	θ_0	v_0
1	1	1	−1	−1	1	−1	−1

Entering

```
> t4:=[[1,2],[1,3],[?1,4],[?1,5]]:
```

```
to_graph:=map(s,t4):
to_show:=map(odeplot,to_graph,[t,theta(t)],0..15,
           color=BLACK):
display(to_show);
```

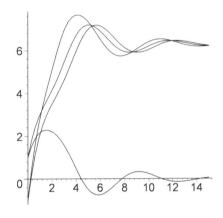

and

```
> t5:=[[?1,2],[?1,3],[1,?4],[1,?5]]:
to_graph:=map(s,t5):
to_show:=map(odeplot,to_graph,[t,theta(t)],0..15,
           color=BLACK):
display(to_show);
```

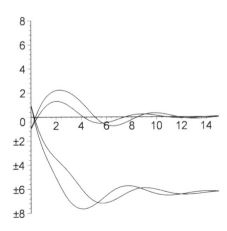

displays the graphs of the solutions that satisfy the initial conditions

θ_0	v_0	θ_0	v_0	θ_0	v_0	θ_0	v_0
1	2	1	3	-1	4	-1	5

CHAPTER 6

Introduction to the Laplace Transform

In previous chapters we have investigated solving the equation

$$a_n y^{(n)} + a_{n-1} y^{(n-1)} + \cdots + a_1 y' + a_0 y = g(x)$$

for y. We have seen that if the coefficients a_n, a_{n-1}, $\ldots$, a_0 are numbers, we can find a general solution of the equation by first solving the characteristic equation of the corresponding homogeneous equation, forming a general solution of the corresponding homogeneous equation, and then finding a particular solution to the nonhomogeneous equation.

If the coefficients a_n, a_{n-1}, $\ldots$, a_0 are not constants, the situation is more difficult. In particular cases, such as when the equation is a Cauchy–Euler equation, similar techniques can be used. In other cases, we might be able to use a series to find a solution. In each of these cases, however, the function $g(x)$ has typically been a smooth function. If $g(x)$ is not a smooth function, such as when $g(x)$ is a piecewise-defined function, solving the equation can become substantially more difficult.

In this chapter, we discuss a technique that transforms the equation

$$a_n y^{(n)} + a_{n-1} y^{(n-1)} + \cdots + a_1 y' + a_0 y = g(x)$$

into an algebraic equation that can often be solved so that a solution to the differential equation can be obtained.

Throughout this chapter we will take advantage of the commands `laplace` and `invlaplace`, which are contained in the **inttrans** package. Maple help supplies a great deal of information about these commands, as indicated in the following screen shots.

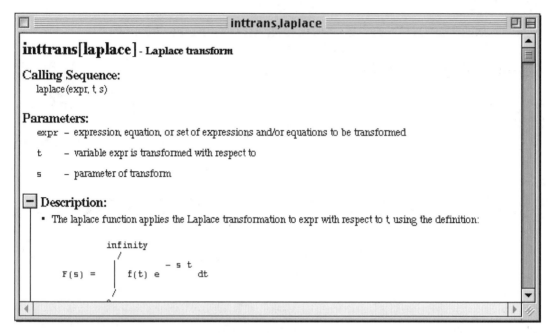

inttrans,laplace

inttrans[laplace] - Laplace transform

Calling Sequence:
laplace(expr, t, s)

Parameters:

expr – expression, equation, or set of expressions and/or equations to be transformed

t   – variable expr is transformed with respect to

s   – parameter of transform

☐ **Description:**

- The laplace function applies the Laplace transformation to expr with respect to t, using the definition:

$$F(s) = \int_{0}^{\infty} f(t)\, e^{-st}\, dt$$

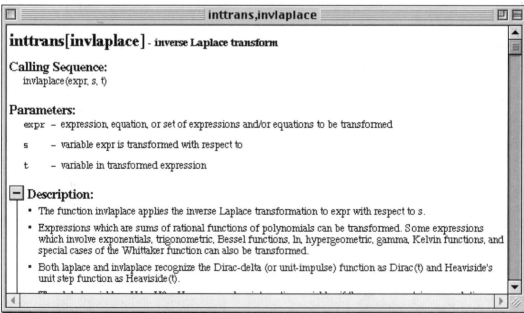

inttrans,invlaplace

inttrans[invlaplace] - inverse Laplace transform

Calling Sequence:
invlaplace(expr, s, t)

Parameters:

expr – expression, equation, or set of expressions and/or equations to be transformed

s   – variable expr is transformed with respect to

t   – variable in transformed expression

☐ **Description:**

- The function invlaplace applies the inverse Laplace transformation to expr with respect to s.

- Expressions which are sums of rational functions of polynomials can be transformed. Some expressions which involve exponentials, trigonometric, Bessel functions, ln, hypergeometric, gamma, Kelvin functions, and special cases of the Whittaker function can also be transformed.

- Both laplace and invlaplace recognize the Dirac-delta (or unit-impulse) function as Dirac(t) and Heaviside's unit step function as Heaviside(t).

Be sure that you have loaded this package by entering `with(inttrans)` during your current Maple work session before completing any example that uses either of these commands.

6.1 Definition of the Laplace Transform

Definition of the Laplace Transform

Definition
Laplace Transform

Let $f(t)$ be a function defined on the interval $[0, \infty)$. The **Laplace transform** of $f(t)$ is the function (of s)

$$\mathcal{L}\{f(t)\} = \int_0^\infty e^{-st} f(t)\,dt.$$

Because the Laplace transform yields a function of s, we often use the notation $\mathcal{L}\{f(t)\} = F(s)$ to denote the Laplace transform of $f(t)$.

EXAMPLE 1: Compute $\mathcal{L}\{f(t)\}$ if $f(t) = 1$.

SOLUTION:

$$\mathcal{L}\{f\}(s) = \int_0^\infty e^{-st}\,dt = \lim_{M \to \infty} \int_0^M e^{-st}\,dt = \lim_{M \to \infty} \left[\frac{e^{-st}}{s}\right]_{t=0}^{t=M} = -\frac{1}{s}\lim_{M \to \infty}\left(e^{-sM} - 1\right)$$

$$= -\frac{1}{s}(0 - 1) = \frac{1}{s}, \qquad s > 0.$$

We can use int to evaluate this integral as well.

```
> int(exp(-s*t),t=0..infinity);
```

$$\lim_{t \to \infty} -\frac{e^{(-st)} - 1}{s}$$

Notice that in order for $\lim_{M \to \infty} e^{-sM} = 0$, we must require that $s > 0$. (Otherwise, the limit does not exist.) After using assume to instruct Maple to assume that s is positive, Maple is able to compute this improper integral with int. Note that the symbol "$\sim$" appearing in the result indicates that we have made assumptions about s.

```
> assume(s>0):
int(exp(-s*t),t=0..infinity);
```

$$\frac{1}{s\sim}$$

Alternatively, we can take advantage of the `laplace` command, which is contained in the **inttrans** package.

```
> with(inttrans):
s:='s':
laplace(1,t,s);
```

$$\frac{1}{s}$$

■

EXAMPLE 2: Compute $\mathscr{L}\{f(t)\}$ if $f(t) = e^{at}$.

SOLUTION:

$$\mathscr{L}\{f(t)\} = \int_0^\infty e^{-st}f(t)dt = \int_0^\infty e^{-st}e^{at}dt = \int_0^\infty e^{-(s-a)t}dt = \lim_{M\to\infty}\left[-\frac{e^{-(s-a)t}}{s-a}\right]_{t=0}^{t=M}$$

$$= -\lim_{M\to\infty}\left(\frac{e^{-(s-a)M}}{s-a} - \frac{1}{s-a}\right) = -\left(0 - \frac{1}{s-a}\right) = \frac{1}{s-a}, \qquad s>a.$$

Notice that we must require $s>a$ so that $\lim_{M\to\infty} e^{-(s-a)M} = 0$. If you loaded the **intrans** package as in Example 1, `laplace` can be used to compute the Laplace transform of this function as well.

```
> a:='a':s:='s':
laplace(exp(a*t),t,s);
```

$$\frac{1}{s-a}$$

■

The formula $\mathscr{L}\{e^{at}\} = 1/(s-a)$, $s>a$ can now be used to avoid using the definition.

EXAMPLE 3: Compute (a) $\mathscr{L}\{e^{-3t}\}$ and (b) $\mathscr{L}\{e^{5t}\}$.

SOLUTION: We have that (a) $\mathscr{L}\{e^{-3t}\} = \frac{1}{s-(-3)} = \frac{1}{s+3}$, $s > -3$, and (b) $\mathscr{L}\{e^{5t}\} = \frac{1}{s-5}$, $s > 5$. Here, we use `map` to apply the function `laplace` to the list of functions $\{e^{-3t}, e^{5t}\}$ to compute both Laplace transforms immediately.

```
> map(laplace,[exp(-3*t),exp(5*t)],t,s);
```

$$\left[\frac{1}{s+3}, \frac{1}{s-5} \right]$$

In most cases, using the definition of the Laplace transform to calculate the Laplace transform of a function is a difficult and time-consuming task.

EXAMPLE 4: Compute (a) $\mathscr{L}\{t^3\}$, (b) $\mathscr{L}\{\sin a t\}$, and (c) $\mathscr{L}\{\cos a t\}$.

SOLUTION: To compute $\mathscr{L}\{t^3\}$ by hand requires application of integration by parts three times. Instead, we proceed with `int`. After computing $\int_0^M t^3 e^{-st} dt$,

```
> s:='s':
step_1:=int(t^3*exp(-s*t),t=0..M);
```

$$step\_1 := -\frac{e^{(-sM)}s^3 M^3 + 3e^{(-sM)}s^2 M^2 + 6e^{(-sM)}sM + 6e^{(-sM)} - 6}{s^4}$$

we instruct Maple to assume that s is positive with `assume` and then use `limit` to calculate $\lim\limits_{M \to \infty} \int_0^M t^3 e^{-st} dt$.

```
> assume(s>0):
limit(step_1,M=infinity);
```

$$6\frac{1}{s\sim^4}$$

The integrals that result when computing $\mathscr{L}\{\sin a t\}$ and $\mathscr{L}\{\cos a t\}$ using the definition of the Laplace transform each require the use of integration by parts twice. Instead, we use `laplace` to compute each Laplace transform. (Be sure that you have loaded the **inttrans** package during your current Maple session by entering `with(inttrans)` before entering the following commands.)

```
> s:='s':
```

```
map(laplace,[sin(a*t),cos(a*t)],t,s);
```

$$\left[\frac{a}{s^2 + a^2}, \frac{s}{s^2 + a^2} \right]$$

∎

As we can see, the definition of the Laplace transform is difficult to apply in most cases. We now discuss the **linearity property** that enables us to use the transforms that we have found thus far to find the Laplace transform of other functions.

Theorem
Linearity
Property

Let a and b be constants, and suppose that $\mathscr{L}\{f(t)\}$ and $\mathscr{L}\{g(t)\}$ exist. Then,

$$\mathscr{L}\{af(t) + bg(t)\} = a\mathscr{L}\{f(t)\} + b\mathscr{L}\{g(t)\}.$$

EXAMPLE 5: Calculate (a) $\mathscr{L}\{6\}$; (b) $\mathscr{L}\{5 - 2e^{-t}\}$.

SOLUTION: Using the results obtained in previous examples, we have for (a)

$$\mathscr{L}\{6\} = 6\mathscr{L}\{1\} = 6 \cdot \frac{1}{s} = \frac{6}{s}$$

and for (b)

$$\mathscr{L}\{5 - 2e^{-t}\} = 5\mathscr{L}\{1\} - 2\mathscr{L}\{e^{-t}\} = 5 \cdot \frac{1}{s} - 2 \cdot \frac{1}{s - (-1)} = \frac{5}{s} - \frac{2}{s+1}.$$

∎

Exponential Order, Jump Discontinuities, and Piecewise Continuous Functions

In calculus, we learn that some improper integrals diverge, which indicates that the Laplace transform may not exist for some functions. Therefore, we present the following definitions and theorems so that we can better understand the types of functions for which the Laplace transform exists.

Definition **Exponential Order**	A function f is of **exponential order b** if there are numbers b, $M > 0$, and $T > 0$ such that $$\|f(t)\| \le Me^{bt}$$ for $t > T$.

In the following sections, we will see that the Laplace transform is particularly useful in solving equations involving piecewise or recursively defined functions.

Definition **Jump Discontinuity**	A function f has a **jump discontinuity** at $t = c$ on the closed interval $[a,b]$ if the one-sided limits $\lim_{t \to c^+} f(t)$ and $\lim_{t \to c^-} f(t)$ are finite, but unequal, values. f has a **jump discontinuity** at $t = a$ if $\lim_{t \to a^+} f(t)$ is a finite value different from $f(a)$. f has a **jump discontinuity** at $t = b$ if $\lim_{t \to b^-} f(t)$ is a finite value different from $f(b)$.

Definition **Piecewise Continuous**	A function f is **piecewise continuous on the finite interval** $[a,b]$ if f is continuous at every point in $[a,b]$ except at finitely many points at which f has a jump discontinuity. A function f is **piecewise continuous on** $[0,\infty)$ if f is piecewise continuous on $[0,N]$ for all $N > 0$.

Theorem **Sufficient Condition** **for Existence of** $\mathscr{L}\{f(t)\}$	Suppose that f is a piecewise continuous function on the interval $[0,\infty)$ and that it is of exponential order b for $t > T$. Then, $\mathscr{L}\{f(t)\}$ exists for $s > b$.

EXAMPLE 6: Find the Laplace transform of $f(t) = \begin{cases} -1, & 0 \le t \le 4 \\ 1, & t > 4 \end{cases}$.

SOLUTION: Because f is a piecewise continuous function on $[0,\infty)$ and of exponential order, $\mathscr{L}\{f(t)\}$ exists. We use the definition and evaluate the integral using the sum of two integrals.

$$\mathscr{L}\{f(t)\} = \int_0^\infty f(t)e^{-st}dt = \int_0^4 -1 \cdot e^{-st}dt = \left[\frac{e^{-st}}{s}\right]_{t=0}^{t=4} + \lim_{M\to\infty}\left[-\frac{e^{-st}}{s}\right]_{t=4}^{t=M}$$

$$= \frac{1}{s}\left(e^{-4s} - 1\right) - \frac{1}{s}\lim_{M\to\infty}\left(e^{-Ms} - e^{-4s}\right) = \frac{1}{s}\left(2e^{-4s} - 1\right).$$

We can define and graph this piecewise-defined function with `piecewise` as illustrated next.

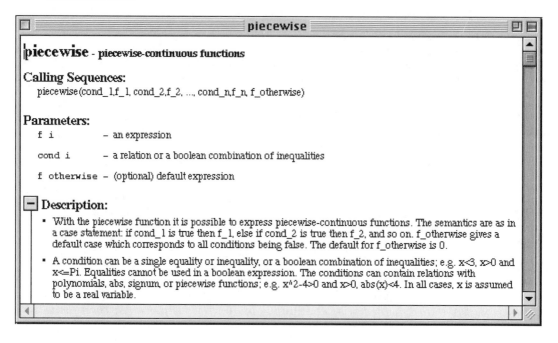

> `f:=piecewise(t>=0 and t<=4,-1,t>4,1);`

$$f := \begin{cases} -1 & -t \le 0 \text{ and } t - 4 \le 0 \\ 1 & 4 < t \end{cases}$$

> `plot(f,t=0..8);`

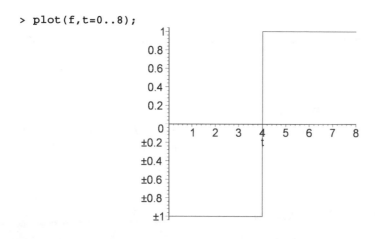

However, the `laplace` command is unable to compute the Laplace transform of f when f is defined in this manner.

> `laplace(f,t,s);`

$$\text{laplace}\left(\left\{\begin{array}{ll} -1 & -t \leq 0 \text{ and } t-4 \leq 0 \\ 1 & 4 < t \end{array}\right., t, s\right)$$

To compute the Laplace transform using Maple, we take advantage of the `Heaviside` function. The `Heaviside` function is defined by

$$\text{Heaviside (t)} = \begin{cases} 0, & \text{if } t < 0 \\ 1, & \text{if } t \geq 0 \end{cases}$$

so f is given by $f(t) = \text{Heaviside (t} - 4) - \text{Heaviside (4} - \text{t)}.$

> `plot(Heaviside(t-4)-Heaviside(4-t),t=0..8);`

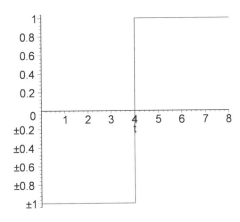

`laplace` is able to compute the Laplace transform of functions defined in terms of `Heavisde`. Thus, to use `laplace` to compute the Laplace transform of $f(t) = \text{Heaviside (t} - 4) - \text{Heaviside (4} - \text{t)}$ we enter

> `laplace(Heaviside(t-4)-Heaviside(4-t),t,s);`

$$2\frac{e^{(-4s)}}{s} - \frac{1}{s}$$

which gives us the same result as that given before.

■

Properties of the Laplace Transform

The definition of the Laplace transform is not easy to apply to most functions. Therefore, we now discuss several properties of the Laplace transform so that numerous transformations can be made without having to use the definition. Most of the properties discussed here follow directly from our knowledge of integrals.

Theorem
Shifting Property

If $\mathscr{L}\{f(t)\} = F(s)$ exists for $s > b$, then

$$\mathscr{L}\{e^{at}f(t)\} = F(s - a).$$

EXAMPLE 7: Find the Laplace transform of (a) $e^{-2t} \cos t$ and (b) $4te^{3t}$.

SOLUTION: (a)   In   this   case, $f(t) = \cos t$ and $a = -2$.   Using $F(s) = \mathscr{L}\{\cos t\} = s/(s^2 + 1)$, we replace each s with $s - a = s + 2$. Therefore,

$$\mathscr{L}\{e^{-2t} \cos t\} = \frac{s + 2}{(s + 2)^2 + 1} = \frac{s + 2}{s^2 + 4s + 5}.$$

(b) Using the linearity property, we have $\mathscr{L}\{4te^{3t}\} = 4\mathscr{L}\{te^{3t}\}$. To apply the shifting property we have $f(t) = t$ and $a = 3$, so we replace s in $F(s) = \mathscr{L}\{t\} = 1/s^2$ by $s - a = s - 3$. Therefore,

$$\mathscr{L}\{4te^{3t}\} = \frac{4}{(s - 3)^2}.$$

Identical results are obtained with `laplace`.

```
> laplace(exp(-2*t)*cos(t),t,s);
```

$$\frac{s + 2}{(s + 2)^2 + 1}$$

```
> laplace(4*t*exp(3*t),t,s);
```

$$4\frac{1}{(s - 3)^2}$$

■

In order to use the Laplace transform to solve differential equations, we will need to be able to compute the Laplace transform of the derivatives of an arbitrary function, provided the Laplace transform of such a function exists.

Theorem **Laplace Transform of** **the First Derivative**	Suppose that f is a piecewise continuous function on the interval $[0,\infty)$ and that it is of exponential order b for $t > T$. Then, for $s > b$ $$\mathscr{L}\{f'(t)\} = s\mathscr{L}\{f(t)\} - f(0).$$

```
> laplace(diff(f(t),t),t,s);
```

$$s\,\text{laplace}(f(t), t, s) - f(0)$$

The following corollary is obtained from the theorem using induction.

Corollary **Laplace Transform of** **the Higher Derivatives**	More generally, if $f^{(i)}(t)$ is a continuous function on $[0,\infty)$ for $i = 0, 1, \ldots, n-1$ and $f^{(n)}(t)$ is piecewise continuous on $[0, \infty)$ and of exponential order b, then for $s > b$ $$\mathscr{L}\left\{f^{(n)}(t)\right\} = s^n\mathscr{L}\{f(t)\} - s^{n-1}f(0) - \cdots - sf^{(n-2)}(0) - f^{(n-1)}(0).$$

We use `laplace` together with `map` to compute the Laplace transform of $f^{(i)}(t)$ for $i = 1, 2, 3$, and 4.

```
> derivs:=[seq(diff(f(t),t$n),n=1..4)];
```

$$derivs := \left[\frac{\partial}{\partial t}f(t), \frac{\partial^2}{\partial t^2}f(t), \frac{\partial^3}{\partial t^3}f(t), \frac{\partial^4}{\partial t^4}f(t)\right]$$

```
> map(laplace,derivs,t,s);
```

$$[s\,\text{laplace}(f(t), t, s) - f(0), s(s\,\text{laplace}(f(t), t, s) - f(0)) - D(f)(0),$$
$$s(s(s\,\text{laplace}(f(t), t, s) - f(0)) - D(f)(0)) - (D^{(2)})(f)(0),$$
$$s(s(s(s\,\text{laplace}(f(t), t, s) - f(0)) - D(f)(0)) - (D^{(2)})(f)(0))$$
$$- (D^{(3)})(f)(0)]$$

We will use this theorem and corollary in solving initial-value problems. However, we can also use them to find the Laplace transform of a function when we know the Laplace transform of the derivative of the function.

EXAMPLE 8: Find $\mathscr{L}\{\sin^2 kt\}$.

SOLUTION: We can use this corollary to find the Laplace transform of $f(t) = \sin^2 kt$. Notice that $f'(t) = 2k \sin kt \cos kt = k \sin 2kt$. Then, because $\mathscr{L}\{f'(t)\} = s\mathscr{L}\{f(t)\} - f(0)$ and $\mathscr{L}\{f'(t)\} = \mathscr{L}\{k \sin 2kt\} = k\frac{2k}{s^2+(2k)^2} = \frac{2k^2}{s^2+4k^2}$, we have $\frac{2k^2}{s^2+4k^2} = s\mathscr{L}\{f(t)\} - 0$. Therefore, $\mathscr{L}\{f(t)\} = \frac{2k^2}{s(s^2+4k^2)}$. As in previous examples, we see that the same results are obtained with `laplace`.

```
> laplace(sin(k*t)^2,t,s);
```

$$2\frac{k^2}{s(s^2 + 4k^2)}$$

∎

Theorem	Suppose that $F(s) = \mathscr{L}\{f(t)\}$ where f is a piecewise continuous
Derivatives of the	function on $[0,\infty)$ and of exponential order b. Then, for $s > b$,
Laplace Transform	$$\mathscr{L}\{t^n f(t)\} = (-1)^n \frac{d^n F}{ds^n}(s).$$

```
> array([seq([n,laplace(t^n*f(t),t,s)],n=1..4)]);
```

$$
\begin{vmatrix}
1 & -\left(\frac{\partial}{\partial s} \text{ laplace}(f(t), t, s)\right) \\
2 & \frac{\partial^2}{\partial s^2} \text{ laplace}(f(t), t, s) \\
3 & -\left(\frac{\partial^3}{\partial s^3} \text{ laplace}(f(t), t, s,)\right) \\
4 & \frac{\partial^4}{\partial s^4} \text{ laplace}(f(t), t, s)
\end{vmatrix}
$$

EXAMPLE 9: Find the Laplace transform of (a) $f(t) = t \cos 2t$, (b) $f(t) = t^2 e^{-3t}$.

SOLUTION: (a) In this case, $n = 1$ and $F(s) = \mathscr{L}\{\cos 2t\} = \frac{s}{s^2+4}$. Then

$$\mathscr{L}\{t\cos 2t\} = (-1)\frac{d}{ds}\left(\frac{s}{s^2+4}\right) = -\frac{(s^2+4) - s \cdot 2s}{(s^2+4)^2} = \frac{s^2-4}{(s^2+4)^2}.$$

```
> simplify(expand(laplace(t*cos(2*t),t,s)));
```

$$\frac{s^2-4}{(s^2+4)^2}$$

(b) Because $n = 2$ and $F(s) = \mathscr{L}\{e^{-3t}\} = 1/(s+3)$, we have

$$\mathscr{L}\{t^2 e^{-3t}\} = (-1)^2\frac{d^2}{ds^2}\left(\frac{1}{s+3}\right) = \frac{2}{s+3)^3}.$$

```
> laplace(t^2*exp(-3*t),t,s);
```

$$2\frac{1}{(s+3)^3}$$

■

EXAMPLE 10: Find $\mathscr{L}\{t^n\}$.

SOLUTION: Using the theorem with $\mathscr{L}\{t^n\} = \mathscr{L}\{t^n \cdot 1\}$, we have $f(t) = 1$. Then, $F(s) = \mathscr{L}\{1\} = 1/s$. Calculating the derivatives of F, we obtain

$$\frac{dF}{ds}(s) = -\frac{1}{s^2}$$

$$\frac{d^s F}{ds^2}(s) = \frac{2}{s^3}$$

$$\frac{d^3 F}{ds^3}(s) = -\frac{3 \cdot 2}{s^4}$$

$$\vdots$$

$$\frac{d^n F}{ds^n}(s) = (-1)^n\frac{n!}{s^{n+1}}$$

Therefore,

$$\mathscr{L}\{t^n\} = \mathscr{L}\{t^n \cdot 1\} = (-1)^n(-1)^n\frac{n!}{s^{n+1}} = (-1)^{2n}\frac{n!}{s^{n+1}} = \frac{n!}{s^{n+1}}.$$

```
> assume(n>0):
laplace(t^n,t,s);
```

$$s^{(-n\sim-1)}\Gamma(n\sim+1)$$

(Recall that for nonnegative integers n, $\Gamma(n+1) = n!$.)

∎

EXAMPLE 11: Compute the Laplace transform of $f(t)$, $f'(t)$, and $f''(t)$ if $f(t) = (3t-1)^3$.

SOLUTION: First, $f(t) = (3t-1)^3 = 27t^3 - 27t^2 + 9t - 1$ and $\mathscr{L}\{t^n\} = n!/s^{n+1}$ so

$$\mathscr{L}\{f(t)\} = 27\frac{3!}{s^4} - 27\frac{2!}{s^3} + 9\frac{1}{s^2} - \frac{1}{s} = \frac{162 - 54s + 9s^2 - s^3}{s^4}$$

```
> f:=t->(3*t-1)^3:
lf:=laplace(f(t),t,s);
```

$$lf := 162\frac{1}{s^4} - 54\frac{1}{s^3} + 9\frac{1}{s^2} - \frac{1}{s}$$

By the previous theorem, $\mathscr{L}\{f'(t)\} = s\,\mathscr{L}\{f(t)\} - f(0)$. Hence,

$$\mathscr{L}\{f'(t)\} = s\frac{162 - 54s + 9s^2 - s^3}{s^4} - f(0) = \frac{162 - 54s + 9s^2 - s^3}{s^3} + 1 = \frac{9(18 - 6s + s^2)}{s^3}$$

```
> lfprime:=laplace(diff(f(t),t),t,s);
```

$$lfprime := 162\frac{1}{s^3} - 54\frac{1}{s^2} + 9\frac{1}{s}$$

```
> expand(s*lf-f(0));
```

$$162\frac{1}{s^3} - 54\frac{1}{s^2} + 9\frac{1}{s}$$

With the corollary, $\mathscr{L}\{f''(t)\} = s^2\,\mathscr{L}\{f(t)\} - sf(0) - f'(0)$ and using $f'(t) = 9(3t-1)^2$,

$$\mathscr{L}\{f''(t)\} = s^2 \frac{162 - 54s + 9s^2 - s^3}{s^4} - sf(0) - f'(0) = \frac{54(3-s)}{s^2}.$$

```
> lfdoubleprime:=laplace(diff(f(t),t$2),t,s);
```

$$lfdoubleprime := 162\frac{1}{s^2} - 54\frac{1}{s}$$

```
> expand(s^2*lf-s*f(0)-D(f)(0));
```

$$162\frac{1}{s^2} - 54\frac{1}{s}$$

∎

Using the properties of the Laplace transform, we can compute the Laplace transform of a large number of frequently encountered functions. The following table lists the Laplace transform of several frequently encountered functions.

$f(t)$	$F(s) = \mathscr{L}\{f(t)\}$	$f(t)$	$F(s) = \mathscr{L}\{f(t)\}$
1	$\frac{1}{s}$, $s>0$	t^n, $n = 1, 2, \ldots$	$\frac{n!}{s^{n+1}}$, $s>0$
e^{at}	$\frac{1}{s-a}$, $s>a$	$t^n e^{at}$, $n = 1, 2, \ldots$	$\frac{n!}{(s-a)^{n+1}}$
$\sin kt$	$\frac{k}{s^2+k^2}$	$e^{at}\sin kt$	$\frac{k}{(s-a)^2+k^2}$
$\cos kt$	$\frac{s}{s^2+k^2}$	$e^{at}\cos kt$	$\frac{s-a}{(s-a)^2+k^2}$
$\sinh kt$	$\frac{k}{s^2-k^2}$	$e^{at}\sin hkt$	$\frac{k}{(s-a)^2-k^2}$
$\cosh kt$	$\frac{s}{s^2-k^2}$	$e^{at}\cos hkt$	$\frac{s-a}{(s-a)^2-k^2}$

6.2 The Inverse Laplace Transform

Definition of the Inverse Laplace Transform

In the previous section, we were concerned with finding the Laplace transform of a given function either through the use of the definition of the Laplace transform or with one of the numerous properties of the Laplace transform. At that time, we discussed the sufficient conditions for the existence of the Laplace transform. In this section, we will reverse this process: given a function $F(s)$ we want to find a function $f(t)$ such that $\mathscr{L}\{f(t)\} = F(s)$.

Definition	The **inverse Laplace transform** of the function $F(s)$ is the unique
Inverse Laplace	continuous function $f(t)$ on $[0,\infty)$ that satisfies $\mathscr{L}\{f(t)\} = F(s)$.
Transform	We denote the inverse Laplace transform of $F(s)$ as

$$f(t) = \mathscr{L}^{-1}\{F(s)\}$$

If the only functions that satisfy this relationship are discontinuous on $[0,\infty)$, we choose a piecewise continuous function on $[0,\infty)$ to be $\mathscr{L}^{-1}\{F(s)\}$

The table of Laplace transforms listed in Section 7.1 is useful in finding the inverse Laplace transform of a given function.

EXAMPLE 1: Find the inverse Laplace transform of (a) $F(s) = 1/(s-6)$, (b) $F(s) = 2/(s^2+4)$, (c) $F(s) = 6/s^4$, and (d) $F(s) = 6\big/(s+2)^4$.

SOLUTION: (a) Because $\mathscr{L}\{e^{6t}\} = 1/(s-6)$, $\mathscr{L}^{-1}\{1/(s-6)\} = e^{6t}$.

(b) $\mathscr{L}\{\sin 2t\} = \frac{2}{s^2+2^2} = \frac{2}{s^2+4}$ so $\mathscr{L}^{-1}\left\{\frac{2}{s^2+4}\right\} = \sin 2t$.

(c) Note that $\mathscr{L}\{t^3\} = 3/s^4 = 6/s^4$ so $\mathscr{L}^{-1}\{6/s^4\} = t^3$.

(d) $F(s) = 6\big/(s+2)^4$ is obtained from $F(s) = 6/s^4$ by substituting $s+2$ for s.
Therefore by the shifting property, $\mathscr{L}\{e^{-2t}t^3\} = 6\big/(s+2)^4$, so
$\mathscr{L}^{-1}\left\{6\big/(s+2)^4\right\} = e^{-2t}\mathscr{L}^{-1}\{6/s^4\} = e^{-2t}t^3$. In the same way that we use
`laplace` to calculate $\mathscr{L}\{f(t)\}$ we use `invlaplace` to calculate $\mathscr{L}^{-1}\{F(s)\}$.

```
> with(inttrans):
invlaplace(1/(s-6),s,t);
```

$$e^{(6t)}$$

Here, we use `map` to apply the function `invlaplace` to the list of functions
$\left\{2\big/(s^2+4),\ 6/s^4,\ 6\big/(s+2)^4\right\}$.

```
> simplify(map(invlaplace,[2/(s^2+4),6/s^4,6/(s+2)^2],s,t));
```

$$[\sin(2t), t^3, 6te^{(-2t)}]$$

Theorem **Linearity Property of** **the Inverse Laplace** **Transform**	Suppose that $\mathscr{L}^{-1}\{F(s)\}$ and $\mathscr{L}^{-1}\{G(s)\}$ exist and are continuous on $[0,\infty)$. Also, suppose that a and b are constants. Then, $$\mathscr{L}^{-1}\{aF(s)+bG(s)\}=a\mathscr{L}^{-1}\{F(s)\}+b\mathscr{L}^{-1}\{G(s)\}$$

EXAMPLE 2: Find the inverse Laplace transform of (a) $F(s)=\frac{1}{s^3}$, (b) $F(s)=-\frac{7}{s^2+16}$, and (c) $F(s)=\frac{5}{s}-\frac{2}{s-10}$.

SOLUTION: (a) $\mathscr{L}^{-1}\left\{\frac{1}{s^3}\right\}=\mathscr{L}^{-1}\left\{\frac{1}{2}\frac{2}{s^3}\right\}=\frac{1}{2}\mathscr{L}^{-1}\left\{\frac{2}{s^3}\right\}=\frac{1}{2}t^2.$

```
> invlaplace(1/s^3,s,t);
```

$$\frac{1}{2}t^2$$

(b) $\mathscr{L}^{-1}\left\{-\frac{7}{s^2-16}\right\}=7\mathscr{L}^{-1}\left\{\frac{1}{s^2+16}\right\}-7\mathscr{L}^{-1}\left\{\frac{1}{4}\frac{4}{s^2-4^2}\right\}=\frac{7}{4}\mathscr{L}^{-1}\left\{\frac{4}{s^2-4^2}\right\}=-\frac{7}{4}\sin 4t.$

```
> simplify(invlaplace(-7/(s^2+16),s,t));
```

$$-\frac{7}{4}\sin(4t)$$

(c) $\mathscr{L}^{-1}\left\{\frac{5}{s}-\frac{2}{s-10}\right\}=5\mathscr{L}^{-1}\left\{\frac{1}{s}\right\}-2\mathscr{L}^{-1}\left\{\frac{1}{s-10}\right\}=5-2e^{10t}.$

```
> invlaplace(5/s-2/(s-10),s,t);
```

$$5-2e^{(10t)}$$

■

Of course, the functions $F(s)$ that are encountered do not have to be of the forms previously discussed. For example, sometimes we must complete the square in the denominator of $F(s)$ before finding $\mathscr{L}^{-1}\{F(s)\}$.

EXAMPLE 3: Determine $\mathscr{L}^{-1}\left\{\frac{s}{s^2+2s+5}\right\}$.

SOLUTION: Notice that all of the forms of $F(s)$ in the table of Laplace transforms involve a term of the form $s^2 + k^2$ in the denominator. However, through shifting, this term is replaced by $(s - a)^2 + k^2$. We obtain a term of this form in the denominator by completing the square. This yields

$$\frac{s}{s^2 + 2s + 5} = \frac{s}{(s^2 + 2s + 1) + 4} = \frac{s}{(s + 1)^2 + 4}.$$

Note that we can take advantage of the `completesquare` command, which is contained in the **student** package, to perform the operation of completing the square encountered here. In the following, we load the **student** package by entering `with(student)`. The commands contained in the **student** package are returned because a semicolon is included at the end of the command instead of a colon.

```
> with(student);
```

> [*D, Diff, Doubleint, Int, Limit, Lineint, Product, Sum, Tripleint,*
>
> *changevar, combine, completesquare, distance, equate, extrema,*
>
> *integrand, intercept, intparts, isolate, leftbox, leftsum, makeproc,*
>
> *maximize, middlebox, middlesum, midpoint, minimize, powsubs,*
>
> *rightbox, rightsum, showtangent, simpson, slope, summand,*
>
> *trapezoid, value]*

Then, entering

```
> completesquare(s^2+2*s+5,s);
```

$$(s + 1)^2 + 4$$

shows us that $s^2 + 2s + 5 = (s + 1)^2 + 4$.
Because the variable appears in the numerator, we must write it in the form $s + 1$ in order to find the inverse Laplace transform. Doing so, we find that

$$\frac{s}{s^2 + 2s + 5} = \frac{s}{(s + 1)^2 + 4} = \frac{(s + 1) - 1}{(s + 1)^2 + 4}.$$

Hence,

$$\mathcal{L}^{-1}\left\{\frac{s}{s^2 + 2s + 5}\right\} = \mathcal{L}^{-1}\left\{\frac{(s + 1) - 1}{(s + 1)^2 + 4}\right\} = \mathcal{L}^{-1}\left\{\frac{(s + 1)}{(s + 1)^2 + 2^2}\right\} - \frac{1}{2}\mathcal{L}^{-1}\left\{\frac{2}{(s + 1)^2 + 2^2}\right\}$$

$$= e^{-t}\cos 2t - \frac{1}{2}e^{-t}\sin 2t.$$

As in previous examples, we see that `invlaplace` quickly finds $\mathcal{L}^{-1}\left\{\frac{s}{s^2+2s+5}\right\}$.

```
> invlaplace(s/(s^2+2*s+5),s,t);
```

$$e^{(-t)}\cos(2t) - \frac{1}{2}e^{(-t)}\sin(2t)$$

■

In other cases, *partial fractions* must be used to obtain terms for which the inverse Laplace transform can be found. Suppose that $F(s) = P(s)/Q(s)$, where $Q(s)$ are polynomials of degree m and n, respectively. If $n>m$, the method of partial fractions can be used to expand $F(s)$. Recall from calculus that there are many possible situations that can be solved through partial fractions. We illustrate three cases in the examples that follow.

Linear Factors (Nonrepeated)

In this case, $Q(s)$ can be written as a product of linear factors, so

$$Q(s) = (s - q_1)(s - q_2)\cdots(s - q_n)$$

where $q_1, q_2, \ldots, q_n$ are distinct numbers. Therefore, $F(s)$ can be written as

$$F(s) = \frac{A_1}{s - q_1} + \frac{A_2}{s - q_2} + \cdots + \frac{A_n}{s - q_n}$$

where $A_1, A_2, \ldots, A_n$ are constants that must be determined.

EXAMPLE 4: Find $\mathscr{L}^{-1}\left\{\frac{3s-4}{s(s-4)}\right\}$.

SOLUTION: In this case, we have distinct linear factors in the denominator. Hence, we write $F(s)$ as

$$\frac{3s - 4}{s(s - 4)} = \frac{A}{s} + \frac{B}{s - 4}.$$

Multiplying both sides of the equation by the denominator $s(s - 4)$, we have

$$3s - 4 = A(s - 4) + Bs = (A + B)s - 4A.$$

Equating the coefficients of s as well as the constant terms, we see that the following system of equations must be satisfied.

$$\begin{cases} A + B = 3 \\ -4A = -4 \end{cases}.$$

Maple can solve the system of equations $\begin{cases} A + B = 3 \\ -4A = -4 \end{cases}$ with `solve` or we can solve the equation $3s - 4 = A(s - 4) + Bs = (A + B)s - 4A$ for A and B with `match`

as shown next.

```
> match(3*s-4=(A+B)*s-4*A,s,'Val'):
Val;
```

$${B = 2, A = 1}$$

Hence, $A = 1$ and $B = 2$. Therefore,

$$\frac{3s - 4}{s(s - 4)} = \frac{A}{s} + \frac{B}{s - 4} = \frac{1}{s} + \frac{2}{s - 4},$$

so

$$\mathscr{L}^{-1}\left\{\frac{3s - 4}{s(s - 4)}\right\} = \mathscr{L}^{-1}\left\{\frac{1}{s} + \frac{2}{s - 4}\right\} = 1 + 2e^{4t}$$

or we can simply use `invlaplace` as shown in the previous examples.

```
> invlaplace((3*s-4)/(s*(s-4)),s,t);
```

$$1 + 2e^{(4t)}$$

In this case, we can use `convert` together with the `parfrac` option to compute the partial fraction decomposition directly

```
> convert((3*s-4)/(s*(s-4)),parfrac,s);
```

$$\frac{1}{s} + 2\frac{1}{s - 4}$$

∎

Repeated Linear Factors

If $s - q$ is a factor of $Q(s)$ of multiplicity k, the terms in the partial fraction expansion of $F(s)$ that correspond to this factor are

$$\frac{A_1}{s-q} + \frac{A_2}{(s-q)^2} + \cdots + \frac{A_k}{(s-q)^k}$$

where $A_1, A_2, \ldots, A_k$ are constants that must be found.

EXAMPLE 5: Calculate $\mathscr{L}^{-1}\left\{\frac{5s^2+20s+6}{s^3+2s^2+s}\right\}$.

SOLUTION: After using `convert` together with the `parfrac` option,

```
> convert((5*s^2+20*s+6)/(s^3+2*s^2+s),parfrac,s);
```

$$6\frac{1}{s} + 9\frac{1}{(s+1)^2} - \frac{1}{s+1}$$

we see that

$$\frac{5s^2+20s+6}{s^3+2s^2+s} = \frac{6}{s} - \frac{1}{s+1} + \frac{9}{(s+1)^2}.$$

Therefore,

$$\mathscr{L}^{-1}\left\{\frac{5s^2+20s+6}{s^3+2s^2+s}\right\} = \mathscr{L}^{-1}\left\{\frac{6}{s} - \frac{1}{s+1} + \frac{9}{(s+1)^2}\right\}$$

$$= 6\mathscr{L}^{-1}\left\{\frac{1}{s}\right\} - \mathscr{L}^{-1}\left\{\frac{1}{s+1}\right\} + 9\mathscr{L}^{-1}\left\{\frac{1}{(s+1)^2}\right\}$$

$$= 6 - e^{-t} + 9te^{-t}.$$

As expected, we obtain the same results using `invlaplace`.

```
> invlaplace((5*s^2+20*s+6)/(s^3+2*s^2+s),s,t);
```

$$6 + 9te^{(-t)} - e^{(-t)}$$

■

Irreducible Quadratic Factors

If $(s-a)^2 + b^2$ is a factor of $Q(s)$ of multiplicity k that cannot be reduced to linear factors, the partial fraction expansion of $F(s)$ corresponding to $(s-a)^2 + b^2$ is

$$\frac{A_1 s + B_1}{(s-a)^2 + b^2} + \frac{A_2 s + B_2}{[(s-a)^2 + b^2]^2} + \cdots + \frac{A_{ks} + B_k}{[(s-a)^2 + b^2]^k}.$$

EXAMPLE 6: Find $\mathscr{L}^{-1}\left\{\frac{2s^3 - 4s - 8}{(s^2 - s)(s^2 + 4)}\right\}$.

SOLUTION: As in the previous example, we use `convert` together with the `parfrac` option

```
> convert((2*s^3-4*s-8)/((s^2-s)*(s^2+4)),parfrac,s);
```

$$2\frac{1}{s} - 2\frac{1}{s-1} + 2\frac{2+s}{s^2+4}$$

to obtain

$$\frac{2s^3 - 4s - 8}{s(s-1)(s^2+4)} = \frac{2}{s} - \frac{2}{s-1} + \frac{2s+4}{s^2+4}.$$

Thus,

$$\mathscr{L}^{-1}\left\{\frac{2s^3 - 4s - 8}{s(s-1)(s^2+4)}\right\} = 2\mathscr{L}^{-1}\left\{\frac{1}{s}\right\} - 2\mathscr{L}^{-1}\left\{\frac{1}{s-1}\right\} + 2\mathscr{L}^{-1}\left\{\frac{s}{s^2+4}\right\} + 2\mathscr{L}^{-1}\left\{\frac{2}{s^2+4}\right\}$$

$$= 2 - 2e^t + 2\cos 2t + 2\sin 2t.$$

```
> invlaplace((2*s^3-4*s-8)/((s^2-s)*(s^2+4)),s,t);
```

$$2 - 2e^t + 2\cos(2t) + 2\sin(2t)$$

■

Laplace Transform of an Integral

We have seen that the Laplace transform of the derivatives of a given function can be found from the Laplace transform of the function. Similarly, the Laplace transform of the integral of a given function can also be obtained from the Laplace transform of the original function.

Theorem	Suppose that $F(s) = \mathcal{L}\{f(t)\}$ where f is a piecewise continuous
Laplace Transform of	function on $[0,\infty)$ and of exponential order b. Then, for $s > b$,
an Integral	

$$\mathcal{L}\left\{\int_0^t f(\alpha)d\alpha\right\} = \frac{\mathcal{L}\{f(t)\}}{s}.$$

The theorem states that

$$\mathcal{L}^{-1}\left\{\frac{\mathcal{L}\{f(t)\}}{s}\right\} = \int_0^t f(\alpha)d\alpha.$$

EXAMPLE 7: Compute $\mathcal{L}^{-1}\left\{\frac{1}{s(s+2)}\right\}$.

SOLUTION: In   this   case, $\frac{1}{s(s+2)} = \frac{1/(s+2)}{s}$,   so $\mathcal{L}\{f(t)\} = \frac{1}{s+2}$.   Therefore,
$f(t) = \mathcal{L}^{-1}\left\{\frac{1}{s+2}\right\} = e^{-2t}$.

With the previous theorem, we then have

$$\mathcal{L}^{-1}\left\{\frac{1}{s(s+2)}\right\} = \int_0^t e^{-2\alpha}d_\alpha = \frac{1}{2}(1 - e^{2t}).$$

Note that the same result is obtained through a partial fraction expansion of $\frac{1}{s(s+2)}$:
Because $\frac{1}{s(s+2)} = \frac{1}{2s} - \frac{1}{2(2+s)}$,

$$\mathcal{L}^{-1}\left\{\frac{1}{s(s+2)}\right\} = \mathcal{L}^{-1}\left\{\frac{1}{2s} - \frac{1}{2(2+s)}\right\} = \frac{1}{2} - \frac{1}{2}e^{-2t}.$$

■

The following theorem is useful in determining whether the inverse Laplace transform of
a function $F(s)$ exists.

| **Theorem** | Suppose that f is a piecewise continuous function on $[0,\infty)$ and |
| | of exponential order b. Then. |

$$\lim_{s\to\infty} F(s) = \lim_{s\to\infty} \mathcal{L}\{f(t)\} = 0.$$

EXAMPLE 8: Determine whether the inverse Laplace transform of the functions may exist for the functions (a) $F(s) = \frac{2s}{s-6}$ and (b) $F(s) = \frac{s^3}{s^2+16}$.

SOLUTION: In both cases, we find $\lim_{s \to \infty} F(s)$. If this value is not zero, then $\mathcal{L}^{-1}\{F(s)\}$ cannot be found.

(a) $\lim_{s \to \infty} F(s) = \lim_{s \to \infty} \frac{2s}{s-6} = 2 \neq 0$, so $\mathcal{L}^{-1}\left\{\frac{2s}{s-6}\right\}$ does not exist.

(b) $\lim_{s \to \infty} F(s) = \lim_{s \to \infty} \frac{s^3}{s^2+16} = \infty \neq 0$.

Thus, $\mathcal{L}^{-1}\left\{\frac{s^3}{s^2+16}\right\}$ does not exist.

■

6.3 Solving Initial-Value Problems with the Laplace Transform

Laplace transforms can be used to solve certain initial-value problems. Typically, when we use Laplace transforms to solve an initial-value problem for a function y, we use the following steps:

1. Compute the Laplace transform of each term in the differential equation.
2. Solve the resulting equation for $\mathcal{L}\{y(t)\}$.
3. Determine y by computing the inverse Laplace transform of $\mathcal{L}\{y(t)\}$.

The advantage of this method is that through the use of the property

$$\mathcal{L}\left\{f^{(n)}(t)\right\} = s^n \mathcal{L}\{f(t)\} - s^{n-1}f(0) - \cdots - sf^{(n-2)}(0) - f^{(n-1)}(0)$$

we change the differential equation to an algebraic equation.

EXAMPLE 1: Solve the initial-value problem $y' - 4y = e^{4t}$, $y(0) = 0$.

SOLUTION: We begin by taking the Laplace transform of both sides of the differential equation and then solving for $\mathcal{L}\{y(t)\} = Y(s)$. Because $\mathcal{L}\{y'\} = sY(s) - y(0) = sY(s)$, we have

$$\mathscr{L}\{y' - 4y\} = \mathscr{L}\{e^{4t}\}$$

$$\mathscr{L}\{y'\} - 4\mathscr{L}\{y\} = \frac{1}{s-4}$$

$$sY(s) - 4Y(s) = \frac{1}{s-4}$$

$$(s-4)Y(s) = \frac{1}{s-4} \Rightarrow Y(s) = \frac{1}{(s-4)^2}.$$

We carry out the same steps with Maple. After computing the Laplace transform of each side of the equation,

```
> with(inttrans):
step_1:=laplace(diff(y(t),t)-4*y(t)=exp(4*t),t,s);
```

$$step\_1 := s\ \mathrm{laplace}(y(t), t, s) - y(0) - 4\ \mathrm{laplace}(y(t), t, s) = \frac{1}{s-4}$$

we apply the initial condition

```
> step_2:=subs(y(0)=0,step_1);
```

$$step\_2 := s\ \mathrm{laplace}(y(t), t, s) - 4\ \mathrm{laplace}(y(t), t, s) = \frac{1}{s-4}$$

and solve the result equation for $\mathscr{L}\{y(t)\} = Y(s)$.

```
> step_3:=solve(step_2,laplace(y(t),t,s));
```

$$step\_3 := \frac{1}{s^2 - 8s + 16}$$

Hence, by using the shifting property with $\mathscr{L}\{t\}\} = 1/s^2$, we have

$$y(t) = \mathscr{L}^{-1}\frac{1}{(s-4)^2} = te^{4t}.$$

Identical results are obtained using `invlaplace`.

```
> Sol:=invlaplace(step_3,s,t);
```

$$Sol := te^{(4t)}$$

We then graph the solution obtained with `plot`.

```
> plot(Sol,t=0..1);
```

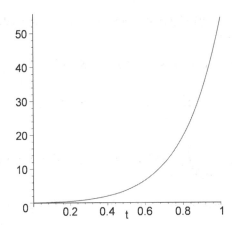

We can also use `dsolve` to solve the initial-value problem directly.

```
> dsolve({diff(y(t),t)-4*y(t)=exp(4*t),y(0)=0},y(t));
```

$$y(t) = te^{(4t)}$$

■

As we can see from the previous example, Laplace transforms are useful in solving nonhomogeneous equations. Hence, problems in Chapter 4 for which the methods of undetermined coefficients and variation of parameters were difficult to apply may be more easily solved through the method of Laplace transforms.

EXAMPLE 2: Use Laplace transforms to solve $y'' + 4y = e^{-t}\cos 2t$ subject to $y(0) = 0$ and $y'(0) = -1$.

SOLUTION: As in Example 1, we proceed by computing the Laplace transform of each side of the equation with `laplace`

```
> step_1:=laplace(diff(y(t),t$2)+4*y(t)=
       exp(-t)*cos(2*t),t,s);
```

$$step\_1 := s(s \, \text{laplace}(y(t), t, s) - y(0)) - D(y)(0) + 4 \, \text{laplace}(y(t), t, s) = \frac{s+1}{(s+1)^2 + 4}$$

and then applying the initial conditions $y(0) = 0$ and $y'(0) = -1$ with `subs`, naming the result `step_2`.

```
> step_2:=subs({y(0)=0,D(y)(0)=-1},step_1);
```

$$step\_2 := s^2 \, \text{laplace}(y(t), t, s) + 1 + 4 \, \text{laplace}(y(t), t, s) = \frac{s+1}{(s+1)^2 + 4}$$

Next, we solve `step_2` for the Laplace transform of *y(t)* and simplify the result, naming the resulting output `step_3`

```
> step_3:=simplify(solve(step_2,laplace(y(t),t,s)));
```

$$step\_3 := \frac{s^2 + s + 4}{s^4 + 2s^3 + 9s^2 + 8s + 20}$$

and use `invlaplace` to compute the inverse Laplace transform of `step_3`, naming the result `Sol`.

```
> Sol:=invlaplace(step_3,s,t);
```

$$Sol := \frac{1}{17} e^{(-t)} \cos(2t) - \frac{4}{17} e^{-t} \sin(2t) - \frac{1}{17} \cos(2t) - \frac{4}{17} \sin(2t)$$

Last, we use `plot` to graph the solution obtained in `Sol` on the interval $[0, 2\pi]$.

```
> plot(Sol,t=0..2*Pi);
```

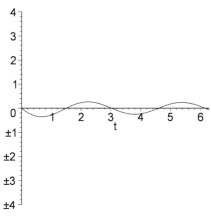

As we have seen in many previous examples, `dsolve` is able to find an equivalent solution to the initial-value problem.

```
> Sol:=dsolve({diff(y(t),t$2)+4*y(t)=exp(-t)*cos(2*t),
    y(0)=0,D(y)(0)=-1},y(t)):
combine(Sol,trig);
```

$$y(t) = \frac{1}{34}(2e^{(-t)}e^t\cos(2t) + \sin(2t)e^{(-t)}e^t - 9\sin(2t) - 2\cos(2t)e^t - 8\sin(2t)e^t)/e^t$$

■

Higher order initial-value problems can often be solved with the method of Laplace transforms as well.

EXAMPLE 3: Solve $y''' + y'' - 6y' = \sin 4t$, $y(0) = 2$, $y'(0) = 0$, $y''(0) = -1$.

SOLUTION: We first note that `dsolve` can quickly find an explicit solution of the initial-value problem.

```
> EQ:=diff(y(t),t$3)+diff(y(t),t$2)-6*diff(y(t),t)=sin(4*t):
> dsolve({EQ,y(0)=2,D(y)(0)=0,(D@@2)(y)(0)=-1},y(t));
```

$$y(t) = \frac{11}{1000}\cos(4t) - \frac{1}{500}\sin(4t) + \frac{17}{8} - \frac{7}{125}e^{(-3t)} - \frac{2}{25}e^{(2t)}$$

Alternatively, we can use Maple to implement the steps encountered when solving the equation using the method of Laplace transforms, as in the previous two examples.

Taking the Laplace transform of both sides of the equation, we find

```
> step_1:=laplace(EQ,t,s);
```

$$step\_1 := s(s(s\,\text{laplace}(y(t),t,s) - y(0)) - D(y)(0)) - (D^{(2)})(y)(0)$$
$$+ s(s(s\,\text{laplace}(y(t),t,s) - y(0)) - D(y)(0) - 6s\,\text{laplace}(y(t),t,s) + 6y(0)$$
$$= 4\frac{1}{s^2 + 16}$$

and then we apply the initial conditions, naming the result `step_2`.

```
> step_2:=subs({y(0)=2,D(y)(0)=0,
      (D@@2)(y)(0)=-1},step_1);
```

$$step\_2 := s^2(s\ \mathrm{laplace}(y(t),t,s)-2)+13+s(s\ \mathrm{laplace}(y(t),t,s)-2)-6s\ \mathrm{laplace}(y(t),t,s)=4\,\frac{1}{s^2+16}$$

Solving for *Y(s)*, we obtain

```
> step_3:=simplify(solve(step_2,laplace(y(t),t,s)));
```

$$step\_3 := \frac{2s^4+19s^2-204+2s^3+32s}{s(s^4+10s^2+s^3+16s-96)}$$

and computing the inverse Laplace transform of `step_3` with `invlaplace` yields the solution to the initial-value problem.

```
> Sol:=invlaplace(step_3,s,t);
```

$$Sol := \frac{11}{1000}\cos(4t)-\frac{1}{500}\sin(4t)+\frac{17}{8}-\frac{7}{125}e^{(-3t)}-\frac{2}{25}e^{(2t)}$$

Last, a graph of the solution is generated with `plot`.

```
> plot(Sol,t=-3/2..3/2);
```

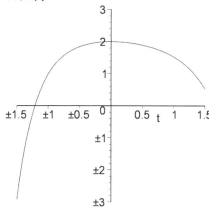

Some initial-value problems that involve differential equations with nonconstant coefficients can also be solved with the method of Laplace transforms as illustrated in the next example. However, Laplace transforms do not provide a general method for solving equations with nonconstant coefficients.

EXAMPLE 4: Solve $\begin{cases} y'' + ty' - 4y = 2 \\ y(0) = y'(0) = 0 \end{cases}$.

SOLUTION: In this case, we see that `dsolve` is able to solve this equation.

```
> dsolve({diff(y(t),t$2)+t*diff(y(t),t)-
    4*y(t)=2,y(0)=0,D(y)(0)=0},y(t));
```

$$y(t) = \frac{1}{6}t^4 + t^2$$

To solve the initial-value problem using the method of Laplace transforms, we take the Laplace transform of both sides of the equation.

```
> step_1:=laplace(diff(y(t),t$2)+t*diff(y(t),t)-
    4*y(t)=2,t,s);
```

$$step\_1 := s(s\ \text{laplace}(y(t), t, s) - y(0)) - D(y)(0) - 5\ \text{laplace}(y(t), t, s) - s\left(\frac{\partial}{\partial s}\ \text{laplace}(y(t), t, s)\right) = 2\frac{1}{s}$$

Next, we apply the initial conditions.

```
> step_2:=subs({y(0)=0,D(y)(0)=0},step_1);
```

$$step\_2 := s^2\ \text{laplace}(y(t), t, s) - 5\ \text{laplace}(y(t), t, s) - s\left(\frac{\partial}{\partial s}\ \text{laplace}(y(t), t, s)\right) = 2\frac{1}{s}$$

For convenience, we replace each occurrence of `laplace(y(t),t,s)`, which represents the Laplace transform of y(t), in `step_2` by Y(s).

```
> step_3:=subs(laplace(y(t),t,s)=Y(s),step_2);
```

$$step\_3 := s^2 Y(s) - 5Y(s) - s\left(\frac{\partial}{\partial s} Y(s)\right) = 2\frac{1}{s}$$

The equation in `step_3` is a first-order (separable) linear differential equation, which we solve with `dsolve`, naming the result `step_4`. (Note that we must have that $\lim_{s \to \infty} Y(s) = 0$ in order for $\mathcal{L}^{-1}\{Y(s)\}$ to exist.)

```
> step_4:=dsolve({step_3,Y(infinity)=0},Y(s));
```

$$step\_4 := Y(s) = 2\frac{s^2 + 2}{s^5}$$

We then use `assign` to name *Y(s)* the result obtained and use `invlaplace` to compute $\mathscr{L}^{-1}\{Y(s)\}$.

```
> assign(step_4):
```

```
step_5:=invlaplace(Y(s),s,t);
```

$$\frac{1}{6}t^4 + t^2$$

■

6.4 Laplace Transforms of Step and Periodic Functions

Piecewise Defined Functions: The Unit Step Function

An important function in modeling many physical situations is the unit step function $\mathscr{U}$.

Definition **Unit Step Function**	The **unit step function** $\mathscr{U}(t - a) = \mathscr{U}_a(t)$ where *a* is a number is defined by $$\mathscr{U}(t - a) = \mathscr{U}_a(t) = \begin{cases} 0, 0 \leq t < a \\ 1, t \geq a \end{cases}.$$

We can use the function `Heaviside` to define the unit step function. The `Heaviside` function is defined by

$$\text{Heaviside (t)} = \begin{cases} 0, \text{ if } t < 0 \\ 1, \text{ if } t \geq 0 \end{cases}.$$

Thus, $\mathscr{U}(t - a) = $ `Heaviside(t-a)`

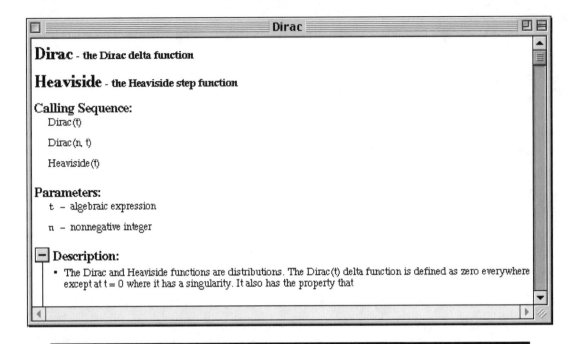

EXAMPLE 1: Graph (a) $2\mathscr{U}(t)$, (b) $\frac{1}{2}\mathscr{U}(t-5)$, and (c) $\mathscr{U}(t-2) - \mathscr{U}(t-8)$.

SOLUTION: (a) Here, $2\mathscr{U}(t) = 2\mathscr{U}(t-0)$, so $2\mathscr{U}(t) = 2$ for $t \geq 0$.

(b) In this case, $\frac{1}{2}\mathscr{U}(t-5) = \begin{cases} 0, & 0 \leq t < 5 \\ 1/2, & t \geq 5 \end{cases}$, so the "jump" occurs at $t = 5$.

(c) Note that

$$\mathscr{U}(t-2) - \mathscr{U}(t-8) = \begin{cases} 0, 0 \leq t < 2 \\ 1, t \geq 2 \end{cases} - \begin{cases} 0, 0 \leq t < 8 \\ 1, t \geq 8 \end{cases} = \begin{cases} 0, 0 \leq t < 2 \text{ or } t \geq 8 \\ 1, 2 \leq t < 8 \end{cases},$$

as we see using `convert` together with the `piecewise` option.

```
> convert(Heaviside(t-2)-Heaviside(t-8),piecewise);
```

$$\begin{cases} 0 & t < 2 \\ undefined & t = 2 \\ 1 & t < 8 \\ undefined & t = 8 \\ 0 & 8 < t \end{cases}$$

These functions are graphed using `plot` and `Heaviside`.

```
> plot({2*Heaviside(t),1/2*Heaviside(t-5),
```

Heaviside(t-2)-Heaviside(t-8)},t=0-1..10);

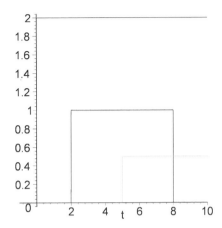

The unit step function is useful in defining functions that are piecewise continuous. For example, we can define the function

$$g(t) = \begin{cases} 0, & 0 \leq t < a \\ h(t), & a \leq t < b \\ 0, & t \geq b \end{cases}$$

as $g(t) = h(t)[\mathcal{U}(t-a) - \mathcal{U}(t-b)]$.

Similarly, a function such as

$$f(t) = \begin{cases} g(t), & 0 \leq t < a \\ h(t), & t \geq a \end{cases}$$

can be written as

$$f(t) = g(t)[\mathcal{U}(t-0) - \mathcal{U}(t-a)] + h(t)\mathcal{U}(t-a) = g(t)[1 - \mathcal{U}(t-a)] + h(t)\mathcal{U}(t-a).$$

The reason for writing piecewise continuous functions in terms of step functions is that we encounter functions of this type in solving initial-value problems. Using our methods in Chapters 4 and 5, we had to solve the problem over each piece of the function. However, the method of Laplace transforms can be used to avoid these complicated calculations.

Theorem

Suppose that $F(s) = \mathcal{L}\{f(t)\}$ exists for $s > b \geq 0$. If a is a positive constant and f is continuous on $[0,\infty)$, then

$$\mathcal{L}\{f(t-a)\mathcal{U}(t-a)\} = e^{-as}F(s).$$

EXAMPLE 2: Find $\mathcal{L}\left\{(t-3)^5 \mathcal{U}(t-3)\right\}$.

SOLUTION: In this case, $a = 3$ and $f(t) = t^5$. Thus,

$$\mathcal{L}\left\{(t-3)^5 \mathcal{U}(t-3)\right\} = e^{-3s}\mathcal{L}\{t^5\} = e^{-3s}\frac{5!}{s^6} = \frac{120}{s^6}e^{-3s}.$$

To use Maple, we note that `(t-3)^5*Heaviside(t-3)` represents $(t-3)^5 \mathcal{U}(t-3)$, so entering

```
> with(inttrans):
laplace((t-3)^5*Heaviside(t-3),t,s);
```

$$120\frac{e^{(-3s)}}{s^6}$$

computes $\mathcal{L}\left\{(t-3)^5 \mathcal{U}(t-3)\right\}$.

■

In most cases, we must calculate

$$\mathcal{L}\{g(t)\mathcal{U}(t-a)\}$$

instead of $\mathcal{L}\{f(t-a)\mathcal{U}(t-a)\}$. To solve this problem, we let $g(t) = f(t-a)$, so $f(t) = g(t+a)$. Therefore,

$$\mathcal{L}\{g(t)\mathcal{U}(t-a)\} = e^{-as}\mathcal{L}\{g(t+a)\}.$$

EXAMPLE 3: Calculate $\mathcal{L}\{\sin t\ \mathcal{U}(t-\pi)\}$.

SOLUTION: In this case, $g(t) = \sin t$ and $a = \pi$. Thus,

$$\mathcal{L}\{\sin t\,\mathcal{U}(t-\pi)\} = e^{-\pi s} = \mathcal{L}\{\sin(t+\pi)\} = e^{-\pi s}\mathcal{L}\{-\sin t\}$$

$$= -e^{-\pi s}\frac{1}{s^2+1} = -\frac{e^{-\pi s}}{s^2+1}.$$

The same result is obtained using `laplace`.

```
> laplace(sin(t)*Heaviside(t-Pi),t,s);
```

$$-\frac{e^{(-s\pi)}}{s^2+1}$$

■

Theorem

> Suppose that $F(s) = \mathscr{L}\{f(t)\}$ exists for $s > b \geq 0$. If a is a positive constant, then
> $$\mathscr{L}^{-1}\{e^{-as}F(s)\} = f(t-a)\mathscr{U}(t-a).$$

EXAMPLE 4: Find (a) $\mathscr{L}^{-1}\{e^{-4s}/s^3\}$ and (b) $\mathscr{L}^{-1}\left\{\frac{e^{-\pi s/2}}{s^2+16}\right\}$.

SOLUTION: (a) If we write the expression e^{-4s}/s^3 in the form $e^{-as}F(s)$, we see that $a = 4$ and $F(s) = 1/s^3$. Hence, $f(t) = \mathscr{L}^{-1}\{1/s^3\} = \frac{1}{2}t^2$ and

$$\mathscr{L}^{-1}\{e^{-4s}/s^3\} = f(t-4)\mathscr{U}(t-4) = \frac{1}{2}(t-4)^2\mathscr{U}(t-4).$$

(b) In this case, $a = \pi/2$ and $F(s) = 1/(s^2+16)$. Then, $f(t) = \mathscr{L}^{-1}\{1/(s^2+16)\} = \frac{1}{4}\sin 4t$ and

$$\mathscr{L}^{-1}\left\{\frac{e^{-\pi s/2}}{s^2+16}\right\} = f(t-\pi/2)\mathscr{U}(t-\pi/2) = \frac{1}{4}\sin 4(t-\pi/2)\mathscr{U}(t-\pi/2)$$

$$= \frac{1}{4}\sin(4t-2\pi)\mathscr{U}(t-\pi/2) = \frac{1}{4}\sin 4t\,\mathscr{U}(t-\pi/2).$$

For each of (a) and (b), the same results are obtained using `invlaplace`, although we must use `factor` to obtain the result given in (a) and `simplify` to simplify the result obtained for (b).

```
> factor(invlaplace(exp(-4*s)/s^3,s,t));
simplify(invlaplace(exp(-Pi*s/2)/(s^2+16),s,t));
```

$$\frac{1}{2}\,\text{Heaviside}(t-4)(t-4)^2$$

$$\frac{1}{4}\,\text{Heaviside}\left(t-\frac{1}{2}\pi\right)\sin(4t)$$

∎

Solving Initial-Value Problems

With the unit step function, we can solve initial-value problems that involve `piecewise-continuous` functions.

EXAMPLE 5: Solve $y'' + 9y = \begin{cases} 1, & 0 \le t < \pi \\ 0, & t \ge \pi \end{cases}$ subject to $y(0) = y'(0) = 0$.

SOLUTION: In order to solve this initial-value problem, we must compute $\mathcal{L}\{f(t)\}$ where $f(t) = \begin{cases} 1, & 0 \le t < \pi \\ 0, & t \ge \pi \end{cases}$. This is a piecewise continuous function, so we write it in terms of the unit step function as

$$f(t) = 1[\mathcal{U}(t-0) - \mathcal{U}(t-\pi)] + 0[\mathcal{U}(t-\pi)] = \mathcal{U}(t) - \mathcal{U}(t-\pi).$$

Then,

$$\mathcal{L}\{f(t)\} = \mathcal{L}\{1 - \mathcal{U}(t-\pi)\} = \frac{1}{s} - \frac{e^{-\pi s}}{s}$$

Hence,

$$\mathcal{L}\{y''\} + 9\mathcal{L}\{y\} = \mathcal{L}\{f(t)\}$$

$$s^2 Y(s) - sy(0) - y'(0) + 9Y(s) = \frac{1}{s} - \frac{e^{-\pi s}}{s}$$

$$(s^2 + 9)Y(s) = \frac{1}{s} - \frac{e^{-\pi s}}{s}$$

$$Y(s) = \frac{1}{s(s^2 + 9)} - \frac{e^{-\pi s}}{s(s^2 + 9)}.$$

The same steps are performed subsequently with Maple. First, we define EQ to be the equation $y'' + 9y = \begin{cases} 1, & 0 \le t < \pi \\ 0, & t \ge \pi \end{cases}$.

```
> f:=t->Heaviside(t)-Heaviside(t-Pi):
EQ:=diff(y(t),t$2)+9*y(t)=f(t):
```

Next, we use `laplace` to compute the Laplace transform of each side of the equation, naming the resulting equation `step_1`,

```
> step_1:=laplace(EQ,t,s);
```

$$step\_1 := s(s\,\text{laplace}(y(t), t, s) - y(0)) - D(y)(0) + 9\,\text{laplace}(y(t), t, s) = \frac{1}{s} - \frac{e^{(-s\pi)}}{s}$$

apply the initial conditions, naming the result `step_2`,

```
> step_2:=subs({y(0)=0,D(y)(0)=0},step_1);
```

$$step\_2 := s^2 \, \text{laplace}(y(t), t, s) + 9 \, \text{laplace}(y(t), t, s) = \frac{1}{s} - \frac{e^{(-s\pi)}}{s}$$

and solve `step_2` for `laplace(y(t),t,s)`, naming the result `step_3`.

> `step_3:=simplify(solve(step_2,laplace(y(t),t,s)));`

$$step\_3 := -\frac{-1 + e^{(-s\pi)}}{s(s^2 + 9)}$$

Then,

$$y(t) = \mathcal{L}^{-1}\{Y(s)\} = \mathcal{L}^{-1}\left\{\frac{1}{s(s^2+9)}\right\} - \mathcal{L}^{-1}\left\{\frac{e^{-\pi s}}{s(s^2+9)}\right\}.$$

Consider $\mathcal{L}^{-1}\left\{\frac{e^{-\pi s}}{s(s^2+9)}\right\}$. In the form of $\mathcal{L}^{-1}\{e^{-as}F(s)\}$, $a = \pi$ and $F(s) = \frac{1}{s(s^2+9)}$. $f(t) = \mathcal{L}^{-1}\{F(s)\}$ can be found either with a partial fraction expansion or with the formula

$$f(t) = \mathcal{L}^{-1}\left\{\frac{1}{s(s^2+9)}\right\} = \int_0^t \mathcal{L}^{-1}\left\{\frac{1}{s^2+9}\right\} d\alpha = \int_0^t \frac{1}{3}\sin 3\alpha \, d\alpha = -\frac{1}{3}\left[\frac{\cos 3\alpha}{3}\right]_0^t = \frac{1}{9} - \frac{1}{9}\cos 3t.$$

Then,

$$\mathcal{L}^{-1}\left\{\frac{e^{-\pi s}}{s(s^2+9)}\right\} = \left[\frac{1}{9} - \frac{1}{9}\cos 3(t-\pi)\right]\mathcal{U}(t-\pi)$$

$$= \left[\frac{1}{9} - \frac{1}{9}\cos(3t - \pi)\right]\mathcal{U}(t-\pi) = \left[\frac{1}{9} + \frac{1}{9}\cos 3t\right]\mathcal{U}(t-\pi).$$

Combining these results yields the solution

$$y(t) = \mathcal{L}^{-1}\{Y(s)\} = \mathcal{L}^{-1}\left\{\frac{1}{s(s^2+9)}\right\} - \mathcal{L}^{-1}\left\{\frac{e^{-\pi s}}{s(s^2+9)}\right\}$$

$$= \frac{1}{9} - \frac{1}{9}\cos 3t - \left[\frac{1}{9} + \frac{1}{9}\cos 3t\right]\mathcal{U}(t-\pi).$$

Notice that we can rewrite this solution as the piecewise-defined function

$$y = \begin{cases} \frac{1}{9} - \frac{1}{9}\cos 3t, & 0 \le t < \pi \\ -\frac{2}{9}\cos 3t, & t < \pi \end{cases}$$

Equivalent results are obtained with `invlaplace`.

> `Sol:=invlaplace(step_3,s,t);`

$$Sol := \frac{1}{9} - \frac{1}{9}\cos(3t) - \frac{1}{9}\text{Heaviside}(t - \pi) - \frac{1}{9}\cos(3t)\text{Heaviside}(t - \pi)$$

```
> convert(Sol,piecewise);
```

$$\begin{cases} \frac{1}{9} - \frac{1}{9}\cos(3t) & t \le \pi \\ -\frac{2}{9}\cos(3t) & \pi < t \end{cases}$$

An equivalent result is obtained using `dsolve`, as shown next, which we then graph with `plot`.

```
> sol:=dsolve({EQ,y(0)=0,D(y)(0)=0},y(t)):
sol:=combine(sol,trig);
```

$$sol := y(t) = \frac{1}{9}\text{Heaviside}(t) - \frac{1}{9}\text{Heaviside}(t - \pi) - \frac{1}{9}\cos(3t)\text{Heaviside}(t) - \frac{1}{9}\cos(3t)\text{Heaviside}(t - \pi)$$

```
> plot(rhs(sol),t=-1..2*Pi);
```

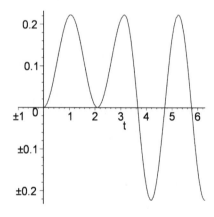

Periodic Functions

Another type of function that is encountered in many areas of applied mathematics is the **periodic** function.

Definition
Periodic Function

A function f is **periodic** if there is a positive number T such that

$$f(t + T) = f(t)$$

for all $t \ge 0$. The minimum value of T that satisfies this equation is called the **period** of f.

Due to the nature of periodic functions, we can simplify the calculation of the Laplace transform of these functions as indicated in the following theorem.

Theorem
Laplace Transform of
Periodic Functions

Suppose that f is a periodic function of period T and that f is piecewise continuous on $[0,\infty)$. Then $\mathscr{L}\{f(t)\}$ exists for $s > 0$ and is given by the definite integral

$$\mathscr{L}\{f(t)\} = \frac{1}{1 - e^{-sT}} \int_0^T e^{-st} f(t)\,dt.$$

EXAMPLE 6: Find the Laplace transform of the periodic function $f(t) = t, 0 \le t < 1$, and $f(t + 1) = f(t)$.

SOLUTION: To graph the periodic function f, we proceed in two steps. If $g(t)$ is defined on the interval $[0,T]$, then the periodic extension of g with period T, f, is defined with the following procedure.

```
f:=proc(t)
        if t>=0 and t<T then g(t) else f(t-T) fi;
        end:
```

Thus, to graph f, we first define $g(t) = t$ and then use the foregoing procedure with $T = 1$ to define f.

```
> g:=t->t:
f:=proc(t)
        if t>=0 and t<1 then g(t) else f(t-1) fi;
        end:
```

Because f has been defined as a procedure, we use operator notation to graph f.

```
> plot(f,0..5);
```

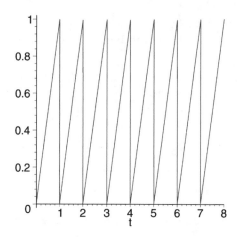

Alternatively, note that

$$f(t) = t(\mathcal{U}(t-1) - \mathcal{U}(t)) + (t-1)(\mathcal{U}(t-2) - \mathcal{U}(t-1)) + (t-2)(\mathcal{U}(t-3) - \mathcal{U}(t-2)) + \cdots$$

$$= t - \mathcal{U}(t-1) - \mathcal{U}(t-2) - \mathcal{U}(t-3) - \mathcal{U}(t-4) - \cdots$$

$$= t - \sum_{n=1}^{\infty} \mathcal{U}(t-n)$$

Then, on the interval [0,N], $f(t) = t - \sum_{n=1}^{N-1} \mathcal{U}(t-n)$. For example, entering

```
> f:=t->t-sum(Heaviside(t-n),n=1..7):
plot(f(t),t=0..8);
```

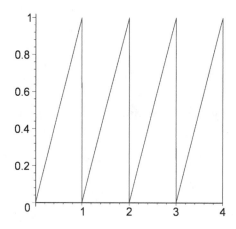

graphs f on the interval [0,8].
To compute $\mathcal{L}\{f(t)\}$, we use integration by parts,

$$\mathscr{L}\{f(t)\} = \frac{1}{1 - e^{-s}} \int_0^1 e^{-st} t\, dt = \frac{1}{1 - e^{-s}} \left\{ \left[-\frac{te^{-st}}{s} \right]_0^1 + \int_0^1 \frac{e^{-st}}{s}\, dt \right\}$$

$$= \frac{1}{1 - e^{-s}} \left\{ -\frac{e^{-s}}{s} - \left[\frac{e^{-st}}{s^2} \right]_0^1 \right\} = \frac{1}{1 - e^{-s}} \left(-\frac{e^{-s}}{s} + \frac{1 - e^{-s}}{s^2} \right) = \frac{1 - (s+1)e^{-s}}{s^2(1 - e^{-s})}$$

or Maple.

```
> simplify(1/(1-exp(-s))*int(t*exp(-s*t),t=0..1));
```

$$\frac{e^{(-s)}s + e^{(-s)} - 1}{(-1 + e^{(-s)})s^2}$$

Alternatively, using $f(t) = t - \sum_{n=1}^{\infty} \mathscr{U}(t - n)$ we have that

$$\mathscr{L}\{f(t)\} = \mathscr{L}\{t\} - \mathscr{L}\left\{ \sum_{n=1}^{\infty} \mathscr{U}(t - n) \right\} = \mathscr{L}\{t\} - \sum_{n=1}^{\infty} \mathscr{L}\{\mathscr{U}(t - n)\}.$$

```
> f:=t->t-sum(Heaviside(t-n),n=1..infinity);
```

$$f := t \rightarrow t - \left(\sum_{n=1}^{\infty} \text{Heaviside}(t - n) \right)$$

```
> laplace(f(t),t,s);
```

$$\frac{1}{s^2} - \frac{e^{(-s)}}{s(1 - e^{(-s)})}$$

■

Laplace transforms can now be used to solve initial-value problems with periodic forcing functions more easily.

EXAMPLE 7: Solve $y'' + y = f(t)$ subject to $y(0) = y'(0) = 0$ if

$f(t) = \begin{cases} \sin t, & 0 \le t < \pi \\ 0, & \pi \le t < 2\pi \end{cases}$ and $f(t + 2\pi) = f(t).$ (f is known as the **half-wave**

rectification of $\sin t$.)

SOLUTION: To graph f, we define $g(t) = \begin{cases} \sin t, & 0 \le t < \pi \\ 0, & \pi \le t < 2\pi \end{cases}$ and then f using proc.

```
> g:=t->sin(t)*(1-Heaviside(t-Pi));
```

$$g := t \rightarrow \sin(t)(1 - \text{Heaviside}(t - \pi))$$

```
> convert(g(t),piecewise);
```

$$\begin{cases} \sin(t) & t \leq \pi \\ 0 & \pi < t \end{cases}$$

```
> f:=proc(t)
      if t>=0 and t<2*Pi then g(t) else f(t-2*Pi) fi;
   end:
> plot(f,0..10*Pi);
```

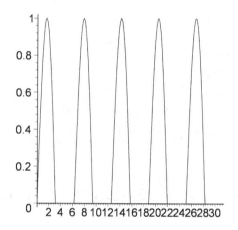

Alternatively, note that

$$f(t) = g(t)(\mathcal{U}(t) - \mathcal{U}(t - 2\pi)) + g(t - 2\pi)(\mathcal{U}(t - 2\pi) - \mathcal{U}(t - 4\pi)) + \cdots$$

$$= \sum_{n=0}^{\infty} g(t - 2n\pi)(\mathcal{U}(t - 2n\pi) - \mathcal{U}(t - 2(n+1)\pi)).$$

Thus, the graph of f on the interval $[0,2k\pi]$, where k represents a positive integer, is obtained by graphing

$$f_k(t) = \sum_{n=0}^{k-1} g(t - 2n\pi)(\mathcal{U}(t - 2n\pi) - \mathcal{U}(t - 2(n+1)\pi))$$

on the interval $[0,2k\pi]$. For convenience, we define `nth_term(n)` to be $g(t - 2n\pi)(\mathcal{U}(t - 2n\pi) - \mathcal{U}(t - 2(n+1)\pi))$.

```
> nth_term:=n->g(t-2*n*Pi)*
        (Heaviside(t-2*n*Pi)-Heaviside(t-2*(n+1)*Pi)):
```

Then, to graph f on the interval $[0,20\pi]$, we first enter

```
> to_graph:=sum(nth_term(n),n=0..9):
```

and then

```
> plot(to_graph,t=0..20*Pi,0..2);
```

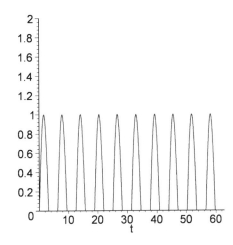

To solve the initial-value problem we must find $\mathscr{L}\{f(t)\}$. Because the period is $T = 2\pi$, we have

$$\mathscr{L}\{f(t)\} = \frac{1}{1-e^{-2\pi s}} \int_0^{2\pi} e^{-st} f(t)\,dt = \frac{1}{1-e^{-2\pi s}} \left[\int_0^{\pi} e^{-st} \sin t\,dt + \int_0^{\pi} e^{-st} \cdot 0\,dt \right]$$

$$= \frac{1}{1-e^{-2\pi s}} \int_0^{\pi} e^{-st} \sin t\,dt.$$

We use `int` to evaluate this integral

```
> s:='s':t:='t':
step_1:=simplify(1/(1-exp(-2*Pi*s))*
        int(exp(-s*t)*sin(t),t=0..Pi));
```

$$step\_1 := -\frac{1}{(s^2+1)(e^{(-\pi s)}-1)}$$

```
> lap_f:=normal(step_1,expanded);
```

$$lap\_f := \frac{e^{(-\pi s)}}{-s^2 + s^2 e^{(\pi s)} - 1 + e^{(\pi s)}}$$

and see that $\mathcal{L}\{f(t)\} = \frac{e^{\pi s}}{(e^{\pi s} - 1)(s^2 + 1)} = \frac{1}{(1 - e^{-\pi s})(s^2 + 1)}$. Alternatively, we can use

$$f(t) = \sum_{n=0}^{\infty} g(t - 2n\pi)(\mathcal{U}(t - 2n\pi) - \mathcal{U}(t - 2(n+1)\pi)).$$

to rewrite f as

$$f(t) = \sum_{n=0}^{\infty} (-1)^n \sin t \,\mathcal{U}(t - n\pi).$$

Then,

$$\mathcal{L}\{f(t)\} = \mathcal{L}\left\{ \sum_{n=0}^{\infty} (-1)^n \sin t \,\mathcal{U}(t - n\pi) \right\} = \sum_{n=0}^{\infty} \mathcal{L}\{(-1)^n \sin t \,\mathcal{U}(t - n\pi)\}.$$

```
> f:=t->sum((-1)^n*sin(t)*Heaviside(t-n*Pi),n=0..infinity);
```

$$f := t \to \sum_{n=0}^{\infty} (-1)^n \sin(t) \text{Heaviside}(t - n\pi)$$

```
> laplace(f(t),t,s);
```

$$\frac{1}{(s^2 + 1)(1 - e^{(-\pi s)})}$$

Taking the Laplace transform of both sides of the equation, applying the initial conditions, and solving for $Y(s)$ then give us

$$\mathcal{L}[y''] + \mathcal{L}\{y\} = \mathcal{L}\{f(t)\}$$

$$s^2 Y(s) - sy(0) - y'(0) + Y(s) = \frac{1}{(1 - e^{-\pi s})(s^2 + 1)}$$

$$Y(s) = \frac{1}{(1 - e^{-\pi s})(s^2 + 1)^2}$$

We perform the same steps with Maple.

```
> step1:=laplace(diff(y(t),t$2)+
          y(t)=f(t),t,s);
```

$$step1 := s(s \,\text{laplace}(y(t), t, s) - y(0)) - D(y)(0) + \text{laplace}(y(t), t, s) = \frac{1}{(s^2 + 1)(1 - e^{(-\pi s)})}$$

```
> step2:=subs({y(0)=0,D(y)(0)=0},step1);
```

$$step2 := s^2 \text{ laplace}(y(t), t, s) + \text{ laplace}(y(t), t, s) = \frac{1}{(s^2 + 1)(1 - e^{(-\pi s)})}$$

```
> step3:=solve(step2,laplace(y(t),t,s));
```

$$step3 := -\frac{1}{-s^4 + s^4 e^{(-\pi s)} - 2s^2 + 2s^2 e^{(-\pi s)} - 1 + e^{(-\pi s)}}$$

```
> factor(step3);
```

$$-\frac{1}{(s^2 + 1)^2 (e^{(-\pi s)} - 1)}$$

Recall from our work with the geometric series that if $|x| < 1$, then

$$\frac{1}{1 - x} = 1 + x + x^2 + x^3 + \cdots = \sum_{n=0}^{\infty} x^n.$$

Because we do not know the inverse Laplace transform of $\frac{1}{(1-e^{-\pi s})(s^2+1)^2}$, we must use a geometric series expansion of $\frac{1}{1-e^{-\pi s}}$ to obtain terms for which we can calculate the inverse Laplace transform. This gives us

$$\frac{1}{1 - e^{-\pi s}} = 1 + e^{-\pi s} + e^{-2\pi s} + e^{-3\pi s} + \cdots = \sum_{n=0}^{\infty} e^{-n\pi s},$$

so

$$Y(s) = (1 + e^{-\pi s} + e^{-2\pi s} + e^{-3\pi s} + \cdots) \frac{1}{(s^2 + 1)^2}$$

$$= \frac{1}{(s^2 + 1)^2} + \frac{e^{-\pi s}}{(s^2 + 1)^2} + \frac{e^{-2\pi s}}{(s^2 + 1)^2} + \frac{e^{-3\pi s}}{(s^2 + 1)^2} + \cdots$$

$$= \sum_{n=0}^{\infty} \frac{e^{-n\pi s}}{(s^2 + 1)^2}.$$

Then,

$$y(t) = \mathcal{L}^{-1}\left\{\frac{1}{(s^2+1)^2} + \frac{e^{-\pi s}}{(s^2+1)^2} + \frac{e^{-2\pi s}}{(s^2+1)^2} + \frac{e^{-3\pi s}}{(s^2+1)^2} + \cdots\right\}$$

$$= \mathcal{L}^{-1}\left\{\frac{1}{(s^2+1)^2}\right\} + \mathcal{L}^{-1}\left\{\frac{e^{-\pi s}}{(s^2+1)^2}\right\} + \mathcal{L}^{-1}\left\{\frac{e^{-2\pi s}}{(s^2+1)^2}\right\} + \mathcal{L}^{-1}\left\{\frac{e^{-3\pi s}}{(s^2+1)^2}\right\} + \cdots$$

$$= \sum_{n=0}^{\infty} \mathcal{L}^{-1}\left\{\frac{e^{-n\pi s}}{(s^2+1)^2}\right\}.$$

Notice that $\mathcal{L}^{-1}\left\{1/(s^2+1)^2\right\}$ is needed to find all of the other terms. Using invlaplace,

> `invlaplace(1/(s^2+1)^2,s,t);`

$$\frac{1}{2}\sin(t) - \frac{1}{2}t\cos(t)$$

we have $\mathcal{L}^{-1}\left\{1/(s^2+1)^2\right\} = \frac{1}{2}(\sin t - t\cos t)$. In fact, we can use invlaplace to compute the inverse Laplace transform of $e^{-n\pi s}/(s^2+1)^2$.

> `invlaplace(exp(-n*Pi*s)/(s^2+1)^2,s,t);`

$$\text{Heaviside}(t - n\pi)\left(\frac{1}{2}\sin(t - n\pi) - \frac{1}{2}(t - n\pi)\cos(t - n\pi)\right)$$

Then,

$$y(t) = \frac{1}{2}\left\{(\sin t - t\cos t) + [\sin(t - \pi) - (t - \pi)\cos(t - \pi)\mathcal{U}(t - \pi)\right.$$

$$+ [\sin(t - 2\pi) - (t - 2\pi)\cos(t - 2\pi)\mathcal{U}(t - 2\pi)$$

$$\left.+ [\sin(t - 3\pi) - (t - 3\pi)\cos(t - 3\pi)]\mathcal{U}(t - 3\pi) + \cdots\right\}$$

$$= \frac{1}{2}\sum_{n=0}^{\infty}[\sin(t - n\pi) - (t - n\pi)\cos(t - n\pi)\mathcal{U}(t - n\pi).$$

To graph $y(t)$ on the interval $[0, k\pi]$, where k represents a positive integer, we note that

$$\frac{1}{2}[\sin(t - n\pi) - (t - n\pi)\cos(t - n\pi)]\mathcal{U}(t - n\pi) = 0$$

for all values of t in $[0, k\pi]$ if $n \geq k$ so we need to graph

$$\frac{1}{2}\sum_{n=0}^{k-1}[\sin(t-n\pi)-(t-n\pi)\cos(t-n\pi)\mathscr{U}(t-n\pi).$$

Thus, to graph $y(t)$ on the interval $[0,5\pi]$, we enter

```
toplot:=sum(invlaplace(exp(-n*Pi*s)/(s^2+1)^2,s,t),n=0..4);
```

$$toplot := \text{Heaviside}(t)\left(\frac{1}{2}\sin(t)-\frac{1}{2}t\cos(t)\right)$$

$$+\,\text{Heaviside}(t-\pi)\left(-\frac{1}{2}\sin(t)+\frac{1}{2}(t-\pi)\cos(t)\right)$$

$$+\,\text{Heaviside}(t-2\pi)\left(\frac{1}{2}\sin(t)-\frac{1}{2}(t-2\pi)\cos(t)\right)$$

$$+\,\text{Heaviside}(t-3\pi)\left(-\frac{1}{2}\sin(t)+\frac{1}{2}(t-3\pi)\cos(t)\right)$$

$$+\,\text{Heaviside}(t-4\pi)\left(\frac{1}{2}\sin(t)-\frac{1}{2}(t-4\pi)\cos(t)\right)$$

```
> plot(toplot,t=-1..5*Pi);
```

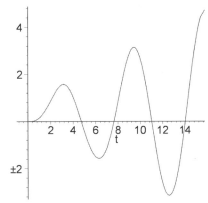

Impulse Functions: The Delta Function

We now consider differential equations of the form $ax''+bx'+cx=f(t)$ where f is "large" over the short interval centered at t_0, $t_0-\alpha<t<t_0+\alpha$, and zero otherwise. Hence, we define the **impulse** delivered by the function $f(t)$ as $I(t)=\int_{t_0-a}^{t_0+a}f(t)\,dt$, or since $f(t)=0$ for t on $(-\infty,t_0-\alpha)\cup(t_0+\alpha,+\infty)$,

$$I(t) = \int_{-\infty}^{+\infty} f(t)\, dt.$$

In order to understand the impulse function better, we let f be defined in the following manner:

$$f(t) = \delta_a(t - t_0) = \begin{cases} \frac{1}{2\alpha}, & t_0 - \alpha < t < t_0 + \alpha \\ 0, & \text{otherwise} \end{cases}$$

To graph $\delta_\alpha(t - t_0)$ for several values of α and $t_0 = 0$, we define `del`.

```
> del:=(t,t0,alpha)->

    piecewise(t>=t0-alpha and t<=t0+alpha,2/(2*alpha),

    t<t0-alpha or t>t0+alpha,0);
```

$$del := (t, t0, \alpha) \rightarrow \text{piecewise}\left(t0 - \alpha \le t \ \textbf{and}\ t \le t0 + \alpha, \frac{1}{\alpha}, t < t0 - \alpha \ \textbf{or}\ t0 + \alpha < t, 0 \right)$$

For example, entering

```
> plot(del(t,0,.25),t=-1..1);
```

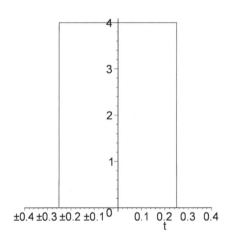

graphs $\delta_{1/4}(t)$ on the interval [-1,1]. Similarly, to graph $\delta_i(t)$ for $i = 0.01, 0.02, 0.03, 0.04$, and 0.05, we first define toplot using seq and then use `plot` to graph this set of functions on the interval [-0.1,0.1].

```
> alphavals:=seq(0.01+0.01*i,i=0..4):

toplot:={seq(del(t,0,alpha),alpha=alphavals)};
```

$$toplot := \left\{ \begin{cases} 25.00000000 & -0.04 - t \leq 0 \text{ and } t - 0.04 \leq 0 \\ 0 & t + 0.04 < 0 \text{ or } 0.04 - t < 0 \end{cases}, \right.$$

$$\begin{cases} 50.00000000 & -0.02 - t \leq 0 \text{ and } t - 0.02 \leq 0 \\ 0 & t + 0.02 < 0 \text{ or } 0.02 - t < 0 \end{cases},$$

$$\begin{cases} 33.33333333 & -0.03 - t \leq 0 \text{ and } t - 0.03 \leq 0 \\ 0 & t + 0.3 < 0 \text{ or } 0.03 - t < 0 \end{cases},$$

$$\begin{cases} 20.00000000 & -0.05 - t \leq 0 \text{ and } t - 0.05 \leq 0 \\ 0 & t + 0.05 < 0 \text{ or } 0.05 - t < 0 \end{cases},$$

$$\left. \begin{cases} 100. & -0.01 - t \leq 0 \text{ and } t - 0.01 \leq 0 \\ 0 & t + 0.01 < 0 \text{ or } 0.01 - t < 0 \end{cases} \right\}$$

```
> plot(toplot,t=-0.1..0.1);
```

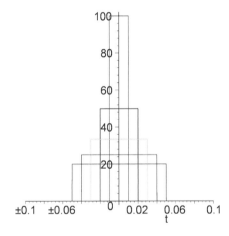

With this definition, the impulse is given by

$$I(t) = \int_{t_0-\alpha}^{t_0+\alpha} f(t)dt = \int_{t_0-\alpha}^{t_0+\alpha} \frac{1}{2\alpha} dt = \frac{1}{2\alpha} \left((t_0 + \alpha) - t_0 - \alpha) \right) = \frac{1}{2\alpha} \cdot 2\alpha = 1.$$

Notice that the value of this integral does not depend on α as long as α is not zero. We now try to create the idealized impulse function by requiring that $\delta_\alpha(t - t_0)$ act on smaller and smaller intervals. From the integral calculation, we have

$$\lim_{\alpha \to 0} I(t) = 1.$$

We also note that

$$\lim_{\alpha \to 0} \delta_\alpha(t - t_0) = 0, \qquad t \neq t_0.$$

We use these properties to now define the **idealized unit impulse function**.

Definition **Unit Impulse Function**	The **idealized unit impulse function (Dirac delta function)** δ satisfies $$\delta(t - t_0) = 0, \qquad t \neq t_0 \int_{-\infty}^{+\infty} \delta(t - t_0)dt = 1.$$

The function `Dirac(t)` represents the Dirac delta function.

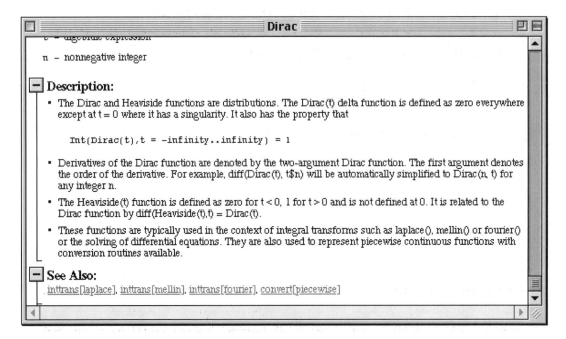

```
b   algebraic expression

n – nonnegative integer
```

Description:

* The Dirac and Heaviside functions are distributions. The Dirac(t) delta function is defined as zero everywhere except at t = 0 where it has a singularity. It also has the property that

```
    Int(Dirac(t),t = -infinity..infinity) = 1
```

* Derivatives of the Dirac function are denoted by the two-argument Dirac function. The first argument denotes the order of the derivative. For example, diff(Dirac(t), t$n) will be automatically simplified to Dirac(n, t) for any integer n.

* The Heaviside(t) function is defined as zero for t < 0, 1 for t > 0 and is not defined at 0. It is related to the Dirac function by diff(Heaviside(t),t) = Dirac(t).

* These functions are typically used in the context of integral transforms such as laplace(), mellin() or fourier() or the solving of differential equations. They are also used to represent piecewise continuous functions with conversion routines available.

See Also:

inttrans[laplace], inttrans[mellin], inttrans[fourier], convert[piecewise]

We now state the following useful theorem involving the unit impulse function.

Theorem	Suppose that g is a bounded and continuous function. Then, $$\int_{-\infty}^{+\infty} \delta(t - t_0)g(t)\, dt = g(t_0).$$

The Laplace transform of $\delta(t - t_0)$ is found by using the function $\delta_\alpha(t - t_0)$ and L'Hôpital's rule.

Theorem

> For $t_0 > 0$,
> $$\mathscr{L}(\delta(t - t_0)) = e^{-st_0}.$$

EXAMPLE 8: Find (a) $\mathscr{L}\{\delta(t - 1)\}$, (b) $\mathscr{L}\{\delta(t - \pi)\}$, and(c) $\mathscr{L}\{\delta(t)\}$.

SOLUTION: (a) In this case, $t_0 = 1$, so $\mathscr{L}\{\delta(t - 1)\} = e^{-s}$.

(b) With $t_0 = \pi$, $\mathscr{L}\{\delta(t - \pi)\} = e^{-s\pi}$.
(c) Because $t_0 = 0$, $\mathscr{L}\{\delta(t)\} = \mathscr{L}\{\delta(t - 0)\} = e^{-s \cdot 0} = 1$.

We obtain the same results using `Dirac` and `laplace` as shown next.

```
> map(laplace,[Dirac(t-1),Dirac(t-Pi),Dirac(t)],t,s);
```

$$[e^{(-s)}, e^{(-\pi s)}, 1]$$

■

EXAMPLE 9: Solve $y'' + y = \delta(t - \pi) + 1$ subject to $y(0) = y'(0) = 0$.

SOLUTION: As in previous examples, we solve this initial-value problem by taking the Laplace transform of both sides of the differential equation

```
> step_1:=laplace(diff(y(t),t$2)+y(t)=Dirac(t-Pi)+1,t,s);
```

$$step\_1 := s(s \, laplace(y(t), t, s) - y(0)) - D(y)(0) + laplace(y(t), t, s) = e^{(-\pi s)} + \frac{1}{s}$$

applying the initial conditions

```
> step_2:=subs({y(0)=0,D(y)(0)=0},step_1);
```

$$step\_2 := s^2 \, \text{laplace}(y(t), t, s) + \text{laplace}(y(t), t, s) = e^{(-\pi s)} + \frac{1}{s}$$

and solving for $Y(s)$.

```
> step_3:=solve(step_2,laplace(y(t),t,s));
```

$$step\_3 := \frac{e^{(-\pi s)}s + 1}{s(s^2 + 1)}$$

We find *y(t)* using `invlaplace`.

```
> Sol:=invlaplace(step_3,s,t);
```

$$Sol := -\sin(t)\text{Heaviside}(t - \pi) + 1 - \cos(t)$$

```
> plot(Sol,t=0..2*Pi);
```

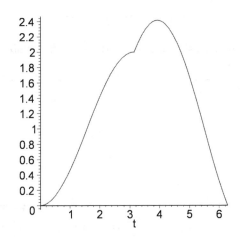

We can use `dsolve` to find the solution to the initial-value problem, although the form is different from that obtained before.

```
> dsolve({diff(y(t),t$2)+y(t)=Dirac(t-Pi)+1,
    y(0)=0,D(y)(0)=0},y(t));
```

$$y(t) = (-\text{Heaviside}(t - \pi) + \sin(t))\sin(t) + \cos(t)^2 - \cos(t)$$

On the other hand, when we use `dsolve` together with the `method=laplace` option, we obtain yet another form of the solution.

```
> dsolve({diff(y(t),t$2)+y(t)=Dirac(t-Pi),
     y(0)=0,D(y)(0)=0},y(t),laplace);
```

$$y(t) = -\sin(t)\text{Heaviside}(t - \pi) + 2\sin\left(\frac{1}{2}t\right)^2$$

∎

The forcing function may involve a combination of functions as illustrated in the following example.

EXAMPLE 10: Solve $y'' + 2y' + y = 1 + \delta(t - \pi) + \delta(t - 2\pi)$ subject to $y(0) = y'(0) = 0$.

SOLUTION: We proceed in exactly the same manner as in Example 9. After computing the Laplace transform of each side of the equation

```
> EQ:=diff(y(t),t$2)+2*diff(y(t),t)+
     y(t)=1+Dirac(t-Pi)+Dirac(t-2*Pi):
step_1:=laplace(EQ,t,s);
```

$$step\_1 := s(s \text{ laplace}(y(t), t, s) - y(0)) - D(y)(0) + 2s \text{ laplace}(y(t), t, s)$$
$$- 2y(0) + \text{ laplace}(y(t), t, s) = \frac{1}{s} + e^{(-\pi s)} + e^{(-2\pi s)}$$

and applying the initial conditions,

```
> step_2:=subs({y(0)=0,D(y)(0)=0},step_1);
```

$$step\_2 := s^2 \text{ laplace}(y(t), t, s) + 2s \text{ laplace}(y(t), t, s) + \text{ laplace}(y(t), t, s) = \frac{1}{s} + e^{(-\pi s)} + e^{(-2\pi s)}$$

we solve for $Y(s)$

```
> step_3:=solve(step_2,laplace(y(t),t,s));
```

$$step\_3 := \frac{1 + e^{(-\pi s)}s + e^{(-2\pi s)}s}{s(s^2 + 2s + 1)}$$

and then compute $y(t) = \mathscr{L}^{-1}\{Y(s)\}$.

```
> Sol:=invlaplace(step_3,s,t);
```

$$Sol := 1 - te^{(-t)} - e^{(-t)} + \text{Heaviside}(t - \pi)e^{(\pi-t)}t$$
$$- \text{Heaviside}(t - \pi)e^{(\pi-t)}\pi + \text{Heaviside}(t - 2\pi)e^{(-t+2\pi)}t$$
$$- 2\text{Heaviside}(t - 2\pi)e^{(-t+2\pi)}\pi$$

Equivalent results are obtained with `dsolve`.

```
> Sol_2:=dsolve({EQ,y(0)=0,D(y)(0)=0},y(t));
```

$$Sol\_2 := y(t) = 1 - te^{(-t)} - e^{(-t)} + \text{Heaviside}(t - \pi)e^{(\pi-t)}t$$
$$- \text{Heaviside}(t - \pi)e^{(\pi-t)}\pi + \text{Heaviside}(t - 2\pi)e^{(-t+2\pi)}t$$
$$- 2\text{Heaviside}(t - 2\pi)e^{(-t+2\pi)}\pi$$

```
> plot(rhs(Sol_2),t=0..4*Pi,0..1.6);
```

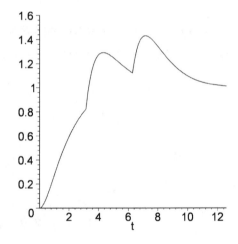

■

6.5 The Convolution Theorem

In many cases, we are required to determine the inverse Laplace transform of a product of two functions. Just as in integral calculus when the integral of the product of two functions did not produce the product of the integrals, neither does the inverse Laplace transform of the product yield the product of the inverse Laplace transforms.

Thus, we state the following theorem.

Theorem
Convolution Theorem

Suppose that $f(t)$ and $g(t)$ are piecewise continuous on $[0,\infty)$ and both of exponential order b. Further suppose that $\mathscr{L}\{f(t)\} = F(s)$ and $\mathscr{L}\{g(t)\} = G(s)$. Then,

$$\mathscr{L}^{-1}\{F(s)G(s)\} = \mathscr{L}^{-1}\{\mathscr{L}\{(f * g)(t)\}\} = (f * g)(t)$$

$$= \int_0^t f(t - v)g(v)dv.$$

Note that $(f * g)(t) = \int_0^t f(t - v)g(v)dv$ is called the **convolution integral**.

EXAMPLE 1: Compute $(f * g)(t)$ if $f(t) = e^{-t}$ and $g(t) = \sin t$. Verify the convolution theorem with these functions.

SOLUTION: We use the definition and integration by parts to obtain

$$(f * g)(t) = \int_0^t f(t - v)g(v)dv = \int_0^t e^{-(-v)} \sin v dv = e^{-t} \int_0^t e^v \sin v dv$$

$$= e^{-t} \left[\frac{1}{2} e^v (\sin v - \cos v) \right]_0^t = \frac{1}{2} e^{-t} [e^t (\sin t - \cos t) - (\sin 0 - \cos 0)]$$

$$= \frac{1}{2} (\sin t - \cos t) + \frac{1}{2} e^{-t}.$$

The same results are obtained with Maple. After defining `convolution`, which computes $(f * g)(t)$,

```
> convolution:=(f,g)->int(f(t-v)*g(v),v=0..t):
```

we define f and g

```
> f:='f':g:='g':
f:=t->exp(-t):g:=t->sin(t):
```

and then use `convolution` to compute $(f * g)(t)$

```
> convolution(f,g);
```

$$-\frac{1}{2}\cos(t) + \frac{1}{2}\sin(t) + \frac{1}{2}e^{(-t)}$$

Note that $(g * f)(t)$ is the same as $(f * g)(t)$.

```
> convolution(g,f);
```

$$-\frac{1}{2}\cos(t) + \frac{1}{2}\sin(t) + \frac{1}{2}e^{(-t)}$$

Now, according to the convolution theorem, $\mathscr{L}\{f(t)\}\mathscr{L}\{g(t)\} = \mathscr{L}\{(f * g)(t)\}$. In this example, we have

$$F(s) = \mathscr{L}\{f(t)\} = \mathscr{L}\{\{e^{-t}\} = \frac{1}{s=1} \quad \text{and} \quad G(s) = \mathscr{L}\{g(t)\} = \mathscr{L}\{\sin t\} = \frac{1}{s^2+1}.$$

Hence, $\mathscr{L}^{-1}\{F(s)G(s)\} = \mathscr{L}^{-1}\{\frac{1}{s+1} \cdot \frac{1}{s^2+1}\}$ should equal $(f * g)(t)$. We compute $\mathscr{L}^{-1}\{\frac{1}{s+1} \cdot \frac{1}{s^2+1}\}$ with `invlaplace`.

```
> with(inttrans):
invlaplace(1/((s+1)*(s^2+1)),s,t);
```

$$-\frac{1}{2}\cos(t) + \frac{1}{2}\sin(t) + \frac{1}{2}e^{(-t)}$$

Hence,

$$\mathscr{L}^{-1}\left\{\frac{1}{s+1} \cdot \frac{1}{s^2+1}\right\} = \frac{1}{2}e^{-t} - \frac{1}{2}\cos t + \frac{1}{2}\sin t$$

which is the same result as that obtained for $(f * g)(t)$.

■

EXAMPLE 2: Use the convolution theorem to find the Laplace transform of $h(t) = \int_0^t \cos(t-v)\sin(v)\,dv$.

SOLUTION: Notice that $h(t) = (f * g)(t)$, where $f(t) = \cos t$ and $g(t) = \sin t$. Therefore, by the convolution theorem, $\mathscr{L}\{(f * g)(t)\} = F(s)G(s)$. Hence,

$$\mathcal{L}\{h(t)\} = \mathcal{L}\{f(t)\}\mathcal{L}\{g(t)\} = \mathcal{L}\{\cos t\}\mathcal{L}\{\sin t\} = \frac{s}{s^2 + 1} \cdot \frac{1}{s^2 + 1} = \frac{s}{(s^2 + 1)^2}.$$

The same result is obtained with `laplace`. (Remember that `Int` represents the inert form of the `int` command.)

```
> h:=t->Int(cos(t-v)*sin(v),v=0..t);
```

$$h := t \rightarrow \int_0^t \cos(t - v) \sin(v) dv$$

```
> laplace(h(t),t,s);
```

$$\frac{s}{(s^2 + 1)^2}$$

■

Integral and Integrodifferential Equations

The convolution theorem is useful in solving numerous problems. In particular, this theorem can be employed to solve **integral equations**, which are equations that involve an integral of the unknown function.

EXAMPLE 3: Use the convolution theorem to solve the integral equation $h(t) = 4t + \int_0^t h(t - v) \sin v \, dv$.

SOLUTION: We first note that the integral in this equation represents $(h * g)(t)$ where $g(t) = \sin t$. Therefore, if we apply the Laplace transform to both sides of the equation, we obtain

$$\mathcal{L}\{h(t)\} = \mathcal{L}\{4t\} + \mathcal{L}\{h(t)\}\mathcal{L}\{\sin t\}$$

or

$$H(s) = \frac{4}{s^2} + H(s)\frac{1}{s^2 + 1},$$

where $\mathcal{L}\{h(t)\} = H(s)$.

The same result is obtained with `laplace`.

```
> h:='h':
```

```
EQ:=h(t)=4*t+Int(h(t-v)*sin(v),v=0..t);
```

$$EQ := h(t) = 4t + \int_0^t h(t-v)\sin(v)dv$$

> `step_1:=laplace(EQ,t,s);`

$$step\_1 := \text{laplace}(h(t),t,s) = 4\frac{1}{s^2} + \frac{\text{laplace}(h(t),t,s)}{s^2+1}$$

Solving for $H(s)$, we have

$$H(s)\left(1 - \frac{1}{s^2+1}\right) = \frac{4}{s^2},$$

so

$$H(s) = \frac{4(s^2+1)}{s^4} = \frac{4}{s^2} + \frac{4}{s^4}.$$

> `step_2:=simplify(solve(step_1,laplace(h(t),t,s)));`

$$step\_2 := 4\frac{s^2+1}{s^4}$$

Then by computing the inverse Laplace transform,

> `Sol:=invlaplace(step_2,s,t);`

$$Sol := \frac{2}{3}t^3 + 4t$$

we find that

$$h(t) = \mathscr{L}^{-1}\left\{\frac{4}{s^2} + \frac{4}{s^4}\right\} = 4t + \frac{2}{3}t^3.$$

∎

Laplace transforms are helpful in solving problems of other types as well. Next, we illustrate how Laplace transforms can be used to solve an **integrodifferential equation**, an equation that involves a derivative as well as an integral of the unknown function.

EXAMPLE 4: Solve $\frac{dy}{dt} + y + \int_0^t y(u)du = 1$ subject to $y(0) = 0$.

SOLUTION: Because we must take the Laplace transform of both sides of this integrodifferential equation, we first compute

$$\mathscr{L}\left\{\int_0^t y(u)du\right\} = \mathscr{L}\{(1*y)(t)\} = \mathscr{L}\{1\}\mathscr{L}\{y\} = \frac{Y(s)}{s}.$$

Hence,

$$\mathscr{L}\left\{\frac{dy}{dt}\right\} + \mathscr{L}\{y\} + \mathscr{L}\left\{\int_0^t y(u)du\right\} = \mathscr{L}\{1\}.$$

$$sY(s) - y(0) + Y(s) + \frac{Y(s)}{s} = \frac{1}{s}$$

$$s^2Y(s) + sY(s) + Y(s) = 1$$

$$Y(s) = \frac{1}{s^2 + s + 1}$$

The same steps are carried out with Maple.

> ```
EQ:=diff(y(t),t)+y(t)+Int(y(u),u=0..t)=1;
```

$$EQ := \left(\frac{\partial}{\partial t} y(t)\right) + y(t) + \int_0^t y(u)du = 1$$

> ```
step_1:=laplace(EQ,t,s);
```

$$step\_1 := s \text{ laplace}(y(t), t, s) - y(0) + \text{laplace}(y(t), t, s) + \frac{\text{laplace}(y(t), t, s)}{s} = \frac{1}{s}$$

> ```
step_2:=subs(y(0)=0,step_1);
```

$$step\_2 := s \text{ laplace}(y(t), t, s) + \text{laplace}(y(t), t, s) + \text{laplace}(y(t), t, s) + \frac{\text{laplace}(y(t), t, s)}{s} = \frac{1}{s}$$

> ```
step_3:=simplify(solve(step_2,laplace(y(t),t,s)));
```

$$step\_3 := \frac{1}{s^2 + s + 1}$$

Because $Y(s) = \frac{1}{s^2+s+1} = \frac{1}{(s+1/2)^2+(\sqrt{3}/2)^2}$, $y(t) = \frac{2}{\sqrt{3}}e^{-t/2}\sin\frac{\sqrt{3}}{2}t$.

The same solution, which is then graphed on the interval [0,3] with `plot`, is found with `invlaplace` and named `Sol`.

```
> Sol:=invlaplace(step_3,s,t);
```

$$Sol := -\frac{1}{3}\sqrt{-3}\left(e^{\left(\left(-\frac{1}{2}+\frac{1}{2}\sqrt{-3}\right)t\right)} - e^{\left(\left(-\frac{1}{2}-\frac{1}{2}\sqrt{-3}\right)t\right)}\right)$$

```
> Sol:=simplify(Sol);
```

$$Sol := \frac{1}{3}I\sqrt{3}\left(-e^{\left(\frac{1}{2}(-1+I\sqrt{3})t\right)} + e^{\left(-\frac{1}{2}(1+I\sqrt{3})t\right)}\right)$$

```
> Sol:=convert(Sol,trig);
```

$$Sol := \frac{1}{3}I\sqrt{3}\left(-\left(\cos h\left(\frac{1}{2}t\right) - \sin h\left(\frac{1}{2}t\right)\right)\left(\cos\left(\frac{1}{2}\sqrt{3}t\right) + I\sin\left(\frac{1}{2}\sqrt{3}t\right)\right)\right.$$
$$\left. +\left(\cos h\left(\frac{1}{2}t\right) - \sin h\left(\frac{1}{2}t\right)\right)\left(\cos\left(\frac{1}{2}\sqrt{3}t\right) - I\sin\left(\frac{1}{2}\sqrt{3}t\right)\right)\right)$$

```
> Sol:=convert(Sol,expsincos);
```

$$Sol := \frac{1}{3}I\sqrt{3}$$
$$\left(\frac{\cos\left(\frac{1}{2}\sqrt{3}t\right) + I\sin\left(\frac{1}{2}\sqrt{3}t\right)}{e^{\left(\frac{1}{2}t\right)}} + \frac{\cos\left(\frac{1}{2}\sqrt{3}t\right) - I\sin\left(\frac{1}{2}\sqrt{3}t\right)}{e^{\left(\frac{1}{2}t\right)}}\right)$$

```
> Sol:=simplify(Sol);
```

$$Sol := \frac{2}{3}e^{\left(-\frac{1}{2}t\right)}\sqrt{3}\sin\left(\frac{1}{2}\sqrt{3}t\right)$$

```
> plot(Sol,t=0..3*Pi);
```

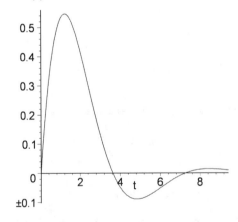

6.6 Applications of Laplace Transforms

Spring–Mass Systems

Laplace transforms are useful in solving the spring–mass systems that were discussed in earlier sections. Although the method of Laplace transforms can be used to solve all problems discussed in the section on applications of higher order equations, this method is most useful in alleviating the difficulties associated with problems that involve piecewise-defined forcing functions. Hence, we investigate the use of Laplace transforms to solve the second order initial-value problem that models the motion of a mass attached to the end of a spring. We found in Chapter 5 that this situation is modeled by the initial-value problem

$$\begin{cases} m\frac{d^2x}{dt^2} + c\,\frac{dx}{dt} + kx = 0 \\ x(0) = \alpha,\ x'(0) = \beta \end{cases}$$

where m represents the mass, c the damping coefficient, and k the spring constant determined by Hooke's law. We demonstrate how the method of Laplace transforms is used to solve initial-value problems of this type if the forcing function is discontinuous.

EXAMPLE 1: Suppose that a mass with $m = 1$ is attached to a spring with spring constant $k = 1$. If there is no resistance due to damping, determine the displacement of the mass if it is released from its equilibrium position and is subjected to the force $f(t) = \begin{cases} \sin t, & ;0 \leq t < \pi/2 \\ 0, & t \geq \pi/2 \end{cases}$

SOLUTION: In this case, the constants are $m = 1$, $c = 0$, and $k = 1$. The initial position is $x(0) = 0$ and the initial velocity is $\frac{dx}{dt}(0) = 0$. Hence, the initial-value problem that models this situation is

$$\begin{cases} \frac{d^2x}{dt^2} + x = \begin{cases} \sin t, 0 \leq t < \pi/2 \\ 0, t \geq \pi/2 \end{cases} \\ x(0) = 0, \frac{dx}{dt}(0) = 0 \end{cases}.$$

Because we will take the Laplace transform of both sides of the differential equation, we write $f(t)$ in terms of the unit step function. This gives us

$$f(t) = \sin t[\mathcal{U}(t - 0) - \mathcal{U}(t - \pi/2)] = (1 - \mathcal{U}(t - \pi/2))\sin t,$$

```
> convert(piecewise(t>=0 and t<Pi/2,sin(t),t>=Pi/2,0),
        Heaviside);
```

$$\sin(t)\,\text{Heaviside}(t) - \sin(t)\,\text{Heaviside}\left(t - \frac{1}{2}\pi\right)$$

which we now graph with `plot`.

```
> plot(sin(t)*(1-Heaviside(t-Pi/2)),t=0..Pi);
```

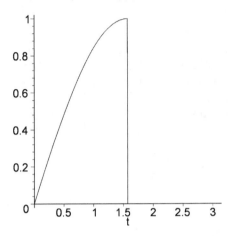

Using the method of Laplace transforms, we compute the Laplace transform of each side of the equation

```
> with(inttrans):
step1:=laplace(diff(x(t),t$2)+x(t)=
            sin(t)*(1-Heaviside(t-Pi/2)),t,s);
```

$$step1 := s(s\,\text{laplace}(x(t),t,s) - x(0)) - D(x)(0) + \text{laplace}(x(t),t,s) =$$
$$\frac{1}{s^2+1} - \frac{e^{\left(-\frac{1}{2}s\pi\right)}s}{s^2+1}$$

apply the initial conditions

```
> step2:=subs({x(0)=0,D(x)(0)=0},step1);
```

$$step2 := s^2\text{laplace}(x(t),t,s) + \text{laplace}(x(t),t,s) = \frac{1}{s^2+1} - \frac{e^{\left(-\frac{1}{2}s\pi\right)}s}{s^2+1}$$

and solve the resulting equation for $x(t) = \mathscr{L}^{-1}\{X(s)\}$.

```
> step3:=solve(step2,laplace(x(t),t,s));
```

$$step3 := -\frac{-1 + e^{\left(-\frac{1}{2}s\pi\right)}s}{s^4 + 2s^2 + 1}$$

The solution is obtained with `invlaplace`.

```
> sol:=invlaplace(step3,s,t);
```

$$sol := -\frac{1}{2}t\cos(t) + \frac{1}{2}\sin(t) + \frac{1}{2}\text{Heaviside}\left(t - \frac{1}{2}\pi\right)\cos(t)t - \frac{1}{4}\text{Heaviside}\left(t - \frac{1}{2}\pi\right)\cos(t)\pi$$

The same result is obtained with `dsolve` together with the option `method=la-place`, which we then graph with `plot`.

```
> sol:=dsolve({diff(x(t),t$2)+x(t)=
       sin(t)*(1-Heaviside(t-Pi/2)),x(0)=0,D(x)(0)=0},x(t),
         method=laplace);
```

$$sol := x(t) = -\frac{1}{2}t\cos(t) + \frac{1}{2}\sin(t) + \frac{1}{2}\text{Heaviside}\left(t - \frac{1}{2}\pi\right)\cos(t)t - \frac{1}{4}\text{Heaviside}\left(t - \frac{1}{2}\pi\right)\cos(t)\pi$$

```
> convert(rhs(sol),piecewise);
```

$$\begin{cases} -\frac{1}{2}t\cos(t) + \frac{1}{2}\sin(t) & t \le \frac{1}{2}\pi \\ -\frac{1}{4}\cos(t)\pi + \frac{1}{2}\sin(t) & \frac{1}{2}\pi < t \end{cases}$$

```
> plot(rhs(sol),t=0..2*Pi);
```

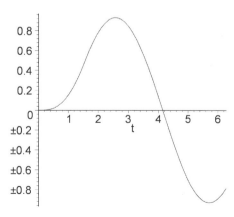

Notice that resonance occurs on the interval $0 \leq t < \pi/2$. Then, for $t \geq \pi/2$, the motion is harmonic. Hence, although the forcing function is zero for $t \geq \pi/2$, the mass continues to follow the path defined by $x(t)$ indefinitely.

■

EXAMPLE 2: Suppose that a mass of $m = 1$ is attached to a spring with spring constant $k = 13$. If the mass is subjected to the resistive force due to damping $F_R = 4dx/dt$, determine the displacement of the mass if it is released from its equilibrium position and is subjected to the force
$f(t) = 2t(1 - \mathcal{U}(t - 1)) + 2\mathcal{U}(t - 1) + 10\delta(t - 3)$.

SOLUTION: In this case, the initial-value problem is

$$\begin{cases} \frac{d^2x}{dt^2} + 4\frac{dx}{dt} + 13x = 2t(1 - \mathcal{U}(t - 1)) + 2\mathcal{U}(t - 1) + 10\delta(t - 3) \\ \qquad\qquad x(0) = 0, \ \frac{dx}{dt}(0) = 0 \end{cases}.$$

We first graph $2t(1 - \mathcal{U}(t - 1)) + 2\mathcal{U}(t - 1)$.

```
> plot(2*t*(1-Heaviside(t-1))+2*Heaviside(t-1),t=0..4,0..4);
```

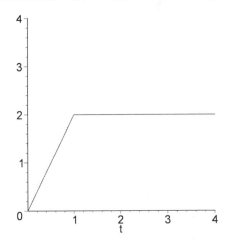

Using the method of Laplace transforms, we take the Laplace transform of each side of the equation

```
> EQ:=diff(x(t),t$2)+4*diff(x(t),t)+13*x(t)=
      2*t*(1-Heaviside(t-1))+2*Heaviside(t-1)+10*Dirac(t-3):
```

```
step1:=laplace(EQ,t,s);
```

$step1 := s(s \, \text{laplace}(x(t), t, s) - x(0)) - D(x)(0) + 4s \, \text{laplace}(x(t), t, s) - 4x(0) + 13\text{laplace}(x(t), t, s) =$

$$2\frac{1}{s^2} - 2\frac{e^{(-s)}}{s^2} + 10e^{(-3s)}$$

apply the initial conditions

```
> step2:=subs({x(0)=0,D(x)(0)=0},step1);
```

$step2 := s^2\text{laplace}(x(t), t, s) + 4s \, \text{laplace}(x(t), t, s) + 13\text{laplace}(x(t), t, s) =$

$$2\frac{1}{s^2} - 2\frac{e^{(-s)}}{s^2} + 10e^{(-3s)}$$

and solve for $X(s) = \mathscr{L}\{x(t)\}$.

```
> step3:=solve(step2,laplace(x(t),t,s));
```

$$step3 := -2\frac{-1 + e^{(-s)} - 5e^{(-3s)}s^2}{s^2(s^2 + 4s + 13)}$$

The solution to the initial-value problem is obtained with `invlaplace`.

```
> sol:=invlaplace(step3,s,t);
```

$$sol := \frac{2}{13}t - \frac{8}{169} + \frac{8}{169}e^{-2t}\cos(3t) - \frac{10}{507}e^{(-2t)}\sin(3t)$$

$$- \frac{2}{13}t\text{Heaviside}(t - 1) + \frac{34}{169}\text{Heaviside}(t - 1)$$

$$- \frac{8}{169}\text{Heaviside}(t - 1)e^{(-2t+2)}\cos(3t - 3)$$

$$+ \frac{10}{507}\text{Heaviside}(t - 1)e^{(-2t+2)}\sin(3t - 3)$$

$$+ \frac{10}{3}\text{Heaviside}(t - 3)e^{(-2t+6)}\sin(3t - 9)$$

An equivalent result is obtained with `dsolve`. The solution is graphed with `plot`.

```
> sol:=dsolve({EQ,x(0)=0,D(x)(0)=0},x(t));
```

$$sol := x(t) = -\frac{8}{169}\cos(3t) = \text{Heaviside}(t-1)\cos(3)e^{(-2t+2)}$$

$$-\frac{10}{507}\cos(3t)\text{Heaviside}(t-1)\sin(3)e^{(-2t+2)}$$

$$-\frac{10}{3}\cos(3t)\sin(9)\text{Heaviside}(t-3)e^{(-2t+6)}+\frac{2}{13}t-\frac{8}{169}$$

$$-\frac{2}{13}t\text{Heaviside}(t-1)+\frac{34}{169}\text{Heaviside}(t-1)$$

$$+\frac{10}{507}\sin(3t)\text{Heaviside}(t-1)\cos(3)e^{(-2t+2)}$$

$$-\frac{8}{169}\sin(3t)\text{Heaviside}(t-1)\sin(3)e^{(-2t+2)}$$

$$+\frac{10}{3}\sin(3t)\cos(9)\text{Heaviside}(t-3)e^{(-2t+6)}$$

$$+\frac{8}{169}e^{-2t}\cos(3t)-\frac{10}{507}e^{(-2t)}\sin(3t)$$

```
> plot(rhs(sol),t=0..6);
```

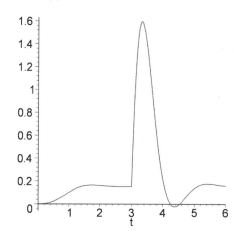

The graph of the solution shows the effect of the impulse delivered at $t = 3$, which is especially evident when we compare this result with the solution of

$$\begin{cases} \frac{d^2x}{dt^2}+4\frac{dx}{dt}+13x = 2t(1-\mathscr{U}(t-1))+2\mathscr{U}(t-1) \\ \quad x(0) = 0,\ \frac{dx}{dt}(0) = 0 \end{cases}.$$

```
> sol2:=dsolve({diff(x(t),t$2)+4*diff(x(t),t)+13*x(t)=
    2*t*(1-Heaviside(t-1))+2*Heaviside(t-1),
    x(0)=0,D(x)(0)=0},x(t));
```

$$sol2 := x(t) = -\frac{8}{169}\cos(3t)\text{Heaviside}(t-1)\cos(3)e^{(-2t+2)}$$

$$-\frac{10}{507}\cos(3t)\text{Heaviside}(t-1)\sin(3)e^{(-2t+2)} + \frac{2}{13}t - \frac{8}{169}$$

$$-\frac{2}{13}t\text{Heaviside}(t-1) + \frac{34}{169}\text{Heaviside}(t-1)$$

$$+\frac{10}{507}\sin(3t)\text{Heaviside}(t-1)\cos(3)e^{(-2t+2)}$$

$$-\frac{8}{169}\sin(3t)\text{Heaviside}(t-1)\sin(3)e^{(-2t+2)}$$

$$+\frac{8}{169}e^{(-2t)}\cos(3t) - \frac{10}{507}e^{(-2t)}\sin(3t)$$

```
> plot(rhs(sol2),t=0..6);
```

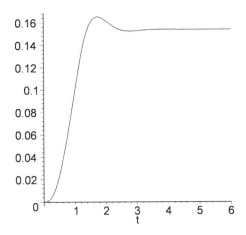

L–R–C Circuits Revisited

Laplace transforms can be used to solve the L–R–C circuit problems that were introduced earlier. Recall that the initial-value problem that is used to find the current is

$$\begin{cases} L\frac{d^2Q}{dt^2} + R\frac{dQ}{dt} + \frac{1}{C}Q = E(t)) \\ Q(0) = Q_0, \ I(0) = \frac{dQ}{dt}(0) = I_0 \end{cases}$$

where L, R, and C represent the inductance, resistance, and capacitance, respectively. Q is the charge of the capacitor and $dQ/dt = I$, where I is the current. $E(t)$ is the voltage supply.

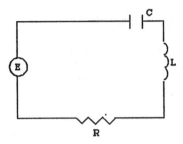

In particular, this method is most useful when the supplied voltage $E(t)$ is a piecewise defined function.

EXAMPLE 3: Suppose that we consider a circuit with a capacitor C, a resistor R,

and a voltage supply $E(t) = \begin{cases} 100, 0 \leq t < 1 \\ 200 - 100t, \ 1 \leq t < 2 \\ 0, \ t \geq 2 \end{cases}$. $L = 0$, find $Q(t)$ and $I(t)$ if

$Q(0) = 0$, $C = 10^{-2}$ farads, and $R = 100\Omega$.

SOLUTION: Because $L = 0$, we can state the first-order initial-value problem as

$$\begin{cases} 100\frac{dQ}{dt} + 100Q = \begin{cases} 100, & 0 \leq t < 1 \\ 200 - 100t, & 1 \leq t < 2 \\ 0, & t \geq 2 \end{cases} \\ \qquad\qquad Q(0) = 0 \end{cases}$$

First, we rewrite $E(t)$ in terms of the unit step functions as

$$E(t) = 100(1 - \mathscr{U}(t - 1)) + (200 - 100t)(\mathscr{U}(t - 1) - \mathscr{U}(t - 2)).$$

```
> convert(piecewise(t>=0 and t
      t>=1 and t=2,0),Heaviside);
```

$$100 \ \text{Heaviside}(t) + 100 \ \text{Heaviside}(t - 1) - 200 \ \text{Heaviside}(t - 2)$$
$$- 100t \ \text{Heaviside}(t - 1) + 100t \ \text{Heaviside}(t - 2)$$

```
> E:=t->100*(1-Heaviside(t-1))+
          (200-100*t)*(Heaviside(t-1)-Heaviside(t-2));
```

$$E := t \to 100 - 100 \ \text{Heaviside}(t - 1)$$
$$+ (200 - 100t)(\text{Heaviside}(t - 1) - \text{Heaviside}(t - 2))$$

```
> plot(E(t),t=0..4);
```

Now, we take the Laplace transform of both sides of the differential equation

```
> step1:=laplace(100*diff(q(t),t)+100*q(t)=E(t),t,s);
```

$$step1 := 100s \, \text{laplace}(q(t), t, s) - 100 \, \text{laplace}(q(t), t, s) =$$
$$100\frac{1}{s} - 100\frac{e^{(-s)}}{s^2} + 100\frac{e^{(-2s)}}{s^2}$$

apply the initial condition

```
> step2:=subs(q(0)=0,step1);
```

$$step2 := 100s \, \text{laplace}(q(t), t, s) + 100 \, \text{laplace}(q(t), t, s) =$$
$$100\frac{1}{s} - 100\frac{e^{(-s)}s^2 + 100e^{(-2s)}}{s^2}$$

and solve for $\mathcal{L}\{Q(t)\}$.

```
> step3:=solve(step2,laplace(q(t),t,s));
```

$$step3 := \frac{s - e^{(-s)} + e^{(-2s)}}{s^2(s+1)}$$

The solution to the initial-value problem is obtained with `invlaplace`.

```
> sol:=invlaplace(step3,s,t);
```

$$sol := 1 - e^{(-t)} - t\text{Heaviside}(t-1) + 2\text{Heaviside}(t-1)$$
$$- \text{Heaviside}(t-1)e^{(1-t)} + t\text{Heaviside}(t-2)$$
$$- 3\text{Heaviside}(t-2) + \text{Heaviside}(t-2)e^{(2-t)}$$

The same result is obtained with `dsolve`.

```
> sol:=dsolve({100*diff(q(t),t)+100*q(t)=E(t),q(0)=0},q(t));
```

$$sol := q(t)1 - e^{(-1)} - t\text{Heaviside}(t-1) + 2\text{Heaviside}(t-1)$$
$$- \text{Heaviside}(t-1)e^{(1-t)} + t\text{Heaviside}(t-2)$$
$$- 3\text{Heaviside}(t-2) + \text{Heaviside}(t-2)e^{(2-t)}$$

We now compute $I = dQ/dt$ and then graph both $Q(t)$ and $I(t)$ on the interval [0,4].

```
> assign(sol):
plot(q(t),t=0..4);
plot(diff(q(t),t),t=0..4);
```

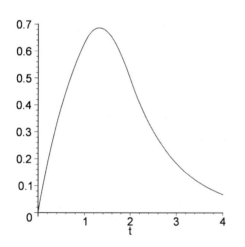
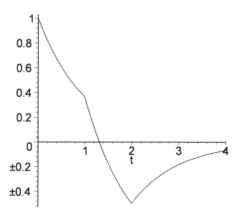

From the graph, we see that after the voltage source is turned off at $t = 2$, the charge approaches zero.

■

EXAMPLE 4: Consider the circuit with no capacitor, $R = 100\,\Omega$, and $L = 100\,\text{H}$ if

$$E(t) = \begin{cases} 100\,\text{V}, 0 \leq t < 1 \\ 0, 1 \leq t < 2 \end{cases} \quad \text{and } E(t+2) = E(t). \text{ Find the current } I(t) \text{ if } I(0) = 0.$$

SOLUTION: The differential equation that models the situation is $100\frac{d^2Q}{dt^2} + 100\frac{dQ}{dt} = E(t)$. Now, $dQ/dt = I$, so we can write this equation as $100\frac{dI}{dt} + 100I = E(t)$. Hence, the initial-value problem is

$$\begin{cases} 100\frac{dI}{dt} + 100I = E(t) \\ I(0) = 0 \end{cases}$$

Notice that $E(t)$ is a periodic function, so we first compute $\mathscr{L}\{E(t)\}$

```
> lape:=simplify(int(100*exp(-s*t),t=0..1)/(1-exp(-2*s)));
```

$$lape := 100\frac{1}{(e^{(-s)} + 1)s}$$

and see that $\mathscr{L}\{E(t)\} = \frac{100}{s(1+e^{-s})}$.
Alternatively, note that we can write $E(t)$ as

$$E(t) = 100 - 100\mathscr{U}(t-1) + 100(\mathscr{U}(t-2) - \mathscr{U}(t-3)) + 100(\mathscr{U}(t-4) - \mathscr{U}(t-5)) + \cdots$$

$$= 100\left(1 + \sum_{n=1}^{\infty}(-1)^n\mathscr{U}(t-n)\right)$$

```
> E:=t->100*(1+sum((-1)^n*Heaviside(t-n),n=1..infinity));
```

$$E := (t) \to 100 + 100\left(\sum_{n=1}^{\infty}(-1)^n\text{Heaviside}(t-n)\right)$$

and then compute the Laplace transform with `laplace`.

```
> simplify(laplace(E(t),t,s));
```

$$100\frac{e^s}{s(e^s + 1)}$$

We now compute the Laplace transform of the left side of the equation. (Note that we use `i` to represent $I(t)$ instead of `I` because `I` represents the imaginary number $\sqrt{-1}$.)

```
> step1:=laplace(100*diff(i(t),t)+100*i(t)=E(t),t,s);
```

$$step1 := 100s \text{ laplace}(i(t), t, s) - 100i(0) + 100 \text{ laplace}(i(t), t, s) =$$

$$100\frac{1}{s} - 100\frac{e^{(-s)}e^s}{s(e^s + 1)}$$

apply the initial condition

```
> step2:=subs(i(0)=0,step1);
```

$$step2 := 100s \text{ laplace}(i(t), t, s) + 100 \text{ laplace}(i(t), t, s) =$$

$$100\frac{1}{s} - 100\frac{e^{(-s)}e^s}{s(e^s + 1)}$$

and solve for $\mathscr{L}\{I(t)\}$.

```
> step3:=solve(step2,laplace(i(t),t,s));
```

$$step3 := -\frac{-e^s - 1 + e^{(-s)}e^s}{s(se^s + s + e^s + 1)}$$

```
> factor(step3);
```

$$-\frac{-e^s - 1 + e^{(-s)}e^s}{s(e^s + 1)(s + 1)}$$

As we did in Section 6.4, we write a power series expansion of $\frac{1}{1+e^{-s}}$. We use $\frac{1}{1+x} = 1 - x + x^2 - x^3 + \cdots$,

$$\frac{1}{1 + e^{-s}} = 1 - e^{-s} + e^{-2s} - e^{-3s} + \cdots.$$

Thus,

$$\mathscr{L}\{I\} = \frac{1}{s(s+1)}(1 - e^{-s} + e^{-2s} - e^{-3s} + \cdots) = \frac{1}{s(s+1)} - \frac{e^{-s}}{s(s+1)} + \frac{e^{-2s}}{s(s+1)} - \frac{e^{-3s}}{s(s+1)} + \cdots.$$

Because, $\mathscr{L}^{-1}\left\{\frac{1}{s(s+1)}\right\} = 1 - e^{-t}$,

```
> invlaplace(1/(s*(s+1)),s,t);
```

$$1 - e^{(-t)}$$

we have that

$$I(t) = (1 - e^{-t}) - (1 - e^{-(t-1)})\mathcal{U}(t-1) + (1 - e^{-(t-2)})\mathcal{U}(t-2) - (1 - e^{-(t-3)})\mathcal{U}(t-3) + \cdots.$$

We can write this function as

$$I(t) = \begin{cases} 1 - e^{-t}, 0 \le t < 1 \\ -e^{-t} + e^{-(t-1)}, 1 \le t < 2 \\ 1 - e^{-t} + e^{-(t-1)} - e^{-(t-2)}, 2 \le t < 3 \\ -e^{-t} + e^{-(t-1)} - e^{-(t-2)} + e^{-(t-3)}, 3 \le t < 4 \\ \vdots \end{cases}$$

To graph $I(t)$ on the interval $[0,n]$, we note that $\mathcal{U}(t - n) = 0$ for $t \le n$ so the graph of $I(t)$ on the interval $[0,n]$ is the same as the graph of

$$(1 - e^{-t}) - (1 - e^{-(t-1)})\mathcal{U}(t-1) + (1 - e^{-(t-2)})\mathcal{U}(t-2) - (1 - e^{-(t-3)})\mathcal{U}(t-3) +$$

$$\cdots + (-1)^{n-1}(1 - e^{-[t-(n-1)]})\mathcal{U}(t - (n-1)).$$

We define `i` using the option `remember`

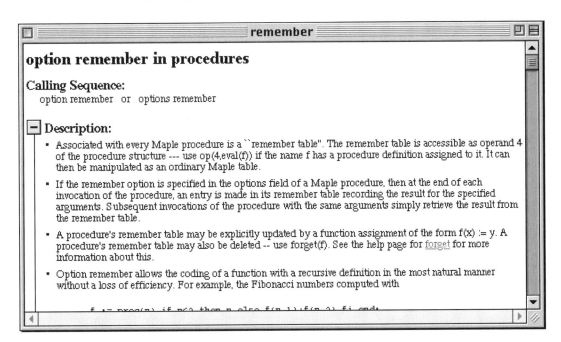

so that Maple "remembers" the values of `i` computed. For example, if `i(3)` is computed, Maple need not recompute `i(0)`, `i(1)`, `i(2)`, and `i(3)` to compute `i(4)`.

```
> i := proc(n) option remember;
      if n=0 then 1-exp(-t) else
          i(n-1)+(-1)^n*(1-exp(-(t-n)))*Heaviside(t-n)
      fi end:
```

Then, to graph $I(t)$ on the interval $[0,5]$ we enter

```
> i(4);
```

$$1 - e^{(-t)} - (1 - e^{(1-t)})\text{Heaviside}(t-1) + (1 - e^{(2-t)})\text{Heaviside}(t-2)$$
$$- (1 - e^{(-t+3)})\text{Heaviside}(t-3) + (1 - e^{(-t+4)})\text{Heaviside}(t-4)$$

and then use `plot`.

```
> plot(i(4),t=0..5);
```

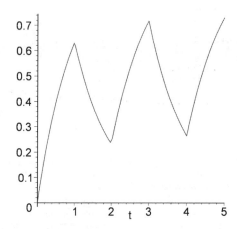

Notice that I increases over the intervals where $E(t) = 100$ and decreases on those where $E(t) = 0$.

■

We can consider the L–R–C circuit in terms of the **integrodifferential equation**

$$L\frac{dI}{dt} + RI + \frac{1}{C}\int_0^t I(\alpha)d\alpha = E(t)$$

which is useful when using the method of Laplace transforms to find the current.

EXAMPLE 5: Find the current $I(t)$ if $L = 1$ henry, $R = 6\,\Omega$, $C = 1/9$ farad, $E(t) = 1$ volt, and $I(0) = 0$.

SOLUTION: In this case, we must solve the initial-value problem

$$\begin{cases} \frac{dI}{dt} + 6I + 9\int_0^t I(\alpha)d\alpha = 1 \\ I(0) = 0 \end{cases}.$$

First, we compute the Laplace transform of each side of the equation,

```
> i:='i':
> step1:=laplace(diff(i(t),t)+6*i(t)+
            9*Int(i(alpha),alpha=0..t)=1,t,s);
```

$$step1 := s\,\text{laplace}(\text{i}(t), t, s) - \text{i}(0) + 6\,\text{laplace}(\text{i}(t), t, s) + 9\frac{\text{laplace}(\text{i}(t), t, s)}{s} = \frac{1}{s}$$

applying the initial condition,

```
> step2:=subs(i(0)=0,step1);
```

$$step2 := s\,\text{laplace}(\text{i}(t), t, s) + 6\,\text{laplace}(\text{i}(t), t, s) + 9\frac{\text{laplace}(\text{i}(t), t, s)}{s} = \frac{1}{s}$$

and solve for $\mathscr{L}\{I(t)\}$.

```
> step3:=solve(step2,laplace(i(t),t,s));
```

$$step3 := \frac{1}{s^2 + 6s + 9}$$

The solution is obtained with `invlaplace`,

```
> sol:=invlaplace(step3,s,t);
```

$$sol := te^{(-3t)}$$

which we graph with `plot`.

```
> plot(sol,t=0..3);
```

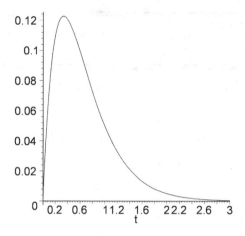

■

Population Problems Revisited

Laplace transforms can used to solve the population problems that were discussed as applications of first-order equations. Laplace transforms are especially useful when dealing with piecewise-defined forcing functions, but they are useful in many other cases as well.

EXAMPLE 6: Let $x(t)$ represent the population of a certain country. The rate at which the population increases and decreases depends on the growth rate of the country as well as the rate at which people are being added to or subtracted from the population by immigration or emigration. Hence, we consider the population problem

$$\begin{cases} x' + kx = 1000(1 + a \sin t) \\ \qquad x(0) = x_0 \end{cases}.$$

Solve this problem using Laplace transforms with $k = 3$, $x_0 = 2000$, and $a = 0.2$, 0.4, 0.6, and 0.8. Plot the solution in each case.

SOLUTION: Using the method of Laplace transforms, we begin by computing the Laplace transform of each side of the equation with `laplace`

```
> step1:=laplace(diff(x(t),t)+3*x(t)=1000*(1+a*sin(t)),t,s);
```

$$step1 := s \, \mathrm{laplace}(x(t), t, s) - x(0) + 3 \, \mathrm{laplace}(x(t), t, s) = 1000\frac{1}{s} + 1000\frac{a}{s^2 + 1}$$

apply the initial condition

```
> step2:=subs(x(0)=2000,step1);
```

$$step2 := s \, \text{laplace}(x(t), t, s) - 2000 + 3 \, \text{laplace}(x(t), t, s) = 1000\frac{1}{s} + 1000\frac{a}{s^2 + 1}$$

and then use `solve` to solve `step2` for $X(s) = \mathscr{L}\{x(t)\}$.

```
> step3:=solve(step2,laplace(x(t),t,s));
```

$$step3 := 1000\frac{2s^3 + 2s + s^2 + 1 + as}{s(s^3 + s + 3s^2 + 3)}$$

To find the solution, we use `invlaplace` and name the result `sol`.

```
> sol:=invlaplace(step3,s,t);
```

$$sol := \frac{1000}{3} + \frac{5000}{3}e^{(-3t)} + 100e^{(-3t)}a - 100a\cos(t) + 300a\sin(t)$$

We use the result to investigate the population for the values of a using `plot`.

```
> avals:=seq(0.2+0.2*i,i=0..3):
toplot:={seq(sol,a=avals)}:
> plot(toplot,t=0..25);
```

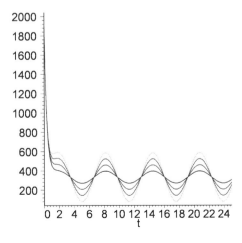

APPLICATION

The Tautochrone

Suppose that from rest, a particle slides down a frictionless curve under the force of gravity. What must the shape of the curve be in order for the time of descent to be independent of the starting position of the particle?

We can determine the shape of the curve using the method of Laplace transforms. Suppose that the particle starts at height y and that its speed is v when it is at a height of z. If m is the mass of the particle and g is the acceleration of gravity, the speed is found by equating the kinetic and potential energies of the particle with

$$\frac{1}{2}mv^2 = mg(y - z)$$

$$v = \sqrt{2g}\sqrt{y - z}.$$

Let σ denote the arc length along the curve from its lowest point to the particle. Then, the time required for the descent is

$$\text{time} = \int_0^{\sigma(y)} \frac{d\sigma}{v} = \int_0^y \frac{1}{v}\frac{d\sigma}{dz} = \int_0^y \frac{1}{v}\phi(z)dz,$$

where $\phi(y) = d\sigma/dy$. The time is constant and $v = \sqrt{2g}\sqrt{y - z}$. so we have

$$\int_0^y \frac{\phi(z)}{\sqrt{y - z}}dz = c_1,$$

where c_1 is a constant. To use a convolution, we multiply by $e^{-sy}dy$ and integrate:

$$\int_0^\infty e^{-sy} \int_0^y \frac{\phi(z)}{\sqrt{y - z}}dzdy = \int_0^\infty e^{-sy}c_1 dy$$

$$\mathscr{L}\{\phi*y^{-1/2}\} = \mathscr{L}\{c_1\}.$$

Using the convolution theorem, we simplify to obtain

$$\mathscr{L}\{\phi\}\mathscr{L}\{y^{-1/2}\} = \frac{c_1}{s}.$$

```
> step1:=laplace(phi(y),y,s)*laplace(y^(-1/2),y,s)=
      laplace(c[1],y,s);
```

$$step1 := \text{laplace}(\phi(y), y, s)\sqrt{\frac{\pi}{s}} = \frac{c_1}{s}$$

Then, $\mathscr{L}\{\phi\} = c_1/\sqrt{\pi s}$.

```
> assume(s>0);
```

```
step2:=simplify(solve(step1,laplace(phi(y),y,s)),radical);
```

$$step2 := \frac{c_1}{\sqrt{\pi s \sim}}$$

We use `invlaplace` to compute $\phi = \mathcal{L}^{-1}\{c_1/\sqrt{\pi s}\} = \frac{c_1}{\pi}y^{-1/2} = ky^{-1/2}$.

```
> step3:=simplify(invlaplace(step2,s,y));
```

$$step3 := \frac{c_1}{\pi\sqrt{y}}$$

Recall that $\phi(yt) = d\sigma/dy$ represents arc length. Then, $\phi(y) = d\sigma/dy = \sqrt{1 + (dx/dy)^2}$ and substitution of $\phi = ky^{-1/2}$ into this equation gives us

$$\sqrt{1 + (dx/dy)^2} = ky^{-1/2} \quad \text{or} \quad 1 + (dx/dy)^2 = k^2/y.$$

We solve this equation for dx/dy to obtain $dx/dy = \sqrt{k^2/y - 1}$. With the substitution $y = k^2 \sin^2 \theta$ we obtain

$$dx = \sqrt{\frac{k^2}{k^2 \sin^2 \theta} - 1}\left(2k^2 \sin\theta \cos\theta \, d\theta\right) = \sqrt{\frac{k^2(1 - \sin^2\theta)}{k^2 \sin^2\theta}}\left(2k^2 \sin\theta \cos\theta \, d\theta\right)$$

$$= \frac{\cos\theta}{\sin\theta}\left(2k^2 \sin\theta \cos\theta \, d\theta\right) = 2k^2 \cos^2\theta \, d\theta$$

and integration results in $x(\theta) = \frac{1}{2}k^2(2\theta + \sin 2\theta) + C_1$. To find C_1, we apply the initial condition:

$$x(0) = 0 \Rightarrow C_1 = 0 \Rightarrow x(\theta) = \frac{1}{2}k^2(2\theta + \sin 2\theta).$$

```
> x:=(theta,k)->int(2*k^2*cos(theta)^2,theta);
```

$$x := (\theta, k) \rightarrow \int 2k^2 \cos(\theta)^2 d\theta$$

```
> x(theta,k);
```

$$2k^2 \left(\frac{1}{2}\cos(\theta)\sin(\theta) + \frac{1}{2}\theta\right)$$

Using the identity $\sin^2 \theta = \frac{1}{2}(1 - \cos 2\theta)$ yields $y(\theta) = k^2 \sin^2 \theta = \frac{1}{2}k^2(1 - \cos 2\theta)$.

```
> y:=(theta,k)->1/2*k^2*(1-cos(2*theta));
```

$$y := (\theta, k) \rightarrow \frac{1}{2}k^2(1 - \cos(2\theta))$$

We use `plot` to graph $\begin{cases} x = x(\theta) \\ y = y(\theta) \end{cases}$, $-\pi/2 \leq \theta \leq 0$ for various values of k and then display

the results as a graphics array using `display` together with the option `insequence=true`.

```
> kvals:=[0.25,0.5,0.75,1,2,3]:

somegraphs:=[seq(plot([x(theta,k),y(theta,k),

    theta=-Pi/2..0]),k=kvals)]:

> with(plots):

toshow:=display(somegraphs,insequence=true):

display(toshow);
```

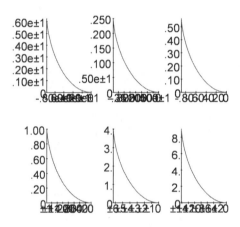

The graphs illustrate that increasing the value of k increases the length of the curve.
The time is independent of the choice of y (that is, the choice of θ). Therefore,

$$\text{time} = \int_0^y \frac{\phi(z)}{\sqrt{y-z}}\,dz = \int_0^y \frac{ky^{-1/2}}{\sqrt{y-z}}\,dz = -2k\left[\sqrt{\frac{y-z}{y}}\right]_0^6 = -2k(-1) = 2k.$$

CHAPTER 7

Systems of Ordinary Differential Equations

7.1 Review of Matrix Algebra and Calculus

Because of their importance in the study of systems of linear equations, we now review matrices and the operations associated with them.

Basic Operations

Definition
$m \times n$ **Matrix**

An **m × n matrix** is an array of the form

$$
\mathbf{A} = (a_{ij}) = \begin{pmatrix} a_{11} & a_{12} & a_{13} & \cdots & a_{1n} \\ a_{21} & a_{22} & a_{23} & \cdots & a_{2n} \\ a_{31} & a_{32} & a_{33} & \cdots & a_{3n} \\ \vdots & \vdots & \vdots & & \vdots \\ a_{m1} & a_{m2} & a_{m3} & \cdots & a_{mn} \end{pmatrix}
$$

with m rows and n columns. This matrix is often denoted by $\mathbf{A} = \left(a_{ij} \right)$.

In Maple, a matrix is simply a list of lists where each list represents a row of the matrix. Therefore, the $m \times n$ matrix

$$\mathbf{A} = (a_{ij}) = \begin{pmatrix} a_{11} & a_{12} & a_{13} & \cdots & a_{1n} \\ a_{21} & a_{22} & a_{23} & \cdots & a_{2n} \\ a_{31} & a_{32} & a_{33} & \cdots & a_{3n} \\ \vdots & \vdots & \vdots & & \vdots \\ a_{m1} & a_{m2} & a_{m3} & \cdots & a_{mn} \end{pmatrix}$$

is entered with the following command.

```
A:=array(
    [[a1,1,a1,2,...,a1,n],[a2,1,a2,2,...,a2,n],...,[am,1,am,2,...,am,n]]
)
```

For example, to use Maple to define m to be the matrix $\begin{pmatrix} a_{11} & a_{12} \\ a_{21} & a_{22} \end{pmatrix}$, enter the following command.

```
> m:=array([[a[1,1],a[1,2]],[a[2,1],a[2,2]]]);
```

$$m := \begin{bmatrix} a_{1,1} & a_{1,2} \\ a_{2,1} & a_{2,2} \end{bmatrix}$$

You can also quickly construct matrices of various sizes using the **Matrix** palette.

For example, to construct a 2×2 matrix, click on the ▦ icon.

We generally call an $n \times 1$ matrix $\begin{pmatrix} v_1 \\ v_2 \\ \vdots \\ v_n \end{pmatrix}$ a **column vector** and a $1 \times n$ matrix $(v_1 \quad v_2 \quad \cdots \quad v_n)$ a **row vector**.

Definition
Transpose

The **transpose** of the $n \times m$ matrix

$$\mathbf{A} = \begin{pmatrix} a_{11} & a_{12} & \cdots & a_{1m} \\ a_{21} & a_{22} & \cdots & a_{2m} \\ \vdots & \vdots & \ddots & \vdots \\ a_{n1} & a_{n2} & \cdots & a_{nm} \end{pmatrix}$$

is the $m \times n$ matrix

$$\mathbf{A}^T = \begin{pmatrix} a_{11} & a_{21} & \cdots & a_{n1} \\ a_{12} & a_{22} & \cdots & a_{n2} \\ \vdots & \vdots & \ddots & \vdots \\ a_{1m} & a_{2m} & \cdots & a_{nm} \end{pmatrix}.$$

Hence, $\mathbf{A}^T = \left(a_{ji} \right)$.

Definition
Scalar Multiplication,
Matrix Addition

Let $A = \left(a_{ij} \right)$ be an $n \times m$ matrix and c a scalar. Then the **scalar multiple of A** by c is the $n \times m$ matrix given by $cA = \left(ca_{ij} \right)$. If $B = \left(b_{ij} \right)$ is also an $n \times m$ matrix, the **sum** of the matrices **A** and **B** is the $n \times m$ matrix $A + B = \left(a_{ij} \right) + \left(b_{ij} \right) = \left(a_{ij} + b_{ij} \right)$.

Note that many commands that are used to perform matrix operations are contained in the `linalg` package.

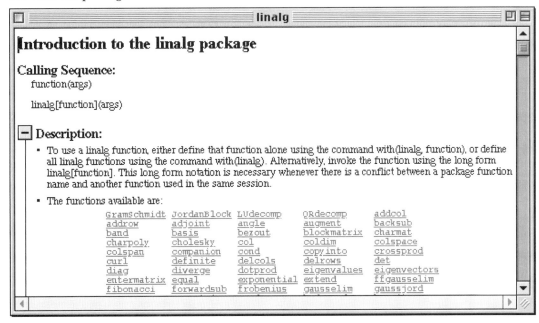

EXAMPLE 1: Compute $3A - 9B$ if $A = \begin{pmatrix} -1 & 4 & -2 \\ 6 & 2 & -10 \end{pmatrix}$ and $B = \begin{pmatrix} 2 & -4 & 8 \\ 7 & 4 & 2 \end{pmatrix}$.

SOLUTION: We have that $3A = \begin{pmatrix} -3 & 12 & -6 \\ 18 & 6 & -30 \end{pmatrix}$
and $-9B = \begin{pmatrix} -18 & 36 & -72 \\ -63 & -36 & -18 \end{pmatrix}$ so

$$3A - 9B = 3A + (-9B) = \begin{pmatrix} -21 & 48 & -78 \\ -45 & -30 & -48 \end{pmatrix}.$$

You should use `array` or `matrix`, which is contained in the **linalg** package, to define matrices. Elementary computations with matrices are carried out using the built-in command `evalm` or commands contained in the **linalg** package.

For example, we use `array` to define **A** and **B** to be the matrices
$A = \begin{pmatrix} -1 & 4 & -2 \\ 6 & 2 & -10 \end{pmatrix}$ and $B = \begin{pmatrix} 2 & -4 & 8 \\ 7 & 4 & 2 \end{pmatrix}$

```
> A:=array([[-1,4,-2],[6,2,-10]]);

  B:=array([[2,-4,8],[7,4,2]]);
```

$$A := \begin{bmatrix} -1 & 4 & -2 \\ 6 & 2 & -10 \end{bmatrix}$$

$$B := \begin{bmatrix} 2 & -4 & 8 \\ 7 & 4 & 2 \end{bmatrix}$$

and then use `evalm` to compute $3A - 9B$.

```
> evalm(3*A-9*B);
```

$$\begin{bmatrix} -21 & 48 & -78 \\ -45 & -30 & -48 \end{bmatrix}$$

■

Definition
Matrix Multiplication

If $\mathbf{A} = \begin{pmatrix} a_{11} & a_{12} & \cdots & a_{1j} \\ a_{21} & a_{22} & \cdots & a_{2j} \\ \vdots & \vdots & \ddots & \vdots \\ a_{n1} & a_{n2} & \cdots & a_{nj} \end{pmatrix}$ is an $n \times j$ matrix and

$\mathbf{B} = \begin{pmatrix} b_{11} & b_{12} & \cdots & b_{1m} \\ b_{21} & b_{22} & \cdots & b_{2m} \\ \vdots & \vdots & \ddots & \vdots \\ b_{j1} & b_{j2} & \cdots & b_{jm} \end{pmatrix}$ is a $j \times m$ matrix,

$\mathbf{AB} = \begin{pmatrix} a_{11} & a_{12} & \cdots & a_{1j} \\ a_{21} & a_{22} & \cdots & a_{2j} \\ \vdots & \vdots & \ddots & \vdots \\ a_{n1} & a_{n2} & \cdots & a_{nj} \end{pmatrix} \begin{pmatrix} b_{11} & b_{12} & \cdots & b_{1m} \\ b_{21} & b_{22} & \cdots & b_{2m} \\ \vdots & \vdots & \ddots & \vdots \\ b_{j1} & b_{j2} & \cdots & b_{jm} \end{pmatrix}$

is the unique matrix $\mathbf{C} = \begin{pmatrix} c_{11} & c_{12} & \cdots & c_{1m} \\ c_{21} & c_{22} & \cdots & c_{2m} \\ \vdots & \vdots & \ddots & \vdots \\ c_{n1} & c_{n2} & \cdots & c_{nm} \end{pmatrix}$

where

$$c_{11} = a_{11}b_{11} + a_{12}b_{12} + \cdots + a_{1j}b_{j1} = \sum_{k=1}^{j} a_{1k}b_{k1},$$

$$c_{12} = a_{11}b_{12} + a_{12}b_{22} + \cdots + a_{1j}b_{j2} = \sum_{k=1}^{j} a_{1k}b_{k2},$$

and

$$c_{uv} = a_{u1}b_{1v} + a_{u2}b_{2v} + \cdots + a_{uj}b_{jv} = \sum_{k=1}^{j} a_{uk}b_{kv}.$$

That is, the element c_{uv}, the entry in the uth row and vth column of $\mathbf{C}$, is obtained by multiplying each member of the uth row of $\mathbf{A}$ by the corresponding entry in the vth column of $\mathbf{B}$ and adding the result.

EXAMPLE 2: Compute $\mathbf{AB}$ if $\mathbf{A} = \begin{pmatrix} 0 & 4 & 5 \\ -5 & -1 & 5 \end{pmatrix}$ and $\mathbf{B} = \begin{pmatrix} -3 & 4 \\ -5 & -4 \\ 1 & -4 \end{pmatrix}$.

SOLUTION: Because $\mathbf{A}$ is a 2×3 matrix and $\mathbf{B}$ is a 3×2 matrix, $\mathbf{AB}$ is the 2×2 matrix:

$$\mathbf{AB} = \begin{pmatrix} 0 & 4 & 5 \\ -5 & -1 & 5 \end{pmatrix} \begin{pmatrix} -3 & 4 \\ -5 & -4 \\ 1 & -4 \end{pmatrix}$$

$$= \begin{pmatrix} 0 \cdot -3 + 4 \cdot -5 + 5 \cdot 1 & 0 \cdot 4 + 4 \cdot -4 + 5 \cdot -4 \\ -5 \cdot -3 + -1 \cdot -5 + 5 \cdot 1 & -5 \cdot 4 + -1 \cdot -4 + 5 \cdot -4 \end{pmatrix} = \begin{pmatrix} -15 & -36 \\ 25 & -36 \end{pmatrix}.$$

After defining $\mathbf{A}$ and $\mathbf{B}$,

```
> A:=array([[0,4,5],[-5,-1,5]]):
B:=array([[-3,4],[-5,-4],[1,-4]]):
```

we use `evalm` together with the `&*` operator, which represents noncommutative multipication, to compute the matrix product $\mathbf{AB}$.

```
> evalm(A
```

$$\begin{vmatrix} -15 & -36 \\ 25 & -36 \end{vmatrix}$$

Alternatively, after loading the **linalg** package, we see that the same result is obtained using `multiply`.

```
> with(linalg):

multiply(A,B);
```

$$\begin{vmatrix} -15 & -36 \\ 25 & -36 \end{vmatrix}$$

■

Definition
Identity Matrix

> The $n \times n$ matrix $\begin{pmatrix} 1 & 0 & 0 & 0 \\ 0 & 1 & 0 & 0 \\ \vdots & \vdots & \ddots & \vdots \\ 0 & 0 & 0 & 1 \end{pmatrix}$ is called the $n \times n$ **identity**
>
> **matrix**, denoted by $\mathbf{I}$ or $\mathbf{I}_n$.

If $\mathbf{A}$ is an $n \times n$ matrix (an $n \times n$ is called a **square matrix**), then $\mathbf{IA} = \mathbf{AI} = \mathbf{A}$. We can define the $n \times n$ identity matrix by entering

```
array(identity,1..n,1..n)
```

```
> seq(array(identity,1..n,1..n),n=3..6);
```

$$\begin{bmatrix} 1 & 0 & 0 \\ 0 & 1 & 0 \\ 0 & 0 & 1 \end{bmatrix}, \begin{bmatrix} 1 & 0 & 0 & 0 \\ 0 & 1 & 0 & 0 \\ 0 & 0 & 1 & 0 \\ 0 & 0 & 0 & 1 \end{bmatrix}, \begin{bmatrix} 1 & 0 & 0 & 0 & 0 \\ 0 & 1 & 0 & 0 & 0 \\ 0 & 0 & 1 & 0 & 0 \\ 0 & 0 & 0 & 1 & 0 \\ 0 & 0 & 0 & 0 & 1 \end{bmatrix}, \begin{bmatrix} 1 & 0 & 0 & 0 & 0 & 0 \\ 0 & 1 & 0 & 0 & 0 & 0 \\ 0 & 0 & 1 & 0 & 0 & 0 \\ 0 & 0 & 0 & 1 & 0 & 0 \\ 0 & 0 & 0 & 0 & 1 & 0 \\ 0 & 0 & 0 & 0 & 0 & 1 \end{bmatrix}$$

Determinants and Inverses

Definition
Determinant

If $\mathbf{A} = (a_{11})$, the **determinant** of $\mathbf{A}$, denoted by $\det(\mathbf{A})$ or $|\mathbf{A}|$, is $\det(\mathbf{A}) = a_{11}$; if $\mathbf{A} = \begin{pmatrix} a_{11} & a_{12} \\ a_{21} & a_{22} \end{pmatrix}$, then

$$\det(\mathbf{A}) = \begin{vmatrix} a_{11} & a_{12} \\ a_{21} & a_{22} \end{vmatrix} = a_{11}a_{22} - a_{12}a_{21}.$$

More generally, if $\mathbf{A} = \begin{pmatrix} a_{11} & a_{12} & \cdots & a_{1n} \\ a_{21} & a_{22} & \cdots & a_{2n} \\ \vdots & \vdots & \ddots & \vdots \\ a_{n1} & a_{n2} & \cdots & a_{nn} \end{pmatrix}$

is an $n \times n$ matrix and $\mathbf{A}_{ij}$ is the $(n-1) \times (n-1)$ matrix obtained by deleting the ith row and jth column from $\mathbf{A}$, then

$$\det(\mathbf{A}) = \begin{vmatrix} a_{11} & a_{12} & \cdots & a_{1n} \\ a_{21} & a_{22} & \cdots & a_{2n} \\ \vdots & \vdots & \ddots & \vdots \\ a_{n1} & a_{n2} & \cdots & a_{nn} \end{vmatrix} = \sum_{j=1}^{n} (-1)^{i+j} a_{ij} \det(\mathbf{A}_{ij})$$

$$= \sum_{j=1}^{n} (-1)^{i+j} a_{ij} |\mathbf{A}_{ij}|.$$

The number $(-1)^{i+j} a_{ij} \det\left(\mathbf{A}_{ij}\right) = (-1)^{i+j} a_{ij} |\mathbf{A}_{ij}|$ is called the **cofactor** of a_{ij}. The **cofactor matrix**, $\mathbf{A}^c$, of $\mathbf{A}$ is the matrix obtained by replacing each element of $\mathbf{A}$ by its cofactor. Hence,

$$\mathbf{A}^c = \begin{pmatrix} |\mathbf{A}_{11}| & -|\mathbf{A}_{12}| & \cdots & (-1)^{n+1}|\mathbf{A}_{1n}| \\ -|\mathbf{A}_{21}| & |\mathbf{A}_{22}| & \cdots & (-1)^{n}|\mathbf{A}_{2n}| \\ \vdots & \vdots & \ddots & \vdots \\ (-1)^{n+1}|\mathbf{A}_{n1}| & (-1)^{n}|\mathbf{A}_{n2}| & \cdots & |\mathbf{A}_{nn}| \end{pmatrix}.$$

EXAMPLE 3: Calculate $|\mathbf{A}|$ if $\mathbf{A} = \begin{pmatrix} -4 & -2 & -1 \\ 5 & -4 & -3 \\ 5 & 1 & -2 \end{pmatrix}$.

SOLUTION:

$$|\mathbf{A}| = \begin{vmatrix} -4 & -2 & -1 \\ 5 & -4 & -3 \\ 5 & 1 & -2 \end{vmatrix} = (-4)\begin{vmatrix} -4 & -3 \\ 1 & -2 \end{vmatrix} + (-1)^3(-2)\begin{vmatrix} 5 & -3 \\ 5 & -2 \end{vmatrix} + (-1)\begin{vmatrix} 5 & -4 \\ 5 & 1 \end{vmatrix}$$

$$= -4((-4)(-2) - (-3)(1)) + 2((5)(-2) - (-3)(5)) - ((5)(1) - (-4)(5)) = -59.$$

Determinants are computed using the det command, which is contained in the linalg package. Next, we load the linalg package.

> with(linalg):

Then, we define **A** and compute the determinant of **A** with det.

> A:=array([[-4,-2,-1],[5,-4,-3],[5,1,-2]]):

det(A);

$$-59$$

∎

Definition **Adjoint and Inverse**	**B** is an **inverse** of the $n \times n$ matrix **A** means that $\mathbf{AB} = \mathbf{BA} = \mathbf{I}$. The **adjoint**, $\mathbf{A}^a$, of an $n \times n$ matrix **A** is the transpose of the cofactor matrix: $\mathbf{A}^a = (\mathbf{A}^c)^T$. If $|\mathbf{A}| \neq 0$ and $\mathbf{B} = \frac{1}{|\mathbf{A}|}\mathbf{A}^a$, then $\mathbf{AB} = \mathbf{BA} = \mathbf{I}$. Therefore, if $|\mathbf{A}| \neq 0$, the inverse of **A** is given by $$\mathbf{A}^{-1} = \frac{1}{|\mathbf{A}|}\mathbf{A}^a.$$

If $\mathbf{A} = \begin{pmatrix} a & b \\ c & d \end{pmatrix}$ and $|\mathbf{A}| = ad - bc \neq 0$, the inverse of **A** is given by

$$\mathbf{A}^{-1} = \frac{1}{ad - bc}\begin{pmatrix} d & -b \\ -c & a \end{pmatrix}$$

as we see using inverse.

```
> with(linalg):
inverse([[a,b],[c,d]]);
```

$$\begin{bmatrix} \frac{d}{ad-bc} & -\frac{b}{ad-bc} \\ -\frac{c}{ad-bc} & \frac{a}{ad-bc} \end{bmatrix}$$

EXAMPLE 4: Find $\mathbf{A}^{-1}$ if $\mathbf{A} = \begin{pmatrix} 2 & -1 \\ -3 & 1 \end{pmatrix}$.

SOLUTION: In this case, $|\mathbf{A}| = \begin{vmatrix} 2 & -1 \\ -3 & 1 \end{vmatrix} = 2 - 3 = -1 \neq 0$, so $\mathbf{A}^{-1}$ exists.

Moreover, $\mathbf{A}^c = \begin{pmatrix} 1 & 3 \\ 1 & 2 \end{pmatrix}$, so $\mathbf{A}^a = \begin{pmatrix} 1 & 1 \\ 3 & 2 \end{pmatrix}$ and $\mathbf{A}^{-1} = \frac{1}{|\mathbf{A}|}\mathbf{A}^a = \begin{pmatrix} -1 & -1 \\ -3 & -2 \end{pmatrix}$.

∎

EXAMPLE 5: Find $\mathbf{A}^{-1}$ if $\mathbf{A} = \begin{pmatrix} -2 & -1 & 1 \\ 2 & 1 & 0 \\ 3 & 1 & -1 \end{pmatrix}$.

SOLUTION: Define $\mathbf{A}$ and then use `inverse` to calculate the inverse.

```
> with(linalg):
A:=array([[-2,-1,1],[2,1,0],[3,1,-1]]):
> inverse(A);
```

$$\begin{bmatrix} 1 & 0 & 1 \\ -2 & 1 & -2 \\ 1 & 1 & 0 \end{bmatrix}$$

∎

The inverse $\mathbf{A}^{-1}$ can be used to solve the linear system of equations $\mathbf{Ax} = \mathbf{b}$. For example, to solve $\begin{pmatrix} 5 & -1 \\ 2 & 3 \end{pmatrix}\begin{pmatrix} x \\ y \end{pmatrix} = \begin{pmatrix} -34 \\ 17 \end{pmatrix}$ in which $\mathbf{A} = \begin{pmatrix} 5 & -1 \\ 2 & 3 \end{pmatrix}$ and $\mathbf{b} = \begin{pmatrix} -34 \\ 17 \end{pmatrix}$, we find

$\mathbf{x} = \mathbf{A}^{-1}\mathbf{b} = \begin{pmatrix} \frac{3}{17} & \frac{1}{17} \\ -\frac{2}{17} & \frac{5}{17} \end{pmatrix}\begin{pmatrix} -34 \\ 17 \end{pmatrix} = \begin{pmatrix} -5 \\ 9 \end{pmatrix}$. We will find several uses for the inverse in solving

systems of differential equations as well. You can also use `linsolve`, which is contained in the **linalg** package, to solve linear systems like this.

```
> with(linalg):
A:=[[5,-1],[2,3]]:
b:=[-34,17]:
> linsolve(A,b);
```

$$[-5, 9]$$

Eigenvalues and Eigenvectors

Definition
Eigenvalues and
Eigenvectors

> A *nonzero* vector $\mathbf{x}$ is an **eigenvector** of the square matrix $\mathbf{A}$ means there is a number λ, called an **eigenvalue** of $\mathbf{A}$, so that
>
> $$\mathbf{Ax} = \lambda\mathbf{x}.$$

If $\mathbf{x}$ is an eigenvector of $\mathbf{A}$ with corresponding eigenvalue λ, then $\mathbf{Ax} = \lambda\mathbf{x}$. Because this equation is equivalent to the equation $(\mathbf{A} - \lambda\mathbf{I})\mathbf{x} = \mathbf{0}$, $\mathbf{x} \neq \mathbf{0}$ is an eigenvector if and only if $\det(\mathbf{A} - \lambda\mathbf{I}) = \mathbf{0}$. We use the commands `eigenvalues` and `eigenvectors` to calculate eigenvalues or both eigenvalues and eigenvectors of a square matrix. Notice that eigenvalues returns a list of the `eigenvalues` of the matrix

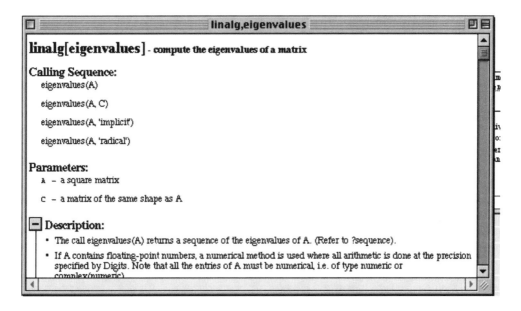

and `eigenvectors` returns the eigenvalues and corresponding eigenvectors of the matrix.

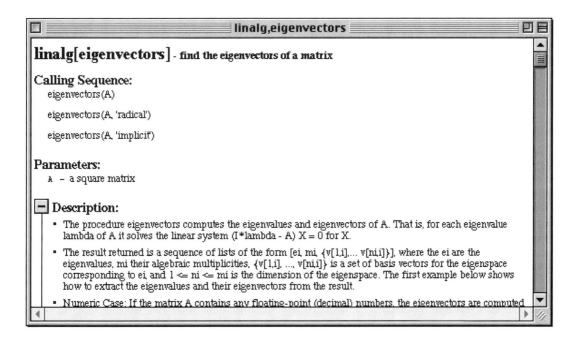

Note that both of these commands are contained in the **linalg** package.

Definition	The equation det $(\mathbf{A} - \lambda\mathbf{I}) = 0$ is called the **characteristic**
Characteristic	**equation** of $\mathbf{A}$; det $(\mathbf{A} - \lambda\mathbf{I})$ is called the **characteristic**
Polynomial	**polynomial** of $\mathbf{A}$; the roots of the characteristic polynomial of
	$\mathbf{A}$ are the eigenvalues of $\mathbf{A}$.

Using pencil-and-paper methods, to find the eigenvectors and corresponding eigenvalues of a square matrix $\mathbf{A}$, we will begin by computing the eigenvalues. The command `charpoly`, which is contained in the **linalg** package, is used to find the characteristic polynomial of a square matrix.

EXAMPLE 6: Calculate the eigenvalues and corresponding eigenvectors of

$$\mathbf{A} = \begin{pmatrix} 4 & -6 \\ 3 & -7 \end{pmatrix}.$$

SOLUTION: The characteristic polynomial of $\mathbf{A} = \begin{pmatrix} 4 & -6 \\ 3 & -7 \end{pmatrix}$ is

$$\begin{vmatrix} 4 - \lambda & -6 \\ 3 & -7 - \lambda \end{vmatrix} = \lambda^2 + 3\lambda - 10 = (\lambda + 5)(\lambda - 2),$$

which is confirmed with `charpoly` and `factor`.

```
> with(linalg):
A:=array([[4,-6],[3,-7]]):
char_A:=charpoly(A,lambda);
```

$$char\_A := \lambda^2 + 3\lambda - 10$$

```
> factor(char_A);
```

$$(\lambda + 5)(\lambda - 2)$$

Because the eigenvalues are found by solving $(\lambda + 5)(\lambda - 2) = 0$, the eigenvalues are $\lambda = -5$ and $\lambda = 2$.

```
> eigs_A:=eigenvalues(A);
```

$$eigs\_A := -5, 2$$

Let $\begin{pmatrix} x_1 \\ y_1 \end{pmatrix}$ denote the eigenvector corresponding to the eigenvalue $\lambda = -5$. Then,

$$\left\{ \begin{pmatrix} 4 & -6 \\ 3 & -7 \end{pmatrix} - (-5) \begin{pmatrix} 1 & 0 \\ 0 & 1 \end{pmatrix} \right\} \begin{pmatrix} x_1 \\ y_1 \end{pmatrix} = \mathbf{0}.$$

Simplifying yields the system of equations $\begin{cases} 9x_1 - 6y_1 = 0 \\ 3x_1 - 2y_1 = 0 \end{cases}$, so $y_1 = \frac{3}{2}x_1$. In the following command,

```
A-eigs_A[1]}^{*}array(identity,1..2,1..2)
```

represents $\begin{pmatrix} 4 & -6 \\ 3 & -7 \end{pmatrix} - (-5) \begin{pmatrix} 1 & 0 \\ 0 & 1 \end{pmatrix}$ and $[0,0]$ represents $\mathbf{0} = \begin{pmatrix} 0 \\ 0 \end{pmatrix}$.

```
> linsolve(A-eigs_A[1]*array(identity,1..2,1..2),[0,0]);
```

$$\left[_{-}t_1, \frac{3}{2} _{-}t_1 \right]$$

Therefore, if x_1 is any real number, then $\begin{pmatrix} x_1 \\ \frac{3}{2}x_1 \end{pmatrix}$ is an eigenvector. In particular, if $x_1 = 2$, then $\begin{pmatrix} 2 \\ 3 \end{pmatrix}$ is an eigenvector of $\mathbf{A} = \begin{pmatrix} 4 & -6 \\ 3 & -7 \end{pmatrix}$ with corresponding eigenvalue $\lambda = -5$. Similarly, if we let $\begin{pmatrix} x_2 \\ y_2 \end{pmatrix}$ denote the eigenvector corresponding to $\lambda = 2$, then $\left\{ \begin{pmatrix} 4 & -6 \\ 3 & -7 \end{pmatrix} - 2 \begin{pmatrix} 1 & 0 \\ 0 & 1 \end{pmatrix} \right\} \begin{pmatrix} x_2 \\ y_2 \end{pmatrix} = \mathbf{0}$, which yields the system $\begin{cases} 2x_2 - 6y_2 = 0 \\ 3x_2 - 9y_2 = 0 \end{cases}$, so $y_2 = \frac{1}{3}x_2$. If $x_2 = 3$, then $\begin{pmatrix} 3 \\ 1 \end{pmatrix}$ is an eigenvector of $\mathbf{A} = \begin{pmatrix} 4 & -6 \\ 3 & -7 \end{pmatrix}$ with corresponding eigenvalue $\lambda = 2$.

```
> linsolve(A-eigs_A[2]*array(identity,1..2,1..2),[0,0]);
```

$$[3 _{-}t_1, _{-}t_1]$$

We confirm these results with Maple.

```
> eigenvectors(A);
```

$$\left[-5, 1, \left\{ \left[1, \frac{3}{2} \right] \right\} \right], [2, 1, \{[3, 1]\}]$$

EXAMPLE 7: Find the eigenvalues and corresponding eigenvectors of

$$A = \begin{pmatrix} 0 & 1 \\ -1 & 0 \end{pmatrix}.$$

SOLUTION: In this case, the characteristic polynomial is $\begin{vmatrix} -\lambda & 1 \\ -1 & -\lambda \end{vmatrix} = \lambda^2 + 1$, so

the eigenvalues are the roots of the equation $\lambda^2 + 1 = 0$. These are the imaginary numbers $\lambda = i$ and $\lambda = -i$ where $i = \sqrt{-1}$.

```
> with(linalg):
A:=[[0,1],[-1,0]]:
eigs_A:=eigenvalues(A);
```

$$eigs\_A := I, -I$$

The corresponding eigenvectors are found by substituting the eigenvalues into the

equation $(A - \lambda I)x = 0$ and solving for x. For $\lambda = i$, this equation is

$\begin{pmatrix} -i & 1 \\ -1 & -i \end{pmatrix}\begin{pmatrix} x_1 \\ y_1 \end{pmatrix} = \begin{pmatrix} 0 \\ 0 \end{pmatrix}$, which is equivalent to the system $\begin{cases} -ix_1 + y_1 = 0 \\ -x_1 - iy_1 = 0 \end{cases}$

Notice that the second equation of this system is a constant multiple of the first

equation. Hence, an eigenvector $\begin{pmatrix} x_1 \\ y_1 \end{pmatrix}$ must satisfy $y_1 = ix_1$. Therefore,

$\begin{pmatrix} x_1 \\ ix_1 \end{pmatrix} = \begin{pmatrix} 1 \\ i \end{pmatrix} x_1$ is an eigenvector for any value of x_1. For example, if $x_1 = 1$,

$\begin{pmatrix} 1 \\ i \end{pmatrix}$ is an eigenvector.

```
> linsolve(A-eigs_A[1]*array(identity,1..2,1..2),[0,0]);
```

$$[-I\_t_1, \_t_1]$$

For $\lambda = -i$, the equation is $\begin{pmatrix} i & 1 \\ -1 & i \end{pmatrix}\begin{pmatrix} x_2 \\ y_2 \end{pmatrix} = \begin{pmatrix} 0 \\ 0 \end{pmatrix}$, which is equivalent to

$\begin{cases} ix_2 + y_2 = 0 \\ -x_2 + iy_2 = 0 \end{cases}$. Because the second equation equals i times the first equation, the eigenvector $\begin{pmatrix} x_2 \\ y_2 \end{pmatrix}$ must satisfy $y_2 = -ix_2$. Hence, $\begin{pmatrix} x_2 \\ -ix_2 \end{pmatrix} = \begin{pmatrix} 1 \\ -i \end{pmatrix} x_2$ is an

eigenvector for any value of x_2. Therefore, if $x_2 = 1$, $\begin{pmatrix} 1 \\ -i \end{pmatrix}$ is an eigenvector.

> `linsolve(A-eigs_A[2]*array(identity,1..2,1..2),[0,0]);`

$$[\_t_1, -I\_t_1]$$

Equivalent results are obtained with `eigenectors`.

> `vecs_A:=eigenvectors(A);`

$$vecs\_A := [I, 1, \{[-I, 1]\}], [-I, 1, \{[I, 1]\}]$$

We confirm that $\begin{pmatrix} i \\ 1 \end{pmatrix}$ is an eigenvector corresponding to $-i$. First, observe that

> `vecs_A[2];`

$$[-I, 1, \{[I, 1]\}]$$

returns the second part of `vecs_A` and

> `vecs_A[2,3,1];`

$$[I, 1]$$

returns the first part of the third part of the second part of `vecs_A`, the
eigenvector $\begin{pmatrix} i \\ 1 \end{pmatrix}$. Similarly,

> `vecs_A[2,1];`

$$-I$$

returns the first part of the second part of `vecs_A`, the eigenvalue $-i$. We see that

$$A\begin{pmatrix} i \\ 1 \end{pmatrix} = -i\begin{pmatrix} i \\ 1 \end{pmatrix}.$$

> `evalm(A &* vecs_A[2,3,1]=vecs_A[2,1]*vecs_A[2,3,1]);`

$$[1, -I] = [1, -I]$$

Similarly, we confirm that $\begin{pmatrix} -i \\ 1 \end{pmatrix}$ is an eigevector corresponding to i.

```
> evalm(A &* vecs_A[1,3,1]=vecs_A[1,1]*vecs_A[1,3,1]);
```

$$[1,I] = [1,I]$$

■

The **complex conjugate** of the complex number $z = a + bi$ is $\bar{z} = a - bi$. Similarly, the

complex conjugate of the vector $\mathbf{x} = \begin{pmatrix} a_1 + b_1 i \\ a_2 + b_2 i \\ \vdots \\ a_n + b_n i \end{pmatrix}$ is the vector $\bar{\mathbf{x}} = \begin{pmatrix} a_1 - b_1 i \\ a_2 - b_2 i \\ \vdots \\ a_n - b_n i \end{pmatrix}$. We will find

that the eigenvectors that correspond to complex conjugate eigenvalues are themselves complex conjugates.

EXAMPLE 8: Calculate the eigenvalues and corresponding eigenvectors of the

matrix $\mathbf{A} = \begin{pmatrix} 1 & 0 & 0 \\ 2 & 3 & 1 \\ 0 & 2 & 4 \end{pmatrix}$.

SOLUTION: We begin by finding the characteristic polynomial of **A** with

```
> A:=[[1,0,0],[2,3,1],[0,2,4]]:
factor(charpoly(A,lambda));
```

$$(\lambda - 1)(\lambda - 2)(\lambda - 5)$$

to see that the eigenvalues of **A** are $\lambda = 1$, $\lambda = 2$, and $\lambda = 5$. The eigenvalues and corresponding eigenvectors are found with `eigenvectors`.

```
> vects_A:=eigenvectors(A);
```

$$vects\_A := \left[1, 1, \left\{ \left[1, \frac{-3}{2}, 1 \right] \right\} \right], [2, 1, \{[0, -1, 1]\}], [5, 1, \{[0, 1, 2]\}]$$

These results are confirmed as follows.

```
> evalm(A &* vects_A[1,3,1]=vects_A[1,1]*vects_A[1,3,1]);
evalm(A &* vects_A[2,3,1]=vects_A[2,1]*vects_A[2,3,1]);
evalm(A &* vects_A[3,3,1]=vects_A[3,1]*vects_A[3,3,1]);
```

$$\left[1, \frac{-3}{2}, 1\right] = \left[1, \frac{-3}{2}, 1\right]$$

$$[0, -2, 2] = [0, -2, 2]$$

$$[0, 5, 10] = [0, 5, 10]$$

■

Definition **Eigenvalue of** **Multiplicity** m	Suppose that $(\lambda - \lambda_1)^m$ where **m** is a positive integer is a factor of the characteristic polynomial of the $n \times n$ matrix **A**, whereas $(\lambda - \lambda_1)^{m+1}$ is not a factor of this polynomial. Then, $\lambda = \lambda_1$ is an **eigenvalue of multiplicity** m.

We often say that the eigenvalue of an $n \times n$ matrix **A** is repeated if it is of multiplicity m where $m \geq 2$ and $m \leq n$. When trying to find the eigenvector(s) corresponding to an eigenvalue of multiplicity m, two situations may be encountered: (1) m linearly independent eigenvectors can be found that correspond to λ; (2) fewer than m linearly independent eigenvectors can be found that correspond to λ.

EXAMPLE 9: Find the eigenvalues and corresponding eigenvectors of (a) $\mathbf{A} = \begin{pmatrix} 1 & -3 & 3 \\ 3 & -5 & 3 \\ 6 & -6 & 4 \end{pmatrix}$ and (b) $\mathbf{B} = \begin{pmatrix} 5 & -4 & 0 \\ 1 & 0 & 2 \\ 0 & 2 & 5 \end{pmatrix}$.

SOLUTION: After defining **A** and **B**,

```
> A:=[[1,-3,3],[3,-5,3],[6,-6,4]]:
B:=[[5,-4,0],[1,0,2],[0,2,5]]:
```

we compute the eigenvalues and corresponding eigenvectors of each matrix with eigenvectors.

```
> eigenvectors(A);
eigenvectors(B);
```

$$[4, 1, \{[1, 1, 2]\}], [-2, 2, \{[1, 0, -1], [0, 1, 1]\}]$$

$$\left[0, 1, \left\{\left[-2, \frac{-5}{2}, 1\right]\right\}\right], [5, 2, \{[-2, 0, 1]\}]$$

For (a), we see that $\lambda = -2$ is an eigenvalue of multiplicity two. Two linearly independent eigenvectors corresponding to $\lambda = -2$ are $\begin{pmatrix} 1 \\ 0 \\ -1 \end{pmatrix}$ and $\begin{pmatrix} 0 \\ 1 \\ 1 \end{pmatrix}$. For (b), we see that $\lambda = 5$ is an eigenvalue of multiplicity two. In this case, we can find only one linearly independent eigenvector corresponding to $\lambda = 5$, $\begin{pmatrix} -2 \\ 0 \\ 1 \end{pmatrix}$.

■

Matrix Calculus

Definition
Derivative and Integral
of a Matrix

The **derivative** of the $n \times m$ matrix

$$A(t) = \begin{pmatrix} a_{11}(t) & a_{12}(t) & \cdots & a_{1m}(t) \\ a_{21}(t) & a_{22}(t) & \cdots & a_{2m}(t) \\ \vdots & \vdots & \ddots & \vdots \\ a_{n1}(t) & a_{n2}(t) & \cdots & a_{nm}(t) \end{pmatrix}, \text{ where } a_{ij}(t) \text{ is}$$

differentiable for all values of i and j, is

$$\frac{d}{dt}A(t) = \begin{pmatrix} \frac{d}{dt}a_{11}(t) & \frac{d}{dt}a_{12}(t) & \cdots & \frac{d}{dt}a_{1m}(t) \\ \frac{d}{dt}a_{21}(t) & \frac{d}{dt}a_{22}(t) & \cdots & \frac{d}{dt}a_{2m}(t) \\ \vdots & \vdots & \ddots & \vdots \\ \frac{d}{dt}a_{n1}(t) & \frac{d}{dt}a_{n2}(t) & \cdots & \frac{d}{dt}a_{nm}(t) \end{pmatrix}.$$

The **integral** of $A(t)$, where $a_{ij}(t)$ is integrable for all values of i and j, is

$$\int A(t)dt = \begin{pmatrix} \int a_{11}(t)\,dt & \int a_{12}(t)\,dt & \cdots & \int a_{1m}(t)\,dt \\ \int a_{21}(t)\,dt & \int a_{22}(t)\,dt & \cdots & \int a_{2m}(t)\,dt \\ \vdots & \vdots & \ddots & \vdots \\ \int a_{n1}(t)\,dt & \int a_{n2}(t)\,dt & \cdots & \int a_{nm}(t)\,dt \end{pmatrix}.$$

EXAMPLE 10: Find $\frac{d}{dt}\mathbf{A}(t)$ and $\int \mathbf{A}(t)\,dt$ if $\mathbf{A}(t) = \begin{pmatrix} \cos 3t & \sin 3t & e^{-t} \\ t & t\sin t^2 & e^t \end{pmatrix}$.

SOLUTION: We find $\frac{d}{dt}\mathbf{A}(t)$ by differentiating each element of $\mathbf{A}(t)$ by using `map` to apply `diff` to each entry of $\mathbf{A}$.

```
> A:=array([[cos(3*t),sin(3*t),exp(-t)],
            [t,t*sin(t^2),exp(t)]]);
```

$$A := \begin{vmatrix} \cos(3t) & \sin(3t) & \mathbf{e}^{(-t)} \\ t & t\sin(t^2) & \mathbf{e}^t \end{vmatrix}$$

```
> map(diff,A,t);
```

$$\begin{vmatrix} -3\sin(3t) & 3\cos(3t) & -\mathbf{e}^{(-t)} \\ 1 & \sin(t^2) + 2t^2\cos(t^2) & \mathbf{e}^t \end{vmatrix}$$

Similarly, we find $\int \mathbf{A}(t)\,dt$ by integrating each element of $\mathbf{A}(t)$ by using `map` to apply `int` to each entry of $\mathbf{A}$.

```
> map(int,A,t);
```

$$\begin{vmatrix} \frac{1}{3}\sin(3t) & -\frac{1}{3}\cos(3t) & -\mathbf{e}^{(-t)} \\ \frac{1}{2}t^2 & -\frac{1}{2}\cos(t^2) & \mathbf{e}^t \end{vmatrix}$$

■

7.2 Systems of Equations: Preliminary Definitions and Theory

Up to this point, we have focused our attention on solving differential equations that involve one dependent variable. However, many physical situations are modeled with more than one equation and involve more than one dependent variable. For example, if we wanted to determine the population of two interacting populations such as foxes and rabbits, we would have two dependent variables that represent the two populations, where these populations depend on one independent variable that represents time. Situations such as this lead to systems of differential equations, which we study in this chapter.

For example, we encountered the nonlinear initial-value problem

$$\begin{cases} x'' + (x^2 - 1)x' + x = 0 \\ x(0) = 1, \ x'(0) = 1 \end{cases}$$

in Chapter 5. If we let $x' = y$, then

$$y' = x'' = -\left[(x^2 - 1)x' + x \right] = (1 - x^2)y - x$$

so the second-order equation $x'' + (x^2 - 1)x' + x = 0$ is equivalent to the system of first-order differential equations

$$\begin{cases} x' = y \\ y' = (1 - x^2)y - x \end{cases}$$

and the initial-value problem is equivalent to the initial-value problem

$$\begin{cases} x' = y \\ y' = (1 - x^2)y - x. \\ x(0) = 1, y(0) = 1 \end{cases}$$

We use `dsolve` together with the `numeric` option to generate a numerical solution to this initial-value problem.

```
> numsol:=dsolve({diff(x(t),t)=y(t),
       diff(y(t),t)=(1-x(t)^2)*y(t)-x(t),x(0)=1,y(0)=1},
           {x(t),y(t)},numeric);
```

$$numsol := \mathbf{proc}(rkf45\_x) \dots \mathbf{end}$$

We can use this result to approximate the solution for various values of t. For example, entering

```
> numsol(1);
```

$$[t = 1, x(t) = 1.298482155900995, y(t) = -0.3670353768372757]$$

shows us that $x(1) \approx 1.29848$ and $y(1) \approx -0.367035$. We use `odeplot`, which is contained in the **plots** package, to graph $x(t)$ and $y(t)$.

```
> with(plots):
odeplot(numsol,[[t,x(t)],[t,y(t)]],0..25);
```

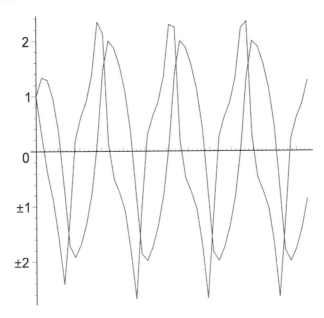

Because we let $x' = y$, notice that $y(t) > 0$ when $x(t)$ is increasing and $y(t) < 0$ when $x(t)$ is decreasing. The observation that these solutions are periodic is further confirmed by a graph of $x(t)$ (the horizontal axis) versus $y(t)$ (the vertical axis) generated with odeplot. We see that as t increases, the solution approaches a certain fixed path, called a *limit cycle*.

```
> odeplot(numsol,[x(t),y(t)],0..25);
```

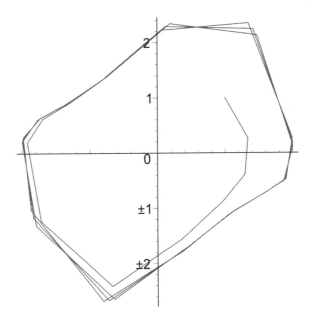

We will find that nonlinear equations are more easily studied when they are written as a system of equations.

Preliminary Theory

Definition System of Ordinary Differential Equations	A **system** of ordinary differential equations is a simultaneous set of equations that involves two or more dependent variables that depend on one independent variable. A **solution** of the system is a set of functions that satisfies each equation on some interval I.

If the differential equations in the system of differential equations are linear equations, we say that the system is a **linear system of differential equations** or a **linear system**.

EXAMPLE 1: Show that $\begin{cases} x(t) = \frac{1}{5}e^{-t}(e^t - \cos 2t - 3\sin 2t) \\ y(t) = -e^{-t}(\cos 2t - \sin 2t) \end{cases}$ is a solution to the

system $\begin{cases} x' - y = 0 \\ y' + 5x + 2y = 1 \end{cases}$

SOLUTION: The set of functions is a solution to the system of equations because

```
> x:='x':y:='y':
x:=t->1/5*exp(-t)*(exp(t)-cos(2*t)-3*sin(2*t)):
y:=t->-exp(-t)*(cos(2*t)-sin(2*t)):
> simplify(diff(x(t),t)-y(t));
```

$$0$$

and

```
> simplify(diff(y(t),t)+5*x(t)+2*y(t));
```

$$1$$

We graph this solution in several different ways. First, we graph the solution

$\begin{cases} x = x(t) \\ y = y(t) \end{cases}$ parametrically. Then, we graph $x(t)$ and $y(t)$ separately.

```
> plot([x(t),y(t),t=0..3*Pi]);
> plot([x(t),y(t)],t=0..3*Pi);
```

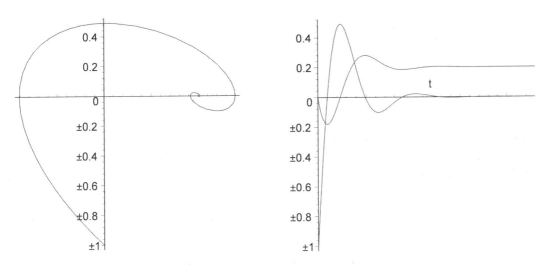

Notice that $\lim\limits_{t\to\infty} x(t) = 1/5$ and $\lim\limits_{t\to\infty} y(t) = 0$. Therefore, in the parametric plot, the points on the curve approach $(1/5, 0)$ as t increases.

■

As with other equations, under reasonable conditions, a solution to a system of differential equations can always be found.

Theorem
Existence and
Uniqueness

Assume that each of the functions $f_1(t, x_1, x_2, \ldots, x_n)$, $f_2(t, x_1, x_2, \ldots, x_n), \ldots, f_n(t, x_1, x_2, \ldots, x_n)$ and the partial derivatives $\partial f_1/\partial x_1$, $\partial f_2/\partial x_2, \ldots, \partial f_n/\partial x_n$ is continuous in a region R containing the point $(t_0, y_1, y_2, \ldots, y_n)$. Then, the initial-value problem

$$\begin{cases} x_1' = f_1(t, x_1, x_2, \ldots, x_n) \\ x_2' = f_2(t, x_1, x_2, \ldots, x_n) \\ \qquad \vdots \\ x_n' = f_n(t, x_1, x_2, \ldots, x_n) \\ x_1(t_0) = y_1,\ x_2(t_0) = y_2,\ \ldots,\ x_n(t_0) = y_n \end{cases}$$

has a unique solution

$$\begin{cases} x_1 = \phi_1(t) \\ x_2 = \phi_2(t) \\ \qquad \vdots \\ x_n = \phi_n(t) \end{cases}$$

on an interval I containing t_0.

EXAMPLE 2: Show that the initial-value problem

$$\begin{cases} dx/dt = 2x - xy \\ dy/dt = -3y + xy \\ x(0) = 2, y(0) = 3/2 \end{cases}$$

has a unique solution.

SOLUTION: In this case, we identify $f_1(t,x,y) = 2x - xy$ and $f_2(t,x,y) = -3y + xy$ with $\partial f_1/\partial x = 2 - y$ $\partial f_2/\partial y = -3 + x$. All four of these functions are continuous on a region containing $(0,2,3/2)$. Thus, by the existence and uniqueness theorem, a unique solution to the initial-value problem exists. In this case, we use dsolve together with the numeric option to approximate the solution to this nonlinear problem.

```
> x:='x':y:='y':
> numsol:=dsolve({diff(x(t),t)=2*x(t)-x(t)*y(t),
      diff(y(t),t)=-3*y(t)+x(t)*y(t),x(0)=2,y(0)=3/2},
          {x(t),y(t)},numeric);
```

$$numsol := \mathbf{proc}(rkf45\_x)\dots\mathbf{end}$$

We can use this result to approximate $x(t)$ and $y(t)$ for various values of t. For example,

```
> numsol(4);
```

$$[t = 4, x(t) = 4.269008041988111, y(t) = 2.624691295424079]$$

shows us that $x(4) \approx 4.26905$ and $y(4) \approx 2.62468$.
Next, we use odeplot to graph $x(t)$ and $y(t)$ and then the parametric equations

$$\begin{cases} x(t) \\ y(t) \end{cases} \text{ for } 0 \le t \le 10.$$

```
> with(plots):
odeplot(numsol,[[t,x(t)],[t,y(t)]],0..10);
> odeplot(numsol,[x(t),y(t)],0..10);
```

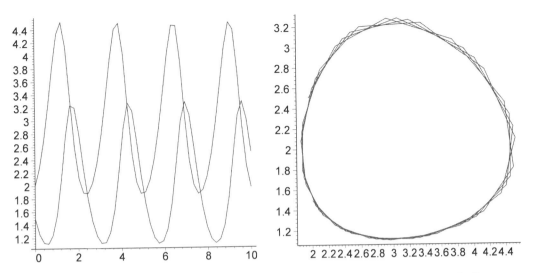

The graphs illustrate that the solution to the initial-value problem is periodic.

■

EXAMPLE 3: Show that the *nonautonomous* initial-value problem

$$\begin{cases} x' = e^{-0.002t} - 0.08x - xy^2 \\ y' = 0.08x - y + xy^2 \\ \qquad x(0) = 0, y(0) = 0 \end{cases}$$

has a unique solution.

SOLUTION: In   this   case, $f_1(t, x, y) = e^{-0.002t} - 0.08x - xy^2$ and
$f_2(t, x, y) = 0.08x - y + xy^2$. The system is nonautonomous because f_1 depends
explicitly on the independent variable t. Because f_1, f_2, $\partial f_1/\partial x$, and $\partial f_2/\partial y$ are
continuous on a region containing $(0, 0, 0)$, a unique solution to the initial-value
problem exists.

```
> x:='x':y:='y':
f[1]:=(x,y,t)->exp(-0.002*t)-0.08*x-x*y^2:
f[2]:=(x,y,t)->0.08*x-y+x*y^2:
> diff(f[1](x,y,t),x);
```

$$-0.08 - y^2$$

```
> diff(f[2](x,y,t),y);
```

$$-1 + 2xy$$

Instead of generating a numerical solution as in Example 2, we use DEplot, which is contained in the **DEtools** package, to graph the parametric equations $\begin{cases} x(t) \\ y(t) \end{cases}$ for $0 \leq t \leq 1000$. Note that the default numerical method used by DEplot to generate a numerical solution is unable to solve the initial-value problem over this interval, so we specify that a more powerful method be used with the option method = dverk78. Notice in the graph that the solution first oscillates and then begins returning to the origin near $t \approx 1000$.

```
> with(DEtools):
DEplot([diff(x(t),t)=exp(-0.002*t)-0.08*x(t)-x(t)*y(t)^2,
        diff(y(t),t)=0.08*x(t)-y(t)+x(t)*y(t)^2],
            [x(t),y(t)],t=0..1000,[[x(0)=0,y(0)=0]],
            scene=[x(t),y(t)],stepsize=0.2,thickness=1,
            linecolor=BLACK,method=dverk78);
```

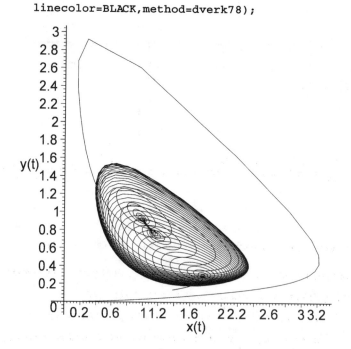

The beginning oscillation and eventual return to the origin are especially evident in a three-dimensional graph of $(x(t), y(t), t)$. To generate a three-dimensional graph of $(x(t), y(t), t)$ for $0 \le t \le 1000$, we use `DEplot3d`, which is also contained in the `DEtools` package.

```
> DEplot3d([diff(x(t),t)=exp(-0.002*t)-0.08*x(t)-x(t)*y(t)^2,
        diff(y(t),t)=0.08*x(t)-y(t)+x(t)*y(t)^2],
            [x(t),y(t)],t=0..1000,[[x(0)=0,y(0)=0]],
            scene=[x(t),y(t),t],stepsize=0.2,thickness=1,
            linecolor=BLACK,method=dverk78);
> DEplot3d([diff(x(t),t)=exp(-0.002*t)-0.08*x(t)-x(t)*y(t)^2,
        diff(y(t),t)=0.08*x(t)-y(t)+x(t)*y(t)^2],
            [x(t),y(t)],t=0..500,[[x(0)=0,y(0)=0]],
            scene=[x(t),y(t),t],stepsize=0.1,thickness=1,
            linecolor=BLACK,method=dverk78);
```

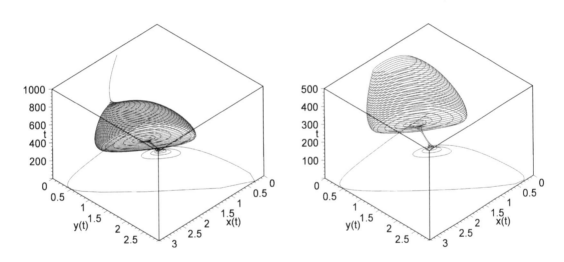

We create a wine glass by graphing the solution for $10 \le t \le 400$.

```
> DEplot3d([diff(x(t),t)=exp(-0.002*t)-0.08*x(t)-x(t)*y(t)^2,
        diff(y(t),t)=0.08*x(t)-y(t)+x(t)*y(t)^2],
            [x(t),y(t)],t=10..400,[[x(0)=0,y(0)=0]],
            scene=[x(t),y(t),t],stepsize=0.1,thickness=1,
            linecolor=BLACK,method=dverk78);
```

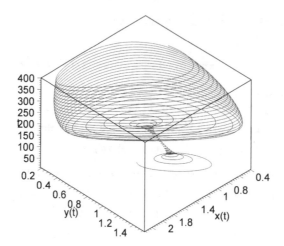

■

We now turn our attention to linear systems.

Linear Systems

We begin our study of linear systems of ordinary differential equations by introducing several definitions along with some convenient notation.

$$\text{Let } \mathbf{X} = \mathbf{X}(t) = \begin{pmatrix} x_1(t) \\ x_2(t) \\ \vdots \\ x_n(t) \end{pmatrix}, \mathbf{A}(t) = \begin{pmatrix} a_{11}(t) & a_{12}(t) & \cdots & a_{1n}(t) \\ a_{21}(t) & a_{22}(t) & \cdots & a_{2n}(t) \\ \vdots & \vdots & \ddots & \vdots \\ a_{n1}(t) & a_{n2}(t) & \cdots & a_{nn}(t) \end{pmatrix}, \text{and } \mathbf{F}(t) = \begin{pmatrix} f_1(t) \\ f_2(t) \\ \vdots \\ f_n(t) \end{pmatrix}.$$

Then, the homogeneous system of first-order linear differential equations

$$\begin{cases} x_1'(t) = a_{11}(t)x_1(t) + a_{12}(t)x_2(t) + \cdots + a_{1n}(t)x_n(t) \\ x_2'(t) = a_{21}(t)x_1(t) + a_{22}(t)x_2(t) + \cdots + a_{2n}(t)x_n(t) \\ \qquad\qquad\qquad \vdots \\ x_n'(t) = a_{n1}(t)x_1(t) + a_{n2}(t)x_2(t) + \cdots + a_{nn}(t)x_n(t) \end{cases}$$

is equivalent to $\mathbf{X}'(t) = \mathbf{A}(t)\mathbf{X}(t)$ and the nonhomogeneous system

$$\begin{cases} x_1'(t) = a_{11}(t)x_1(t) + a_{12}(t)x_2(t) + \cdots + a_{1n}(t)x_n(t) + f_1(t) \\ x_2'(t) = a_{21}(t)x_1(t) + a_{22}(t)x_2(t) + \cdots + a_{2n}(t)x_n(t) + f_2(t) \\ \qquad\qquad\qquad \vdots \\ x_n'(t) = a_{n1}(t)x_1(t) + a_{n2}(t)x_2(t) + \cdots + a_{nn}(t)x_n(t) + f_n(t) \end{cases}$$

is equivalent to $\mathbf{X}'(t) = \mathbf{A}(t)\mathbf{X}(t) + \mathbf{F}(t)$.

EXAMPLE 4: (a) Write the homogeneous system $\begin{cases} x' = -5x + 5y \\ y' = -5x + y \end{cases}$ in matrix form.

(b) Write the nonhomogeneous system $\begin{cases} x' = x + 2y - \sin t \\ y' = 4x - 3y + t^2 \end{cases}$ in matrix form.

SOLUTION: (a) The homogeneous system $\begin{cases} x' = -5x + 5y \\ y' = -5x + y \end{cases}$ is equivalent to the

system $\begin{pmatrix} x' \\ y' \end{pmatrix} = \begin{pmatrix} -5 & 5 \\ -5 & 1 \end{pmatrix} \begin{pmatrix} x \\ y \end{pmatrix}$. (b) The nonhomogeneous system

$\begin{cases} x' = x + 2y - \sin t \\ y' = 4x - 3y + t^2 \end{cases}$ is equivalent to $\begin{pmatrix} x' \\ y' \end{pmatrix} = \begin{pmatrix} 1 & 2 \\ 4 & -3 \end{pmatrix} \begin{pmatrix} x \\ y \end{pmatrix} + \begin{pmatrix} -\sin t \\ t^2 \end{pmatrix}$.

■

The equation

$$y^{(n)}(t) + a_{n-1}(t)y^{(n-1)}(t) + \cdots + a_2(t)y''(t) + a_1(t)y'(t) + a_0(t)y(t) = f(t),$$

discussed in previous chapters, can be written as a system of first-order equations as well. Let $x_1 = y$, $x_2 = dx_1/dt = y'$, $x_3 = dx_2/dt = y''$, ... $x_{n-1} = dx_{n-2}/dt = y^{(n-2)}$, $x_n = dx_{n-1}/dt = y^{(n-1)}$. Then, the equation $y^{(n)}(t) + a_{n-1}(t)y^{(n-1)}(t) + \cdots + a_2(t)y''(t) + a_1(t)y'(t) + a_0(t)y(t) = f(t)$ is equivalent to the system

$$\begin{cases} x_1' = x_2 \\ x_2' = x_3 \\ \quad \vdots \\ x_{n-1}' = x_n \\ x_n' = -a_{n-1}y^{(n-1)} - \cdots - a_2y'' - a_1y' - a_0y + f(t) \\ \quad = -a_{n-1}x_n - \cdots - a_2x_3 - a_1x_2 - a_0x_1 + f(t) \end{cases}$$

which can be written in matrix form as

$$
\begin{pmatrix} x_1' \\ x_2' \\ \vdots \\ x_{n-1}' \\ x_n' \end{pmatrix} = \begin{pmatrix} 0 & 1 & 0 & 0 & 0 \\ 0 & 0 & 1 & 0 & 0 \\ \vdots & \vdots & \vdots & \ddots & \vdots \\ 0 & 0 & 0 & 0 & 1 \\ -a_0 & -a_1 & -a_2 & \cdots & -a_n \end{pmatrix} \begin{pmatrix} x_1 \\ x_2 \\ \vdots \\ x_{n-1} \\ x_n \end{pmatrix} + \begin{pmatrix} 0 \\ 0 \\ 0 \\ 0 \\ f(t) \end{pmatrix}.
$$

EXAMPLE 5: Write the equation $y'' + 5y' + 6y = \cos t$ as a system.

SOLUTION: We let $x_1 = y$ and $x_2 = x_1' = y'$. Then, $x_2' = y'' = \cos t - 6y$
$-5y' = \cos t - 6x_1 - 5x_2$ so $\begin{cases} x_1' = x_2 \\ x_2' = \cos t - 6x_1 - 5x_2 \end{cases}$, which can be written in

matrix form as $\begin{pmatrix} x_1' \\ x_2' \end{pmatrix} = \begin{pmatrix} 0 & 1 \\ -6 & -5 \end{pmatrix} \begin{pmatrix} x_1 \\ x_2 \end{pmatrix} + \begin{pmatrix} 0 \\ \cos t \end{pmatrix}$.

■

At this point, given a system of ordinary differential equations, our goal is to construct either an explicit, numerical, or graphical solution of the system of equations.

We now state the following theorems and terminology, which are used in establishing the fundamentals of solving systems of differential equations. In each case, we assume that the matrix $\mathbf{A}(t)$ in the systems $\mathbf{X}'(t) = \mathbf{A}(t)\mathbf{X}(t) + \mathbf{F}(t)$ and $\mathbf{X}'(t) = \mathbf{A}(t)\mathbf{X}(t)$ is an $n \times n$ matrix.

Definition **Solution Vector**	A **solution vector** of the system $\mathbf{X}'(t) = \mathbf{A}(t)\mathbf{X}(t) + \mathbf{F}(t)$ on the interval I is an $n \times 1$ matrix of the form $$\mathbf{X}(t) = \begin{pmatrix} x_1(t) \\ x_2(t) \\ \vdots \\ x_n(t) \end{pmatrix}$$ where the $x_i(t)$ are differentiable functions that satisfy $\mathbf{X}'(t) = \mathbf{A}(t)\mathbf{X}(t) + \mathbf{F}(t)$ on I.

Let $\mathbf{X}'(t) = \mathbf{A}(t)\mathbf{X}(t)$ where $X(t) = \begin{pmatrix} x_1(t) \\ x_2(t) \\ \vdots \\ x_n(t) \end{pmatrix}$ and $\mathbf{A}(t) = \begin{pmatrix} a_{11}(t) & a_{12}(t) & \cdots & a_{1n}(t) \\ a_{21}(t) & a_{22}(t) & \cdots & a_{2n}(t) \\ \vdots & \vdots & \ddots & \vdots \\ a_{n1}(t) & a_{n2}(t) & \cdots & a_{nn}(t) \end{pmatrix}$

where $a_{ij}(t)$ is continuous for all $1 \leq j \leq n$ and $1 \leq i \leq n$. Let $\{\Phi_i\}_{i=1}^{m} = \left\{ \begin{pmatrix} \Phi_{1i} \\ \Phi_{2i} \\ \vdots \\ \Phi_{ni} \end{pmatrix} \right\}_{i=1}^{m}$ be a set of m solutions of $\mathbf{X}'(t) = \mathbf{A}(t)\mathbf{X}(t)$.

We define linear dependence and independence of the set of vectors $\{\Phi_i\}_{i=1}^{m} = \left\{ \begin{pmatrix} \Phi_{1i} \\ \Phi_{2i} \\ \vdots \\ \Phi_{ni} \end{pmatrix} \right\}_{i=1}^{m}$ in the same way as we define linear dependence and independence of sets of functions. The set $\{\Phi_i\}_{i=1}^{m} = \left\{ \begin{pmatrix} \Phi_{1i} \\ \Phi_{2i} \\ \vdots \\ \Phi_{ni} \end{pmatrix} \right\}_{i=1}^{m}$ is **linearly dependent** on an interval

I means that there is a set of constants $\{c_i\}_{i=1}^{m}$ not all zero such that $\sum\limits_{i=1}^{m} c_i \Phi_i = 0$; otherwise, the set is **linearly independent**.

We can determine whether a set of vectors is linearly independent or linearly dependent

Definition
Fundamental Set of
Solutions

Any set $\{\Phi_i\}_{i=1}^{m} = \left\{ \begin{pmatrix} \Phi_{1i} \\ \Phi_{2i} \\ \vdots \\ \Phi_{ni} \end{pmatrix} \right\}_{i=1}^{m}$ of n linearly independent

solution vectors of $\mathbf{X}'(t) = \mathbf{A}(t)\mathbf{X}(t)$ on an interval I is called a **fundamental set of solutions** on I.

by computing the Wronskian.

Theorem

The set $\{\Phi_i\}_{i=1}^{m} = \left\{ \begin{pmatrix} \Phi_{1i} \\ \Phi_{2i} \\ \vdots \\ \Phi_{ni} \end{pmatrix} \right\}_{i=1}^{m}$ is linearly independent if and only

only if the **Wronskian**

$$W(\Phi_1, \Phi_2, \ldots \Phi_n) = \det \begin{pmatrix} \Phi_{11} & \Phi_{12} & \cdots & \Phi_{1n} \\ \Phi_{21} & \Phi_{22} & \cdots & \Phi_{2n} \\ \vdots & \vdots & \ddots & \vdots \\ \Phi_{n1} & \Phi_{n2} & \cdots & \Phi_{nn} \end{pmatrix} \neq 0.$$

EXAMPLE 6: Which of the following is a fundamental set of solutions for

$$\begin{pmatrix} x' \\ y' \end{pmatrix} = \begin{pmatrix} -2 & -8 \\ 1 & 2 \end{pmatrix} \begin{pmatrix} x \\ y \end{pmatrix}?$$

(a)

$$\left\{ \begin{pmatrix} \cos 2t \\ \sin 2t \end{pmatrix}, \begin{pmatrix} \sin 2t \\ \cos 2t \end{pmatrix} \right\}$$

(b)

$$\left\{ \begin{pmatrix} -2\sin 2t + 2\cos 2t \\ \sin 2t \end{pmatrix}, \begin{pmatrix} 4\cos 2t \\ \sin 2t - \cos 2t \end{pmatrix} \right\}$$

SOLUTION: We first remark that the equation $\begin{pmatrix} x' \\ y' \end{pmatrix} = \begin{pmatrix} -2 & -8 \\ 1 & 2 \end{pmatrix} \begin{pmatrix} x \\ y \end{pmatrix}$ is equivalent to the system

$$\begin{cases} x' = -2x - 8y \\ y' = x + 2y \end{cases}$$

(a) Differentiating, we see that

$$\begin{pmatrix} \cos 2t \\ \sin 2t \end{pmatrix}' = \begin{pmatrix} -2\sin 2t \\ 2\cos 2t \end{pmatrix} \neq \begin{pmatrix} -2\cos 2t - 8\sin 2t \\ \cos 2t + 2\sin 2t \end{pmatrix}$$

which shows us that $\begin{pmatrix} \cos 2t \\ \sin 2t \end{pmatrix}$ is not a solution of the system.

```
> A:=[[-2,-8],[1,2]]:
v1:=[cos(2*t),sin(2*t)]:
> map(diff,v1,t);
```
$$[-2\sin(2t), 2\cos(2t)]$$

```
> evalm(A &* v1);
```
$$[-2\cos(2t) - 8\sin(2t), \cos(2t) + 2\sin(2t)]$$

Therefore, $\left\{ \begin{pmatrix} \cos 2t \\ \sin 2t \end{pmatrix}, \begin{pmatrix} \sin 2t \\ \cos 2t \end{pmatrix} \right\}$ is not a fundamental set of solutions.

(b) First we verify that $\begin{pmatrix} -2\sin 2t + 2\cos 2t \\ \sin 2t \end{pmatrix}$ is a solution to the system.

```
> v2:=[-2*sin(2*t)+2*cos(2*t),sin(2*t)]:
> map(diff,v2,t);
```
$$[-4\cos(2t) - 4\sin(2t), 2\cos(2t)]$$

```
> evalm(A &* v2);
```

$$[-4\cos(2t) - 4\sin(2t), 2\cos(2t)]$$

Next, we see that $\begin{pmatrix} 4\cos 2t \\ \sin 2t - \cos 2t \end{pmatrix}$ is a solution to the system.

```
> v3:=[4*cos(2*t),sin(2*t)-cos(2*t)]:
```

```
> map(diff,v3,t);
```

$$[-8\sin(2t), 2\cos(2t) + 2\sin(2t)]$$

```
> evalm(A &* v3);
```

$$[-8\sin(2t), 2\cos(2t) + 2\sin(2t)]$$

To see that these vectors are linearly independent, we compute the Wronskian.

```
> with(linalg):
```

```
m1:=transpose(array([v2,v3]));
```

$$m1 := \begin{bmatrix} -2\sin(2t) + 2\cos(2t) & 4\cos(2t) \\ \sin(2t) & \sin(2t) - \cos(2t) \end{bmatrix}$$

```
> simplify(det(m1));
```

$$-2$$

Thus, the set $\left\{ \begin{pmatrix} -2\sin 2t + 2\cos 2t \\ \sin 2t \end{pmatrix}, \begin{pmatrix} 4\cos 2t \\ \sin 2t - \cos 2t \end{pmatrix} \right\}$ is linearly independent

and, consequently, a fundamental set of solutions.

■

The following theorem implies that a fundamental set of solutions cannot contain more than n vectors, because the solutions could not be linearly independent.

Theorem

> Any $n + 1$ nontrivial solutions of $\mathbf{X}'(t) = \mathbf{A}(t)\mathbf{X}(t)$ are linearly dependent.

Finally, we state the following theorems, which indicate that a fundamental set of solutions can always be found and a general solution can be constructed.

Theorem

> There is a set of n nontrivial linearly independent solutions of $\mathbf{X}'(t) = \mathbf{A}(t)\mathbf{X}(t)$

Theorem

Let $\{\boldsymbol{\Phi}_i\}_{i=1}^{n} = \left\{ \begin{pmatrix} \boldsymbol{\Phi}_{1i} \\ \boldsymbol{\Phi}_{2i} \\ \vdots \\ \boldsymbol{\Phi}_{ni} \end{pmatrix} \right\}_{i=1}^{n}$ be a set of n linearly independent

solutions of $\mathbf{X}'(t) = \mathbf{A}(t)\mathbf{X}(t)$. Then every solution of $\mathbf{X}'(t) = \mathbf{A}(t)\mathbf{X}(t)$ is a linear combination of these solutions. Hence, a **general solution** of $\mathbf{X}'(t) = \mathbf{A}(t)\mathbf{X}(t)$ is

$$\mathbf{X}(t) = c_1\boldsymbol{\Phi}_1(t) + c_2\boldsymbol{\Phi}_2(t) + \cdots + c_n\boldsymbol{\Phi}_n(t).$$

Definition
Fundamental Matrix
General Solution

Let $\{\boldsymbol{\Phi}_i\}_{i=1}^{n} = \left\{ \begin{pmatrix} \boldsymbol{\Phi}_{1i} \\ \boldsymbol{\Phi}_{2i} \\ \vdots \\ \boldsymbol{\Phi}_{ni} \end{pmatrix} \right\}_{i=1}^{n}$ be a set of n linearly independent

solutions of $\mathbf{X}'(t) = \mathbf{A}(t)\mathbf{X}(t)$. Then

$$\boldsymbol{\Phi}(t) = (\boldsymbol{\Phi}_1 \quad \boldsymbol{\Phi}_2 \quad \cdots \quad \boldsymbol{\Phi}_n) = \begin{pmatrix} \boldsymbol{\Phi}_{11} & \boldsymbol{\Phi}_{12} & \cdots & \boldsymbol{\Phi}_{1n} \\ \boldsymbol{\Phi}_{21} & \boldsymbol{\Phi}_{22} & \cdots & \boldsymbol{\Phi}_{2n} \\ \vdots & \vdots & \ddots & \vdots \\ \boldsymbol{\Phi}_{n1} & \boldsymbol{\Phi}_{n2} & \cdots & \boldsymbol{\Phi}_{nn} \end{pmatrix}$$

is called a **fundamental matrix** of the system $\mathbf{X}'(t) = \mathbf{A}(t)\mathbf{X}(t)$. Thus, a **general solution** can be written as $\mathbf{X}(t) = \boldsymbol{\Phi}(t)\mathbf{C}$

where $\mathbf{C} = \begin{pmatrix} c_1 \\ c_2 \\ \vdots \\ c_n \end{pmatrix}$.

EXAMPLE 7: Show that $\boldsymbol{\Phi}(t) = \begin{pmatrix} e^{-2t} & -3e^{5t} \\ 2e^{-2t} & e^{5t} \end{pmatrix}$ is a fundamental matrix for the

system $\mathbf{X}'(t) = \begin{pmatrix} 4 & -3 \\ -2 & -1 \end{pmatrix} \mathbf{X}(t)$. Use the matrix to find a general solution of

$\mathbf{X}'(t) = \begin{pmatrix} 4 & -3 \\ -2 & -1 \end{pmatrix} \mathbf{X}(t).$

SOLUTION: Because

$$\begin{pmatrix} e^{-2t} \\ 2e^{-2t} \end{pmatrix}' = \begin{pmatrix} -2e^{-2t} \\ -4e^{-2t} \end{pmatrix} = \begin{pmatrix} 4 & -3 \\ -2 & -1 \end{pmatrix}\begin{pmatrix} e^{-2t} \\ 2e^{-2t} \end{pmatrix} \text{ and } \begin{pmatrix} -3e^{5t} \\ e^{5t} \end{pmatrix}' = \begin{pmatrix} -15e^{5t} \\ 5e^{5t} \end{pmatrix} = \begin{pmatrix} 4 & -3 \\ -2 & -1 \end{pmatrix}\begin{pmatrix} -3e^{5t} \\ e^{5t} \end{pmatrix}$$

both $\begin{pmatrix} e^{-2t} \\ 2e^{-2t} \end{pmatrix}$ and $\begin{pmatrix} -3e^{5t} \\ e^{5t} \end{pmatrix}$ are solutions of the system $\mathbf{X}'((t) = \begin{pmatrix} 4 & -3 \\ -2 & -1 \end{pmatrix}\mathbf{X}(t)$.

Alternatively, we show that $\mathbf{\Phi}'(t)$ and $\begin{pmatrix} 4 & -3 \\ -2 & -1 \end{pmatrix}\mathbf{\Phi}(t)$ are the same.

```
> A:=array([[4,-3],[-2,-1]]):
Phi:=array([[exp(-2*t),-3*exp(5*t)],
           [2*exp(-2*t),exp(5*t)]]);
```

$$\Phi := \begin{bmatrix} \mathbf{e}^{(-2t)} & -3\mathbf{e}^{(5t)} \\ 2\mathbf{e}^{(-2t)} & \mathbf{e}^{(5t)} \end{bmatrix}$$

```
> map(diff,Phi,t);
```

$$\begin{bmatrix} -2\mathbf{e}^{(-2t)} & -15\mathbf{e}^{(5t)} \\ -4\mathbf{e}^{(-2t)} & 5\mathbf{e}^{(5t)} \end{bmatrix}$$

```
> evalm(A &* Phi);
```

$$\begin{bmatrix} -2\mathbf{e}^{(-2t)} & -15\mathbf{e}^{(5t)} \\ -4\mathbf{e}^{(-2t)} & 5\mathbf{e}^{(5t)} \end{bmatrix}$$

The solutions are linearly independent because the Wronskian is not the zero function.

```
> with(linalg):
simplify(det(Phi));
```

$$7\mathbf{e}^{(3t)}$$

A general solution is given by

$$\mathbf{X}(t) = \mathbf{\Phi}(t)\mathbf{C} = \begin{pmatrix} e^{-2t} & -3e^{5t} \\ 2e^{-2t} & e^{5t} \end{pmatrix}\begin{pmatrix} c_1 \\ c_2 \end{pmatrix} = \begin{pmatrix} c_1 e^{-2t} - 3c_2 e^{5t} \\ 2c_1 e^{-2t} + c_2 e^{5t} \end{pmatrix} = c_1\begin{pmatrix} e^{-2t} \\ 2e^{-2t} \end{pmatrix} + c_2\begin{pmatrix} -3e^{5t} \\ e^{5t} \end{pmatrix}.$$

■

In the next sections, we will see that we are often able to use commands such as `dsolve` to solve systems of differential equations either exactly or numerically. In addtion, the **DEtools** package contains the `matrixDE` command that can often be used to solve systems of the form $\mathbf{X}' = \mathbf{A}(t)\mathbf{X} + \mathbf{F}(t)$.

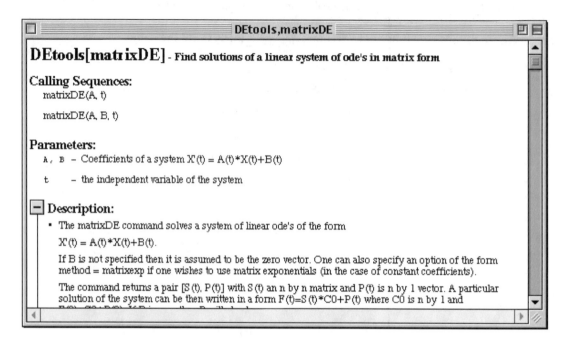

DEtools[matrixDE] - Find solutions of a linear system of ode's in matrix form

Calling Sequences:

matrixDE(A, t)

matrixDE(A, B, t)

Parameters:

A, B – Coefficients of a system X'(t) = A(t)*X(t)+B(t)

t – the independent variable of the system

Description:

* The matrixDE command solves a system of linear ode's of the form

  X'(t) = A(t)*X(t)+B(t).

  If B is not specified then it is assumed to be the zero vector. One can also specify an option of the form method = matrixexp if one wishes to use matrix exponentials (in the case of constant coefficients).

  The command returns a pair [S(t), P(t)] with S(t) an n by n matrix and P(t) is n by 1 vector. A particular solution of the system can be then written in a form F(t)=S(t)*C0+P(t) where C0 is n by 1 and

7.3 Homogeneous Linear Systems with Constant Coefficients

Now that we have covered the necessary terminology, we can turn our attention to solving

linear systems with constant coefficients. Let $\mathbf{A} = \begin{pmatrix} a_{11} & a_{12} & \cdots & a_{1n} \\ a_{21} & a_{22} & \cdots & a_{2n} \\ \vdots & \vdots & \ddots & \vdots \\ a_{n1} & a_{n2} & \cdots & a_{nn} \end{pmatrix}$ be an $n \times n$ real

matrix and let $\{\lambda_k\}_{k=1}^{n}$ be the eigenvalues and $\{\mathbf{v}_k\}_{k=1}^{n}$ the corresponding eigenvectors of $\mathbf{A}$. Then a general solution of the system $\mathbf{X}' = \mathbf{A}\mathbf{X}$ is determined by the eigenvalues and corresponding eigenvectors of $\mathbf{A}$. For the moment, we consider the cases in which the eigenvalues of $\mathbf{A}$ are distinct and real or the eigenvalues of $\mathbf{A}$ are distinct and complex. We will consider the case in which $\mathbf{A}$ has repeated eigenvalues (eigenvalues of multiplicity greater than one) separately.

Distinct Real Eigenvalues

If the eigenvalues $\{\lambda_k\}_{k=1}^{n}$ of $\mathbf{A}$ are distinct, a general solution of $\mathbf{X}' = \mathbf{AX}$ can be written as

$$\mathbf{X}(t) = c_1 \mathbf{v}_1 \mathbf{e}^{\lambda_1 t} + c_2 \mathbf{v}_2 \mathbf{e}^{\lambda_2 t} + \cdots + c_n \mathbf{v}_n \mathbf{e}^{\lambda_n t},$$

where $\{\mathbf{v}_k\}_{k=1}^{n}$ are eigenvectors corresponding to $\{\lambda_k\}_{k=1}^{n}$. We investigate this claim by assuming that $\mathbf{X} = \mathbf{v}e^{\lambda t}$ is a solution of $\mathbf{X}' = \mathbf{AX}$. Then, $\mathbf{X}' = \lambda \mathbf{v}e^{\lambda t}$ must satisfy the differential equation which implies that

$$\lambda \mathbf{v}e^{\lambda t} = \mathbf{A}\mathbf{v}e^{\lambda t}.$$

Now, because $\mathbf{Iv} = \mathbf{v}$, we make the following substitution so that the expression can be simplified.

$$\lambda \mathbf{Iv}e^{\lambda t} = \mathbf{A}\mathbf{v}e^{\lambda t}$$

$$(\mathbf{A} - \lambda \mathbf{I})\mathbf{v}e^{\lambda t} = \mathbf{0}.$$

Then, because $e^{\lambda t} \neq 0$, we have

$$(\mathbf{A} - \lambda \mathbf{I})\mathbf{v} = \mathbf{0}.$$

In order for this system of equations to have a solution other than $\mathbf{v} = \mathbf{0}$,

$$|\mathbf{A} - \lambda \mathbf{I}| = 0.$$

We recall that a solution λ to this equation is an eigenvalue of $\mathbf{A}$ and a nonzero vector $\mathbf{v}$ satisfying $(\mathbf{A} - \lambda \mathbf{I})\mathbf{v} = \mathbf{0}$ is an eigenvector that corresponds to λ. Hence, if $\mathbf{A}$ has n distinct eigenvalues $\{\lambda_k\}_{k=1}^{n}$, we can find a set of n linearly independent eigenvectors $\{\mathbf{v}_k\}_{k=1}^{n}$. From these eigenvalues and corresponding eigenvectors, we form n linearly independent solutions

$$\mathbf{X}_1 = \mathbf{v}_1 e^{\lambda_1 t}, \mathbf{X}_2 = \mathbf{v}_2 e^{\lambda_2 t}, \ldots, \mathbf{X}_n = \mathbf{v}_n e^{\lambda_n t}.$$

Therefore, if $\mathbf{A}$ is an $n \times n$ matrix with n distinct real eigenvalues $\{\mathbf{v}_k\}_{k=1}^{n}$, a general solution of $\mathbf{X}' = \mathbf{AX}$ is the linear combination of the set of solutions $\{\mathbf{X}_1, \mathbf{X}_2, \ldots, \mathbf{X}_n\}$,

$$\mathbf{X}(t) = c_1 \mathbf{v}_1 e^{\lambda_1 t} + c_2 \mathbf{v}_2 e^{\lambda_2 t} + \cdots + c_n \mathbf{v}_n e^{\lambda_n t} = \sum_{i=1}^{n} c_i \mathbf{v}_i e^{\lambda_i t}.$$

EXAMPLE 1: Solve $\begin{cases} x' = 5x - y \\ y' = 3y \end{cases}$.

SOLUTION: In matrix form the system is $\mathbf{X}' = \begin{pmatrix} 5 & -1 \\ 0 & 3 \end{pmatrix}\mathbf{X}$. We find the

eigenvalues of $\mathbf{A} = \begin{pmatrix} 5 & -1 \\ 0 & 3 \end{pmatrix}$ with `eigenvectors`. The results indicate that

the eigenvalues are $\lambda_1 = 3$ and $\lambda_2 = 5$ with corresponding eigenvectors $\mathbf{v}_1 = \begin{pmatrix} 1 \\ 2 \end{pmatrix}$

and $\mathbf{v}_2 = \begin{pmatrix} 1 \\ 0 \end{pmatrix}$, respectively.

```
> with(linalg):
A:=array([[5,-1],[0,3]]):
vecs_A:=eigenvectors(A);
```

$$vecs\_A := [3, 1, \{[1, 2]\}], [5, 1, \{[1, 0]\}]$$

Therefore, a general solution of the system $\mathbf{X}' = \begin{pmatrix} 5 & -1 \\ 0 & 3 \end{pmatrix}\mathbf{X}$ is

$$\mathbf{X} = c_1\mathbf{v}_1 e^{\lambda_1 t} + c_2\mathbf{v}_2 e^{\lambda_2 t} = c_1\begin{pmatrix} 1 \\ 2 \end{pmatrix}e^{3t} + c_2\begin{pmatrix} 1 \\ 0 \end{pmatrix}e^{5t} = \begin{pmatrix} e^{3t} & e^{5t} \\ 2e^{3t} & 0 \end{pmatrix}\begin{pmatrix} c_1 \\ c_2 \end{pmatrix}.$$

We can write the general solution obtained here as $\begin{cases} x = c_1 e^{3t} + c_2 e^{5t} \\ y = 2c_1 e^{3t} \end{cases}$. We can use `matrixDE` to find a general solution as well.

```
> matrixDE(A,t);
```

$$\left[\left[\begin{array}{cc} e^{(3t)} & e^{(5t)} \\ 2e^{(3t)} & 0 \end{array}\right], [0, 0]\right]$$

The result means that a general solution of the system is $\mathbf{X} = \begin{pmatrix} e^{3t} & e^{5t} \\ 2e^{3t} & 0 \end{pmatrix}\begin{pmatrix} c_1 \\ c_2 \end{pmatrix}$.

Remember that the system $\mathbf{X}' = \begin{pmatrix} 5 & -1 \\ 0 & 3 \end{pmatrix}\mathbf{X}$ is the same as the system

$\begin{cases} x' = 5x - y \\ y' = 3y \end{cases}$, which we form in the following using `equate`, which is contained

in the **student** package.

```
> with(student):
sys:=equate(array([[diff(x(t),t)],[diff(y(t),t)]]),
      A &* array([[x(t)],[y(t)]]));
```

$$sys := \left\{ \frac{\partial}{\partial t}x(t) = 5x(t) - y(t), \frac{\partial}{\partial t}y(t) = 3y(t) \right\}$$

We can then use `dsolve` to solve `sys`, obtaining a result equivalent to that obtained previously.

```
> gensol:=dsolve(sys,{x(t),y(t)});
```

$$gensol := \{y(t) = \mathbf{e}^{(3t)}\_C2, x(t) = \mathbf{e}^{(5t)}\_C1 - \frac{1}{2}\_C2\mathbf{e}^{(5t)} + \frac{1}{2}\mathbf{e}^{(3t)}\_C2\}$$

We then assign $x(t)$ and $y(t)$ the results obtained in `gensol` with `assign`.

```
> assign(Gen_Sol):
```

We can graph the solution parametrically for various values of c_1 and c_2 with `plot`. First, we use `seq` to generate a list corresponding to replacing c_1 and c_2 in

$$\begin{cases} x = x(t) \\ y = y(t) \end{cases} \quad \text{by} -2, -1, 0, 1, \text{and } 2.$$

```
> toplot:={seq(seq(subs({_C1=i,_C2=j},
            [x(t),y(t),t=-1..1]),i=-2..2),j=-2..2)};
```

$$toplot := \{\left[-\frac{5}{2}\mathbf{e}^{(5t)} + \frac{1}{2}\mathbf{e}^{(3t)}, \mathbf{e}^{(3t)}, t = -1..1\right],$$

$$[\mathbf{e}^{(3t)}, 2\mathbf{e}^{(3t)}, t = -1..1], [2\mathbf{e}^{(5t)}, 0, t = -1..1],$$

$$\left[\frac{3}{2}\mathbf{e}^{(5t)} + \frac{1}{2}\mathbf{e}^{(3t)}, \mathbf{e}^{(3t)}, t = -1..1\right], [0, 0, t = -1..1],$$

$$[\mathbf{e}^{(5t)}, 0, t = -1..1], \left[\frac{1}{2}\mathbf{e}^{(5t)} + \frac{1}{2}\mathbf{e}^{(3t)}, \mathbf{e}^{(3t)}, t = -1..1\right],$$

$$[-2\mathbf{e}^{(5t)} + \mathbf{e}^{(3t)}, 2\mathbf{e}^{(3t)}, t = -1..1],$$

$$[-3\mathbf{e}^{(5t)} + \mathbf{e}^{(3t)}, 2\mathbf{e}^{(3t)}, t = -1..1],$$

$$[-\mathbf{e}^{(5t)} - \mathbf{e}^{(3t)}, -2\mathbf{e}^{(3t)}, t = -1..1], [-\mathbf{e}^{(3t)}, -2\mathbf{e}^{(3t)}, t = -1..1],$$

$$[\mathbf{e}^{(5t)} - \mathbf{e}^{(3t)}, -2\mathbf{e}^{(3t)}, t = -1..1],$$

$$[2\mathbf{e}^{(5t)} - \mathbf{e}^{(3t)}, -2\mathbf{e}^{(3t)}, t = -1..1],$$

$$\left[-\frac{3}{2}\mathbf{e}^{(5t)} + \frac{1}{2}\mathbf{e}^{(3t)}, \mathbf{e}^{(3t)}, t = -1..1\right],$$

$$\left[-\frac{3}{2}\mathbf{e}^{(5t)} - \frac{1}{2}\mathbf{e}^{(3t)}, -\mathbf{e}^{(3t)}, t = -1..1\right],$$

$$\left[-\frac{1}{2}\mathbf{e}^{(5t)} - \frac{1}{2}\mathbf{e}^{(3t)}, -\mathbf{e}^{(3t)}, t = -1..1\right],$$

$$\left[\frac{1}{2}\mathbf{e}^{(5t)} - \frac{1}{2}\mathbf{e}^{(3t)}, -\mathbf{e}^{(3t)}, t = -1..1\right], [-\mathbf{e}^{(5t)}, 0, t = -1..1],$$

$$[3\mathbf{e}^{(5t)} - \mathbf{e}^{(3t)}, -2\mathbf{e}^{(3t)}, t = -1..1],$$

$$\left[\frac{5}{2}\mathbf{e}^{(5t)} - \frac{1}{2}\mathbf{e}^{(3t)}, -\mathbf{e}^{(3t)}, t = -1..1\right],$$

$$\left[-\frac{1}{2}e^{(5t)}+\frac{1}{2}e^{(3t)}, e^{(3t)}, t=-1..1\right], [-2e^{(5t)}, 0, t=-1..1],$$

$$\left[\frac{3}{2}e^{(5t)}-\frac{1}{2}e^{(3t)}, -e^{(3t)}, t=-1..1\right],$$

$$[e^{(5t)}+e^{(3t)}, 2e^{(3t)}, t=-1..1], [-e^{(5t)}+e^{(3t)}, 2e^{(3t)}, t=-1..1]\}$$

Next, we use `plot` to graph the list of parametric functions in `toplot` and name the resulting graphics object `pp1`.

```
> pp1:=plot(toplot,view=[-5..5,-5..5],color=BLACK):
```

To show the graphs of the solutions together with the direction field associated with the system, we use `DEplot` to graph the direction field associated with the system on the rectangle $[-5,5] \times [-5,5]$, naming the resulting graphics object `pvf`.

```
> x:='x':y:='y':
pvf:=DEplot(sys,[x(t),y(t)],t=-1..1,

        x=-5..5,y=-5..5,color=GRAY):
```

`display`, which is contained in the **plots** package, is then used to display the graphs together.

```
> with(plots):
display({pvf,pp1});
```

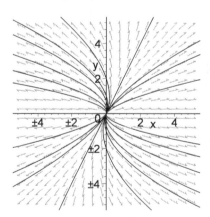

Notice that each curve corresponds to the parametric plot of the pair $\begin{cases} x = x(t) \\ y = y(t) \end{cases}$.

Because both eigenvalues are positive, all solutions move away from the origin as t increases. The arrows on the vectors in the direction field show this behavior.

■

Complex Conjugate Eigenvalues

If $\mathbf{A}$ has complex conjugate eigenvalues $\lambda_1 = \alpha + \beta i$ and $\lambda_2 = \alpha - \beta i$ and corresponding eigenvectors $\mathbf{v}_1 = \mathbf{a} + \mathbf{b}i$ and $\mathbf{v}_2 = \mathbf{a} - \mathbf{b}i$, then one solution of $\mathbf{X}' = \mathbf{AX}$ is

$$\mathbf{X} = \mathbf{v}_1 e^{\lambda t} = (\mathbf{a} + \mathbf{b}i)e^{(\alpha + \beta i)t} = e^{\alpha t}(\mathbf{a} + \mathbf{b}i)e^{i\beta t} = e^{\alpha t}(\mathbf{a} + \mathbf{b}i)(\cos \beta t + \sin \beta t)$$

$$= e^{\alpha t}(\mathbf{a}\cos \beta t - \mathbf{b}\sin \beta t) + ie^{\alpha t}(\mathbf{a}\sin \beta t + \mathbf{b}\cos \beta t)$$

$$= \mathbf{X}_1(t) + i\mathbf{X}_2(t).$$

Now, because $\mathbf{X}$ is a solution of the system, $\mathbf{X}' = \mathbf{AX}$, we have $\mathbf{X}_1'(t) + i\mathbf{X}_2'(t) = \mathbf{AX}_1(t) + i\mathbf{AX}_2(t)$. Equating the real and imaginary parts of this equation yields $\mathbf{X}_1'(t) = \mathbf{AX}_1(t)$ and $\mathbf{X}_2'(t) = \mathbf{AX}_2(t)$. Therefore, $\mathbf{X}_1(t)$ and $\mathbf{X}_2(t)$ are solutions of $\mathbf{X}' = \mathbf{AX}$, so any linear combination of $\mathbf{X}_1(t)$ and $\mathbf{X}_2(t)$ is also a solution. We can show that $\mathbf{X}_1(t)$ and $\mathbf{X}_2(t)$ are linearly independent, so this linear combination forms a portion of a general solution of $\mathbf{X}' = \mathbf{AX}$.

Theorem

> If $\mathbf{A}$ has complex conjugate eigenvalues $\lambda_1 = \alpha + \beta i$ and $\lambda_2 = \alpha - \beta i$ and corresponding eigenvectors $\mathbf{v}_1 = \mathbf{a} + \mathbf{b}i$ and $\mathbf{v}_2 = \mathbf{a} - \mathbf{b}i$, then two linearly independent solutions of $\mathbf{X}' = \mathbf{AX}$ are $\mathbf{X}_1(t) = e^{\alpha t}(\mathbf{a}\cos \beta t - \mathbf{b}\sin \beta t)$ and $\mathbf{X}_2(t) = e^{\alpha t}(\mathbf{a}\sin \beta t + \mathbf{b}\cos \beta t)$.

Notice that in the case of complex conjugate eigenvalues, we are able to obtain two linearly independent solutions by knowing one of the eigenvalues and the eigenvector that corresponds to it.

EXAMPLE 2: Find a general solution of $\mathbf{X}' = \begin{pmatrix} 3 & -2 \\ 4 & -1 \end{pmatrix}\mathbf{X}$.

SOLUTION: In this case, $\mathbf{A} = \begin{pmatrix} 3 & -2 \\ 4 & -1 \end{pmatrix}$. We find the eigenvalues and corresponding eigenvectors with `eigenvectors`.

```
> with(linalg):

A:=array([[3,-2],[4,-1]]):

vecs_A:=eigenvectors(A);
```

$$vecs\_A := \left[1 + 2I, 1, \left\{ \left[\frac{1}{2} + \frac{1}{2}I, 1 \right] \right\} \right], \left[1 - 2I, 1, \left\{ \left[\frac{1}{2} - \frac{1}{2}I, 1 \right] \right\} \right]$$

These results mean that an eigenvector that corresponds to $\lambda_1 = 1 + 2i$ is
$\mathbf{v}_1 = \begin{pmatrix} 1 + i \\ 2 \end{pmatrix} = \begin{pmatrix} 1 \\ 2 \end{pmatrix} + i \begin{pmatrix} 1 \\ 0 \end{pmatrix}$. Therefore, in the notation used in the theorem,

$$\mathbf{a} = \begin{pmatrix} 1 \\ 2 \end{pmatrix} \quad \text{and} \quad \mathbf{b} = \begin{pmatrix} 1 \\ 0 \end{pmatrix}.$$

Hence, with $\alpha = 1$ and $\beta = 2$, a general solution is

$$\mathbf{X}(t) = c_1 e^t \left[\begin{pmatrix} 1 \\ 2 \end{pmatrix} \cos 2t - \begin{pmatrix} 1 \\ 0 \end{pmatrix} \sin 2t \right] + c_2 e^t \left[\begin{pmatrix} 1 \\ 2 \end{pmatrix} \sin 2t + \begin{pmatrix} 1 \\ 0 \end{pmatrix} \cos 2t \right]$$

$$= \begin{pmatrix} e^t[(c_1 + c_2) \cos 2t + (c_2 - c_1) \sin 2t] \\ e^t[2c_1 \cos 2t + 2c_2 \sin 2t] \end{pmatrix}.$$

A slightly different form of the general solution is found with `matrixDE`. Entering

```
> matrixDE(A,t);
```

$$\left[\left| \left| \begin{matrix} e^t \cos(2t) & e^t \sin(2t) \\ e^t \cos(2t) + e^t \sin(2t) & e^t \sin(2t) - e^t \cos(2t) \end{matrix} \right| , [0, 0] \right| \right]$$

shows that a general solution is $\mathbf{X} = \begin{pmatrix} e^t \cos 2t & e^t \sin 2t \\ e^t(\cos 2t + \sin 2t) & e^t(\sin 2t - \cos 2t) \end{pmatrix} \begin{pmatrix} c_1 \\ c_1 \end{pmatrix}$;

a fundamental matrix for the system is

$$\mathbf{\Phi} = \begin{pmatrix} e^t \cos 2t & e^t \sin 2t \\ e^t(\cos 2t + \sin 2t) & e^t(\sin 2t - \cos 2t) \end{pmatrix}.$$

Using `dsolve` together with `collect`, we obtain a general solution of the system, although the form is slightly different from the first solution obtained.

```
> gensol:=dsolve({diff(x(t),t)=3*x(t)-2*y(t),
          diff(y(t),t)=4*x(t)-y(t)},{x(t),y(t)});
```

$gensol := \{y(t) = e^t(2 \sin(2t)\_C1 + \_C2 \cos(2t) - \sin(2t)\_C2),$

$\quad x(t) = e^t(\_C1 \cos(2t) + \sin(2t)\_C1 - \sin(2t)\_C2)\}$

```
> collect(gensol,{_C1,_C2});
```

$\{x(t) = -e^t \sin(2t)\_C2 + e^t(\cos(2t) + \sin(2t))\_C1,$

$\quad y(t) = e^t(\cos(2t) - \sin(2t))\_C2 + 2e^t \sin(2t)\_C1\}$

We can graph the solution for various values of the arbitrary constants using plot, as in the previous example, or we can take advantage of the DEplot command, which is contained in the **DEtools** package. After defining init_conds to be the set of ordered triples $(0,0.25i,0.25j)$ for $i = -1, 0$, and 1 and $j = -1, 0$, and 1 and loading the **DEtools** package, we use DEplot to graph the direction field associated with the system and then use DEplot to graph the solutions that satisfy the initial conditions specified in init_conds. In the second DEplot command, the option stepsize $= 0.1$ instructs Maple to use a step size of 0.1, which is smaller than the default value of $\frac{3-(-3)}{20} = \frac{3}{10}$, and helps ensure that the resulting graphs appear smooth, and the option arrows $=$ NONE instructs Maple not to display the direction field associated with the system.

```
> init_conds:={seq(seq([0,0.25*i,0.25*j],i=-1..1),j=-1..1)}:

with(DEtools):

DEplot({diff(x(t),t)=3*x(t)-2*y(t),
       diff(y(t),t)=4*x(t)-y(t)},
           [x(t),y(t)],t=-3..3,x=-15..15,y=-15..15);
DEplot({diff(x(t),t)=3*x(t)-2*y(t),
       diff(y(t),t)=4*x(t)-y(t)},
       [x(t),y(t)],t=-5..5,init_conds,x=-15..15,y=-15..15,
       stepsize=0.1,arrows=NONE,linecolor=BLACK,thickness=1);
```

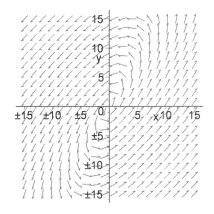 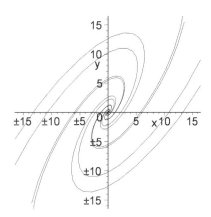

Notice the spiraling motion of the vectors in the direction field. This is due to terms in the solution formed by a product of exponential and trigonometric functions.

■

Initial-value problems can be solved through the use of eigenvalues and eigenvectors as well.

EXAMPLE 3: Solve $\begin{cases} x' = -\frac{1}{2}x - y + 64z \\ y' = -\frac{1}{4}y - 16z \\ z' = y - \frac{1}{4}z \\ x(0) = 1,\ y(0) = -1,\ z(0) = 0 \end{cases}$.

SOLUTION: In matrix form, the system is equivalent to $\mathbf{X}' = \mathbf{A}\mathbf{X}$, where

$\mathbf{A} = \begin{pmatrix} -1/2 & -1 & 64 \\ 0 & -1/4 & -16 \\ 0 & 1 & -1/4 \end{pmatrix}$. The eigenvalues and corresponding eigenvectors

of $\mathbf{A}$ are found with `eigenvectors`.

> `with(linalg):`

`A:=array([[-1/2,-1,64],[0,-1/4,-16],[0,1,-1/4]]):`

`vecs_A:=eigenvectors(A);`

$$vecs_A := \left[\frac{-1}{2}, 1, \{[1,0,0]\} \right],$$

$$\left[-\frac{1}{4} + 4I, 1, \{[-16I, 4I, 1]\} \right],$$

$$\left[-\frac{1}{4} - 4I, 1, \{[16I, -4I, 1]\} \right]$$

These results mean that the eigenvalue $\lambda_1 = -1/2$ has corresponding eigenvector

$\mathbf{v}_1 = \begin{pmatrix} 1 \\ 0 \\ 0 \end{pmatrix}$, so one solution of the system is $\mathbf{X}_1 = \mathbf{v}_1 e^{\lambda_1 t} = \begin{pmatrix} 1 \\ 0 \\ 0 \end{pmatrix} e^{-t/2}$. An

eigenvector corresponding to $\lambda_2 = -1/4 + 4i$ is $\mathbf{v}_2 = \begin{pmatrix} -16i \\ 4i \\ 1 \end{pmatrix}$

$= \begin{pmatrix} 0 \\ 0 \\ 1 \end{pmatrix} + i \begin{pmatrix} -16 \\ 4 \\ 0 \end{pmatrix} = \mathbf{a} + \mathbf{b}i$. Thus, two linearly independent solutions that

correspond to the complex conjugate pair of eigenvalues are

$$\mathbf{X}_2 = e^{-t/4} \left[\begin{pmatrix} 0 \\ 0 \\ 1 \end{pmatrix} \cos 4t - \begin{pmatrix} -16 \\ 4 \\ 0 \end{pmatrix} \sin 4t \right] = \begin{pmatrix} 16e^{-t/4} \sin 4t \\ -4e^{-t/4} \sin 4t \\ e^{-t/4} \cos 4t \end{pmatrix}$$

and

$$\mathbf{X}_3 = e^{-t/4} \left[\begin{pmatrix} 0 \\ 0 \\ 1 \end{pmatrix} \sin 4t + \begin{pmatrix} -16 \\ 4 \\ 0 \end{pmatrix} \cos 4t \right] = \begin{pmatrix} -16e^{-t/4} \cos 4t \\ 4e^{-t/4} \cos 4t \\ e^{-t/4} \sin 4t \end{pmatrix}.$$

Hence, a general solution is

$$\mathbf{X} = c_1\mathbf{X}_1 + c_2\mathbf{X}_2 + c_3\mathbf{X}_3 = c_1 \begin{pmatrix} 1 \\ 0 \\ 0 \end{pmatrix} e^{-t/2} + c_2 \begin{pmatrix} 16e^{-t/4} \sin 4t \\ -4e^{-t/4} \sin 4t \\ e^{-t/4} \cos 4t \end{pmatrix} + c_3 \begin{pmatrix} -16e^{-t/4} \cos 4t \\ 4e^{-t/4} \cos 4t \\ e^{-t/4} \sin 4t \end{pmatrix}$$

$$= \begin{pmatrix} +c_1 e^{-1/2} + 16e^{-t/4}(c_2 \sin 4t - c_3 \cos 4t) \\ 4e^{-t/4} \sin 4t + c_3 \cos 4t) \\ e^{-t/4}(\sin 4t + c_2 \cos 4t) \end{pmatrix}$$

```
> x:=t->c[1]*exp(-t/2)+
      16*exp(-t/4)*(c[2]*sin(4*t)-c[3]*cos(4*t)):
y:=t->4*exp(-t/4)*(-c[2]*sin(4*t)+c[3]*cos(4*t)):
z:=t->exp(-t/4)*(c[3]*sin(4*t)+c[2]*cos(4*t)):
```

We solve the initial-value problem by applying the initial condition
$\mathbf{X}(0) = \begin{pmatrix} 1 \\ -1 \\ 0 \end{pmatrix}$

```
> sysofeqs:={x(0)=1,y(0)=-1,z(0)=0};
```

$$sysofeqs := \{c_1 - 16c_3 = 1, c_2 = 0, 4c_3 = -1\}$$

and solving the resulting system of equations for c_1, c_2, and c_3.

```
> cvals:=solve(sysofeqs);
```

$$cvals : \{c_2 = 0, c_3 = \frac{-1}{4}, c_1 = -3\}$$

Substitution of these values into the general solution yields the solution to the initial-value problem.

```
> assign(cvals):
```

We graph $x(t)$, $y(t)$, and $z(t)$ with `plot`

```
> plot({x(t),y(t),z(t)},t=0..3*Pi);
```

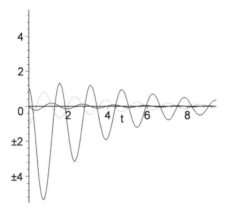

and a parametric plot of $\begin{cases} x = x(t) \\ y = y(t) \\ z = z(t) \end{cases}$ in three dimensions with `spacecurve`,

which is contained in the **plots** package.

```
> with(plots):
spacecurve([x(t),y(t),z(t)],t=0..3*Pi,
          axes=BOXED,numpoints=200,color=BLACK);
```

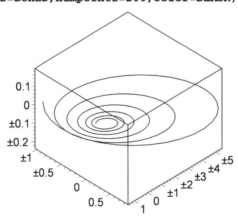

As in previous examples, we see that `dsolve` is able to find a general solution of the system as well as solve the initial-value problem.

```
> x:='x':y:='y':z:='z':
gensol:=dsolve({diff(x(t),t)=-x(t)/2-y(t)+64*z(t),
```

```
diff(y(t),t)=-y(t)/4-16*z(t),
        diff(z(t),t)=y(t)-z(t)/4},{x(t),y(t),z(t)});
```

$$gensol := \left\{ x(t) = e^{\left(-\frac{1}{2}t\right)}\_C1 + 4e^{\left(-\frac{1}{4}t\right)}\cos(4t)\_C2 \right.$$

$$+ 4\_C2e^{\left(-\frac{1}{2}t\right)} + 16e^{\left(-\frac{1}{4}t\right)}\sin(4t)\_C3,$$

$$z(t) = \frac{1}{4}e^{\left(-\frac{1}{4}t\right)}(\sin(4t)\_C2 + 4\cos(4t)\_C3),$$

$$\left. y(t) = -e^{\left(-\frac{1}{4}t\right)}(\cos(rt)\_C2 + 4\sin(4t)\_C3) \right\}$$

```
> partsol:=dsolve({diff(x(t),t)=-x(t)/2-y(t)+64*z(t),
        diff(y(t),t)=-y(t)/4-16*z(t),
        diff(z(t),t)=y(t)-z(t)/4,
            x(0)=1,y(0)=-1,z(0)=0},{x(t),y(t),z(t)});
```

$$partsol := \left\{ x(t) = -3e^{\left(-\frac{1}{2}t\right)} + 4e^{\left(-\frac{1}{4}t\right)}\cos(4t), . \right.$$

$$\left. z(t) = -\frac{1}{4}e^{\left(-\frac{1}{4}t\right)}\sin(4t), y(t) = -e^{\left(-\frac{1}{4}t\right)}\cos(4t) \right\}$$

∎

Alternative Method for Solving Initial-Value Problems

An alternative method can be used to solve initial-value problems. Let $\mathbf{\Phi}(t)$ be a fundamental matrix for the system of equations $\mathbf{X}' = \mathbf{AX}$. Then, a general solution is $\mathbf{X}(t) = \mathbf{\Phi}(t)\mathbf{C}$, where $\mathbf{C}$ is a constant vector. If the initial condition $\mathbf{X}(0) = \mathbf{X}_0$ is given, then

$$\mathbf{X}(0) = \mathbf{\Phi}(0)\mathbf{C}$$

$$\mathbf{X}_0 = \mathbf{\Phi}(0)\mathbf{C}$$

$$\mathbf{C} = \mathbf{\Phi}^{-1}(0)\mathbf{X}_0$$

Therefore, the solution to the initial-value problem is $\mathbf{X}(t) = \mathbf{\Phi}(t)\mathbf{\Phi}^{-1}(0)\mathbf{X}_0$.

EXAMPLE 4: Use a fundamental matrix to solve the initial-value problem

$$\mathbf{X}' = \begin{pmatrix} 1 & 1 \\ 4 & -2 \end{pmatrix}\mathbf{X} \text{ subject to } \mathbf{X}(0) = \begin{pmatrix} 1 \\ -2 \end{pmatrix}.$$

SOLUTION: We first remark that we can use `dsolve` to solve the initial-value problem directly.

```
> dsolve({diff(x(t),t)=x(t)+y(t),
       diff(y(t),t)=4*x(t)-2*y(t),x(0)=1,y(0)=-2},
          {x(t),y(t)});
```

$$\{x(t) = \frac{3}{5}e^{(-3t)} + \frac{2}{5}e^{(2t)}, y(t) = \frac{2}{5}e^{2t} - \frac{12}{5}e^{(-3t)}\}$$

The eigenvalues and corresponding eigenvectors of $\mathbf{A} = \begin{pmatrix} 1 & 1 \\ 4 & -2 \end{pmatrix}$ are found with `eigenvectors`.

```
> with(linalg):
A:=array([[1,1],[4,-2]]):
vecs_A:=eigenvects(A);
```

$$vecs\_A := [2, 1, \{[1, 1]\}], [-3, 1, \{[1, -4]\}]$$

Hence, the eigenvalues are $\lambda_1 = -3$ and $\lambda_2 = 2$ and corresponding eigenvectors are $\mathbf{v}_1 = \begin{pmatrix} 1 \\ -4 \end{pmatrix}$ and $\mathbf{v}_2 = \begin{pmatrix} 1 \\ 1 \end{pmatrix}$, respectively. A fundamental matrix is then given by $\mathbf{\Phi}(t) = \begin{pmatrix} e^{-3t} & e^{2t} \\ -4e^{-3t} & e^{2t} \end{pmatrix}$.

```
> with(DEtools):
step1:=matrixDE(A,t);
```

$$step1 := \left[\left[\begin{matrix} e^{(-3t)} e^{(2t)} \\ -4e^{(-3t)} e^{(2t)} \end{matrix} \right], [0, 0] \right]$$

```
> Phi:=step1[1];
```

$$\mathbf{\Phi} := \left[\begin{matrix} e^{(-3t)} & e^{(2t)} \\ -4e^{(-3t)} & e^{(2t)} \end{matrix} \right]$$

We calculate $\mathbf{\Phi}^{-1}(0)$ with `inverse`, which is contained in the **linalg** package.

```
> eval(inverse(Phi),t=0);
```

$$\begin{bmatrix} \frac{1}{5} & \frac{-1}{5} \\ \frac{4}{5} & \frac{1}{5} \end{bmatrix}$$

Hence, the solution to the initial-value problem is $\mathbf{X}(t) = \boldsymbol{\Phi}(t)\boldsymbol{\Phi}^{-1}(0)\mathbf{X}_0$.

```
> sol:=evalm(Phi &* eval(inverse(Phi),t=0) &* [[1],[-2]]);
```

$$sol := \begin{bmatrix} \frac{3}{5}\mathbf{e}^{(-3t)} + \frac{2}{5}\mathbf{e}^{(2t)} \\ -\frac{12}{5}\mathbf{e}^{(-3t)} + \frac{2}{5}\mathbf{e}^{(2t)} \end{bmatrix}$$

As in the previous examples, we graph $x(t)$ and $y(t)$ together and parametrically.

```
> plot([sol[1,1],sol[2,1]],t=-1..3,view=[-1..3,-2..2]);
> plot([sol[1,1],sol[2,1],t=-1..3]);
```

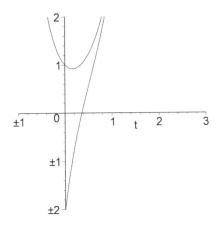
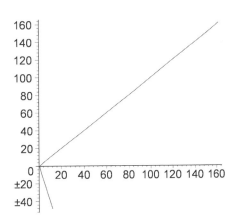

Repeated Eigenvalues

We now consider the case of repeated eigenvalues, which is a little more complicated than the other cases because two situations can arise. As we illustrated in Section 7.1, an eigenvalue of multiplicity m can have m corresponding linearly independent eigenvectors or it can have fewer than m corresponding linearly independent eigenvectors. In the case of m linearly independent eigenvectors, a general solution is found in the same manner as in the case of n distinct eigenvalues.

EXAMPLE 5: Solve $X' = \begin{pmatrix} 1 & -3 & 3 \\ 3 & -5 & 3 \\ 6 & -6 & 4 \end{pmatrix} X$.

SOLUTION: The eigenvalues and corresponding eigenvectors of

$A = \begin{pmatrix} 1 & -3 & 3 \\ 3 & -5 & 3 \\ 6 & -6 & 4 \end{pmatrix}$ are found with `eigenvectors`.

```
> with(linalg):
A:=[[1,-3,3],[3,-5,3],[6,-6,4]]:
> eigenvectors(A);
```

$$[4, 1, \{[1, 1, 2]\}], [-2, 2, \{[1, 1, 0], [-1, 0, 1]\}]$$

From the results, we see that the eigenvalue $\lambda_1 = \lambda_2 = -2$ of multiplicity two has

two corresponding linearly independent eigenvectors, $v_1 = \begin{pmatrix} -1 \\ 0 \\ 1 \end{pmatrix}$ and

$v_2 = \begin{pmatrix} 1 \\ 1 \\ 0 \end{pmatrix}$. An eigenvector corresponding to $\lambda_3 = 4$ is $v_3 = \begin{pmatrix} 1 \\ 1 \\ 2 \end{pmatrix}$. A general

solution is then

$$X = c_1 v_1 e^{\lambda_1 t} + c_2 v_2 e^{\lambda_2 t} + c_3 v_3 e^{\lambda_3 t} = c_1 \begin{pmatrix} -1 \\ 0 \\ 1 \end{pmatrix} e^{-2t} + c_2 \begin{pmatrix} 1 \\ 1 \\ 0 \end{pmatrix} e^{-2t} + c_3 \begin{pmatrix} 1 \\ 1 \\ 2 \end{pmatrix} e^{4t}$$

$$= \begin{pmatrix} (c_1 - c_2)e^{-2t} + c_3 e^{4t} \\ c_1 e^{-2t} + c_3 e^{4t} \\ c_2 e^{-2t} + 2c_3 e^{4t} \end{pmatrix}.$$

Of course, `dsolve` can be used to find a general solution of the system as well, although the form is slightly different from that obtained before.

```
> simplify(dsolve({diff(x(t),t)=x(t)-3*y(t)+3*z(t),
        diff(y(t),t)=3*x(t)-5*y(t)+3*z(t),
```

```
diff(z(t),t)=6*x(t)-6*y(t)+4*z(t)},
    {x(t),y(t),z(t)}));
```

$$\{x(t) = -\frac{1}{2}$$

$$(-\_C1e^{(-6t)} - \_C1 + \_C2 - \_C2e^{(-6t)} - \_C3 + \_C3e^{(-6t)})e^{(4t)},$$

$$y(t) = -\frac{1}{2}$$

$$(-\_C1 + \_C1e^{(-6t)} - 3\_C2e^{(-6t)} + \_C2 - \_C3 + \_C3e^{(-6t)})$$

$$\mathbf{e}^{(4t)}, z(t) = (\_C1 - \_C1e^{(-6t)} - \_C2 + \_C2e^{(-6t)} + \_C3e^{(4t)}\}$$

∎

Because an eigenvalue of multiplicity two can have only one corresponding eigenvector, let us first restrict our attention to a system with the repeated eigenvalue $\lambda_1 = \lambda_2$, which has only one corresponding eigenvector $\mathbf{v}_1$. We obtain one solution to the system $\mathbf{X}_1 = \mathbf{v}_1 e^{\lambda_1 t}$ that corresponds to λ_1. We now seek a second linearly independent solution corresponding to λ_1 in a manner similar to that considered in the case of repeated characteristic roots of higher order equations. In this case, however, we suppose that the second linearly independent solution corresponding to λ_1 is of the form

$$\mathbf{X}_2 = (\mathbf{v}_2 t + \mathbf{w}_2)e^{\lambda_1 t}$$

In order to find $\mathbf{v}_2$ and $\mathbf{w}_2$, we substitute $\mathbf{X}_2$ into $\mathbf{X}' = \mathbf{A}\mathbf{X}$. Because $\mathbf{X}_2' = \lambda_1(\mathbf{v}_2 t + \mathbf{w}_2)e^{\lambda_1 t} + \mathbf{v}_2 e^{\lambda_1 t}$, we have

$$\mathbf{X}_2' = \mathbf{A}\mathbf{X}_2$$

$$\lambda_1(\mathbf{v}_2 t + \mathbf{w}_2)e^{\lambda_1 t} + \mathbf{v}_2 e^{\lambda_1 t} = \mathbf{A}(\mathbf{v}_2 t + \mathbf{w}_2)e^{\lambda_1 t}$$

$$\lambda_1\mathbf{v}_2 t + (\lambda_1\mathbf{w}_2 + \mathbf{v}_2) = \mathbf{A}\mathbf{v}_2 t + \mathbf{A}\mathbf{w}_2$$

Equating coefficients yields $\lambda_1\mathbf{v}_2 = \mathbf{A}\mathbf{v}_2$ and $\lambda_1\mathbf{w}_2 + \mathbf{v}_2 = \mathbf{A}\mathbf{w}_2$. The equation $\lambda_1\mathbf{v}_2 = \mathbf{A}\mathbf{v}_2$ indicates that $\mathbf{v}_2$ is an eigenvector that corresponds to λ_1, so $\mathbf{v}_2 = \mathbf{v}_1$. We simplify the equation $\lambda_1\mathbf{w}_2 + \mathbf{v}_2 = \mathbf{A}\mathbf{w}_2$:

$$\lambda_1\mathbf{w}_2 + \mathbf{v}_2 = \mathbf{A}\mathbf{w}_2$$

$$\mathbf{v}_2 = \mathbf{A}\mathbf{w}_2 - \lambda_1\mathbf{w}_2$$

$$\mathbf{v}_2 = (A - \lambda_1 I)\mathbf{w}_2.$$

Hence, $\mathbf{w}_2$ satisfies the equation

$$(\mathbf{A} - \lambda_1\mathbf{I})\mathbf{w}_2 = \mathbf{v}_1.$$

Therefore, a second linearly independent solution corresponding to the eigenvalue λ_1 has the form

$$X_2 = (v_1 t + w_2)e^{\lambda_1 t}$$

where w_2 satisfies $(A - \lambda_1 I)w_2 = v_1$

EXAMPLE 6: Find a general solution of $X' = \begin{pmatrix} -8 & -1 \\ 16 & 0 \end{pmatrix} X$.

SOLUTION: We first note that `dsolve` can find a general solution of the system immediately.

```
> dsolve({D(x)(t)=-8*x(t)-y(t),D(y)(t)=16*x(t)},
          {x(t),y(t)});
```

$\{x(t) = -e^{(-4t)}(-\_C1 + 4t\_C1 + t\_C2),$

$\quad y(t) = e^{(-4t)}(16t\_C1 + \_C2 + 4t\_C2)\}$

We find the eigenvalues and corresponding eignevectors of $A = \begin{pmatrix} -8 & -1 \\ 16 & 0 \end{pmatrix}$ with `eigenvectors`.

```
> with(linalg):
A:=array([[-8,-1],[16,0]]):
vecs_A:=eigenvects(A);
```

$$vecs\_A := [-4, 2, \{[1, -4]\}]$$

Hence, $\lambda_1 = \lambda_2 = -4$ with corresponding eigenvector $v_1 = \begin{pmatrix} 1 \\ -4 \end{pmatrix}$, so one solution to the system is $X_1 = \begin{pmatrix} 1 \\ -4 \end{pmatrix} e^{-4t}$. Therefore, to find $w_2 = \begin{pmatrix} x_2 \\ y_2 \end{pmatrix}$ in a second linearly independent solution $X_2 = (v_1 t + w_2)e^{\lambda_1 t}$, we solve $(A - \lambda_1 I)w_2 = v_1$, which in this case is

$$\begin{pmatrix} -4 & -1 \\ 16 & 4 \end{pmatrix} \begin{pmatrix} x_2 \\ y_2 \end{pmatrix} = \begin{pmatrix} 1 \\ -4 \end{pmatrix},$$

with `linsolve`. (Notice that `A-vecs_A[1]*array(identity,1..2,1..2)` represents the matrix $A - \lambda_1 I$.)

```
> sol_vec:=linsolve(
        evalm(A-vecs_A[1]*array(identity,1..2,1..2)),
            vecs_A[3,1]);
```

$$sol\_vec := [\_t_1, -4\_t_1 - 1]$$

The result means that for any value of t_1, $x_2 = t_1$ and $y_2 = -4t_1 - 1 = -4x_2 - 1$. For example, if we choose $t_1 = 0$, $x_2 = 0$ and $y_2 = -1$.

```
> subs(_t[1]=0,eval(sol_vec));
```

$$[0, -1]$$

With $\mathbf{w}_2 = \begin{pmatrix} 0 \\ -1 \end{pmatrix}$, a second linearly independent solution is

$$\mathbf{X}_2 = \left(\begin{pmatrix} 1 \\ -4 \end{pmatrix} t + \begin{pmatrix} 0 \\ -1 \end{pmatrix} \right) e^{-4t}.$$

Hence, a general solution is

$$\mathbf{X} = c_1 \begin{pmatrix} 1 \\ -4 \end{pmatrix} e^{-4t} + c_2 \left(\begin{pmatrix} 1 \\ -4 \end{pmatrix} t + \begin{pmatrix} 0 \\ -1 \end{pmatrix} \right) e^{-4t} = \begin{pmatrix} c_1 + c_2 t \\ (-4c_1 - c_2) - 4tc_2 \end{pmatrix} e^{-4t}.$$

After loading the **DEtools** package, we use DEplot to graph the direction field associated with the system on the rectangle $[-2, 2] \times [-2, 2]$ along with the graphs of the solutions that satisfy the initial conditions $\begin{cases} x(0) = 0 \\ y(0) = i \end{cases}$, $i = -2, -5/3, \ldots, 5/3$, and 2.

```
> with(DEtools):
ivals:=seq(-2+1/3*i,i=0..12):
inits:=[seq([0,0,i],i=ivals)]:
> DEplot({D(x)(t)=-8*x(t)-y(t),
        D(y)(t)=16*x(t)},[x(t),y(t)],
        t=-4..4,inits,x=-2..2,y=-2..2,
            stepsize=0.1,color=GRAY,
            linecolor=BLACK,thickness=1);
```

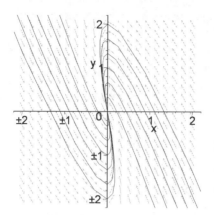

Notice that the behavior of these solutions differs from those of the other systems solved earlier in the section. This is due to the repeated eigenvalues.

■

A similar method is carried out in the case of three equal eigenvalues $\lambda_1 = \lambda_2 = \lambda_3$ When we encounter this situation, we assume that

$$\mathbf{X}_1 = \mathbf{v}_1 e^{\lambda_1 t}, \mathbf{X}_2 = (\mathbf{v}_2 t + \mathbf{w}_2)e^{\lambda_1 t}, \text{ and } \mathbf{X}_3 = \left(\mathbf{v}_1 \tfrac{1}{2}t^2 + \mathbf{w}_2 t + \mathbf{u}_3\right)e^{\lambda_1 t}.$$

Substitution of these solutions into the system of differential equations yields the following system of equations, which is solved for the unknown vectors $\mathbf{v}_2$, $\mathbf{w}_2$, $\mathbf{v}_3$, $\mathbf{w}_3$, and $\mathbf{u}_3$:

$$\begin{cases} \lambda_1 \mathbf{v}_2 = \mathbf{A}\mathbf{v}_2 \\ (\mathbf{A} - \lambda_1 \mathbf{I})\mathbf{w}_2 = \mathbf{v}_2 \\ \lambda_1 \mathbf{v}_3 = \mathbf{A}\mathbf{v}_3 \\ (\mathbf{A} - \lambda_1 \mathbf{I})\mathbf{w}_3 = \mathbf{v}_3 \\ (\mathbf{A} - \lambda_1 \mathbf{I})\mathbf{u}_3 = \mathbf{w}_3 \end{cases}.$$

Similarly to the previous case, $\mathbf{v}_3 = \mathbf{v}_2 = \mathbf{v}_1$, $\mathbf{w}_2 = \mathbf{w}_3$, and the vector $\mathbf{u}_3$ is found by solving the system

$$(\mathbf{A} - \lambda_1 \mathbf{I})\mathbf{u}_3 = \mathbf{w}_3.$$

Hence, the three solutions have the form

$$\mathbf{X}_1 = \mathbf{v}_1 e^{\lambda_1 t}, \mathbf{X}_2 = (\mathbf{v}_1 t + \mathbf{w}_2)e^{\lambda_1 t}, \text{ and } \mathbf{X}_3 = \left(\mathbf{v}_1 \tfrac{1}{2}t^2 + \mathbf{w}_2 t + \mathbf{u}_3\right)e^{\lambda_1 t}.$$

Notice that this method is easily generalized for instances in which the multiplicity of the repeated eigenvalue is greater than three.

EXAMPLE 7: Solve $\mathbf{X}' = \begin{pmatrix} 1 & 1 & 1 \\ 2 & 1 & -1 \\ -3 & 2 & 4 \end{pmatrix} \mathbf{X}$.

SOLUTION: The eigenvalues and corresponding eigenvectors of $\mathbf{A} = \begin{pmatrix} 1 & 1 & 1 \\ 2 & 1 & -1 \\ -3 & 2 & 4 \end{pmatrix}$ are found with `eigenvectors`.

```
> with(linalg):
A:=array([[1,1,1],[2,1,-1],[-3,2,4]]):
vecs_A:=eigenvects(A);
```

$$vecs\_A := [2, 3, \{[0, 1, -1]\}]$$

Hence, $\lambda_1 = \lambda_2 = \lambda_3 = 2$ with corresponding eigenvector $\mathbf{v}_1 = \begin{pmatrix} 0 \\ 1 \\ -1 \end{pmatrix}$, so one solution is $\mathbf{X}_1 = \mathbf{v}_1 e^{2t} = \begin{pmatrix} 0 \\ 1 \\ -1 \end{pmatrix} e^{2t}$. The vector $\mathbf{w}_2 = \begin{pmatrix} x_2 \\ y_2 \\ z_2 \end{pmatrix}$ in a second linearly independent solution of the form $\mathbf{X}_2 = (\mathbf{v}_1 t + \mathbf{w}_2)e^{2t}$ is found by solving the system

$(\mathbf{A} - \lambda \mathbf{I})\mathbf{w}_2 = \mathbf{v}_1$. As in Example 6, we can solve this system with `linsolve`.

```
> vec_2:=linsolve(
        evalm(A-vecs_A[1]*array(identity,1..3,1..3)),
            vecs_A[3,1]);
```

$$vec\_2 := [1, 1 - \_t_1, \_t_1]$$

Choosing $t_1 = 0$, we obtain

```
> w[2]:=subs(_t[1]=0,eval(vec_2));
```

$$w_2 := [1, 1, 0]$$

$\mathbf{w}_2 = \begin{pmatrix} 1 \\ 1 \\ 0 \end{pmatrix}$, so $\mathbf{X}_2 = \left(\begin{pmatrix} 0 \\ 1 \\ -1 \end{pmatrix} t + \begin{pmatrix} 1 \\ 1 \\ 0 \end{pmatrix} \right) e^{2t}$. Finally, we must determine the vector $\mathbf{u}_3 = \begin{pmatrix} x_3 \\ y_3 \\ z_3 \end{pmatrix}$ in a third linearly independent solution

$\mathbf{X}_3 = \left(\mathbf{v}_1\frac{1}{2}t^2 + \mathbf{w}_2 t + \mathbf{u}_3\right)e^{\lambda_1 t}$ by solving the system $(\mathbf{A} - \lambda\mathbf{I})\mathbf{u}_3 = \mathbf{w}_2$.

```
> vec_3:=linsolve(
        evalm(A-vecs_A[1]*array(identity,1..3,1..3)),
        w[2]);
```

$$vec\_3 := [2, 3 - \_t_1, \_t_1]$$

Choosing $t_1 = 0$, we obtain

```
> u[3]:=subs(_t[1]=0,eval(vec_2));
```

$$u_3 := [2, 3, 0]$$

$\mathbf{u}_3 = \begin{pmatrix} 2 \\ 3 \\ 0 \end{pmatrix}$, so a third linearly independent solution is

$\mathbf{X}_3 = \left(\begin{pmatrix} 0 \\ 1 \\ -1 \end{pmatrix}\frac{t^2}{2} + \begin{pmatrix} 1 \\ 1 \\ 0 \end{pmatrix}t + \begin{pmatrix} 2 \\ 3 \\ 0 \end{pmatrix}\right)e^{2t}$. A general solution is then given by

$\mathbf{X} = c_1\mathbf{X}_1 + c_2\mathbf{X}_2 + c_3\mathbf{X}_3$

$$= c_1\begin{pmatrix} 0 \\ 1 \\ -1 \end{pmatrix}e^{2t} + c_2\left(\begin{pmatrix} 0 \\ 1 \\ -1 \end{pmatrix}t + \begin{pmatrix} 1 \\ 1 \\ 0 \end{pmatrix}\right)e^{2t} + c_3\left(\begin{pmatrix} 0 \\ 1 \\ -1 \end{pmatrix}\frac{t^2}{2} + \begin{pmatrix} 1 \\ 1 \\ 0 \end{pmatrix}t + \begin{pmatrix} 2 \\ 3 \\ 0 \end{pmatrix}\right)e^{2t}$$

$$= \begin{pmatrix} c_2e^{2t} + c_3(t+2)e^{2t} \\ c_1e^{2t} + c_2(t+1)e^{2t} + c_3\left(\frac{t^2}{2} + t + 3\right)e^{2t} \\ -c_1e^{2t} - c_2te^{2t} - c_3\left(\frac{t^2}{2}\right)e^{2t} \end{pmatrix}.$$

As we have seen in previous examples, we obtain equivalent results with `dsolve`.

```
> simplify(dsolve({D(x)(t)=x(t)+y(t)+z(t),
        D(y)(t)=2*x(t)+y(t)-z(t),
        D(z)(t)=-3*x(t)+2*y(t)+4*z(t)},
        {x(t),y(t),z(t)}));
```

$\{x(t)\mathbf{e}^{(2t)}(\_C1 - \_C1t + t\_C2 + t\_C3), y(t) = \dfrac{1}{2}$

$\mathbf{e}^{(2t)}(4\_C1t - \_C1t^2 + 2\_C2 - 2t\_C2 + \_C2t^2$

$- 2t\_C3 + \_C3t^2), z(t) = -\dfrac{1}{2}\mathbf{e}^{(2t)}(6\_C1t - \_C1t^2$

$- 4t\_C2 + \_C2t^2 - 2\_C3 + \_C3t^2 - 4t\_C3)\}$

■

7.4 Nonhomogeneous First-Order Systems: Undetermined Coefficients, Variation of Parameters, and the Matrix Exponential

In Chapter 4, we learned how to solve nonhomogeneous differential equations through the use of undetermined coefficients and variation of parameters. Here we approach the solution of systems of nonhomogeneous equations using those methods.

Let $\mathbf{X} = \mathbf{X}(t) = \begin{pmatrix} x_1(t) \\ x_2(t) \\ \vdots \\ x_n(t) \end{pmatrix}$, $\mathbf{A} = \begin{pmatrix} a_{11} & a_{12} & \cdots & a_{1n} \\ a_{21} & a_{22} & \cdots & a_{2n} \\ \vdots & \vdots & \ddots & \vdots \\ a_{n1} & a_{n2} & \cdots & a_{nn} \end{pmatrix}$, $\mathbf{F}(t) = \begin{pmatrix} f_1(t) \\ f_2(t) \\ \vdots \\ f_n(t) \end{pmatrix}$, and $\mathbf{\Phi}(t)$ be a

fundamental matrix of the system $\mathbf{X}' = \mathbf{A}\mathbf{X}$. Then a general solution of the homogeneous

system $\mathbf{X}' = \mathbf{A}\mathbf{X}$ is $\mathbf{X} = \mathbf{\Phi}(t)\,\mathbf{C}$ where $\mathbf{C} = \begin{pmatrix} c_1 \\ c_2 \\ \vdots \\ c_n \end{pmatrix}$ is an $n \times 1$ constant vector. To find a general

solution of $\mathbf{X}' = \mathbf{A}\mathbf{X} + \mathbf{F}(t)$, we note that if $\mathbf{X}_p$ is a particular solution of the equation then all other solutions of the equation can be written in the form $\mathbf{X} = \mathbf{\Phi}(t)\,\mathbf{C} + \mathbf{X}_p$

Undetermined Coefficients

We use the method of undetermined coefficients to find a particular solution of a nonhomogeneous system in much the same way as we approached nonhomogeneous higher order equations in Chapter 4. The main difference is that the coefficients are *constant vectors* when we work with systems.

EXAMPLE 1: Solve $\begin{cases} x' = 2x + y + \sin 3t \\ y' = -8x - 2y \\ x(0) = 0, y(0) = 1 \end{cases}$.

SOLUTION: Let $\mathbf{X} = \begin{pmatrix} x \\ y \end{pmatrix}$. Then, in matrix form, the system is equivalent to

$\mathbf{X}' = \begin{pmatrix} 2 & 1 \\ -8 & -2 \end{pmatrix}\mathbf{X} + \begin{pmatrix} \sin 3t \\ 0 \end{pmatrix}$. We find a general solution of the corresponding

homogeneous system $\mathbf{X}' = \begin{pmatrix} 2 & 1 \\ -8 & -2 \end{pmatrix}\mathbf{X}$ with `dsolve`

```
> homsol:=dsolve({diff(x(t),t)=2*x(t)+y(t),
        diff(y(t),t)=-8*x(t)-2*y(t)},{x(t),y(t)});
```

$$homsol := \{x(t) = \_C1\cos(2t) + \sin(2t)\_C1 + \frac{1}{2}\sin(2t)\_C2,$$

$$y(t) = -4\sin(2t)\_C1 + \_C2\cos(2t) - \sin(2t)\_C2\}$$

or `matrixDE`.

```
> with(DEtools):
A:=array([[2,1],[-8,-2]])
homsol:=matrixDE(A,t);
```

$$homsol :=$$
$$\left[\left[\begin{matrix} \cos(2t) & \sin(2t) \\ -2\cos(2t) - 2\sin(2t) & -2\sin(2t) + 2\cos(2t) \end{matrix}\right], [0,0]\right]$$

These results indicate that a general solution of the corresponding homogeneous system is

$$\mathbf{X}_h = \begin{pmatrix} \cos 2t & \sin 2t \\ -2\cos 2t - 2\sin 2t & -2\sin 2t + 2\cos t2 \end{pmatrix}\begin{pmatrix} c_1 \\ c_2 \end{pmatrix}.$$

```
> xh:=evalm(homsol[1] &* vector([c[1],c[2]]));
```

$$xh := [\cos(2t)c_1 + \sin(2t)c_2,$$

$$(-2\cos(2t) - 2\sin(2t))c_1 + (-2\sin(2t) + 2\cos(2t))c_2]$$

Thus, we search for a particular solution of the nonhomogeneous system of the

form $\mathbf{X}_p = \mathbf{a}\sin 3t + \mathbf{b}\cos 3t$, where $\mathbf{a} = \begin{pmatrix} a_1 \\ a_2 \end{pmatrix}$ and $\mathbf{b} = \begin{pmatrix} b_1 \\ b_2 \end{pmatrix}$. After defining

$\mathbf{X}_p = \mathbf{a}\sin 3t + \mathbf{b}\cos 3t$, we substitute $\mathbf{X}_p$ into the nonhomogeneous system.

```
> xp:=evalm(vector([a[1],a[2]])*sin(3*t)+
            vector([b[1],b[2]])*cos(3*t));
```

$$xp := [\sin(3t)a_1 + \cos(3t)b_1, \sin(3t)a_2 + \cos(3t)b_2]$$

```
> step1:=evalm(map(diff,xp,t)=
            A &* xp+vector([sin(3*t),0]));
```

$$step1 := [3\cos(3t)a_1 - 3\sin(3t)b_1, 3\cos(3t)a_2 - \sin(3t)b_2]$$
$$= [2\sin(3t)a_1 + 2\cos(3t)b_1, \sin(3t)a_2 + \cos(3t)b_2 + \sin(3t),$$
$$- [8\sin(3t)a_1 - 8\cos(3t)b_1 - 2\sin(3t)a_2 - 2\cos(3t)b_2]$$

The result represents a system of equations that is true for all values of t. In particular, substituting $t = 0$ yields

```
> eq1:=eval(step1,t=0);
```

$$eq1 := [3a_1, 3a_2] = [2b_1 + b_2, -8b_1 - 2b_2]$$

which is equivalent to the system of equations
$$\begin{cases} 3a_1 = 2b_1 + b_2 \\ 3a_2 = -2(4b_1 + b_2) \end{cases}.$$

```
> with(student):
sys1:=equate(lhs(eq1),rhs(eq1));
```

$$sys1 := \{3a_2 = -8b_1 - 2b_2, 3a_1 = 2b_1 + b_2\}$$

Similarly, substituting $t = \pi/2$ results in

```
> eq2:=eval(step1,t=Pi/2);
```

$$eq2 := [3b_1, 3b_2] = [-2a_1 - 1 - a_2, 8a_1 + 2a_2]$$

which is equivalent to the system of equations

$$\begin{cases} 3b_1 = -1 - 2a_1 - a_2 \\ 3b_2 = -2(-4a_1 - a_2) \end{cases}.$$

```
> sys2:=equate(lhs(eq2),rhs(eq2));
```

$$sys2 := \{3b_1 = -2a_1 - 1 - a_2, 3b_2 = 8a_1 + 2a_2\}$$

We now use `solve` to solve these four equations for a_1, a_2, b_1, and b_2.

```
> vals:=solve('union'(sys1,sys2));
```

$$vals := \left\{ a_2 = \frac{8}{5}, b_1 = \frac{-3}{5}, a_1 = \frac{-2}{5}, b_2 = 0 \right\}$$

and substitute into $\mathbf{X}_p$ to obtain a particular solution to the nonhomogeneous system.

```
> xp:=subs(vals,eval(xp));
```

$$xp := \left[-\tfrac{2}{5}\sin(3t) - \tfrac{3}{5}\cos(3t), \tfrac{8}{5}\sin(3t)cr \right]$$

A general solution to the nonhomogeneous system is then given by $\mathbf{X} = \mathbf{X}_h + \mathbf{X}_p$.

```
> x:=evalm(xh+xp);
```

$$x := \left[\cos(2t)c_1 + \sin(2t)c_2 - \frac{2}{5}\sin(3t) - \frac{3}{5}\cos(3t), \right.$$
$$(-2\cos(2t) - 2\sin(2t))c_1 + (-2\sin(2t) + 2\cos(2t))c_2$$
$$\left. + \frac{8}{5}\sin(3t) \right]$$

To solve the initial-value problem, we apply the initial condition and solve for the unknown constants.

```
> step2:=eval(eval(x),t=0);
```

$$step2 := \left[c_1 - \frac{3}{5}, -2c_1 + 2c_2 \right]$$

```
> step3:=equate(eval(eval(x),t=0),[0,1]);
```

$$step3 := \left\{ -2c_1 + 2c_2 = 1, c_1 - \frac{3}{5} = 0 \right\}$$

```
> cvals:=solve(step3);
```

$$cvals := \left\{ c_2 = \frac{11}{10}, c_1 = \frac{3}{5} \right\}$$

We obtain the solution by substituting these values back into the general solution.

```
> x:=subs(cvals,eval(x));
```

$$x := \left[\frac{3}{5}\cos(2t) + \frac{11}{10}\sin(2t) - \frac{2}{5}\sin(3t) - \frac{3}{5}\cos(3t), \right.$$

$$\left. \cos(2t) - \frac{17}{5}\sin(2t) + \frac{8}{5}\sin(3t) \right]$$

We confirm this result by graphing $x(t)$ and $y(t)$ as well as parametrically.

```
> plot(x,t=0..4*Pi);
```

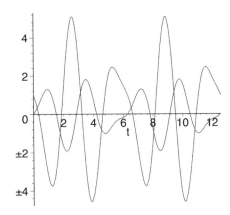

```
> plot([
       3/5*cos(2*t)+11/10*sin(2*t)-2/5*sin(3*t)-3/5*cos(3*t),
       cos(2*t)-17/5*sin(2*t)+8/5*sin(3*t),t=0..4*Pi]);
```

Finally, we note that dsolve is able to find a general solution of the nonhomogeneous system.

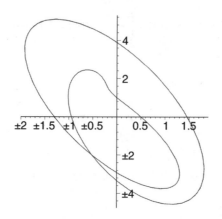

```
> x:='x':
dsolve({diff(x(t),t)=2*x(t)+y(t)+sin(3*t),
            diff(y(t),t)=-8*x(t)-2*y(t)},{x(t),y(t)});
```

$$\{y(t) = -4\sin(2t)\_C1 + \_C2\cos(2t) - \sin(2t)\_C2 + \frac{8}{5}\sin(3t),$$

$$x(t) =$$

$$\_C1\cos(2t) + \sin(2t)\_C1 + \frac{1}{2}\sin(2t)\_C2 - \frac{3}{5}\cos(3t) - \frac{2}{5}\sin(3t)$$

$$\}$$

as well as solve the initial-value problem.

```
> dsolve({diff(x(t),t)=2*x(t)+y(t)+sin(3*t),
         diff(y(t),t)=-8*x(t)-2*y(t),x(0)=0,y(0)=1},
         {x(t),y(t)});
```

$$\{y(t) = \cos(2t) - \frac{17}{5}\sin(2t) + \frac{8}{5}\sin(3t),$$

$$x(t) = \frac{3}{5}\cos(2t)\frac{11}{10}\sin(2t) - \frac{2}{5}\sin(3t) - \frac{3}{5}\cos(3t)\}$$

On the other hand, after defining $\mathbf{F}(t) = \begin{pmatrix} \sin 3t \\ 0 \end{pmatrix}$,

```
> F:=vector([sin(3*t),0]);
```

$$F := [\sin(3t), 0]$$

we can use `matrixDE` to find a fundamental matrix for the corresponding homogeneous system as well as a particular solution to the nonhomogeneous equation. The results of entering

```
> simplify(matrixDE(A,F,t));
```

$$\left[\left[\left|\begin{array}{cc} \cos(2t) & \sin(2t) \\ -2\cos(2t) - 2\sin(2t) & -2\sin(2t) + 2\cos(2t) \end{array}\right|\right.\right.,$$

$$\left.\left.\left[-\frac{2}{5}\sin(3t) - \frac{3}{5}\cos(3t), \frac{32}{5}\sin(t)\cos(t)^2 - \frac{8}{5}\sin(t)\right]\right]\right]$$

show us that a fundamental matrix for the corresponding homogeneous system is

$$\Phi(t) = \begin{pmatrix} \cos 2t & \sin 2t \\ -2\cos 2t - 2\sin 2t & -2\sin 2t + 2\cos 2t \end{pmatrix} \text{ and a particular solution to the}$$

nonhomogeneous system is $X_p(t) = \begin{pmatrix} -\frac{2}{5}\sin 3t - \frac{3}{5}\cos 3t \\ \frac{32}{5}\sin t \cos^2 t - \frac{8}{5}\sin 2 \end{pmatrix}.$

■

Variation of Parameters

In much the same way that we derived the method of variation of parameters for solving higher order differential equations, we assume that a particular solution of the nonhomogeneous system can be expressed in the form

$$X_p(t) = \Phi(t)V(t) \quad \text{where } V(t) = \begin{pmatrix} v_1(t) \\ v_2(t) \\ \vdots \\ v_n(t) \end{pmatrix}.$$

Notice that $X_p' = \Phi(t)V'(t) + \Phi'(t)V(t)$. Then if $X' = AX + F(t)$, we have

$$\Phi(t)V'(t) + \Phi'(t)V(t) = A\Phi(t)V(t) + F(t).$$

However, the fundamental matrix $\Phi(t)$ satisfies $X' = AX$, so $\Phi'(t) = A\Phi(t)$. Hence,

$$\Phi(t)V'(t) + A\Phi(t)V(t) = A\Phi(t)V(t) + F(t)$$

$$\Phi(t)V'(t) = F(t).$$

Multiplying both sides of this equation by $\Phi^{-1}(t)$ yields

$$\Phi^{-1}(t)\Phi(t)\mathbf{V}'(t) = \Phi^{-1}(t)\mathbf{F}(t)$$

$$\mathbf{V}'(t) = \Phi^{-1}(t)\mathbf{F}(t).$$

Therefore, $\mathbf{V}(t) = \int \Phi^{-1}(t)\mathbf{F}(t)dt$, so a particular solution of the nonhomogeneous system is

$$\mathbf{X}_p(t) = \Phi(t) \int \Phi^{-1}(t)\,\mathbf{F}(t)\,dt,$$

and a general solution of the system is

$$\mathbf{X}(t) = \Phi(t)\,\mathbf{C} + \mathbf{X}_p(t) = \Phi(t)\,\mathbf{C} + \Phi(t) \int \Phi^{-1}(t)\,\mathbf{F}(t)\,dt.$$

EXAMPLE 2: Solve $\mathbf{X}' = \begin{pmatrix} -5 & 3 \\ 2 & -10 \end{pmatrix} \mathbf{X} + \begin{pmatrix} e^{-2t} \\ 1 \end{pmatrix}$.

SOLUTION: In order to apply variation of parameters, we first calculate a fundamental matrix for the associated homogeneous system $\mathbf{X}' = \begin{pmatrix} -5 & 3 \\ 2 & -10 \end{pmatrix} \mathbf{X}$.

Using `dsolve`, we find a general solution of the corresponding homogeneous system

```
> homsol:=dsolve({D(x)(t)=-5*x(t)+3*y(t),
            D(y)(t)=2*x(t)-10*y(t)},{x(t),y(t)});
```

homsol := {$x(t) =$

$\dfrac{1}{7}\_C1e^{-11t} + \dfrac{6}{7}\_C1e^{(-4t)} + \dfrac{3}{7}\_C2e^{(-4t)} - \dfrac{3}{7}\_C2e^{(-11t)}$,

$y(t) =$

$\dfrac{2}{7}\_C1e^{(-4t)} - \dfrac{2}{7}\_C1e^{(-11t)} + \dfrac{6}{7}\_C2e^{(-11t)} + \dfrac{1}{7}\_C2e^{(-4t)}$}

or we can use `matrixDE` to find a fundamental matrix directly.

```
> with(DEtools):
A:=array([[-5,3],[2,-10]]):
homsol:=matrixDE(A,t);
```

$$homsol := \left[\left[\begin{array}{cc} \mathbf{e}^{(-4t)} & \mathbf{e}^{(-11t)} \\ \frac{1}{3}\mathbf{e}^{(-4t)} & -2\mathbf{e}^{(-11t)} \end{array} \right], [0,0] \right]$$

Then, a general solution of this system is $\mathbf{X} = \left(\begin{array}{cc} e^{-4t} & e^{-11t} \\ \frac{1}{3}e^{-4t} & -2e^{-11t} \end{array} \right) \left(\begin{array}{c} c_1 \\ c_2 \end{array} \right)$, so a

fundamental matrix is given by $\mathbf{\Phi}(t) = \left(\begin{array}{cc} e^{-4t} & e^{-11t} \\ \frac{1}{3}e^{-4t} & -2e^{-11t} \end{array} \right)$.

> `Phi:=homsol[1]:`

We find $\mathbf{\Phi}^{-1}(t)$ with `inverse`.

> `with(linalg):`

`inversePhi:=inverse(Phi);`

$$inversePhi := \left[\begin{array}{cc} \frac{6}{7}\frac{1}{\mathbf{e}^{(-4t)}} & \frac{3}{7}\frac{1}{\mathbf{e}^{(-4t)}} \\ \frac{1}{7}\frac{1}{\mathbf{e}^{(-11t)}} & -\frac{3}{7}\frac{1}{\mathbf{e}^{(-11t)}} \end{array} \right]$$

We now compute $\mathbf{\Phi}^{-1}(t)\mathbf{F}(t)$

> `f:=array([[exp(-2*t)],[1]]);`

$$f := \left[\begin{array}{c} \mathbf{e}^{(-2t)} \\ 1 \end{array} \right]$$

> `step1:=simplify(evalm(inversePhi&*f ));`

$$step1 := \left[\begin{array}{c} \frac{6}{7}\mathbf{e}^{(2t)} + \frac{3}{7}\mathbf{e}^{(4t)} \\ \frac{1}{7}\mathbf{e}^{(9t)} - \frac{3}{7}\mathbf{e}^{(11t)} \end{array} \right]$$

and integrate the result to obtain $\mathbf{V}(t) = \int \mathbf{\Phi}^{-1}(t)\mathbf{F}(t)dt$.

> `v:=map(int,step1,t);`

$$v := \left[\begin{array}{c} \frac{3}{7}\mathbf{e}^{(2t)} + \frac{3}{28}\mathbf{e}^{(4t)} \\ \frac{1}{63}\mathbf{e}^{(9t)} - \frac{3}{77}\mathbf{e}^{(11t)} \end{array} \right]$$

By variation of parameters, we have the particular solution $\mathbf{X}_p(t) = \mathbf{\Phi}(t) \int \mathbf{\Phi}^{-1}(t)\mathbf{F}(t)dt$.

```
> xp:=simplify(evalm(Phi&*v));
```

$$xp := \begin{vmatrix} \frac{1}{396}\mathbf{e}^{(-2t)}\left(176 - 27\mathbf{e}^{(2t)}\right) \\ \frac{1}{396}\mathbf{e}^{(-2t)}\left(44 + 45\mathbf{e}^{(2t)}\right) \end{vmatrix}$$

Therefore, a general solution is given by $\mathbf{X}(t) = \mathbf{\Phi}(t)\mathbf{C} + \mathbf{X}_p(t)$.

```
> x:=simplify(evalm(Phi&*array([[c[1]],[c[2]]])+xp));
```

$$x := \begin{vmatrix} \mathbf{e}^{(-4t)}c_1 + \mathbf{e}^{(-11t)}c_2 + \frac{4}{9}\mathbf{e}^{(-2t)} + \frac{3}{44} \\ \frac{1}{3}\mathbf{e}^{(-4t)}c_1 - 2\mathbf{e}^{(-11t)}c_2 + \frac{1}{9}\mathbf{e}^{(-2t)} + \frac{5}{44} \end{vmatrix}$$

We graph $x(t)$ and $y(t)$ for several values of c_1 and c_2.

```
> cvals:=seq(-0.5+0.5*i,i=0..2):
xs:={seq(seq(x[1,1],c[1]=cvals),c[2]=cvals)};
plot(xs,t=0..2,view=[0..3,-1.5..1.5]);
> ys:={seq(seq(x[2,1],c[1]=cvals),c[2]=cvals)};
plot(ys,t=0..2,view=[0..3,-1.5..1.5]);
```

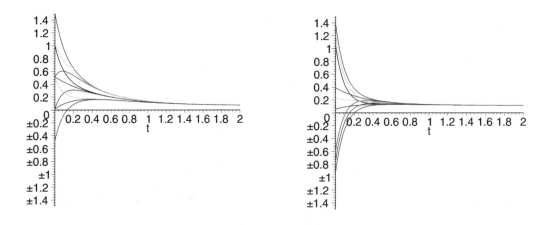

We see that `dsolve` can find a general solution of the nonhomogeneous system as well.

```
> x:='x':
```

```
gensol:=dsolve({D(x)(t)=-5*x(t)+3*y(t)+exp(-2*t),
                D(y)(t)=2*x(t)-10*y(t)+1},{x(t),y(t)});
```

$gensol := \{x(t) = \frac{1}{7}\_C1e^{(-11t)} + \frac{6}{7}\_C1e^{(-4t)} + \frac{3}{7}\_C2e^{-4t)}$

$-\frac{3}{7}\_C2e^{(-11t)} + \frac{3}{44} + \frac{4}{9}e^{(-2t)}, y(t) = \frac{2}{7}\_C1e^{(-4t)}$

$-\frac{2}{7}\_C1e^{(-11t)} + \frac{6}{7}\_C2e^{(-11t)} + \frac{1}{7}\_C2e^{(-4t)} + \frac{5}{44}$

$+\frac{1}{9}e^{(-2t)}\}$

To see that these functions satisfy the system, we define $x(t)$ and $y(t)$ to be these functions with `assign`.

> `assign(gensol):`

Then, we subtract the right side of each equation from the left side, evaluate, and simplify the results.

> `simplify(diff(x(t),t)-(-5*x(t)+3*y(t)+exp(-2*t)));`

$$0$$

> `simplify(diff(y(t),t)-(2*x(t)-10*y(t)+1));`

$$0$$

The result is 0; these functions form a general solution of the system.

■

EXAMPLE 3: Solve $\begin{cases} x' = -3x + 2y + e^{-t}\sec t \\ y' = -10x + 5y + e^{-t}\csc t \\ x(\pi/4) = 3, y(\pi/4) = -1 \end{cases}$

SOLUTION: To implement the method of variation of parameters, we proceed in the same manner as in Example 2. First, we find a fundamental matrix for the corresponding homogeneous system with `matrixDE`.

```
> A:=array([[-3,2],[-10,5]]):
with(DEtools):
homsol:=matrixDE(A,t);
```

$$homsol :=$$

$$\left[\left[\begin{matrix} \mathbf{e}^{(t)}\cos(2t) & \mathbf{e}^{(t)}\sin(2t) \\ 2\mathbf{e}^{(t)}\cos(2t) - \mathbf{e}^{(t)}\sin(2t) & 2\mathbf{e}^{(t)}\sin(2t) + \mathbf{e}^{(t)}\cos(2t) \end{matrix}\right], [0,0]\right]$$

This result means that a fundamental matrix is given by

$$\Phi(t) = \left(\begin{matrix} e^t\cos 2t & e^t\sin 2t \\ e^t(2\cos 2t - \sin 2t) & e^t(2\sin 2t + \cos 2t) \end{matrix}\right).$$

```
> Phi:=homsol[1]:
```

Next, we compute $\mathbf{X}_p(t) = \Phi(t)\int\Phi^{-1}(t)\mathbf{F}(t)dt$.

```
> f:=array([[exp(t)*sec(t)],[exp(t)*csc(t)]]);
```

$$f := \left|\begin{matrix} \mathbf{e}^t\sec(t) \\ \mathbf{e}^t\csc(t) \end{matrix}\right|$$

```
> with(linalg):
step1:=simplify(evalm(inverse(Phi)&*f));
```

$$step1 := \left|\begin{matrix} \dfrac{4\cos(t)\sin(t)-1}{\cos(t)} \\ -\dfrac{4\sin(t)\cos(t)^2-2\sin(t)-\cos(t)}{\cos(t)\sin(t)} \end{matrix}\right|$$

```
> step2:=map(int,step1,t);
```

$$step2 := \left|\begin{matrix} -4\cos(t) - \ln(\sec(t)+\tan(t)) \\ -4\sin(t) + 2\ln(\sec(t)+\tan(t)) + \ln(\csc(t)-\cot(t)) \end{matrix}\right|$$

```
> xp:=simplify(evalm(Phi&*step2));
```

$xp :=$

$$\left[-4e^t \cos(2t) \cos(t) - e^t \cos(2t) \ln\left(\frac{1+\sin(t)}{\cos(t)}\right)\right.$$

$$- 4e^t \sin(2t) \sin(t) + 2e^t \sin(2t) \ln\left(\frac{1+\sin(t)}{\cos(t)}\right)$$

$$\left.+e^t \sin(2t) \ln\left(\frac{\sin(t)}{\cos(t)+1}\right)\right]$$

$$\left[-8e^t \cos(2t) \cos(t) + 4e^t \sin(2t) \cos(t)\right.$$

$$+ 5e^t \sin(2t) \ln\left(\frac{1+\sin(t)}{\cos(t)}\right) - 8e^t \sin(2t) \sin(t)$$

$$+ 2e^t \sin(2t) \ln\left(\frac{\sin(t)}{\cos(t)+1}\right) - 4e^t \cos(2t)\sin(t)$$

$$\left.+e^t \cos(2t) \ln\left(\frac{\sin(t)}{\cos(t)+1}\right)\right]$$

Finally, we form a general solution of the nonhomogeneous system.

```
> x:=simplify(evalm(Phi&*array([[c[1]],[c[2]]])+xp));
```

$x :=$

$$\left[e^t \cos(2t)c_1 + e^t \sin(2t)c_2 - 4e^t \cos(2t)\cos(t)\right.$$

$$- e^t \cos(2t) \ln\left(\frac{1+\sin(t)}{\cos(t)}\right) - 4e^t \sin(2t) \sin(t)$$

$$\left.+2e^t \sin(2t) \ln\left(\frac{1+\sin(t)}{\cos(t)}\right) + e^t \sin(2t) \ln\left(\frac{\sin(t)}{\cos(t)+1}\right)\right]$$

$$\left[2e^t \cos(2t)c_1 - c_1 e^t \sin(2t) + 2e^t \sin(2t)c_2 + c_2 e^t \cos(2t)\right.$$

$$- 8e^t \cos(2t) \cos(t) + 4e^t \sin(2t) \cos(t)$$

$$+ 5e^t \sin(2t) \ln\left(\frac{1+\sin(t)}{\cos(t)}\right) - 8e^t \sin(2t) \sin(t)$$

$$+ 2e^t \sin(2t) \ln\left(\frac{\sin(t)}{\cos(t)+1}\right) - 4e^t \cos(2t) \sin(t)$$

$$\left.+e^t \cos(2t) \ln\left(\frac{\sin(t)}{\cos(t)+1}\right)\right]$$

To solve the initial-value problem, we substitute $t = \pi/4$ into the general solution

```
> initcond:=eval(eval(x),t=Pi/4);
```

$$initcond :=$$

$$\left[e^{\left(\frac{1}{4}\pi\right)} c_2 - 2e^{\left(\frac{1}{4}\pi\right)} \sqrt{2} + 2e^{\left(\frac{1}{4}\pi\right)} \ln\left(\left(1 + \frac{1}{2}\sqrt{2}\right)\sqrt{2} \right) . \right.$$

$$\left. + e^{\left(\frac{1}{4}\pi\right)} \ln\left(\frac{1}{2} \frac{\sqrt{2}}{1 + \frac{1}{2}\sqrt{2}} \right) \right]$$

$$\left[-c_1 e^{\left(\frac{1}{4}\pi\right)} + 2e^{\left(\frac{1}{4}\pi\right)} c_2 - 2e^{\left(\frac{1}{4}\pi\right)} \sqrt{2} . \right.$$

$$\left. + 5e^{\left(\frac{1}{4}\pi\right)} \ln\left(\left(1 + \frac{1}{2}\sqrt{2}\right)\sqrt{2} \right) + 2e^{\left(\frac{1}{4}\pi\right)} \ln\left(\frac{1}{2} \frac{\sqrt{2}}{1 + \frac{1}{2}\sqrt{2}} \right) \right]$$

and solve $\begin{cases} x(\pi/4) = 3 \\ y(\pi/4) = -1 \end{cases}$ for c_1 and c_2.

```
> with(student):
sys:=equate(initcond,[[3],[-1]]);
```

$$sys := \left\{ -c_1 e^{\left(\frac{1}{4}\pi\right)} + 2e^{\left(\frac{1}{4}\pi\right)} c_2 - 2e^{\left(\frac{1}{4}\pi\right)} \sqrt{2} \right.$$

$$+ 5e^{\left(\frac{1}{4}\pi\right)} \ln\left(\left(1 + \frac{1}{2}\sqrt{2}\right)\sqrt{2} \right) + 2e^{\left(\frac{1}{4}\pi\right)} \ln\left(\frac{1}{2} \frac{\sqrt{2}}{1 + \frac{1}{2}\sqrt{2}} \right) = 1,$$

$$e^{\left(\frac{1}{4}\pi\right)} c_2 - 2e^{\left(\frac{1}{4}\pi\right)} \sqrt{2} + 2e^{\left(\frac{1}{4}\pi\right)} \ln\left(\left(1 + \frac{1}{2}\sqrt{2}\right)\sqrt{2} \right)$$

$$\left. + e^{\left(\frac{1}{4}\pi\right)} \ln\left(\frac{1}{2} \frac{\sqrt{2}}{1 + \frac{1}{2}\sqrt{2}} \right) = 3 \right\}$$

```
> cvals:=solve(sys,{c[1],c[2]});
```

$$cvals := \left\{ c_2 = -\frac{1}{2}\left(-4e^{\left(\frac{1}{4}\pi\right)} \sqrt{2} + 2e^{\left(\frac{1}{4}\pi\right)} \ln\left(2\left(1 + \frac{1}{2}\sqrt{2}\right)^2 \right) \right. \right.$$

$$\left. + e^{\left(\frac{1}{4}\pi\right)} \ln\left(\frac{1}{2} \frac{1}{\left(1 + \frac{1}{2}\sqrt{2}\right)^2} \right) - 6 \right) / e^{\left(\frac{1}{4}\pi\right)},$$

$$c_1 = \frac{1}{2} \frac{4e^{\left(\frac{1}{4}\pi\right)} \sqrt{2} + e^{\left(\frac{1}{4}\pi\right)} \ln\left(2\left(1 + \frac{1}{2}\sqrt{2}\right)^2 \right) + 14}{e^{\left(\frac{1}{4}\pi\right)}} \right\}$$

This result is rather complicated, so we compute more meaningful approximations with `evalf`.

```
> numcvals:=evalf(cvals);
```

$$numcvals := \{ c_1 = 6.901367605, c_2 = 3.314867920 \}$$

The solution to the initial-value problem is obtained by substiuting these numbers back into the general solution.

```
> x:=evalf(subs(cvals,eval(x)));
```

$$x :=$$

$$\Big[6.901367605 e^t \cos(2.t) + 3.314867920 e^t \sin(2.t).$$

$$- 4.e^t \cos(2.t) \cos(t) - 1.e^t \cos(2.t) \ln\left(\frac{1.+\sin(t)}{\cos(t)}\right)$$

$$- 4.e^t \sin(2.t) \sin(t) + 2.e^t \sin(2.t) \ln\left(\frac{1.+\sin(t)}{\cos(t)}\right)$$

$$+ e^t \sin(2.t) \ln\left(\frac{\sin(t)}{\cos(t)+1.}\right) \Big]$$

$$\Big[17.11760313 e^t \cos(2.t) - 0.271631766 e^t \sin(2.t).$$

$$- 8.e^t \cos(2.t) \cos(t) + 4.e^t \sin(2.t) \cos(t)$$

$$+ 5.e^t \sin(2.t) \ln\left(\frac{1.+\sin(t)}{\cos(t)}\right) - 8.e^t \sin(2.t) \sin(t)$$

$$+ 2.e^t \sin(2.t) \ln\left(\frac{\sin(t)}{\cos(t)+1.}\right) - 4.e^t \cos(2.t) \sin(t)$$

$$+ e^t \cos(2.t) \ln\left(\frac{\sin(t)}{\cos(t)+1.}\right) \Big]$$

We confirm that the initial conditions are satisfied by graphing $x(t)$ and $y(t)$ on the interval $(0, \pi/2)$.

```
> plot(convert(x,set),t=0..Pi/2);
```

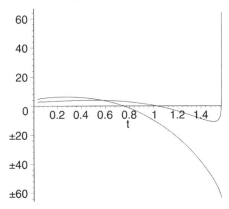

The Matrix Exponential

Definition
Matrix Exponential

> If $\mathbf{A}t$ is $n \times n$, the **matrix exponential** is defined by
>
> $$e^{\mathbf{A}t} = \exp(\mathbf{A}t) = \mathbf{I} + \mathbf{A}t + \frac{1}{2!}\mathbf{A}^2 t^2 + \cdots = \sum_{n=0}^{\infty}\frac{1}{n!}\mathbf{A}^n t^n.$$

Use the command `exponential`, which is contained in the **linalg** package, to compute the matrix exponential of a matrix.

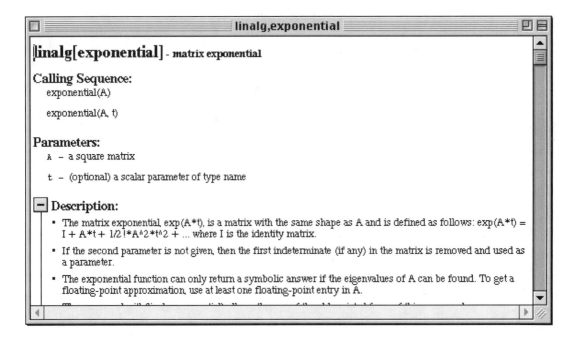

linalg[exponential] - matrix exponential

Calling Sequence:
  exponential(A)

  exponential(A, t)

Parameters:
  A – a square matrix

  t – (optional) a scalar parameter of type name

Description:
- The matrix exponential, exp(A*t), is a matrix with the same shape as A and is defined as follows: exp(A*t) = I + A*t + 1/2!*A^2*t^2 + ... where I is the identity matrix.
- If the second parameter is not given, then the first indeterminate (if any) in the matrix is removed and used as a parameter.
- The exponential function can only return a symbolic answer if the eigenvalues of A can be found. To get a floating-point approximation, use at least one floating-point entry in A.

For example, here we use `exponential` to calculate $e^{\left(\begin{smallmatrix} t & 0 \\ -2t & 3t \end{smallmatrix}\right)}$.

```
> with(linalg):
exponential([[t,0],[-2*t,3*t]]);
```

$$\begin{bmatrix} \mathbf{e}^t & 0 \\ -\mathbf{e}^{(3t)} + \mathbf{e}^t & \mathbf{e}^{(3t)} \end{bmatrix}$$

Differentiating the series $e^{\mathbf{A}t} = \exp(\mathbf{A}t) = \mathbf{I} + \mathbf{A}t + \frac{1}{2!}\mathbf{A}^2 t^2 + \cdots = \sum_{n=0}^{\infty} \frac{1}{n!}\mathbf{A}^n t^n$ term by term shows us that $\frac{d}{dt}e^{\mathbf{A}t} = \mathbf{A} \cdot e^{\mathbf{A}t}$, so $\mathbf{x} = e^{\mathbf{A}t}$ satisfies the differential equation $\mathbf{x}' = \mathbf{A}\mathbf{x}$. We can use the matrix exponential $e^{\mathbf{A}t}$ to solve the linear first-order system $\mathbf{X}' = \mathbf{A}\mathbf{X} + \mathbf{F}(t)$ in much the same way that we used the integrating factor $e^{\int p(x)\,dx}$ to solve the linear first-order equation $y' + p(x)y = q(x)$. Moreover, $e^{\mathbf{A}t}$ is a fundamental matrix for the homogeneous system; $\left(e^{\mathbf{A}t}\right)^{-1} = e^{-\mathbf{A}t}$; and if $t = 0$, $e^{\mathbf{A}t} = \mathbf{I}$.

To solve the system $\mathbf{X}' = \mathbf{A}\mathbf{X} + \mathbf{F}(t)$, we first rewrite it as $\mathbf{X}' - \mathbf{A}\mathbf{X} = \mathbf{F}(t)$. Now, multiply both sides of the equation by $e^{-\mathbf{A}t}$ and integrate:

$$e^{-\mathbf{A}t}(\mathbf{X}' - \mathbf{A}\mathbf{X}) = e^{-\mathbf{A}t}\mathbf{F}(t)$$

$$\frac{d}{dt}\left(e^{-\mathbf{A}t}\mathbf{X}\right) = e^{-\mathbf{A}t}\mathbf{F}(t)$$

$$e^{-\mathbf{A}t}\mathbf{X} = \int e^{-\mathbf{A}t}\mathbf{F}(t)\,dt + \mathbf{C}$$

$$\mathbf{X} = e^{\mathbf{A}t}\int e^{-\mathbf{A}t}\mathbf{F}(t)\,dt + e^{\mathbf{A}t}\mathbf{C},$$

where $\mathbf{C}$ is an arbitrary constant *vector*.

If, in addition, we are given the initial condition $\mathbf{X}(t_0) = \mathbf{X}_0$, the solution to the initial-value problem is

$$\mathbf{X} = \int_{t_0}^{t} e^{\mathbf{A}(t-s)}\mathbf{F}(s)\,ds + e^{\mathbf{A}(t-t_0)}\mathbf{X}_0.$$

EXAMPLE 4: Solve $\mathbf{X}' = \begin{pmatrix} 2 & 5 \\ -4 & -2 \end{pmatrix}\mathbf{X} + \begin{pmatrix} \cos 4t \\ \sin 4t \end{pmatrix}$.

SOLUTION: Here, $\mathbf{A} = \begin{pmatrix} 2 & 5 \\ -4 & -2 \end{pmatrix}$. We compute $e^{\mathbf{A}t}$ with `exponential`.

```
> A:=array([[2,5],[-4,-2]]):
with(linalg):
expa:=exponential(A*t);
```

$$expa := \begin{vmatrix} \cos(4t) + \frac{1}{2}\sin(4t) & \frac{5}{4}\sin(4t) \\ -\sin(4t) & \cos(4t) - \frac{1}{2}\sin(4t) \end{vmatrix}$$

A general solution of the system is then given by $\mathbf{X} = e^{\mathbf{A}t}\int e^{-\mathbf{A}t}\mathbf{F}(t)\,dt + e^{\mathbf{A}t}\mathbf{C}$. First, we compute $e^{-\mathbf{A}t}\mathbf{F}(t)$, $\int e^{-\mathbf{A}t}\mathbf{F}(t)\,dt$, and $e^{\mathbf{A}t}\int e^{-\mathbf{A}t}\mathbf{F}(t)\,dt$.

```
> invexpa:=simplify(inverse(expa));
```

$$invexpa := \left[\begin{array}{cc} \cos(4t) - \frac{1}{2}\sin(4t) & -\frac{5}{4}\sin(4t) \\ \sin(4t) & \cos(4t) + \frac{1}{2}\sin(4t) \end{array} \right]$$

```
> f:=array([[cos(4*t)],[sin(4*t)]]):
step1:=simplify(evalm(invexpa&*f ));
```

$$step1 := \left[\begin{array}{c} \frac{9}{4}\cos(4t)^2 - \frac{1}{2}\sin(4t)\cos(4t) - \frac{5}{4} \\ 2\sin(4t)\cos(4t) + \frac{1}{2} - \frac{1}{2}\cos(4t)^2 \end{array} \right]$$

```
> step2:=map(int,step1,t);
```

$$step2 := \left[\begin{array}{c} \frac{9}{32}\sin(4t)\cos(4t) - \frac{1}{8}t - \frac{1}{16}\sin(4t)^2 \\ \frac{1}{4}\sin(4t)^2 + \frac{1}{4}t - \frac{1}{16}\sin(4t)\cos(4t) \end{array} \right]$$

```
> step3:=simplify(evalm(expa&*step2));
```

$$step3 := \left[\begin{array}{c} -\frac{1}{8}\cos(4t)t + \frac{1}{4}\sin(4t)t + \frac{9}{32}\sin(4t) \\ \frac{1}{4}\cos(4t)t - \frac{1}{16}\sin(4t) \end{array} \right]$$

Then, we form our general solution.

```
> gensol:=simplify(
      evalm(step3+expa&*array([[c[1]],[c[2]]])));
```

$$gensol :=$$

$$\left[-\frac{1}{8}\cos(4t)t + \frac{1}{4}\sin(4t)t + \frac{9}{32}\sin(4t) + c_1 \cos(4t) + \frac{1}{2}\sin(4t)c_1 \right.$$

$$\left. + \frac{5}{4}\sin(4t)c_2 \right]$$

$$\left[\frac{1}{4}\cos(4t)t - \frac{1}{16}\sin(4t) - \sin(4t)c_1 + c_2 \cos(4t) - \frac{1}{2}\sin(4t)c_2 \right]$$

To graph the solution parametrically for various values of the arbitrary constant, we define `step1` to be a list of the form $[x(t), y(t), t = -3\pi..3\pi]$, the form required to graph $\begin{cases} x = x(t) \\ y = y(t) \end{cases}$, $-3\pi \leq t \leq 3\pi$, parametrically with `plot`.

```
> step1:=[gensol[1,1],gensol[2,1],t=-3*Pi..3*Pi]:
```

We then use `plot` to graph $\begin{cases} x = x(t) \\ y = y(t) \end{cases}$, $-3\pi \leq t \leq 3\pi$, parametrically for $c_1 = -1$,

0, and 1 and $c_2 = -1$, 0, and 1. The resulting array of graphs is named `graphs`

```
> graphs:=array([seq(
         [seq(plot(step1,color=BLACK),c[1]=-1..1)],
              c[2]=-1..1)]):
```

and displayed using `display` together with the option `insequence = true`.

```
> with(plots):
display(graphs,insequence=true);
```

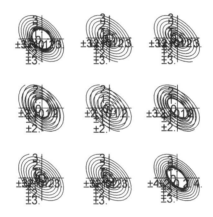

If a system of differential equations contains derivatives of order greater than one, we can often rewrite it as a system of first-order equations. (In the next section, we will also see that Laplace transforms can often be used to solve systems of this type.)

EXAMPLE 5: Solve $\begin{cases} -2\frac{d^2x}{dt^2} - 2\frac{dy}{dt} = 0 \\ \frac{d^2y}{dt^2} + y - \frac{dx}{dt} = \cos t \\ x(0) = 2, \ x'(0) = 1, \ y(0) = 1, \ y'(0) = 2 \end{cases}$.

SOLUTION: To rewrite the system as a system of first-order equations, we let $z = dx/dt$ and $w = dy/dt$. Then, $dz/dt = d^2x/dt^2$ and $dw/dt = d^2y/dt^2$. Substituting into the first equation we have $-2dz/dt - 2w = 0$, so $dz/dt = -w$. Similarly, substituting into the second equation yields $dw/dt + y - z = \cos t$, so $dw/dt = -y + z + \cos t$. Therefore, the original system is equivalent to the system of first-order equations

$$\begin{cases} dx/dt = z \\ dy/dt = w \\ dz/dt = -w \\ dw/dt = -y + z + \cos t \\ x(0) = 2, \ y(0) = 1, \ z(0) = 1, \ w(0) = 2 \end{cases}$$

In matrix form, the initial-value problem is equivalent to $\mathbf{X}' = \mathbf{A}\mathbf{X} + \mathbf{F}(t)$,

$$\mathbf{X}(0) = \begin{pmatrix} 2 \\ 1 \\ 1 \\ 2 \end{pmatrix}, \text{ where } \mathbf{X} = \begin{pmatrix} x \\ y \\ z \\ w \end{pmatrix}, \ \mathbf{A} = \begin{pmatrix} 0 & 0 & 1 & 0 \\ 0 & 0 & 0 & 1 \\ 0 & 0 & 0 & -1 \\ 0 & -1 & 1 & 0 \end{pmatrix}, \text{ and } \mathbf{F}(t) = \begin{pmatrix} 0 \\ 0 \\ 0 \\ \cos t \end{pmatrix}.$$

Using the exponential matrix, the solution to the initial-value problem is given by

$$\mathbf{X} = \int_0^t e^{\mathbf{A}(t-s)} \mathbf{F}(s)\,ds + e^{\mathbf{A}t} \begin{pmatrix} 2 \\ 1 \\ 1 \\ 2 \end{pmatrix}.$$ First, we define $\mathbf{A}$ and then use `exponential` to compute $e^{\mathbf{A}t}$.

```
> A:=array([[0,0,1,0],[0,0,0,1],[0,0,0,-1],[0,-1,1,0]]):
with(linalg):
expa:=exponential(A*t);
```

$$expa :=$$
$$\left[1, -\frac{1}{4}\sin(\sqrt{2}t)\sqrt{2} + \frac{1}{2}t, \frac{1}{2}t + \frac{1}{4}\sin(\sqrt{2}t)\sqrt{2}, \frac{1}{2}\cos(\sqrt{2}t) - \frac{1}{2}\right]$$
$$\left[0, \frac{1}{2} + \frac{1}{2}\cos(\sqrt{2}t), -\frac{1}{2}\cos(\sqrt{2}t) + \frac{1}{2}, \frac{1}{2}\sin(\sqrt{2}t)\sqrt{2}\right]$$
$$\left[0, -\frac{1}{2}\cos(\sqrt{2}t) + \frac{1}{2}, \frac{1}{2} + \frac{1}{2}\cos(\sqrt{2}t), -\frac{1}{2}\sin(\sqrt{2}t)\sqrt{2}\right]$$
$$\left[0, -\frac{1}{2}\sin(\sqrt{2}t)\sqrt{2}, \frac{1}{2}\sin(\sqrt{2}t)\sqrt{2}, \cos(\sqrt{2}t)\right]$$

The matrix $e^{\mathbf{A}(t-s)}$ is obtained by replacing each occurrence of t in $e^{\mathbf{A}t}$ by $t - s$.

```
> expats:=eval(eval(expa),t=t-s):
```

Next, we compute $e^{\mathbf{A}(t-s)}\mathbf{F}(s)$ and integrate the result.

```
> f:=array([[0],[0],[0],[cos(t)]]);
```

$$f := \begin{bmatrix} 0 \\ 0 \\ 0 \\ \cos(t) \end{bmatrix}$$

```
> toint:=simplify(evalm(expats&*f ));
```

$$toint := \begin{bmatrix} \frac{1}{2}\cos(\sqrt{2}(t-s))\cos(t) - \frac{1}{2}\cos(t) \\ \frac{1}{2}\sin(\sqrt{2}(t-s))\sqrt{2}\cos(t) \\ -\frac{1}{2}\sin(\sqrt{2}(t-s))\sqrt{2}\cos(t) \\ \cos(\sqrt{2}(t-s))\cos(t) \end{bmatrix}$$

```
> step2:=simplify(map(int,toint,s=0..t));
```

$$step2 := \begin{bmatrix} -\frac{1}{2}\cos(t)t + \frac{1}{4}\sin(\sqrt{2}t)\sqrt{2}\cos(t) \\ \frac{1}{2}\cos(t) - \frac{1}{2}\cos(\sqrt{2}t)\cos(t) \\ -\frac{1}{2}\cos(t)t + \frac{1}{2}\cos(\sqrt{2}t)\cos(t) \\ \frac{1}{2}\sin(\sqrt{2}t)\sqrt{2}\cos(t) \end{bmatrix}$$

Finally, we form the solution to the initial-value problem. Note that the first and second rows correspond to x and y, respectively.

```
> x0:=array([[2],[1],[1],[2]]):
  sol:=simplify(evalm(step2+expa&*x0));
```

$$sol := \begin{bmatrix} -\frac{1}{2}\cos(t)t + \frac{1}{4}\sin(\sqrt{2}t)\sqrt{2}\cos(t) + 1 + t + \cos(\sqrt{2}t) \\ \frac{1}{2}\cos(t) - \frac{1}{2}\cos(\sqrt{2}t)\cos(t) + 1 + \sin(\sqrt{2}t)\sqrt{2} \\ -\frac{1}{2}\cos(t) + \frac{1}{2}\cos(\sqrt{2}t)\cos(t) + 1 - \sin(\sqrt{2}t)\sqrt{2} \\ \frac{1}{2}\sin(\sqrt{2}t)\sqrt{2}\cos(t) + 2\cos(\sqrt{2}t) \end{bmatrix}$$

We confirm that the initial conditions are satisfied by graphing $x(t)$, $y(t)$, and

$$\begin{cases} x = x(t) \\ y = y(t) \end{cases}.$$

```
> plot([sol[1,1],sol[2,1]],t=0..12,view=[0..12,-2..10]);
```

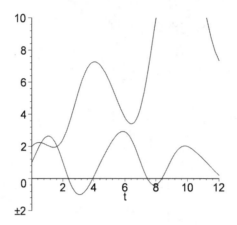

```
> plot([sol[1,1],sol[2,1],t=0..12],view=[0..12,-6..6]);
```

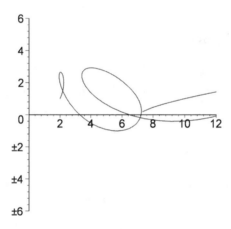

Last, we remark that `dsolve` is able to solve this initial-value problem as well.

```
> sol:=dsolve({-2*diff(x(t),t$2)-2*diff(y(t),t)=0,
       diff(y(t),t$2)+y(t)-diff(x(t),t)=cos(t),
          x(0)=2,D(x)(0)=1,y(0)=1,D(y)(0)=2},{x(t),y(t)});
```

$$sol := \{x(t) = 1 + t + \frac{1}{2}\sin(\sqrt{2}t)\sqrt{2} + \cos(\sqrt{2}t) - \sin(t),$$

$$y(t) = -\cos(\sqrt{2}t) + 1 + \sin(\sqrt{2}t)\sqrt{2} + \cos(t)\}$$

■

7.5 Laplace Transforms

In many cases, Laplace transforms can be used to solve initial-value problems that involve a system of linear differential equations. This method is applied in much the same way as it was in solving initial-value problems involving higher order differential equations. In the case of systems of differential equations, however, a system of algebraic equations is obtained after taking the Laplace transform of each equation. After solving the algebraic system for the Laplace transform of each of the unknown functions, the inverse Laplace transform is used to find each unknown function in the solution of the system. As in Chapter 6, be sure to load the **inttrans** package by entering `with(inttrans)` before completing any example in this section.

EXAMPLE 1: Solve $\mathbf{X}' = \begin{pmatrix} 0 & 1 \\ 1 & 0 \end{pmatrix} \mathbf{X} + \begin{pmatrix} \sin t \\ 2\cos t \end{pmatrix}$ subject to $\mathbf{X}(0) = \begin{pmatrix} 2 \\ 0 \end{pmatrix}$.

SOLUTION: Let $X(t) = \begin{pmatrix} x(t) \\ y(t) \end{pmatrix}$. Then, we can rewrite this problem as

$$\begin{cases} x' = y + \sin t \\ y' = x + 2\cos t \\ x(0) = 2, y(0) = 0 \end{cases}.$$

```
> Eqs:={diff(x(t),t)=y(t)+sin(t),
          diff(y(t),t)=x(t)+2*cos(t)};
```

$$Eqs := \{\frac{\partial}{\partial t} x(t) = y(t) + \sin(t), \frac{\partial}{\partial t} y(t) = x(t) + 2\cos(t)\}$$

Taking the Laplace transform of both sides of each equation yields the system

$$\begin{cases} sX(s) - x(0) = Y(s) + \frac{1}{s^2+1} \\ sY(s) - y(0) = X(s) + \frac{2s}{s^2+1} \end{cases}$$

```
> step_1:=laplace(Eqs,t,s);
```

$$step\_1 := \{s \, \mathrm{laplace}(x(t), t, s) - x(0) = \mathrm{laplace}(y(t), t, s) + \frac{1}{s^2 + 1},$$

$$s \, \mathrm{laplace}(y(t), t, s) - y(0) = \mathrm{laplace}(x(t), t, s) + 2\frac{s}{s^2 + 1}\}$$

and applying the initial condition results in

$$\begin{cases} sX(s) - Y(s) = \frac{1}{s^2+1} + 2 \\ -X(s) + sY(s) = \frac{2s}{s^2+1} \end{cases}.$$

```
> step_2:=subs({x(0)=2,y(0)=0},step_1);
```

$step\_2 := \{s \ \text{laplace}(y(t), t, s) = \text{laplace}(x(t), t, s) + 2\dfrac{s}{s^2+1},$

$\qquad s \ \text{laplace}(x(t), t, s) - 2 = \text{laplace}(y(t), t, s) + \dfrac{1}{s^2+1}\}$

We now use `solve` to solve this system of algebraic equations for $X(s)$ and $Y(s)$.

```
> step_3:=solve(step_2,
            {laplace(x(t),t,s),laplace(y(t),t,s)});
```

$step\_3 := \{$

$\qquad \text{laplace}(x(t), t, s) = \dfrac{(2s^2+5)s}{(s^2+1)(s^2-1)}, \ \ \text{laplace}(y(t), t, s) = \dfrac{4s^2+3}{s^4-1}$

$\qquad \}$

We find $x(t)$ and $y(t)$ with `invlaplace`.

```
> Sol:=invlaplace(step_3,s,t);
```

$Sol :=$

$$\left\{ x(t) = \frac{7}{4}e^t + \frac{7}{4}e^{(-t)} - \frac{3}{2}\cos(t), y(t) = \frac{7}{4}e^t - \frac{7}{4}e^{(-t)} + \frac{1}{2}\sin(t) \right\}$$

Last, we graph $x(t)$, $y(t)$, and $\begin{cases} x = x(t) \\ y = y(t) \end{cases}$.

```
> assign(Sol):
plot([x(t),y(t)],t=-3..3,view=[-2..3,-1..4]);
plot([x(t),y(t),t=-3..3],view=[0..10,-5..5]);
```

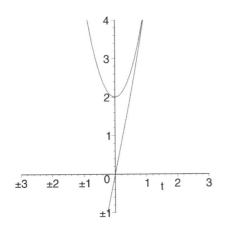

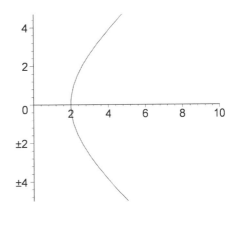

In some cases, systems that involve higher order differential equations can be solved with Laplace transforms.

EXAMPLE 2: Solve $\begin{cases} x'' = -2x - 4y - \cos t \\ y'' = -x - 2y + \sin t \\ x(0) = 0,\ x'(0) = 0,\ y(0) = 0,\ y'(0) = 0 \end{cases}$.

SOLUTION: After defining the system of equations in `sys`, we take the Laplace transform of each equation.

```
> x:='x':y:='y':
sys:={diff(x(t),t$2)=-2*x(t)-4*y(t)-cos(t),
            diff(y(t),t$2)=-x(t)-2*y(t)+sin(t)};
```

$$sys := \left\{ \frac{\partial^2}{\partial t^2} x(t) = -2x(t) - 4y(t) - \cos(t), \right.$$

$$\left. \frac{\partial^2}{\partial t^2} y(t) = -x(t) - 2y(t) + \sin(t) \right\}$$

```
> step_1:=laplace(sys,t,s);
```

$$step\_1 := \{s(s\ \text{laplace}(x(t), t, s) - x(0)) - D(x)(0) =$$

$$- 2\ \text{laplace}(x(t), t, s) - 4\ \text{laplace}(y(t), t, s) - \frac{s}{s^2 + 1},$$

$$s(s\ \text{laplace}(y(t), t, s) - y(0)) - D(y)(0) =$$

$$- \text{laplace}(x(t), t, s) - 2\ \text{laplace}(y(t), t, s) + \frac{1}{s^2 + 1}\}$$

We then apply the initial conditions and solve the resulting *algebraic* system of equations for $X(s)$ and $Y(s)$.

```
> step_2:=subs({x(0)=0,D(x)(0)=0,y(0)=0,D(y)(0)=0},step_1);
```

$$step\_2 := \{\ \text{laplace}(x(t), t, s)s^2 =$$

$$- 2\ \text{laplace}(x(t), t, s) - 4\ \text{laplace}(y(t), t, s) - \frac{s}{s^2 + 1},$$

$$\text{laplace}(y(t), t, s)s^2 =$$

$$\text{laplace}(x(t), t, s) - 2\ \text{laplace}(y(t), t, s) + \frac{1}{s^2 + 1}\}$$

```
> step_3:=solve(step_2,
       {laplace(x(t),t,s),laplace(y(t),t,s)});
```

$$step\_3 := \left\{\ \text{laplace}(x(t), t, s) = -\frac{s^3 + 2s + 4}{(s^2 + 1)(s^2 + 4)s^2},\right.$$

$$\left.\text{laplace}(y(t), t, s) = \frac{s + 2 + s^2}{s^2(s^4 + 5s^2 + 4)}\right\}$$

Finally, we use `invlaplace` to compute $x(t) = \mathscr{L}^{-1}\{X(s)\}$ and $y(t) = \mathscr{L}^{-1}\{Y(s)\}$.

```
> Sol:=invlaplace(step_3,s,t);
```

$$Sol := \{$$

$$y(t) = \frac{1}{2}t + \frac{1}{4} + \frac{1}{12}\cos(2t) - \frac{1}{12}\sin(2t) - \frac{1}{3}\cos(t) - \frac{1}{3}\sin(t),$$

$$x(t) = -t - \frac{1}{2} + \frac{1}{3}\cos(t) + \frac{4}{3}\sin(t) + \frac{1}{6}\cos(2t) - \frac{1}{6}\sin(2t)\}$$

We see that the initial conditions are satisfied by graphing $x(t)$, $y(t)$, and

$$\begin{cases} x = x(t) \\ y = y(t) \end{cases}.$$

```
> assign(Sol):
plot([x(t),y(t)],t=-Pi..4*Pi,
            view=[-Pi..4*Pi,-3*Pi..2*Pi]);
plot([x(t),y(t),t=-Pi..4*Pi],view=[-12..3,-3..12]);
```

As we saw in Chapter 6, Laplace transform methods are especially useful in solving problems that involve piecewise-defined, periodic, or impulse functions.

EXAMPLE 3: Solve $\begin{cases} x' = y + 3\delta(t - \pi) \\ y' = -x + 6\delta(t - 2\pi) \\ x(0) = 1, \ y(0) = -1 \end{cases}.$

SOLUTION: We proceed in exactly the same manner as in Examples 1 and 2. After defining the system of equations,

```
> x:='x':y:='y':
sys:={diff(x(t),t)=y(t)+3*Dirac(t-Pi),
            diff(y(t),t)=-x(t)+6*Dirac(t-2*Pi)};
```

$sys :=$

$$\left\{ \frac{\partial}{\partial t}x(t) = y(t) + 3 \ \mathrm{Dirac}(t - \pi), \frac{\partial}{\partial t}y(t) = -x(t) + 6 \ \mathrm{Dirac}(t - 2\pi) \right\}$$

we use `laplace` to compute the Laplace transform of each equation

```
> step_1:=laplace(sys,t,s);
```

$$step\_1 := \{s \ \mathrm{laplace}(x(t), t, s) - x(0) = \mathrm{laplace}(y(t), t, s) + 3e^{(-\pi s)},$$

$$s \ \mathrm{laplace}(y(t), t, s) - y(0) = -\mathrm{laplace}(x(t), t, s) + 6e^{(-2\pi s)}\}$$

and apply the initial conditions.

```
> step_2:=subs({x(0)=1,y(0)=-1},step_1);
```

$$step\_2 := \{s \ \mathrm{laplace}(x(t), t, s) - 1 = \mathrm{laplace}(y(t), t, s) + 3e^{(-\pi s)},$$

$$s \ \mathrm{laplace}(y(t), t, s) + 1 = -\mathrm{laplace}(x(t), t, s) + 6e^{(-2\pi s)}\}$$

We then solve the resulting algebraic system of equations for $X(s)$ and $Y(s)$ and use `invlaplace` to compute $x(t)$ and $y(t)$.

```
> step_3:=solve(step_2,
              {laplace(x(t),t,s),laplace(y(t),t,s)});
```

$$step\_3 := \left\{ \mathrm{laplace}(x(t), t, s) = \frac{s + 3se^{(-\pi s)} - 1 + 6e^{(-2\pi s)}}{s^2 + 1}, \right.$$

$$\left. \mathrm{laplace}(y(t), t, s) = -\frac{s - 6se^{(-2\pi s)} + 1 + 3e^{(-\pi s)}}{s^2 + 1} \right\}$$

```
> Sol:=invlaplace(step_3,s,t);
```

$Sol := \{x(t) = \cos(t) - 3 \ \mathrm{Heaviside}(t - \pi) \cos(t) - \sin(t)$

$\quad + 6 \ \mathrm{Heaviside}(t - 2\pi) \sin(t), y(t) = -\cos(t)$

$\quad + 6 \ \mathrm{Heaviside}(t - 2\pi) \cos(t), -\sin(t) + 3 \ \mathrm{Heaviside}(t - \pi) \sin(t)\}$

We see that the initial conditions are satisfied by graphing $x(t)$, $y(t)$, and

$$\begin{cases} x = x(t) \\ y = y(t) \end{cases} .$$

```
> assign(Sol):
plot([x(t),y(t)],t=0..4*Pi,view=[0..4*Pi,-2*Pi..2*Pi]);
plot([x(t),y(t),t=0..4*Pi],view=[-2*Pi..2*Pi,-2*Pi..2*Pi]);
```

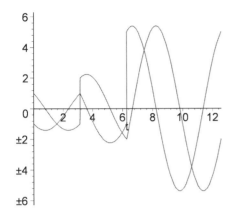
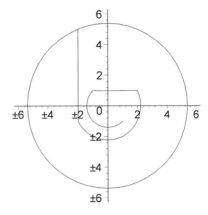

We see that dsolve is also able to solve this initial-value problem.

```
> x:='x':y:='y':
dsolve({diff(x(t),t)=y(t)+3*Dirac(t-Pi),
          diff(y(t),t)=-x(t)+6*Dirac(t-2*Pi),
              x(0)=1,y(0)=-1},{x(t),y(t)});
```

$$\{x(t) = \cos(t) - 3\,\text{Heaviside}(t - \pi)\cos(t) - \sin(t)$$
$$+ 6\,\text{Heaviside}(t - 2\pi)\sin(t), y(t) = -\cos(t)$$
$$+ 6\,\text{Heaviside}(t - 2\pi)\cos(t) - \sin(t) + 3\,\text{Heaviside}(t - \pi)\sin(t)\}$$

∎

EXAMPLE 4: Solve $\begin{cases} x' = -17y + f(t) \\ y' = \frac{1}{4}x - y - f(t) \\ x(0) = 0, y(0) = 0 \end{cases}$, where $f(t) = \begin{cases} 1 + t, & 0 \le t < 1 \\ 3, & t \ge 1 \end{cases}$.

SOLUTION: We define and graph f using piecewise and plot.

```
> f:=piecewise(t>=0 and t<1,1+t,t>=1,3);
```

$$f := \begin{cases} 1+t & -t \leq 0 \text{ and } t-1 < 0 \\ 3 & 1 \leq t \end{cases}$$

```
> plot(f,t=0..4);
```

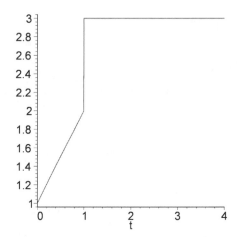

To apply the method of Laplace transforms, we rewrite f in terms of the unit step function:

$$f(t) = \begin{cases} 1+t, 0 \leq t < 1 \\ 3, t \geq 1 \end{cases} = (1+t)(1 - \mathcal{U}(t-1)) + 3\mathcal{U}(t-1).$$

```
> fh:=convert(f,Heaviside);
```

$$fh := \text{Heaviside}(t) + 2\,\text{Heaviside}(t-1) + t\,\text{Heaviside}(t)$$
$$- t\,\text{Heaviside}(t-1)$$

Next, we define the system of equations.

```
sys:={diff(x(t),t)=-17*y(t)+fh,
       diff(y(t),t)=1/4*x(t)-y(t)-fh};
```

$$sys := \left\{ \frac{\partial}{\partial t} y(t) = \frac{1}{4}x(t) - y(t) - \text{Heaviside}(t) - 2\,\text{Heaviside}(t-1) \right.$$

$$- t\,\text{Heaviside}(t) + t\,\text{Heaviside}(t-1), \frac{\partial}{\partial t}x(t) = -17y(t)$$

$$+ \text{Heaviside}(t) + 2\,\text{Heaviside}(t-1) + t\,\text{Heaviside}(t)$$

$$\left. - t\,\text{Heaviside}(t-1) \right\}$$

Then, we compute the Laplace transform of each equation,

```
> step1:=laplace(sys,t,s);
```

$$step1 := \left\{ s \; \text{laplace}(x(t), t, s) - x(0) = . \right.$$

$$- 17 \; \text{laplace}(t(t), t, s) + \frac{1}{s} + \frac{e^{(-s)}}{s} + \frac{1}{s^2} - \frac{e^{(-s)}}{s^2},$$

$$s \; \text{laplace}(t(t), t, s) - y(0) =$$

$$\left. \frac{1}{4} \text{laplace}(x(t), t, s) - \text{laplace}(y(t), t, s) - \frac{1}{s} - \frac{e^{(-s)}}{s} - \frac{1}{s^2} + \frac{e^{(-s)}}{s^2} \right\}$$

apply the initial conditions,

```
> step2:=subs({x(0)=0,y(0)=0},step1);
```

$$step2 := \left\{ s \; \text{laplace}(x(t), t, s) = . \right.$$

$$- 17 \; \text{laplace}(y(t), t, s) + \frac{1}{s} + \frac{e^{(-s)}}{s} + \frac{1}{s^2} - \frac{e^{(-s)}}{s^2},$$

$$s \; \text{laplace}(y(t), t, s) =$$

$$\left. \frac{1}{4} \text{laplace}(x(t), t, s) - \text{laplace}(y(t), t, s) - \frac{1}{s} - \frac{e^{(-s)}}{s} - \frac{1}{s^2} + \frac{e^{(-s)}}{s^2} \right\}$$

and solve the resulting equations for $X(s)$ and $Y(s)$.

```
> step3:=solve(step2,{laplace(x(t),t,s),laplace(y(t),t,s)});
```

$$step3 := \{$$

$$\text{laplace}(x(t), t, s) = 4 \frac{18e^s - 18 + 19se^s + 17s + s^2e^s + s^2}{s^2 e^s (4s^2 + 17 + 4s)},$$

$$\text{laplace}(y(t), t, s) = - \frac{3se^s - s - e^s + 1 + 4s^2 e^s + 4s^2}{s^2 e^s (4s^2 + 17 + 4s)} \right\}$$

The solution is obtained with `invlaplace`.

```
> sol:=invlaplace(step3,s,t);
```

$$sol := \left\{ x(t) = \frac{72}{17}t + \frac{1004}{289} - \frac{1004}{289}\mathbf{e}^{\left(-\frac{1}{2}t\right)}\cos(2t) \right.$$

$$- \frac{1437}{578}\mathbf{e}^{\left(-\frac{1}{2}t\right)}\sin(2t) - \frac{72}{17}t\,\text{Heaviside}(t-1)$$

$$+ \frac{2668}{289}\text{Heaviside}(t-1)$$

$$- \frac{1444}{289}\text{Heaviside}(t-1)\mathbf{e}^{\left(-\frac{1}{2}t+\frac{1}{2}\right)}\cos(2t-2)$$

$$+ \frac{791}{578}\text{Heaviside}(t-1)\mathbf{e}^{\left(-\frac{1}{2}t+\frac{1}{2}\right)}\sin(2t-2), y(t) = \frac{55}{289}$$

$$+ \frac{55}{289}\mathbf{e}^{\left(-\frac{1}{2}t\right)}\cos(2t) - \frac{557}{1156}\mathbf{e}^{\left(-\frac{1}{2}t\right)}\sin(2t)$$

$$+ \frac{106}{289}\text{Heaviside}(t-1)$$

$$- \frac{89}{289}\text{Heaviside}(t-1)\mathbf{e}^{\left(-\frac{1}{2}t+\frac{1}{2}\right)}\cos(2t-2)$$

$$- \frac{633}{1156}\text{Heaviside}(t-1)\mathbf{e}^{\left(-\frac{1}{2}t+\frac{1}{2}\right)}\sin(2t-2) + \frac{1}{17}t$$

$$\left. - \frac{1}{17}t\,\text{Heaviside}(t-1) \right\}$$

Last, we confirm that the initial conditions are satisfied by graphing $x(t)$, $y(t)$, and

$$\begin{cases} x = x(t) \\ y = y(t) \end{cases}.$$

```
> assign(sol):
plot([x(t),y(t)],t=0..2);
plot([x(t),y(t),t=0..2]);
```

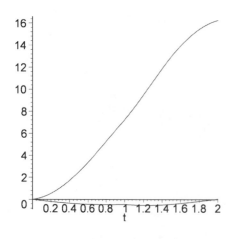
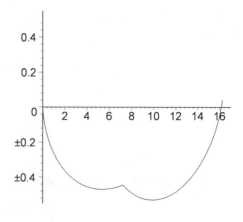

In this case, we obtain a slightly different, but equivalent, form when we use `dsolve` to solve the initial-value problem.

```
> x:='x': y:='y':
altsol:=dsolve({diff(x(t),t)=-17*y(t)+fh,
      diff(y(t),t)=1/4*x(t)-y(t)-fh,
         x(0)=0,y(0)=0},{x(t),y(t)});
```

$$altsol := \left\{ y(t) = -\frac{89}{289} \text{Heaviside}(t-1)e^{\left(-\frac{1}{2}t+\frac{1}{2}\right)} \cos(2t-2) \right.$$

$$-\frac{633}{1156} \text{Heaviside}(t-1)e^{\left(-\frac{1}{2}t+\frac{1}{2}\right)} \sin(2t-2)$$

$$- ovr117t \, \text{Heaviside}(t-1) + \frac{1}{17}t \, \text{Heaviside}(t) - \frac{55}{289} \text{Heaviside}(t)$$

$$+\frac{106}{289} \text{Heaviside}(t-1) + \frac{55}{289} e^{\left(-\frac{1}{2}t\right)} \cos(2t)\text{Heaviside}(t)$$

$$-\frac{557}{1156} e^{\left(-\frac{1}{2}t\right)} \sin(2t)\text{Heaviside}(t), x(t) =$$

$$-\frac{1444}{289} \text{Heaviside}(t-1)e^{\left(-\frac{1}{2}t+\frac{1}{2}\right)} \cos(2t-2)$$

$$+\frac{791}{578} \text{Heaviside}(t-1)e^{\left(-\frac{1}{2}t+\frac{1}{2}\right)} \sin(2t-2)$$

$$-\frac{72}{17}t \, \text{Heaviside}(t-1) + \frac{72}{17}t \, \text{Heaviside}(t) + \frac{1004}{289} \text{Heaviside}(t)$$

$$+\frac{2668}{289} \text{Heaviside}(t-1) - \frac{1004}{289} e^{\left(-\frac{1}{2}t\right)} \cos(2t)\text{Heaviside}(t)$$

$$\left. -\frac{1437}{578} e^{\left(-\frac{1}{2}t\right)} \sin(2t)\text{Heaviside}(t) \right\}$$

■

EXAMPLE 5: Solve $\begin{cases} x' + 2x + 3y = 0 \\ y' - x + 6y = f(t) \\ x(0) = 1, y(0) = 0 \end{cases}$, where $f(t) = \begin{cases} 0, 0 \leq t < 1 \\ 1, 1 \leq t < 2 \\ 2, 2 \leq t < 3 \end{cases}$ and $f(t) = f(t-3)$, $t \geq 3$.

SOLUTION: We begin by defining and graphing f.

```
> g:=t->piecewise(t>=0 and t<1,0,

            t>=1 and t<2,1,t>=2 and t<3,2):
f:=proc(t)

        if t>=0 and t<3 then g(t) else f(t-3) fi;

        end:
```

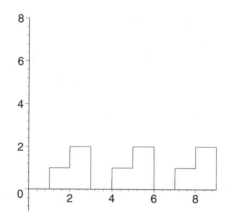

Alternatively, note that

$$f(t) = g(t)(\mathscr{U}(t) - \mathscr{U}(t-3)) + g(t-3)(\mathscr{U}(t-3) - \mathscr{U}(t-6)) + \cdots$$

$$= \sum_{n=0}^{\infty} g(t - 3n)(\mathscr{U}(t - 3n) - \mathscr{U}(t - 3(n+1))),$$

where $g(t) = \begin{cases} 0, 0 \le t < 1 \\ 1, 1 \le t < 2 \\ 2, 2 \le t < 3 \end{cases}$ Thus, the graph of f on the interval $[0, 3k]$, where k

represents a positive integer, is obtained by graphing

$$f_k(t) = \sum_{n=0}^{k-1} g(t - 3n)(\mathscr{U}(t - 3n) - \mathscr{U}(t - 3(n+1)))$$

on the interval $[0, 2k\pi]$. For convenience, we define `nth_term(n)` to be $g(t - 3n)(\mathscr{U}(t - 3n) - \mathscr{U}(t - 3(n+1)))$.

```
> nth_term:=n->g(t-3*n)*

        (Heaviside(t-3*n)-Heaviside(t-3*(n+1))):;
```

Then, to graph f on the interval $[0, 30]$, we first enter

```
> to_graph:=sum(nth_term(n),n=0..9):
```

and then

```
> plot(to_graph,t=-1..31,view=[-1..31,-11..21]);
```

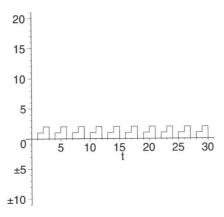

In Section 6.4 we learned that if f is a periodic function with period T, the Laplace transform of f is given by

$$\mathscr{L}\{f(t)\} = \frac{1}{1 - e^{-sT}} \int_0^T e^{-st} f(t) dt.$$

We use `int` and `simplify` to find $\mathscr{L}\{f(t)\}$, naming the result `lapf`.

```
> lapf:=simplify((int(exp(-s*t),t=1..2)+
          int(2*exp(-s*t),t=2..3))/(1-exp(-3*s)));
```

$$lapf := \frac{(1 + 2e^{(-s)})e^{(-s)}}{(e^{(-2s)} + e^{(-s)} + 1)s}$$

Alternatively,

$$\mathscr{L}\{f(t)\} = \mathscr{L}\left\{ \sum_{n=0}^{\infty} g(t - 3n)(\mathscr{U}(t - 3n) - \mathscr{U}(t - 3(n + 1))) \right\}$$

$$= \sum_{n=0}^{\infty} \mathscr{L}\{g(t - 3n)(\mathscr{U}(t - 3n) - \mathscr{U}(t - 3(n + 1)))\}.$$

```
> h:=convert(g(t),Heaviside);
```

$$h := \text{Heaviside}(t - 1) + \text{Heaviside}(t - 2) - 2\,\text{Heaviside}(t - 3)$$

```
> hn:=subs(t=t-3*n,h);
```

$$hn := \text{Heaviside}(t - 3n - 1) + \text{Heaviside}(t - 3n - 2)$$
$$- 2\text{Heaviside}(t - 3n - 3)$$

```
> termn:=simplify(hn*
         (Heaviside(t-3*n)-Heaviside(t-3*(n+1))));
```

$$termn := \text{Heaviside}(t - 3n - 1) + \text{Heaviside}(t - 3n - 2)$$
$$- 2\,\text{Heaviside}(t - 3n - 3)$$

```
> f:=t->sum(Heaviside(t-3*n-1)+
      Heaviside(t-3*n-2)-2*Heaviside(t-3*n-3),
         n=0..infinity);
```

$$f := t \rightarrow \sum_{n=0}^{\infty} (\text{Heaviside}(t - 3n - 1) + \text{Heaviside}(t - 3n - 2)$$
$$- 2\,\text{Heaviside}(t - 3n - 3))$$

```
> with(inttrans):
simplify(laplace(f(t),t,s));
```

$$\frac{(1 + 2e^{(-s)})e^{(-s)}}{(e^{(-2s)} + e^{(-s)} + 1)s}$$

After defining the system,

```
> sys:={diff(x(t),t)+2*x(t)+3*y(t)=0,
            diff(y(t),t)-x(t)+6*y(t)=f(t)}:
```

we take the Laplace transform of both sides of each equation,

```
> step1:=laplace(sys,t,s);
```

$$step1 := \{s\,\text{laplace}(y(t), t, s) - y(0) - \text{laplace}(x(t), t, s)$$
$$+ 6\,\text{laplace}(y(t), t, s) = \frac{\frac{e^{(-s)}}{s} + \frac{e^{(-2s)}}{s} - 2\frac{e^{(-3s)}}{s}}{1 - e^{(-3s)}},$$

$$\text{laplace}(x(t), t, s) - x(0) + 2 \text{ laplace}(x(t), t, s)$$

$$+3 \text{ laplace}(y(t), t, s) = 0\}$$

apply the initial conditions,

```
> step2:=subs({x(0)=1,y(0)=0},step1);
```

$$step2 := \{s \text{ laplace}(x(t), t, s) - 1 + \text{ laplace}(x(t), t, s)$$

$$+ 3 \text{ laplace}(y(t), t, s) = 0,$$

$$s \text{ laplace}(y(t), t, s) - \text{laplace}(x(t), t, s) + 6 \text{ laplace}(y(t), t, s) =$$

$$\frac{\frac{e^{(-s)}}{s} + \frac{e^{(-2s)}}{s} - 2\frac{e^{(-3s)}}{s}}{1 - e^{(-3s)}}\}$$

and solve for $X(s)$ and $Y(s)$.

```
> step3:=factor(solve(step2,
             {laplace(x(t),t,s),laplace(y(t),t,s)}));
```

$$step3 := \{ \text{ laplace}(x(t), t, s) = (-s^2 e^s + s^2 e^{(-3s)} e^s - 6se^s$$

$$+ 6se^{(-3s)} e^s + 3 + 3e^{(-2s)} e^s - 6e^{(-3s)} e^s)/$$

$$(se^s(s+5)(s+3)(-1 + e^{(-3s)})), \text{ laplace}(y(t), t, s)$$

$$= -(s + se^{(-2s)} e^s - 3se^{(-3s)} e^s + se^s + 2 + 2e^{(-2s)} e^s$$

$$- 4e^{(-3s)} e^s)/(se^s(s+5)(s+3)(-1 + e^{(-3s)}))\}$$

Note that `invlaplace` cannot be used to compute $\mathscr{L}^{-1}\{X(s)\}$ and $\mathscr{L}^{-1}\{Y(s)\}$. Note that

$$X(s) = \frac{3(2 + e^{-s})}{s(s+3)(s+5)(1 + e^{-s} + e^{-2s})} + \frac{s^2 + 6s - 6}{s(s+3)(s+5)}$$

and `invlaplace` quickly calculates $\mathscr{L}^{-1}\left\{\frac{s^2+6s-6}{s(s+3)(s+5)}\right\}$.

```
> invlaplace((s^2+6*s-6)/(s*(s+3)*(s+5)),s,t);
```

$$\frac{2}{5} + \frac{5}{2}e^{(-3t)} - \frac{11}{10}e^{(-5t)}$$

To calculate $\mathscr{L}^{-1}\left\{\frac{3(2+e^{-s})}{(1+e^{-2s}+e^{-s})s(s+3)(s+5)}\right\}$, we first rewrite the fraction:

$$\frac{3(2+e^{-s})}{(1+e^{-2s}+e^{-s})s(s+3)(s+5)} = \frac{2+e^{-s}}{1+e^{-s}+e^{-2s}} \cdot \frac{3}{s(s+3)(s+5)}$$

$$= \frac{2+e^{-s}}{1+e^{-s}+e^{-2s}} \cdot \frac{1-e^{-s}}{1-e^{-s}} \cdot \frac{3}{s(s+3)(s+5)}$$

$$= \left(\frac{2}{1-e^{-3s}} + \frac{e^{-s}}{1-e^{-3s}} + \frac{e^{-2s}}{1-e^{-3s}}\right) \cdot \frac{3}{s(s+3)(s+5)}$$

and then use the geometric series $\frac{1}{1-x} = \sum_{n=0}^{\infty} x^n$:

$$\left(\frac{2}{1-e^{-3s}} + \frac{e^{-s}}{1-e^{-3s}} + \frac{e^{-2s}}{1-e^{-3s}}\right) \frac{3}{s(s+3)(s+5)}$$

$$= \left(2\sum_{n=0}^{\infty} e^{-3ns} + \sum_{n=0}^{\infty} e^{-(3n+1)s} + \sum_{n=0}^{\infty} e^{-(3n+2)s}\right) \cdot \frac{3}{s(s+3)(s+5)}.$$

Notice that $\mathscr{L}^{-1}\left\{\frac{3}{s(s+3)(s+5)}\right\} = \frac{1}{5} + \frac{3}{10}e^{-5t} - \frac{1}{2}e^{-3t}$. We name this function $g(t)$ for later use.

```
> invlaplace(3/(s*(s+3)*(s+5)),s,t):
```

$$\frac{1}{5} + \frac{3}{10}e^{(-5t)} - \frac{1}{2}e^{(-3t)}$$

```
> g:=t->1/5+3/10*exp(-5*t)-1/2*exp(-3*t):
```

In Section 6.4 we learned that $\mathscr{L}^{-1}\{e^{-as}F(s)\} = f(t-a)\mathscr{U}(t-a)$. Thus,

$$\mathscr{L}^{-1}\left\{\left(2\sum_{n=0}^{\infty} e^{-3ns} + \sum_{n=0}^{\infty} e^{-(3n+1)s} + \sum_{n=0}^{\infty} e^{-(3n+2)s}\right) \cdot \frac{3}{s(s+3)(s+5)}\right\}$$

$$= 2\sum_{n=0}^{\infty} g(t-3n)\mathscr{U}(t-3n) + \sum_{n=0}^{\infty} g(t-(3n+1))\mathscr{U}(t-(3n+1))$$

$$+ \sum_{n=0}^{\infty} g(t-(3n+2))\mathscr{U}(t(3n+2))$$

and

$$x(t) = -\frac{2}{5} - \frac{11}{10}e^{-5t} + \frac{5}{2}e^{-3t} + 2\sum_{n=0}^{\infty} g(t-3n)\mathscr{U}(t-3n)$$

$$+ \sum_{n=0}^{\infty} g(t-(3n+1))\mathscr{U}(t-(3n+1)) + \sum_{n=0}^{\infty} g(t-(3n+2))\mathscr{U}(t-(3n+2)).$$

We find $y(t)$ in the same way. First, notice that

$$Y(s) = -\frac{(s+2)(2+e^{-s})}{s(s+3)(s+5)(1+e^{-s}+e^{-2s})} + \frac{3s+4}{s(s+3)(s+5)}$$

We use `invlaplace` to see that

$$\mathscr{L}^{-1}\left\{\frac{3s+4}{s(s+3)(s+5)}\right\} = \frac{4}{15} - \frac{11}{10}e^{-5t} + \frac{5}{6}e^{-3t}.$$

and

$$\mathscr{L}^{-1}\left\{-\frac{s+2}{s(s+3)(s+5)}\right\} = -\frac{2}{15} + \frac{3}{10}e^{-5t} - \frac{1}{6}e^{-3t}$$

We name the second result $h(t)$ for later use.

```
> invlaplace((3*s+4)/(s*(s+3)*(s+5)),s,t);
```

$$\frac{4}{15} + \frac{5}{6}e^{(-3t)} - \frac{11}{10}e^{(-5t)}$$

```
> invlaplace(-(s+2)/(s*(s+3)*(s+5)),s,t):
```

$$\frac{2}{15} - \frac{1}{6}e^{(-3t)} + \frac{3}{10}e^{(-5t)}$$

```
> h:=t->-2/15-1/6*exp(-3*t)+3/10*exp(-5*t):
```

To calculate $y(t) = \mathscr{L}^{-1}\{Y(s)\}$, we use the results we obtained when calculating $x(t) = \mathscr{L}^{-1}\{X(s)\}$.

$$\mathscr{L}^{-1}\left\{\frac{2+e^{-s}}{1+e^{-2s}+e^{-s}} \cdot \frac{-(s+2)}{s(s+3)(s+5)}\right\} =$$

$$\mathscr{L}^{-1}\left\{\left(2\sum_{n=0}^{\infty}e^{-3ns} + \sum_{n=0}^{\infty}e^{-(3n+1)s} + \sum_{n=0}^{\infty}e^{-(3n+2)s}\right) \cdot \frac{-(s+2)}{s(s+3)(s+5)}\right\}$$

$$= 2\sum_{n=0}^{\infty}h(t-3ns)\mathscr{U}(t-3n) + \sum_{n=0}^{\infty}h(t-(3n+1))\mathscr{U}(t-(3n+1))$$

$$+ \sum_{n=0}^{\infty}h(t-(3n+2))\mathscr{U}(t-(3n+2))$$

and

$$y(t) = \frac{4}{15} - \frac{11}{10}e^{-5t} + \frac{5}{6}e^{-3t} + 2\sum_{n=0}^{\infty} h(t-3n)\mathcal{U}(t-3n)$$

$$+ \sum_{n=0}^{\infty} h(t-(3n+1))\mathcal{U}(t-(3n+1)) + \sum_{n=0}^{\infty} h(t-(3n+2))\mathcal{U}(t-(3n+2)).$$

We use Maple to graph the solution on the interval [0,6].

```
> xapprox:=-2/5-11/10*exp(-5*t)+5/2*exp(-3*t)+
      sum(2*g(t-3*n)*Heaviside(t-3*n)+
         g(t-(3*n+1))*Heaviside(t-(3*n+1))+
            g(t-(3*n+2))*Heaviside(t-(3*n+2)),n=0..1);
> yapprox:=4/15-11/10*exp(-5*t)+5/6*exp(-3*t)+
      sum(2*h(t-3*n)*Heaviside(t-3*n)+
         h(t-(3*n+1))*Heaviside(t-(3*n+1))+
            h(t-(3*n+2))*Heaviside(t-(3*n+2)),n=0..1);
> plot([xapprox,yapprox],t=-1..6,view=[0..6,-3..3]);
plot([xapprox,yapprox,t=0.001..6]);
```

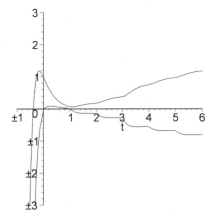
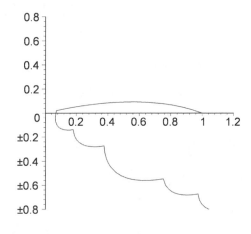

■

7.6 Numerical Methods

Because it may be difficult or even impossible to construct an explicit solution to some systems of differential equations, we now turn our attention to discussing some numerical methods that are used to construct numerical solutions to systems of differential equations.

Built-In Methods

As we saw in Section 7.2, we can use several of the commands contained in the **DEtools** package to provide us with numerical information about solutions to systems of differential equations. The following examples illustrate the use of some of the other commands contained in the **DEtools** package that can help us obtain information about a system of differential equations in addition to showing how to use dsolve together with the numeric option to generate a numerical solution of a system of differential equations.

EXAMPLE 1: Consider the nonlinear system of equations

$$\begin{cases} x' = \mu x + y - x(x^2 + y^2) \\ y' = \mu y - x - y(x^2 + y^2) \end{cases}.$$

(a) Graph the direction field associated with the system for $\mu = 2, 1, 1/4$, and $-1/2$. (b) For each value of μ in (a), approximate the solution that satisfies the initial conditions $x(0) = 0$ and $y(0) = 1/2$. Use each numerical solution to approximate $x(5)$ and $y(5)$.

SOLUTION: After loading the **DEtools** package, we define the function dfield. Given μ, dfield (μ) graphs the direction field associated with the system

$$\begin{cases} x' = \mu x + y - x(x^2 + y^2) \\ y' = \mu y - x - y(x^2 + y^2) \end{cases} \quad \text{on the rectangle } [-2, 2] \times [-2, 2].$$

```
> dfield:=mu->DEplot(

        [diff(x(t),t)=mu*x(t)+y(t)-x(t)*(x(t)^2+y(t)^2),

        diff(y(t),t)=mu*y(t)-x(t)-y(t)*(x(t)^2+y(t)^2)],

            [x(t),y(t)],t=-2..2,x=-2..2,y=-2..2,

            color=BLACK):
```

We use dfield to graph the direction field associated with the system for $\mu = 2, 1, 1/4$, and $-1/2$.

```
> pvfa:=dfield(2):
```

```
pvfb:=dfield(1):
```

```
pvfc:=dfield(1/4):
```

```
pvfd:=dfield(-1/2):
```

display together with insequence = true is used to display all four graphs together. The direction field indicates that the behavior of the solutions strongly depends on the value of μ.

```
> with(plots):
```

```
display(array([[pvfa,pvfb],[pvfc,pvfd]]),

        insequence=true);
```

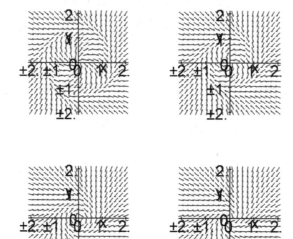

Now, we use dsolve together with the numeric option to generate a numerical approximation to the initial-value problem if $\mu = 2$.

```
> mu:=2:
```

```
sola:=dsolve({

        diff(x(t),t)=mu*x(t)+y(t)-x(t)*(x(t)^2+y(t)^2),

        diff(y(t),t)=mu*y(t)-x(t)-y(t)*(x(t)^2+y(t)^2),

            x(0)=0,y(0)=1/2},{x(t),y(t)},numeric):
```

We see that $x(5) \approx -1.35612$ and $y(5) \approx 0.40116$.

```
> sola(5);
```

$$[t = 5, x(t) = 1.356123696400036, y(t) = 0.4011589549456984]$$

Alternatively, we can use `dsolve` to return these values directly by including the `value` option in the dsolve command. For example, in the following, we use `value` to compute the value of $x(t)$ and $y(t)$ if $t = 2, 5$, and 7.

```
> solb:=dsolve( {
        diff(x(t),t)=mu*x(t)+y(t)-x(t)*(x(t)^2+y(t)^2),
        diff(y(t),t)=mu*y(t)-x(t)-y(t)*(x(t)^2+y(t)^2),
            x(0)=0,y(0)=1/2},{x(t),y(t)},numeric,
                value=array([2,5,7]));
                [t,x(t),y(t)]
```

$$solb := \begin{vmatrix} \begin{bmatrix} 2 & 1.284433554 & -0.5878307493 \\ 5 & -1.356123695 & 0.4011589566 \\ 7 & 0.9291194091 & 1.066178756 \end{bmatrix} \end{vmatrix}$$

We use `odeplot` to graph $x(t)$ and $y(t)$ for $0 \leq t \leq 10$. Notice that the solution appears to become periodic.

```
> with(plots):
```

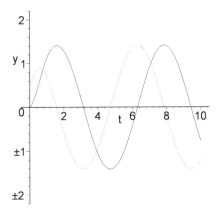

```
odeplot(sola, [[t,x(t)],[t,y(t)]],0..10,numpoints=100);
```

This is further confirmed by graphing the solution parametrically and showing it together with the direction field.

```
> ppa:=odeplot(sola,[x(t),y(t)],0..10,
            numpoints=100,thickness=2,color=BLACK):
display(pvfa,ppa);
```

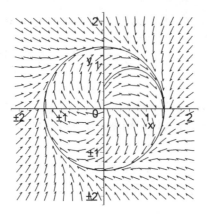

For the remaining values of μ, we define the function `numsol`. Given μ, `numsol` (μ) generates a numerical solution to the initial-value problem.

```
> mu:='mu':
numsol:=mu->dsolve({
        diff(x(t),t)=mu*x(t)+y(t)-x(t)*(x(t)^2+y(t)^2),
        diff(y(t),t)=mu*y(t)-x(t)-y(t)*(x(t)^2+y(t)^2),
            x(0)=0,y(0)=1/2},{x(t),y(t)},numeric):
```

We use `numsol` to solve each initial-value problem. Note that Maple does not display any output because we have included a colon at the end of each command.

```
> solb:=numsol(1):
solc:=numsol(1/4):
sold:=numsol(-1/2):
```

As before, we approximate the value of each solution if $t = 5$.

```
> [solb(5),solc(5),sold(5)];
```

$$[[t = 5, \mathrm{x}(t) = -0.9588589679007017, \mathrm{y}(t) = 0.2836428695421192],$$

$$[t = 5, \mathrm{x}(t) = -0.4794621386293579, \mathrm{y}(t) = 0.1418310847540857],$$

$$[t = 5, \mathrm{x}(t) = -0.03217070862099356, \mathrm{y}(t) = 0.009516514390043265]]$$

For each numerical solution, we use `odeplot` to graph $x(t)$ and $y(t)$ for $0 \le t \le 10$. Notice that the solutions corresponding to positive values of μ appear to become periodic whereas the solution corresponding to the negative value of μ appears to tend toward zero.

```
> pb:=odeplot(solb,[[t,x(t)],[t,y(t)]],0..10,numpoints=100):
      pc:=odeplot(solc,[[t,x(t)],[t,y(t)]],0..10,numpoints=100):
      pd:=odeplot(sold,[[t,x(t)],[t,y(t)]],0..10,numpoints=100):
> display(array([pb,pc,pd]),insequence=true);
```

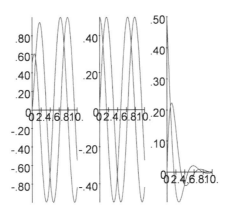

These results are further confirmed when we graph each solution parametrically and display the graphs with the direction fields generated in (a).

```
> ppb:=odeplot(solb,[x(t),y(t)],0..10,numpoints=100,
            thickness=2,color=BLACK):
ppc:=odeplot(solc,[x(t),y(t)],0..10,numpoints=100,
            thickness=2,color=BLACK):
ppd:=odeplot(sold,[x(t),y(t)],0..10,numpoints=100,
            thickness=2,color=BLACK):
> graphb:=display(pvfb,ppb):
      graphc:=display(pvfc,ppc):
      graphd:=display(pvfd,ppd):
> display(array([graphb,graphc,graphd]),insequence=true);
```

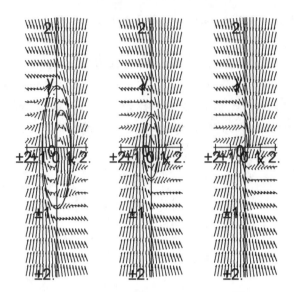

In the cases corresponding to the positive values of μ, we see in the direction field that all solutions appear to tend to a closed curve. Can we find the curve in each case If $\mu = 2$, we see that the solution that satisfies $x(0) = 0$ and $y(0) \approx 1.41$ will be periodic. Similarly, if $\mu = 1/4$, we see that the solution that satisfies $x(0) = 0$ and $y(0) \approx 0.482$ will be periodic. On the other hand, if $\mu = 1$, we need not approximate the solution. From the graph, we see that the solution that satisfies $x(0) = 0$ and $y(0) = 1$ will be periodic. It is relatively easy to verify that the solution that

satisfies these initial conditions is $\begin{cases} x = \sin t \\ y = \cos t \end{cases}$.

■

APPLICATION

Controlling the Spread of a Disease

If a person becomes immune to a disease after recovering from it and births and deaths in the population are not taken into account, then the percentage of persons susceptible to becoming infected with the disease, $S(t)$; the percentage of people in the population infected with the disease, $I(t)$; and the percentage of the population recovered and immune to the disease, $R(t)$, can be modeled by the system

$$\begin{cases} S' = -\lambda SI \\ I' = \lambda SI - \gamma I \\ R' = \gamma I \\ S(0) = S_0, \ I(0) = I_0, \ R(0) = 0 \end{cases}$$

Because $S(t) + I(t) + R(t) = 1$, once we know S and I, we can compute R with $R(t) = 1 - S(t) - I(t)$.

This model is called an **SIR model without vital dynamics** because once a person has had the disease the person becomes immune to the disease and because births and deaths are not taken into consideration. This model might be used to model diseases that are **epidemic** in a population: the diseases that persist in a population for short periods of time (less than 1 year). Such diseases typically include influenza, measles, rubella, and chickenpox.

If $S_0 < \gamma/\lambda$, $I'(0) = \lambda S_0 I_0 - \gamma I_0 < \lambda \frac{\gamma}{\lambda} I_0 - \gamma I_0 = 0$. Thus, the rate of infection immediately begins to decrease; the disease dies out. On the other hand, if $S_0 > \gamma/\lambda$, $I'(0) = \lambda S_0 I_0 - \gamma I_0 > \lambda \frac{\gamma}{\lambda} I_0 - \gamma I_0 = 0$, so the rate of infection first increases; an epidemic results.

Although we cannot find explicit formulas for S, I, and R as functions of t, we can, for example, solve for I in terms of S. The equation $\frac{dI}{dS} = -\frac{(\lambda S - \gamma)I}{\lambda S I} = -1 + \frac{\rho}{S}$, $\rho = \gamma/\lambda$ is separable:

$$\frac{dI}{dS} = -1 + \frac{\rho}{S} \Rightarrow dI = \left(-1 + \frac{\rho}{S}\right)dS \Rightarrow I = -S + \rho \ln S + C$$

and applying the initial condition results in

$$I_0 = -S_0 + \rho \ln S_0 + C \Rightarrow C = I_0 + S_0 - \rho \ln S_0$$

so $I = -S + \rho \ln S + I_0 + S_0 - \rho \ln S_0 \Rightarrow I + S - \rho \ln S = I_0 + S_0 - \rho \ln S_0$.

When diseases persist in a population for long periods of time, births and deaths must be taken into consideration. If a person becomes immune to a disease after recovering from it and births and deaths in the population are taken into account, then the percentage of persons susceptible to becoming infected with the disease, $S(t)$; and the percentage of people in the population infected with the disease, $I(t)$, can be modeled by the system

$$\begin{cases} S' = -\lambda S I + \mu - \mu S \\ I' = \lambda S I - \gamma I - \mu I \\ S(0) = S_0, \ I(0) = I_0 \end{cases}$$

This model is called an **SIR model with vital dynamics** because once a person has had the disease the person becomes immune to the disease and because births and deaths are taken into consideration. This model might be used to model diseases that are **endemic** to a population: the diseases that persist in a population for long periods of time (10 or 20 years). Smallpox is an example of a disease that was endemic until it was eliminated in 1977.

We use `solve` to see that the solutions to the system of equations

$$\begin{cases} -\lambda S I + \mu - \mu S = 0 \\ \lambda S I - \gamma I - \mu I = 0 \end{cases} \quad \text{are } S = 1, I = 0, \text{ and } S = (\gamma + \mu)/\lambda, I = \frac{\mu[\lambda - (\gamma + \mu)]}{\lambda(\gamma + \mu)}.$$

```
> l:='l':m:='m':g:='g':

eq1:=-l*s(t)*i(t)+m-m*s(t):

eq2:=l*s(t)*i(t)-g*i(t)-m*i(t):

> eqpts:=solve({eq1=0,eq2=0},{s(t),i(t)});
```

$$eqpts := \{i(t) = 0, s(t) = 1\}, \left\{ s(t) = \frac{g + m}{l}, i(t) = -\frac{m(-l + g + m)}{l(g + m)} \right\}$$

These two points are called *equilibrium points*.

Because $S(t) + I(t) + R(t) = 1$, it follows that $S(t) + I(t) \le 1$.

The following table shows the average infectious period and typical contact numbers for several diseases during certain epidemics.

Disease	Infectious Period (Average) $1/\gamma$	γ	Typical Contact Number σ
Measles	6.5	0.153846	14.9667
Chickenpox	10.5	0.0952381	11.3
Mumps	19	0.0526316	8.1
Scarlet fever	17.5	0.0571429	8.5

Let us assume that the average lifetime, $1/\mu$, is 70 so that $\mu = 0.0142857$.

For each of the diseases listed in the previous table, we use the formula $\sigma = \lambda/(\gamma + \mu)$ to calculate the daily contact rate λ.

Disease	λ
Measles	2.51638
Chickenpox	1.23762
Mumps	0.54203
Scarlet fever	0.607143

Diseases such as those listed can be controlled once an effective and inexpensive vaccine has been developed. Because it is virtually impossible to vaccinate everybody against a disease, we would like to know what percentage of a population needs to be vaccinated to eliminate a disease. A population of people has **herd immunity** to a disease means that enough people are immune to the disease so that if it is introduced into the population, it will not spread throughout the population. In order to have herd immunity, an infected person must infect less than one uninfected person during the time the person is infectious. Thus, we must have $\sigma S < 1$.

Because $I + S + R = 1$, when $S = 1 - R$ and, consequently, herd immunity is achieved when

$$\sigma(1 - R) < 1$$

$$\sigma - \sigma R < 1$$

$$-\sigma R < 1 - \sigma$$

$$R > \frac{\sigma - 1}{\sigma} = 1 - \frac{1}{\sigma}.$$

For each of the diseases listed, we estimate the minimum percentage of a population that needs to be vaccinated to achieve herd immunity.

Disease	Minimum Value of R to Achieve Herd Immunity
Measles	0.933186
Chickenpox	0.911505
Mumps	0.876544
Scarlet fever	0.882354

Using the values in the previous tables, for each disease we graph the direction field and several solutions $\begin{cases} S(t) \\ I(t) \end{cases}$ parametrically. First, we define two lists of ordered pairs and join the lists together in `initconds`. The points in `initconds1` are "close" to the S axis, and the points in `initconds2` are close to the I axis. We will graph the solutions that satisfy these initial conditions.

```
> initconds1:=seq([s(0)=j/10,i(0)=0.01],j=1..9):
  initconds2:=seq([s(0)=1-j/10,i(0)=j/10],j=1..9);
  initconds:=[initconds1,initconds2];
```

For measles, we proceed as follows. After loading the **DEtools** package, we define μ, γ, σ, and λ. For these values, we graph the direction field associated with the system along with the solutions that satisfy the initial conditions specified in `initconds` on the rectangle $[0, 1] \times [0, 1]$. Because $S(t) + I(t) \le 1$, we are concerned only with solutions of the system that are below the line $S + I = 1$.

```
with(DEtools):
> m:=0.0142857:
g:=0.153846:
sigma:=14.9667:
l:=sigma*(g+m):
> DEplot([diff(s(t),t)=eq1,diff(i(t),t)=eq2],
        {s(t),i(t)},t=0..20,initconds,
```

```
        s=0..1,i=0..1,color=GRAY,

            thickness=1,linecolor=BLACK,stepsize=0.1);
```

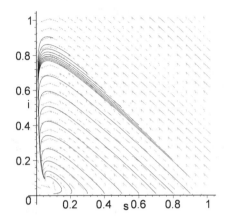

For the remaining three diseases, we change the values of μ, γ, σ, and λ and reenter the code. Here are the results for chickenpox.

```
> m:=0.0142857:

g:=0.0952381:

sigma:=11.3:

l:=sigma*(g+m):

> DEplot([diff(s(t),t)=eq1,diff(i(t),t)=eq2],

        {s(t),i(t)},t=0..20,initconds,

        s=0..1,i=0..1,color=GRAY,

            thickness=1,linecolor=BLACK,stepsize=0.1);
```

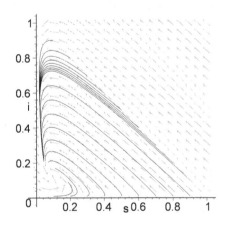

Similar results are obtained for mumps.

```
> m:=0.0142857:
g:=0.0526316:
sigma:=8.1:
l:=sigma*(g+m):
> DEplot([diff(s(t),t)=eq1,diff(i(t),t)=eq2],
        {s(t),i(t)},t=0..40,initconds,
        s=0..1,i=0..1,color=GRAY,
            thickness=1,linecolor=BLACK,stepsize=0.1);
```

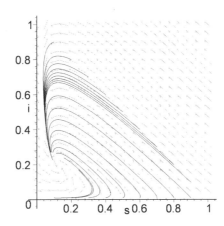

Last, we generate graphs for scarlet fever. In all four cases, we see that all solutions approach the equilibrium point, which indicates that although the epidemic runs its course, the disease is never completely removed from the population.

```
> m:=0.0142857:
g:=0.0571429:
sigma:=8.5:
l:=sigma*(g+m):
> DEplot([diff(s(t),t)=eq1,diff(i(t),t)=eq2],
        {s(t),i(t)},t=0..40,initconds,
        s=0..1,i=0..1,color=GRAY,
            thickness=1,linecolor=BLACK,stepsize=0.1);
```

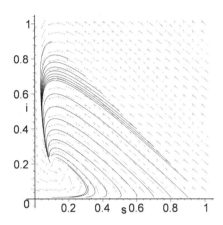

Sources: Herbert W. Hethcote, "Three Basic Epidemiological Models," *Applied Mathematical Ecology,* edited by Simon A. Levin, Thomas G. Hallan, and Louis J. Gross, Springer-Verlag (1989), pp. 119–143. Roy M. Anderson and Robert M. May, "Directly Transmitted Infectious Diseases: Control by Vaccination," *Science,* Volume 215 (February 26, 1982), pp. 1053–1060. J. D. Murray, *Mathematical Biology,* Springer-Verlag (1990), pp. 611–618.

Maple can be used to generate numerical and graphical solutions of systems that involve more than two differential equations as well.

EXAMPLE 2: Under certain assumptions, the **FitzHugh–Nagumo equation**, which arises in the study of the impulses in a nerve fiber, can be written as the system of ordinary differential equations

$$\begin{cases} \frac{dV}{d\xi} = W \\ \frac{dW}{d\xi} = F(V) + R - uW, \\ \frac{dR}{d\xi} = \frac{\varepsilon}{u}(bR - V - a) \end{cases}$$

where $F(V) = \frac{1}{3}V^3 - V$. (a) Graph the solution to the FitzHugh–Nagumo equation that satisfies the initial conditions $V(0) = 1$, $W(0) = 0$, $R(0) = 1$ if $\varepsilon = 0.08$, $a = 0.7$, $b = 0$, and $u = 1$. (b) Graph the solution that satisfies the initial conditions $V(0) = 1$, $W(0) = 0.5$, $R(0) = 0.5$ if $\varepsilon = 0.08$, $a = 0.7$, $b = 0.8$, and $u = 0.6$.

SOLUTION: We begin by defining the FitzHugh–Nagumo system in FHN and then the constants $\varepsilon = 0.08$, $a = 0.7$, $b = 0$, and $u = 1$. In this case, we use

lowercase letters to avoid any ambiguity with built-in Maple functions such as W,
which represents the omega function.

```
> FHN:=[diff(v(xi),xi)=w,
              diff(w(xi),xi)=1/3*v^3-v+r-u*w,
              diff(r(xi),xi)=epsilon/u*(b*r-v-a)];
```

$$FHN := \left[\frac{\partial}{\partial \xi} v(\xi) = w, \frac{\partial}{\partial \xi} w(\xi) = \frac{1}{3} v^3 - v + r - uw, \right.$$

$$\left. \frac{\partial}{\partial \xi} r(\xi) = \frac{\varepsilon(br - v - a)}{u} \right]$$

For (a), we define the values of the constants

```
> epsilon:=0.08:
b:=0:
a:=0.7:
u:=1:
FHN;
```

$$\left[\frac{\partial}{\partial \xi} v(\xi) = w, \frac{\partial}{\partial \xi} w(\xi) = \frac{1}{3} v^3 - v + r - w, \frac{\partial}{\partial \xi} r(\xi) = -0.056 - 0.08v \right]$$

Next, we use DEplot3d to graph the solution that satisfies
$V(0) = 1$, $W(0) = 0$, $R(0) = 1$ for $0 \le \xi \le 30$. Note that the ordered quadruple
$[0,1,0,1]$ corresponds to the initial conditions $V(0) = 1$, $W(0) = 0$, $R(0) = 1$.
Several solutions could be graphed together by including additional ordered
quadruples corresponding to various initial conditions. The option stepsize is
included to help ensure that the resulting graph is smooth.

```
> with(DEtools):
DEplot3d(FHN,[v,w,r],xi=0..30,
                {[0,1,0,1]},stepsize=0.1,linecolor=BLACK);
```

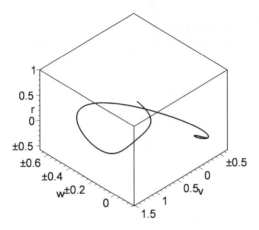

We then graph the functions V, W, and R individually with DEplot and show the three graphs together with display.

```
> pv:=DEplot(FHN,[v,w,r],xi=0..30,
              {[0,1,0,1]},stepsize=0.1,scene=[xi,v],
              linecolor=BLACK):
pw:=DEplot(FHN,[v,w,r],xi=0..30,
              {[0,1,0,1]},stepsize=0.1,scene=[xi,w],
              linecolor=GRAY):
pr:=DEplot(FHN,[v,w,r],xi=0..30,
              {[0,1,0,1]},stepsize=0.1,scene=[xi,r],
              linecolor=BLACK,thickness=1):
> with(plots):
display({pv,pw,pr});
```

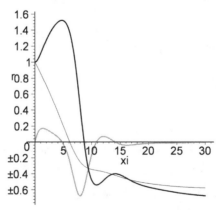

For (b), we redefine the constants $\varepsilon = 0.08$, $a = 0.7$, $b = 0.8$, and $u = 0.6$ and then use `DEplot3d` in the same manner as before to generate various graphs of the solutions that satisfy $V(0) = 1$, $W(0) = 0.5$, $R(0) = 0.5$.

```
> epsilon:=0.08:
b:=0.8:
a:=0.7:
u:=0.6:
FHN;
```

$$\left[\frac{\partial}{\partial \xi} v(\xi) = w, \frac{\partial}{\partial \xi} w(\xi) = \frac{1}{3} v^3 - v + r - 0.6w, \right.$$

$$\left. \frac{\partial}{\partial \xi} r(\xi) = 0.1066666666 \, r - 0.1333333333 \, v - 0.09333333331 \right]$$

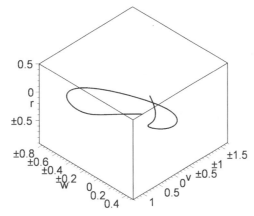

```
> DEplot3d(FHN,[v,w,r],xi=0..15,{[0,1,0.5,0.5]},
            stepsize=0.1,linecolor=BLACK);
```

Individual graphs of V, W, and R are generated with `DEplot` and then shown together with `display`.

```
pv:=DEplot(FHN,[v,w,r],xi=0..15,{[0,1,0.5,0.5]},
            stepsize=0.1,scene=[xi,v],linecolor=BLACK):
pw:=DEplot(FHN,[v,w,r],xi=0..15,{[0,1,0.5,0.5]},
            stepsize=0.1,scene=[xi,w],linecolor=GRAY):
pr:=DEplot(FHN,[v,w,r],xi=0..15,{[0,1,0.5,0.5]},
            stepsize=0.1,scene=[xi,r],linecolor=BLACK,
                thickness=1):
> display({pv,pw,pr});
```

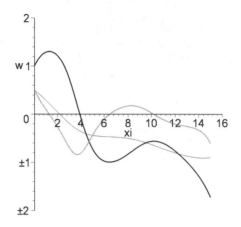

Euler's Method

Euler's method for approximation, which was discussed for first-order equations, may be extended to include systems of first-order equations. Therefore, the initial-value problem

$$\begin{cases} dx/dt = f(t,x,y) \\ dy/dt = g(t,x,y) \\ x(t_0) = x_0, \ y(t_0) = y_0 \end{cases}$$

is approximated at each step by the recursive relationship based on the Taylor expansion of x and y up to order h:

$$\begin{cases} x_{n+1} = x_n + hf(t_n, \ x_n, \ y_n) \\ y_{n+1} = y_n + hg(t_n, \ x_n, \ y_n) \end{cases}$$

where $t_n = t_0 + nh, n = 0, \ 1, \ 2, \ \ldots$.

EXAMPLE 3: Use Euler's method with $h = 0.1$ to approximate the solution of the initial-value problem

$$\begin{cases} dx/dt = x - y + 1 \\ dy/dt = x + 3y + e^{-t} \\ x(0) = 0, y(0) = 1 \end{cases} \ .$$

Compare these results with those of the exact solution of the system of equations.

SOLUTION: In this case, $f(x,y) = x - y + 1$, $g(x,y) = x + 3y + e^{-t}$, $t_0 = 0$, $x_0 = 0$, and $y_0 = 1$, so we use the formulas

$$\begin{cases} x_{n+1} = x_n + h\,(x_n - y_n + 1) \\ y_{n+1} = y_n + h\,(x_n + 3y_n + e^{-t_n}) \end{cases}$$

where $t_n = 0.1 \cdot n$, $n = 0., 1, 2, \ldots$.
If $n = 0$, then

$$\begin{cases} x_1 = x_0 + h(x_0 - y_0 + 1) = 0 \\ y_1 = y_0 + h(x_0 + 3y_0 + e^{-t_0}) = 1.4. \end{cases}$$

To use Maple to implement the Euler method, we define the functions f and g as well as the increment size, the formula for incrementing t, the recursive formulas for determining subsequent values of x and y, and the initial values of x and y. Notice that using the option `remember` in defining these functions causes previous values to be retained so that subsequent values may be based on them.

```
> f:='f':g:='g':t:='t':h:='h':x:='x':y:='y':

f:=(t,x,y)->x-y+1:

g:=(t,x,y)->x+3*y+exp(-t):

h:=0.1:

t:=n->n*h:

x:=proc(n) option remember;

        x(n-1)+h*f(t(n-1),x(n-1),y(n-1))

    end:

x(0):=0:

y:=proc(n) option remember;

        y(n-1)+h*g(t(n-1),x(n-1),y(n-1))

    end:

y(0):=1:
```

The exact solution of this problem is found to be

$$\begin{cases} x(t) = -\frac{3}{4} - \frac{1}{9}e^{-t} + \frac{31}{36}e^{2t} - \frac{11}{6}te^{2t} \\ y(t) = \frac{1}{4} - \frac{2}{9}e^{-t} + \frac{35}{36}e^{2t} + \frac{11}{6}te^{2t} \end{cases}$$

with `dsolve`. Note that we use capital letters to avoid conflict with the definitions of x and y entered previously.

```
> Sol:=dsolve({diff(X(t),t)=X(t)-Y(t)+1,

        diff(Y(t),t)=X(t)+3*Y(t)+exp(-t),X(0)=0,

    Y(0)=1},{X(t),Y(t)});
```

```
assign(Sol):
```

$$Sol := \left\{ Y(t) = \frac{1}{4} = \frac{2}{9}e^{(-t)} + \frac{11}{6}te^{(2t)} + \frac{35}{36}e^{(2t)}, \right.$$

$$\left. X(t) = \frac{3}{4} - \frac{1}{9}e^{(-t)} - \frac{11}{6}te^{(2t)} + \frac{31}{36}e^{(2t)} \right\}$$

We then use `array`, `seq`, `evalf`, and subs to display the results obtained with this method and compare them with the actual function values. The first column gives the value of t; the second column represents the approximate value of x and should be compared with the third column, which gives the exact value of x; and the fourth and fifth columns give the approximate and exact values of y, respectively.

```
> array([seq([t(i),x(i),evalf(subs(t=t(i),X(t))),y(i),
        evalf(subs(t=t(i),Y(t)))],i=0..10)]);
```

0,	0,	0,	1,	1.000000000
0.1,	0,	−0.0226978438,	1.4,	1.460323761
0.2,	−0.04,	−0.1033456498,	1.910483742,	2.065447345
0.3,	−0.1350483742,	−0.2654317312,	2.561501940,	2.859043457
0.4,	−0.3047034056,	−0.5401053309,	3.390529507,	3.896823685
0.5,	−0.5742266969,	−0.9684079527,	4.444250023,	5.249747751
0.6,	−0.9760743689,	−1.604118114,	5.780755426,	7.008061926
0.7,	−1.551757348,	−2.517371685,	7.472255781,	9.286376522
0.8,	−2.354158661,	−3.799261744,	9.608415310,	12.23004487
0.9,	−3.450416058,	−5.567674073,	12.30045693,	16.02317121
1.0,	−4.925503357,	−7.974680032,	15.68620937,	20.89865639

We also graph the approximation with the actual solution.

```
> xs:=[seq([t(i),x(i)],i=0..10)];
p1:=plot(xs,style=POINT,color=BLACK):
p2:=plot(X(t),t=0..1,color=BLACK):
with(plots):
display({p1,p2});
> ys:=[seq([t(i),y(i)],i=0..10)]:
p3:=plot(ys,style=POINT,color=BLACK):
```

```
p4:=plot(Y(t),t=0..1,color=BLACK):
display({p3,p4});
```

Because the accuracy of this approximation diminishes as t increases, we attempt to improve the approximation by decreasing the increment size. We do this next by entering the value $h = 0.05$ and repeating the procedure that was just followed.

```
> h:='h':x:='x':y:='y':
h:=0.05:
t:=n->n*h:
x:=proc(n) option remember;
     x(n-1)+h*f(t(n-1),x(n-1),y(n-1))
     end:
x(0):=0:
y:=proc(n) option remember;
     y(n-1)+h*g(t(n-1),x(n-1),y(n-1))
     end:
y(0):=1:
> array([seq([t(i),x(i),evalf(subs(t=t(i),X(t))),y(i),
     evalf(subs(t=t(i),Y(t)))],i=0..20)]);
```

$$
\begin{bmatrix}
0, & 0, & 0, & 1, & 1.000000000 \\
0.05, & 0, & -0.0053245353, & 1.20, & 1.214394744 \\
0.10, & -0.0100, & -0.0226978438, & 1.427561471, & 1.460323761 \\
0.15, & -0.03187807355, & -0.0544669738, & 1.686437563, & 1.742305463 \\
0.20, & -0.06779385540, & -0.1033456498, & 1.980844693, & 2.065447345 \\
0.25, & -0.1202257828, & -0.1724651306, & 2.315518242, & 2.435520534 \\
0.30, & -0.1920129841, & -0.2654317312, & 2.695774728, & 2.859043457 \\
0.35, & -0.2864023697, & -0.3863918327, & 3.127581199, & 3.343375766 \\
0.40, & -0.4071015482, & -0.5401053309, & 3.617632665, & 3.896823685 \\
0.45, & -0.5583382589, & -0.7320285712, & 4.173438490, & 4.528758223 \\
0.50, & -0.7449270964, & -0.9684079527, & 4.803418758, & 5.249747751 \\
0.55, & -0.9723443891, & -1.256385532, & 5.517011750, & 6.071706638 \\
0.60, & -1.246812196, & -1.604118114, & 6.324793784, & 7.008061926 \\
0.65, & -1.575392495, & -2.020911486, & 7.238612824, & 8.073940117 \\
0.70, & -1.966092761, & -2.517371685, & 8.271737412, & 9.286376522 \\
0.75, & -2.427984270, & -3.105575390, & 9.439022651, & 10.66454984 \\
0.80, & -2.971334616, & -3.799261744, & 10.75709516, & 12.23004487 \\
0.85, & -3.607756105, & -4.614048311, & 12.24455915, & 14.00714688 \\
0.90, & -4.350371868, & -5.567674073, & 13.92222596, & 16.02317121 \\
0.95, & -5.214001760, & -6.680272722, & 15.81336974, & 18.30883219 \\
1.00, & -6.215370335, & -7.974680032, & 17.94401216, & 20.89865639
\end{bmatrix}
$$

```
> xs:=[seq([t(i),x(i)],i=0..20)]:

p1:=plot(xs,style=POINT,color=BLACK):

display({p1,p2});

> ys:=[seq([t(i),y(i)],i=0..20)]:

p3:=plot(ys,style=POINT,color=BLACK):

display({p3,p4});
```

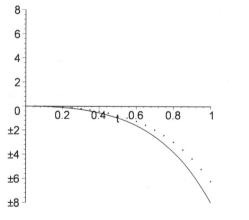
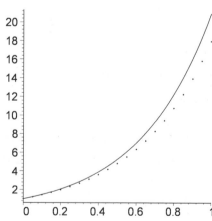

Notice that the approximations are more accurate with the smaller value of h. We also see this in the graphs that compare the approximation with the exact solution.

■

Runge–Kutta Method

Because we would like to be able to improve the approximation without using such a small value for h, we seek to improve the method. As with first-order equations, the Runge–Kutta method can be extended to systems. In this case, the recursive formula at each step is

$$\begin{cases} x_{n+1} = x_n + \frac{1}{6}h(k_1 + 2k_2 + 2k_3 + k_4) \\ y_{n+1} = y_n + \frac{1}{6}h(m_1 + 2m_2 + 2m_3 + m_4) \end{cases}$$

where

$$k_1 = f(t_n, x_n, y_n)$$

$$k_2 = f\left(t_n + \frac{1}{2}h, x_n + \frac{1}{2}hk_1, y_n + \frac{1}{2}hm_1\right)$$

$$k_3 = f\left(t_n + \frac{1}{2}h, , x_n + \frac{1}{2}hk_2, y_n + \frac{1}{2}hm_2\right)$$

$$k_4 = f(t_n + h, x_n + hk_3, y_n + hm_3)$$

$$m_1 = g(t_n, x_n, y_n)$$

$$m_2 = g\left(t_n + \frac{1}{2}h, x_n + \frac{1}{2}hk_1, y_n + \frac{1}{2}hm_1\right)$$

$$m_3 = g\left(t_n + \frac{1}{2}h, , x_n + \frac{1}{2}hk_2, y_n + \frac{1}{2}hm_2\right)$$

$$m_4 = g(t_n + h, x_n + hk_3, y_n + hm_3).$$

EXAMPLE 4: Use the Runge–Kutta method to approximate the solution of the initial-value problem from Example 3

$$\begin{cases} dx/dt = x - y + 1 \\ dy/dt = x + 3y + e^{-t} \\ x(0) = 0, \ y(0) = 1 \end{cases}$$

using $h = 0.1$. Compare these results with those of the exact solution of the system of equations as well as those obtained with Euler's method.

SOLUTION: Because $f(x, y) = x - y + 1$, $g(x, y) = x + 3y + e^{-t}$, $t_0 = 0$, $x_0 = 0$, and $y_0 = 1$, we use the formulas

$$\begin{cases} x_{n+1} = x_n + \frac{1}{6}h(k_1 + 2k_2 + 2k_3 + k_4) \\ y_{n+1} = y_n + \frac{1}{6}h(m_1 + 2m_2 + 2m_3 + m_4) \end{cases}$$

where

$$k_1 = f(t_n, x_n, y_n) = x_n - y_n + 1$$

$$m_1 = g(t_n, x_n, y_n) = x_n + 3y_n + e^{-t_n}$$

$$k_2 = \left(x_n + \tfrac{1}{2}hk_1\right) - \left(y_n + \tfrac{1}{2}hm_1\right) + 1$$

$$m_2 = \left(x_n + \tfrac{1}{2}hk_1\right) + 3\left(y_n + \tfrac{1}{2}hm_1\right) + e^{-(t_n + h/2)}$$

$$k_3 = \left(x_n + \tfrac{1}{2}hk_2\right) - \left(y_n + \tfrac{1}{2}hm_2\right) + 1$$

$$m_3 = \left(x_n + \tfrac{1}{2}hk_2\right) + 3\left(y_n + \tfrac{1}{2}hm_2\right) + e^{-(t_n + h/2)}$$

$$k_4 = (x_n + hk_3) - (y_n + hm_3) + 1$$

$$m_4 = (x_n + hk_3) + 3(y_n + hm$$

For example, if $n = 0$, then

$$k_1 = x_0 - y_0 + 1 = 0 - 1 + 1 = 0,$$

$$m_1 = x_0 + 3y_0 + e^{-t_0} = 0 + 3 + 1 = 4,$$

$$k_2 = \left(x_0 + \tfrac{1}{2}hk_1\right) - \left(y_0 + \tfrac{1}{2}hm_1\right) + 1 = -1 - \tfrac{1}{2} \cdot 4 \cdot 0.1 + 1 = -0.2,$$

$$m_2 = \left(x_0 + \tfrac{1}{2}hk_1\right) + 3\left(y_0 + \tfrac{1}{2}hm_1\right) + e^{-(t_0 + h/2)} = 3\left(1 + \tfrac{1}{2} \cdot 4 \cdot 0.1\right) + e^{-0.05} \approx 4.55123,$$

$$k_3 = \left(x_0 + \tfrac{1}{2}hk_2\right) - \left(y_0 + \tfrac{1}{2}hm_2\right) + 1 = \tfrac{1}{2} \cdot 0.1 \cdot 0.2 - 1 - \tfrac{1}{2} \cdot 0.1 \cdot 4.55123 + 1 \approx -0.23756,$$

$$m_3 = \left(x_0 + \tfrac{1}{2}hk_2\right) + 3\left(y_0 + \tfrac{1}{2}hm_2\right) + e^{-(t_0 + h/2)}$$

$$= \tfrac{1}{2} \cdot 0.1 \cdot 0.2 + 3\left(1 + \tfrac{1}{2} \cdot 0.1 \cdot 4.55123\right) + e^{-0.05} \approx 4.62391,$$

$$k_4 = (x_0 + hk_3) - (y_0 + hm_3) + 1 = 0.1 \cdot -0.23756 - 1 + 0.1 \cdot 4.62391 + 1 \approx -0.48615, \text{ and}$$

$$m_4 = (x_0 + hk_3) + 3(y_0 + hm_3) + e^{-(t_0 + h)}$$

$$= 0.1 \cdot -0.23756 + 3(1 + 0.1 \cdot 4.62391) + e^{-0.1} \approx 5.26826.$$

Therefore,

$$x_1 = x_0 + \tfrac{1}{6} \cdot 0.1 \cdot (k_1 + 2k_2 + 2k_3 + k_4)$$

$$= 0 + \tfrac{1}{6} \cdot 0.1 \cdot (0 + 2 \cdot -0.2 + 2 \cdot -0.23756 - 0.48615) \approx -0.0226878$$

and

$$y_1 = y_0 + \tfrac{1}{6} \cdot 0.1 \cdot (m_1 + 2m_2 + 2m_3 + m_4)$$

$$= 1 + \tfrac{1}{6} \cdot 0.1 \cdot (4 + 2 \cdot 4.55123 + 2 \cdot 4.62391 + 5.26826) \approx 1.46031.$$

We show the results obtained with this method and compare them with the exact values.

To use Maple to implement the Runge–Kutta method, we begin by defining the appropriate functions.

```
> f:=(t,x,y)->x-y+1:

g:=(t,x,y)->x+3*y+exp(-t):

h:=0.1:

t:=n->n*h:
```

The recursive formulas for xrk and yrk are defined using the option remember so that Maple "remembers" the values of xrk and yrk computed. Thus, previously computed values are retained and subsequent values may be based on them.

```
> xrk:='xrk':yrk:='yrk':
xrk:=proc(n)
        local k1,m1,k2,m2,k3,m3,k4,m4;
        option remember;
        k1:=f(t(n-1),xrk(n-1),yrk(n-1));
        m1:=g(t(n-1),xrk(n-1),yrk(n-1));
        k2:=f(t(n-1)+h/2,xrk(n-1)+h*k1/2,yrk(n-1)+h*m1/2);
        m2:=g(t(n-1)+h/2,xrk(n-1)+h*k1/2,yrk(n-1)+h*m1/2);
        k3:=f(t(n-1)+h/2,xrk(n-1)+h*k2/2,yrk(n-1)+h*m2/2);
        m3:=g(t(n-1)+h/2,xrk(n-1)+h*k2/2,yrk(n-1)+h*m2/2);
        k4:=f(t(n-1)+h,xrk(n-1)+h*k3,yrk(n-1)+h*m3);
        m4:=g(t(n-1)+h,xrk(n-1)+h*k3,yrk(n-1)+h*m3);
        xrk(n-1)+h/6*(k1+2*k2+2*k3+k4)
        end:
xrk(0):=0:
> yrk:=proc(n)
        local k1,m1,k2,m2,k3,m3,k4,m4;
        option remember;
        k1:=f(t(n-1),xrk(n-1),yrk(n-1));
        m1:=g(t(n-1),xrk(n-1),yrk(n-1));
        k2:=f(t(n-1)+h/2,xrk(n-1)+h*k1/2,yrk(n-1)+h*m1/2);
        m2:=g(t(n-1)+h/2,xrk(n-1)+h*k1/2,yrk(n-1)+h*m1/2);
        k3:=f(t(n-1)+h/2,xrk(n-1)+h*k2/2,yrk(n-1)+h*m2/2);
        m3:=g(t(n-1)+h/2,xrk(n-1)+h*k2/2,yrk(n-1)+h*m2/2);
        k4:=f(t(n-1)+h,xrk(n-1)+h*k3,yrk(n-1)+h*m3);
        m4:=g(t(n-1)+h,xrk(n-1)+h*k3,yrk(n-1)+h*m3);
        yrk(n-1)+h/6*(m1+2*m2+2*m3+m4)
        end:
yrk(0):=1:
```

In the same way as in the previous example, the actual solution of this problem is determined with `dsolve`. Note that we use capital letters to avoid conflict with the definitions of x and y entered previously.

```
> X:='X':Y:='Y':
Sol:=dsolve({diff(X(t),t)=X(t)-Y(t)+1,
       diff(Y(t),t)=X(t)+3*Y(t)+exp(-t),X(0)=0,
       Y(0)=1},{X(t),Y(t)},method=laplace):
assign(Sol):
```

We show the results obtained with this method and compare them with the exact values using the same format as the two previous tables in Example 3 using Euler's method.

```
> array([seq([t(i),xrk(i),evalf(subs(t=t(i),X(t))),yrk(i),
       evalf(subs(t=t(i),Y(t)))],i=0..10)]);
```

0,	0,	0,	1,	1.000000000
0.1,	−0.02268784122,	−0.0226978438,	1.460309033,	1.460323761
0.2,	−0.1033199101,	−0.1033456498,	2.065409956,	1.460323761
0.3,	−0.2653822034,	−0.2654317312,	2.858972430,	2.859043457
0.4,	−0.5400208432,	−0.5401053309,	3.896703967,	3.896823685
0.5,	−0.9682731495,	−0.9684079527,	5.249558887,	5.249747751
0.6,	−1.603912060,	−1.604118114,	7.007776299,	7.008061926
0.7,	−2.517066041,	−2.517371685,	9.285957081,	9.286376522
0.8,	−3.798818366,	−3.799261744,	12.22944215,	12.23004487
0.9,	−5.567041916,	−5.567674073,	16.02231953,	16.02317121
1.0,	−7.973791088,	−7.974680032,	20.89746888,	20.89865639

Notice that the Runge–Kutta method is much more accurate than Euler's method. In fact, the Runge–Kutta method with $h = 0.1$ is more accurate than Euler's method with $h = 0.05$. We also observe the accuracy of the approximation in the graphs that compare the approximation with the exact solution.

```
> xs:=[seq([t(i),xrk(i)],i=0..20)]:
p1:=plot(xs,style=POINT,color=BLACK):
p2:=plot(X(t),t=0..1,color=BLACK):
with(plots):
display({p1,p2},view=[0..1,0..-8]);
> ys:=[seq([t(i),yrk(i)],i=0..20)]:
p3:=plot(ys,style=POINT,color=BLACK):
```

```
p4:=plot(Y(t),t=0..1,color=BLACK):

display({p3,p4},view=[0..1,0..22]);
```

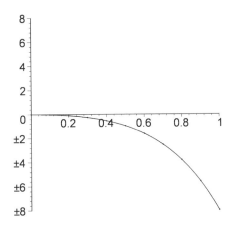
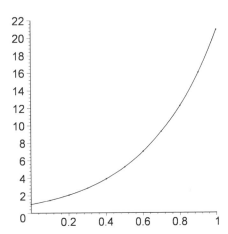

■

7.7 Nonlinear Systems, Linearization, and Classification of Equilibrium Points

We now turn our attention to systems of differential equations of the form

$$\begin{cases} dx/dt = f(x,y) \\ dy/dt = g(x,y). \end{cases}$$

This system is **autonomous**, because f and g do not depend explicitly on the independent variable t.

Definition
Equilibrium Point

A point (x_0, y_0) is an **equilibrium point** of the system

$$\begin{cases} dx/dt = f(x,y) \\ dy/dt = g(x,y) \end{cases} \quad \text{if } f(x_0,y_0) = 0 \text{ and } g(x_0,y_0) = 0.$$

Before discussing nonlinear systems, we first investigate properties of systems of the form

$$\begin{cases} dx/dt = ax + by \\ dy/dt = cx + dy \end{cases}$$

where $\begin{vmatrix} a & b \\ c & d \end{vmatrix} = ad - bc \neq 0$, which have only one equilibrium point: (0,0). We have solved

many systems of this type by using the eigenvalues and corresponding eigenvectors of

$$\mathbf{A} = \begin{pmatrix} a & b \\ c & d \end{pmatrix}.$$

Real Distinct Eigenvalues

If λ_1 and λ_2 are real eigenvalues of $\mathbf{A} = \begin{pmatrix} a & b \\ c & d \end{pmatrix}$ where $\lambda_2 < \lambda_1$, with corresponding

eigenvectors $\mathbf{v}_1$ and $\mathbf{v}_2$, respectively, a general solution of the system is

$$\mathbf{X} = \begin{pmatrix} x \\ y \end{pmatrix} = c_1 \mathbf{v}_1 e^{\lambda_1 t} + c_2 \mathbf{v}_2 e^{\lambda_2 t} = e^{\lambda_1 t} \left[c_1 \mathbf{v}_1 + c_2 \mathbf{v}_2 e^{(\lambda_2 - \lambda_1)t} \right].$$

(a) Suppose that both eigenvalues are negative. If we assume that $\lambda_2 < \lambda_1 < 0$, then $\lambda_2 - \lambda_1 < 0$. This means that $e^{(\lambda_2 - \lambda_1)t}$ is very small for large values of t, so $\mathbf{X} \approx c_1 \mathbf{v}_1 e^{\lambda_1 t}$ is small for large values of t. If $c_1 \neq 0$, then $\lim_{t \to \infty} \mathbf{X} = \mathbf{0}$ in one of the directions determined by $\mathbf{v}_1$ or $-\mathbf{v}_1$. If $c_1 = 0$, then $\mathbf{X} = c_2 \mathbf{v}_2 e^{\lambda_2 t}$. Again, because $\lambda_2 < 0$, $\lim_{t \to \infty} \mathbf{X} = \mathbf{0}$ in the directions determined by $\mathbf{v}_2$ and $-\mathbf{v}_2$. In this case, (0,0) is a **stable node**.

(b) Suppose that both eigenvalues are positive. If $0 < \lambda_2 < \lambda_1$, then $e^{\lambda_1 t}$ and $e^{\lambda_2 t}$ both become unbounded as t increases. If $c_1 \neq 0$, then $\mathbf{X}$ becomes unbounded in the direction of either $\mathbf{v}_1$ or $-\mathbf{v}_1$. If $c_1 = 0$, then $\mathbf{X}$ becomes unbounded in the directions given by $\mathbf{v}_2$ and $-\mathbf{v}_2$. In this case, (0,0) is an **unstable node**.

(c) Suppose that the eigenvalues have opposite sign. Then, if $\lambda_2 < 0 < \lambda_1$ and $c_1 \neq 0$, $\mathbf{x}$ becomes unbounded in the direction of either $\mathbf{v}_1$ or $-\mathbf{v}_1$ as it did in (b). However, if $c_1 = 0$, then due to the fact that $\lambda_2 < 0$, $\lim_{t \to \infty} \mathbf{X} = \mathbf{0}$ along the line determined by $\mathbf{v}_2$. If the initial point $\mathbf{x}(0)$ is not on the line determined by $\mathbf{v}_2$, then the line given by $\mathbf{v}_1$ is an asymptote for the solution. We say that (0,0) is a **saddle point** in this case.

EXAMPLE 1: Classify the equilibrium point (0,0) of the systems: (a)

$$\begin{cases} x' = 5x + 3y \\ y' = -4x - 3y \end{cases} ; \text{(b)} \begin{cases} x' = x - 2y \\ y' = 3x - 4y \end{cases} ; \text{(c)} \begin{cases} x' = -x - 2y \\ y' = 3x + 4y \end{cases} .$$

SOLUTION: (a) We find the eigenvalues and corresponding eigenvectors of

$$\mathbf{A} = \begin{pmatrix} 5 & 3 \\ -4 & -3 \end{pmatrix}$$ with `eigenvectors`.

```
> A:=array([[5,3],[-4,-3]]):
with(linalg):
vecsA:=eigenvectors(A);
```

$$vecsA := [-1, 1, \{[1, -2]\}], \left[3, 1, \left\{\left[\frac{-3}{2}, 1\right]\right\}\right]$$

Because these eigenvalues have opposite sign, (0,0) is a saddle point. Eigenvectors corresponding to $\lambda_1 = -1$ and $\lambda_2 = 3$ are $\mathbf{v}_1 = \begin{pmatrix} -1 \\ 2 \end{pmatrix}$ and $\mathbf{v}_2 = \begin{pmatrix} -3 \\ 2 \end{pmatrix}$, respectively. Hence the solution becomes unbounded in the directions associated with the positive eigenvalue, $\mathbf{v}_2 = \begin{pmatrix} -3 \\ 2 \end{pmatrix}$ and $-\mathbf{v}_2 = \begin{pmatrix} 3 \\ -2 \end{pmatrix}$. Along the line determined by $\mathbf{v}_1 = \begin{pmatrix} -1 \\ 2 \end{pmatrix}$, the solution approaches (0,0).

We see this when we graph various solutions and display the results together with the direction field associated with the system. First, we use `dsolve` to solve the initial-value problem

$$\begin{cases} dx/dt = 5x + 3y \\ dy/dt = -4x - 3y \\ x(0) = x_0, \ y(0) = y_0 \end{cases}.$$

```
> sol:=dsolve({diff(x(t),t)=5*x(t)+3*y(t),
        diff(y(t),t)=-4*x(t)-3*y(t),
            x(0)=x0,y(0)=y0}, {x(t),y(t)});
```

$$sol := \{x(t) = -\frac{1}{2}x0e^{(-t)} + \frac{3}{2}x0e^{(3t)} + \frac{3}{4}y0e^{(3t)} - \frac{3}{4}y0e^{(-t)},$$

$$y(t) = -x0e^{(3t)} + x0e^{(-t)} + \frac{3}{2}y0e^{(-t)} - \frac{1}{2}y0e^{(3t)}\}$$

Then, we use `seq` and `subs` to create a list of ordered triples $\{x(t), y(t), t = -3..3\}$, corresponding to the solution for various initial conditions.

```
> assign(sol):
xvals:=seq(-1+0.2*i,i=0..10):
> toplota:={seq(subs({x0=j,y0=0},[x(t),y(t),t=-3..3]),
```

```
            j=xvals)}:
toplotb:={seq(subs({x0=j,y0=-4/3*j},[x(t),y(t),t=-3..3]),
            j=xvals)}:
```

These functions are then graphed with `plot`.

```
> p1:=plot(toplota,color=BLACK,view=[-1..1,-1..1]):
p2:=plot(toplotb,color=BLACK,view=[-1..1,-1..1]):
> p3:=plot({-2*x,-2*x/3},x=-1..1,color=BLACK,style=POINT):
```

We graph the direction field associated with the system with `DEplot`.

```
> with(DEtools):
x:='x':y:='y':
p4:=DEplot({diff(x(t),t)=5*x(t)+3*y(t),
        diff(y(t),t)=-4*x(t)-3*y(t)},{x(t),y(t)},
            t=-3..3,x=-1..1,y=-1..1,color=GRAY):
```

Last, all graphs are displayed together with `display`.

```
> with(plots):
display({p1,p2,p3,p4});
```

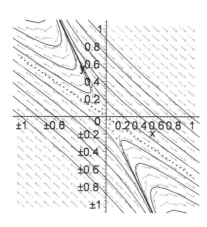

(b) In this case, the eigenvalues $\lambda_1 = -1$ and $\lambda_2 = -2$ are both negative.

```
> eigenvectors([[1,-2],[3,-4]]);
```

$$\left[-2, 1, \left\{\left[1, \frac{3}{2}\right]\right\}\right], \left[-1, 1, \{[1, 1]\}\right]$$

Hence, (0,0) is a stable node. Corresponding eigenvectors are $\mathbf{v}_1 = \begin{pmatrix} 1 \\ 1 \end{pmatrix}$ and $\mathbf{v}_2 = \begin{pmatrix} 2 \\ 3 \end{pmatrix}$. Therefore, the solutions approach (0,0) along the lines given by these vectors, $y = x$ and $y = 3x/2$. We see this in the graph of the direction field and graphs of several solutions to the system. We graph the direction field associated with the system. First, we use `seq` to generate several lists of ordered pairs.

```
> ivals:=seq(-1+2*i/9,i=0..9):
initconds1:=seq([x(0)=-1,y(0)=i],i=ivals):
initconds2:=seq([x(0)=1,y(0)=i],i=ivals):
initconds3:=seq([x(0)=i,y(0)=1],i=ivals):
initconds4:=seq([x(0)=i,y(0)=-1],i=ivals):
> initconds:=[initconds1,initconds2,
        initconds3,initconds4]:
```

Then, we use `DEplot` to graph the direction field associated with the system along with the solutions that satisfy the initial conditions in `initconds` for $-3 \le t \le 3$. The solution curves are graphed in black (`linecolor=BLACK`) and the arrows in the direction field are graphed in gray (`color=GRAY`).

```
> p1:=DEplot({diff(x(t),t)=x(t)-2*y(t),
        diff(y(t),t)=3*x(t)-4*y(t)},{x(t),y(t)},
        t=-3..3,initconds,x=-1..1,y=-1..1,color=GRAY,
            linecolor=BLACK,thickness=1,stepsize=0.05):
```

The lines $y = x$ and $y = 3x/2$ are graphed with `plot`.

```
> p2:=plot({x,3*x/2},x=-1..1,color=BLACK,style=POINT):
```

All graph are displayed together with `display`.

```
> display({p1,p2});
```

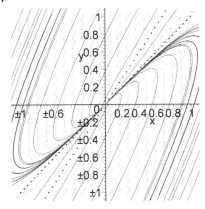

(c) Because the eigenvalues $\lambda_1 = 2$ and $\lambda_2 = 1$ are both positive, (0,0) is an unstable node.

```
> eigenvectors([[-1,-2],[3,4]]);
```

$$\left[2, 1, \left\{\left[1, -\frac{3}{2}\right]\right\}\right], [1, 1, \{[-1, 1]\}]$$

Note that the corresponding eigenvectors are $\mathbf{v}_1 = \begin{pmatrix} 2 \\ -3 \end{pmatrix}$ and $\mathbf{v}_2 = \begin{pmatrix} 1 \\ -1 \end{pmatrix}$,

respectively. Hence, the solutions become unbounded along the lines determined by these vectors, $y = -3x/2$ and $y = -x$. As before, we see this in the graph of the direction field and various solutions of the system.

```
>ivals:=seq(evalf(i/24),i=0..24):
initconds:=[seq([x(0)=0.5*i*cos(2*Pi*i),
        y(0)=0.5*i*sin(2*Pi*i)],i=ivals)]:with(DEtools):
> p1:=DEplot({diff(x(t),t)=-x(t)-2*y(t),
        diff(y(t),t)=3*x(t)+4*y(t)},[x(t),y(t)],
        t=-3..3,initconds,x=-1..1,y=-1..1,color=GRAY,
            linecolor=BLACK,thickness=1,stepsize=0.05):
> p2:=plot({-x,-3*x/2},x=-1..1,color=BLACK,style=POINT):
> display({p1,p2});
```

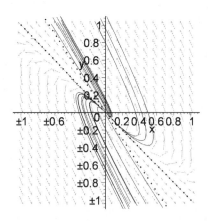

■

Repeated Eigenvalues

We recall from our previous experience with repeated eigenvalues of a 2×2 system that the eigenvalue can have two linearly independent eigenvectors associated with it or only one eigenvector associated with it. Hence, we investigate the behavior of solutions in this case by considering both of these possibilities.

(a) Suppose that the eigenvalue $\lambda = \lambda_1 = \lambda_2$ has two corresponding linearly independent eigenvectors $\mathbf{v}_1$ and $\mathbf{v}_2$. Then, a general solution is

$$\mathbf{X} = c_1 \mathbf{v}_1 e^{\lambda t} + c_2 \mathbf{v}_2 e^{\lambda t} = (c_1 \mathbf{v}_1 + c_2 \mathbf{v}_2) e^{\lambda t}.$$

Hence, if $\lambda > 0$, then $\mathbf{X}$ becomes unbounded along the line determined by the vector $c_1 \mathbf{v}_1 + c_2 \mathbf{v}_2$ where c_1 and c_2 are arbitrary constants. In this case, we call the equilibrium point a **degenerate unstable node** (or an **unstable star**). On the other hand, if $\lambda < 0$, then $\mathbf{X}$ approaches $(0,0)$ along these lines, and we call $(0,0)$ a **degenerate stable node** (or **stable star**). Note that the name "star" was selected because of the shape of the solutions.

(b) Suppose that $\lambda = \lambda_1 = \lambda_2$ has only one corresponding eigenvector $\mathbf{v}_1$. Hence, a general solution is

$$\mathbf{X} = c_1 \mathbf{v}_1 e^{\lambda t} + c_2 [\mathbf{v}_1 t + \mathbf{w}_2] e^{\lambda t} = (c_1 \mathbf{v}_1 + c_2 \mathbf{w}_2) e^{\lambda t} + c_2 \mathbf{v}_1 t e^{\lambda t}$$

where $(\mathbf{A} - \lambda \mathbf{I}) \mathbf{w}_2 = \mathbf{v}_1$. If we write this solution as $\mathbf{X} = t e^{\lambda t} \left[\frac{1}{t} (c_1 \mathbf{v}_1 + c_2 \mathbf{w}_2) + c_2 \mathbf{v}_1 \right]$, we can more easily investigate the behavior of this solution. If $\lambda < 0$, $\lim\limits_{t \to \infty} t e^{\lambda t} = 0$ and $\lim\limits_{t \to \infty} \left[\frac{1}{t} (c_1 \mathbf{v}_1 + c_2 \mathbf{w}_2) + c_2 \mathbf{v}_1 \right] = c_2 \mathbf{v}_1$. Hence, the solutions approach $(0,0)$ along the line determined by $\mathbf{v}_1$, and we call $(0,0)$ a **degenerate stable node**. If $\lambda > 0$, the solutions become unbounded along this line, and we say that $(0,0)$ is a **degenerate unstable node**.

EXAMPLE 2: Classify the equilibrium point $(0,0)$ in the systems: (a)

$$\begin{cases} x' = x + 9y \\ y' = -x - 5y \end{cases} \; ; \text{(b)} \; \begin{cases} x' = 2x \\ y' = 2y \end{cases} .$$

SOLUTION: (a) Using `eigenvectors`,

```
> with(linalg):
A:=array([[1,9],[-1,-5]]):
eigenvectors(A);
```

$$[-2, 2, \{[-3, 1]\}]$$

we see that $\lambda_1 = \lambda_2 = -2$ and that there is only one corresponding eigenvector. Therefore, because $\lambda = -2 < 0$, $(0,0)$ is a degenerate stable node. Notice that in the graph of several members of the family of solutions of this system along with the direction field, which we generate using the same technique as in Example 1 (b),

the solutions approach (0,0) along the line in the direction of $\mathbf{v}_1 = \begin{pmatrix} -3 \\ 1 \end{pmatrix}$ $y = -x/3$.

```
> ivals:=seq(-1+2*i/9,i=0..9):
  initconds1:=seq([x(0)=-1,y(0)=i],i=ivals):
  initconds2:=seq([x(0)=1,y(0)=i],i=ivals):
  initconds3:=seq([x(0)=i,y(0)=1],i=ivals):
  initconds4:=seq([x(0)=i,y(0)=-1],i=ivals):
> initconds:=[initconds1,initconds2,
              initconds3,initconds4]:
> with(DEtools):
> p1:=DEplot({diff(x(t),t)=x(t)+9*y(t),
        diff(y(t),t)=-x(t)-5*y(t)},[x(t),y(t)],
        t=-3..3,initconds,x=-1..1,y=-1..1,color=GRAY,
            linecolor=BLACK,thickness=1,stepsize=0.05):
> p2:=plot(-x/3,x=-1..1,color=BLACK,style=POINT):
> with(plots):
  display({p1,p2});
```

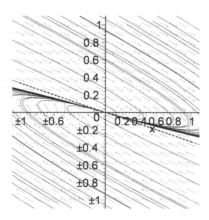

(b) We have $\lambda_1 = \lambda_2 = 2$ and two linearly independent vectors, $\mathbf{v}_1 = \begin{pmatrix} 1 \\ 0 \end{pmatrix}$ and $\mathbf{v}_2 = \begin{pmatrix} 0 \\ 1 \end{pmatrix}$ (Note: The choice of these two vectors does not change the value of the solution, because of the form of the general solution in this case.).

```
> A:=array([[2,0],[0,2]]):
eigenvectors(A);
```

$$[2, 2, \{[0, 1], [1, 0]\}]$$

Because $\lambda = 2 > 0$, we classify (0,0) as a degenerate unstable node (or star). Some of these solutions along with the direction field are graphed in the same manner as in Example 1 (c). Notice that they become unbounded in the direction of any vector in the xy-plane because $\mathbf{v}_1 = \begin{pmatrix} 1 \\ 0 \end{pmatrix}$ and $\mathbf{v}_2 = \begin{pmatrix} 0 \\ 1 \end{pmatrix}$.

```
> ivals:=seq(evalf(i/24),i=0..24):
> initconds:=[seq([x(0)=0.5*cos(2*Pi*i),
            y(0)=0.5*sin(2*Pi*i)],i=ivals)]:
> DEplot({diff(x(t),t)=2*x(t),diff(y(t),t)=2*y(t)},
       [x(t),y(t)],t=-3..3,initconds,
       x=-1..1,y=-1..1,color=GRAY,linecolor=BLACK,
          thickness=1,stepsize=0.05);
```

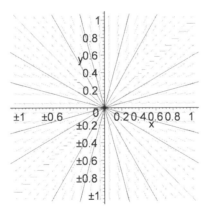

Complex Conjugate Eigenvalues

We have seen that if the eigenvalues of the system of differential equations are $\lambda_1 = \alpha + \beta i$ and $\lambda_2 = \alpha - \beta i$ with corresponding eigenvectors $\mathbf{v}_1 = \mathbf{a} + \mathbf{b}i$ and $\mathbf{v}_2 = \mathbf{a} - \mathbf{b}i$, two linearly independent solutions of the system are

$$\mathbf{X}_1 = e^{\alpha t}(\mathbf{a}\cos\beta t - \mathbf{b}\sin\beta t) \quad \text{and} \quad \mathbf{X}_2 = e^{\alpha t}(\mathbf{b}\cos\beta t + \mathbf{a}\sin\beta t).$$

Hence, a general solution is $\mathbf{X} = c_1\mathbf{X}_1 + c_2\mathbf{X}_2$, so there are constants A_1, A_2, B_1, and B_2 such that x and y are given by

$$\mathbf{X} = \begin{pmatrix} x \\ y \end{pmatrix} = \begin{pmatrix} A_1 e^{\alpha t}\cos\beta t + A_2 e^{\alpha t}\sin\beta t \\ B_1 e^{\alpha t}\cos\beta t + B_2 e^{\alpha t}\sin\beta t \end{pmatrix}.$$

(a) If $\alpha = 0$, the solution is

$$\mathbf{X} = \begin{pmatrix} x \\ y \end{pmatrix} = \begin{pmatrix} A_1\cos\beta t + A_2\sin\beta t \\ B_1\cos\beta t + B_2\sin\beta t \end{pmatrix}.$$

Hence, both x and y are periodic. If fact, if $A_2 = B_1 = 0$, then

$$\mathbf{X} = \begin{pmatrix} x \\ y \end{pmatrix} = \begin{pmatrix} A_1\cos\beta t \\ B_2\sin\beta t \end{pmatrix}.$$

In rectangular coordinates this solution is

$$\frac{x^2}{A_1{}^2} + \frac{y^2}{B_2{}^2} = 1$$

where the graph is either a circle or an ellipse centered at (0,0) depending on the value of A_1 and B_2. Hence, (0,0) is classified as a **center**. Note that the motion around these circles or ellipses is either clockwise or counterclockwise for all solutions.

(b) If $\alpha \neq 0$, then $e^{\alpha t}$ is present in the solution. This term causes the solution to spiral around the equilibrium point. If $\alpha > 0$, then the solution spirals away from (0,0), so we classify (0,0) as an **unstable spiral**. Otherwise, if $\alpha < 0$, the solution spirals toward (0,0), so we say that (0,0) is a **stable spiral**.

EXAMPLE 3: Classify the equilibrium point (0,0) in each of the following systems:

(a) $\begin{cases} x' = -y \\ y' = x \end{cases}$; (b) $\begin{cases} x' = \frac{1}{2}x - \frac{153}{32}y \\ y' = 2x - y \end{cases}$.

SOLUTION: (a) The eigenvalues are found to be $\lambda_{1,2} = \pm i$.

```
> with(linalg):

eigenvectors([[0,-1],[1,0]]);
```

$$[I, 1, \{[I, 1]\}], [-I, 1, \{[-I, 1]\}]$$

Because these eigenvalues have zero real part (and, hence, are purely imaginary), (0,0) is a center. Several solutions along with the direction field are graphed next.

```
> sol:=dsolve({diff(x(t),t)=-y(t),
        diff(y(t),t)=x(t),x(0)=0,y(0)=i},{x(t),y(t)});
```

$$sol := \{x(t) = -\sin(t)i, y(t) = \cos(t)i\}$$

```
> assign(sol):
> ivals:=seq(i/14,i=0..14):
 toplot:={seq(eval([x(t),y(t)],t=0..2*Pi]),
            i=ivals)}:
> p1:=plot(toplot,color=BLACK):
> x:='x':y:='y':
p2:=DEplot({diff(x(t),t)=-y(t),
        diff(y(t),t)=x(t)},[x(t),y(t)],
        t=-3..3,x=-1..1,y=-1..1,color=GRAY,
            arrows=LARGE):
> with(plots):
display({p1,p2});
```

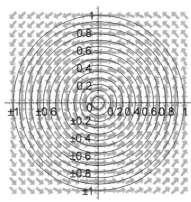

(b) The eigenvalues are found to be $\lambda_{1,2} = -1/4 \pm 3i$.

```
> eigenvectors([[1/2,-153/32],[2,-1]]);
```

$$\left[\left| \frac{1}{4} + 3I, 1, \left\{ \left[\frac{3}{8} + \frac{3}{2}I, 1 \right] \right\} \right|, \left| \frac{1}{4} - 3I, 1, \left\{ \left[\frac{3}{8} - \frac{3}{2}I, 1 \right] \right\} \right| \right]$$

Thus, $(0,0)$ is a stable spiral, because $\alpha = -1/4 < 0$. Several solutions along with the direction field are graphed in the same way as we have done before.

```
> sol:=dsolve({diff(x(t),t)=1/2*x(t)-153/32*y(t),
        diff(y(t),t)=2*x(t)-y(t),
            x(0)=1,y(0)=i},{x(t),y(t)});
```

$$sol := \left\{ x(t) = -\frac{1}{32}e^{\left(-\frac{1}{4}t\right)}(-32\cos(3t) - 8\sin(3t) + 51\sin(3t)i), \right.$$

$$\left. y(t) = -\frac{1}{12}e^{\left(-\frac{1}{4}t\right)}(-8\sin(3t) - 12i\cos(3t) + 3\sin(3t)i) \right\}$$

```
> assign(sol):
ivals:=seq(-1+i/2,i=0..4):
toplot:={seq(eval([x(t),y(t),t=0..4*Pi]),i=ivals)}:
p1:=plot(toplot,color=BLACK):
> x:='x':y:='y':
> p2:=DEplot({diff(x(t),t)=1/2*x(t)-153/32*y(t),
        diff(y(t),t)=2*x(t)-y(t)},[x(t),y(t)],
            t=-3..3,x=-1..1,y=-1..1,color=GRAY,arrows=LARGE):
> display({p1,p2});
```

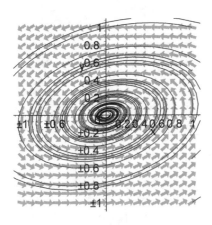

Nonlinear Systems

When working with nonlinear systems, we can often gain a great deal of information concerning the system by making a linear approximation near each equilibrium point of the nonlinear system and solving the linear system. Although the solution to the linearized system only approximates the solution to the nonlinear system, the general behavior of solutions to the nonlinear system near each equilibrium is the same as that of the corresponding linear system in most cases. The first step toward approximating a nonlinear system near each equilibrium point is to find the equilibrium points of the system and the matrix for linearization near each point as defined in the following.

Recall from multivariate calculus that if $z = F(x, y)$ is a differentiable function, the tangent plane to the surface S given by the graph of $z = F(x, y)$ at the point (x_0, y_0) is

$$z = F_x(x_0, y_0)(x - x_0) + F_y(x_0, y_0)(y - y_0) + F(x_0, y_0).$$

Hence, near each equilibrium point (x_0, y_0) of the nonlinear system

$$\begin{cases} dx/dt = f(x, y) \\ dy/dt = g(x, y) \end{cases},$$

the system can be approximated with

$$\begin{cases} dx/dt = f_x(x_0, y_0)(x - x_0) + f_y(x_0, y_0)(y - y_0) + f(x_0, y_0) \\ dy/dt = g_x(x_0, y_0)(x - x_0) + g_y(x_0, y_0)(y - y_0) + g(x_0, y_0) \end{cases}.$$

Then, because $f(x_0, y_0) = 0$ and $g(x_0, y_0) = 0$, the approximate system is

$$\begin{cases} dx/dt = f_x(x_0, y_0)(x - x_0) + f_y(x_0, y_0)(y - y_0) \\ dy/dt = g_x(x_0, y_0)(x - x_0) + g_y(x_0, y_0)(y - y_0) \end{cases},$$

which can be written in matrix form as

$$\begin{pmatrix} dx/dt \\ dy/dt \end{pmatrix} = \begin{pmatrix} f_x(x_0, y_0) & f_y(x_0, y_0) \\ g_x(x_0, y_0) & g_y(x_0, y_0) \end{pmatrix} \begin{pmatrix} x - x_0 \\ y - y_0 \end{pmatrix}.$$

Note that we often call this system **the linearized system corresponding to the nonlinear system** because we have removed the nonlinear terms from the original system. Now that the system is approximated by a system of the form $\begin{cases} dx/dt = ax + by \\ dy/dt = cx + dy \end{cases}$, an equilibrium point (x_0, y_0) of the system $\begin{cases} dx/dt = f(x, y) \\ dy/dt = g(x, y) \end{cases}$ is classified by the eigenvalues of the matrix

$$J(x_0, y_0) = \begin{pmatrix} f_x(x_0, y_0) & f_y(x_0, y_0) \\ g_x(x_0, y_0) & g_y(x_0, y_0) \end{pmatrix}$$

which is called the **Jacobian matrix**. Of course, this linearization must be carried out for each equilibrium point. After determining the matrix for linearization for each equilibrium point, the eigenvalues for the matrix must be found. Then, we classify each equilibrium point according to the following criteria.

Classification of Equilibrium Points

Let (x_0, y_0) be an equilibrium point of the system $\begin{cases} dx/dt = f(x,y) \\ dy/dt = g(x,y) \end{cases}$ and let λ_1 and λ_2 be the eigenvalues of the matrix

$$J(x_0, y_0) = \begin{pmatrix} f_x(x_0, y_0) & f_y(x_0, y_0) \\ g_x(x_0, y_0) & g_y(x_0, y_0) \end{pmatrix}.$$

(a) Suppose that λ_1 and λ_2 are real. If $\lambda_1 > \lambda_2 > 0$, then (x_0, y_0) is an **unstable node**; if $\lambda_2 < \lambda_1 < 0$, then (x_0, y_0) is a **stable node**; and if $\lambda_2 < 0 < \lambda_1$, then (x_0, y_0) is a **saddle**.

(b) Suppose that $\lambda_1 = \alpha + \beta i$ and $\lambda_2 = \alpha - \beta i$ where $\beta \neq 0$. If $\alpha < 0$, (x_0, y_0) is a **stable spiral**; if $\alpha > 0$, (x_0, y_0) is an **unstable spiral**; and if $\alpha = 0$, (x_0, y_0) may be a center, unstable spiral, or stable spiral. Hence, we can draw no conclusion.

We will not discuss the case in which the eigenvalues are the same or one eigenvalue is zero. For analyzing nonlinear systems, we state the following useful theorem:

Theorem: Suppose that (x_0, y_0) is an equilibrium point of the nonlinear system $\begin{cases} dx/dt = f(x,y) \\ dy/dt = g(x,y) \end{cases}$. Then, the relationships in the following table hold for the classification of (x_0, y_0) in the nonlinear system and that in the associated linearized system.

Associated Linearized System	Nonlinear System
Stable node	Stable node
Unstable node	Unstable node
Stable spiral	Stable spiral
Unstable spiral	Unstable spiral
Saddle	Saddle
Center	No conclusion

EXAMPLE 4: Find and classify the equilibrium points of $\begin{cases} dx/dt = 1 - y \\ dy/dt = x^2 - y^2 \end{cases}$.

SOLUTION: We begin by finding the equilibrium points of this nonlinear system by solving

$$\begin{cases} 1 - y = 0 \\ x^2 - y^2 = 0 \end{cases}.$$

```
> x:='x':y:='y':
f:=(x,y)->1-y:
g:=(x,y)->x^2-y^2:
Eq_pts:=solve({f(x,y)=0,g(x,y)=0});
```

$$Eq\_pts := \{y = 1, x = 1\}, \{y = 1, x = -1\}$$

Because $f(x,y) = 1 - y$ and $g(x,y) = x^2 - y^2$, $f_x(x,y) = 0$, $f_y(x,y) = -1$, $g_x(x,y) = 2x$, and $g_y(x,y) = -2y$, so the Jacobian matrix is $J(x,y) = \begin{pmatrix} 0 & -1 \\ 2x & -2y \end{pmatrix}$,

which is formed with `jacobian`. (Note that `jacobian` is contained in the **linalg** package.)

```
> with(linalg):
jac:=jacobian([f(x,y),g(x,y)],[x,y]);
```

$$jac := \begin{bmatrix} 0 & -1 \\ 2x & -2y \end{bmatrix}$$

Next, we obtain the linearized system about each equilibrium point. The eigenvalues of each matrix are computed with `eigenvalues`.
For (1,1), we obtain $\lambda_1 = -1 + i$ and $\lambda_2 = -1 - i$. Because these eigenvalues are complex valued with negative real part, we classify (1,1) as a stable spiral. For $(-1,1)$, we obtain $\lambda_1 = -1 + \sqrt{3} > 0$ and $\lambda_2 = -1 - \sqrt{3} < 0$, so $(-1,1)$ is a saddle.

```
> eigenvalues(subs(Eq_pts[1],eval(jac)));
eigenvalues(subs(Eq_pts[2],eval(jac)));
```

$$-1 + I, -1 - I$$
$$-1 + \sqrt{3}, -1 - \sqrt{3}$$

We graph several solutions to this nonlinear system together with the direction field associated with the nonlinear system using `DEplot`. We can see how the solutions move toward and move away from the equilibrium points by observing the arrows on the vectors in the direction field.

```
> with(DEtools):
> ivals:=seq(-1+i/8,i=0..24):
jvals:=seq(-3/2+3/14*j,j=0..14):
initconds1:=seq([0,-3/2,i],i=ivals);
initconds2:=seq([0,j,2],j=jvals);
> initconds:=[initconds1,initconds2]:
> DEplot([diff(x(t),t)=f(x,y),diff(y(t),t)=g(x,y)],
        [x(t),y(t)],t=0..15,initconds,x=-3/2..3/2,y=-1..2,
        stepsize=0.1,arrows=LARGE,color=GRAY,
            linecolor=BLACK,thickness=1);
```

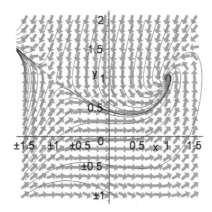

■

CHAPTER 8

Applications of Systems of Ordinary Differential Equations

8.1 Mechanical and Electrical Problems with First-Order Linear Systems

L–R–C Circuits with Loops

As indicated in Chapter 5, an electrical circuit can be modeled with an ordinary differential equation with constant coefficients. In this section, we illustrate how a circuit involving loops can be described as a system of linear ordinary differential equations with constant coefficients. This derivation is based on the following principles.

Kirchhoff's Current Law:	The current entering a point of the circuit equals the current leaving the point.

Kirchhoff's Voltage Law:	The sum of the changes in voltage around each loop in the circuit is zero.

As was the case in Chapter 5, we use the following standard symbols for the components of the circuit: $I(t) = $ current where $I(t) = \frac{dQ}{dt}(t)$, $Q(t) = $ charge, $R = $ resistance, $C = $ capacitance, $V = $ voltage, and $L = $ inductance.

The relationships corresponding to the drops in voltage in the various components of the circuit that were stated in Chapter 5 are also given in the following table.

Circuit Element	Voltage Drop
Inductor	L dI/dt
Resistor	RI
Capacitor	Q/C
Voltage Source	− V(t)

L–R–C Circuit with One Loop

In determining the drops in voltage around the circuit, we consistently add the voltages in the clockwise direction. The positive direction is directed from the negative symbol toward the positive symbol associated with the voltage source. In summing the voltage drops encountered in the circuit, a drop across a component is added to the sum if the positive direction through the component agrees with the clockwise direction. Otherwise, this drop is subtracted. In the case of the following L–R–C circuit with one loop involving each type of component, the current is equal around the circuit by Kirchhoff's current law as illustrated in the following figure.

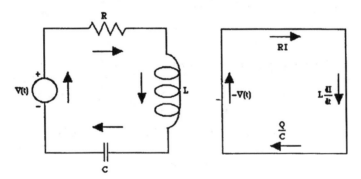

A Simple L–R–C Circuit

Also, by Kirchhoff's voltage law we have the sum

$$RI + L\frac{dI}{dt} + \frac{1}{C}Q - V(t) = 0.$$

Solving this equation for dI/dt and using the relationship between I and Q, $dQ/dt = I$, we have the following system of differential equations with initial conditions on charge and current, respectively.

$$\begin{cases} \frac{dQ}{dt} = I \\ \frac{dI}{dt} = -\frac{1}{LC}Q - \frac{R}{L}I + \frac{V(t)}{L} \\ Q(0) = Q_0, I(0) = I_0 \end{cases}$$

EXAMPLE 1: Determine the charge and current in an L–R–C circuit with $L = 1$, $R = 2$, $C = 4/3$, and $V(t) = e^{-t}$ if $Q(0) = Q_0$ and $I(0) = I_0$.

SOLUTION: We begin by modeling the circuit with the system of differential equations

$$\begin{cases} dQ/dt = I \\ dI/dt = -\frac{3}{4}Q - 2I + e^{-t} \end{cases},$$

which can be written in matrix form as

$$\begin{pmatrix} dQ/dt \\ dI/dt \end{pmatrix} = \begin{pmatrix} 0 & 1 \\ -3/4 & -2 \end{pmatrix}\begin{pmatrix} Q \\ I \end{pmatrix} + \begin{pmatrix} 0 \\ e^{-t} \end{pmatrix}.$$

We solve the initial-value problem with dsolve, naming the result sol.

```
> sol:=dsolve({diff(q(t),t)=i(t),
        diff(i(t),t)=-3/4*q(t)-2*i(t)+exp(-t),
            q(0)=q0,i(0)=i0},{q(t),i(t)});
```

$$sol := \left\{ i(t) = \frac{3}{2}(-4 + i0)e^{\left(-\frac{3}{2}t\right)} - \frac{1}{2}(-4 + i0)e^{\left(-\frac{1}{2}t\right)} \right.$$

$$- \frac{3}{4}(4 + q0)e^{\left(-\frac{1}{2}t\right)} + \frac{3}{4}(4 + q0)e^{\left(-\frac{3}{2}t\right)} + 4e^{(-t)}, q(t) =$$

$$(-4 + i0)e^{\left(-\frac{1}{2}t\right)} - (-4 + i0)e^{\left(-\frac{3}{2}t\right)} - \frac{1}{2}(4 + q0)e^{\left(-\frac{3}{2}t\right)}$$

$$\left. + \frac{3}{2}(4 + q0)e^{\left(-\frac{1}{2}t\right)} - 4e^{(-t)} \right\}$$

The result indicates that $\lim\limits_{t \to \infty} Q(t) = \lim\limits_{t \to \infty} I(t) = 0$ regardless of the values of Q_0 and I_0. This is confirmed by graphing $Q(t)$ (in black) and $I(t)$ (in gray) together for various values of Q_0 and I_0 (we choose $Q_0 = I_0 = 1$) as well as $\mathbf{X}(t) = \begin{pmatrix} Q(t) \\ I(t) \end{pmatrix}$ parametrically.

```
> assign(sol):
plot(subs({q0=1,i0=1},[q(t),i(t)]),t=0..10,
        color=[BLACK,GRAY]);
plot(subs({q0=1,i0=1},[q(t),i(t)]),t=0..10]), colour=BLACK
```

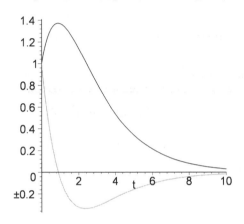
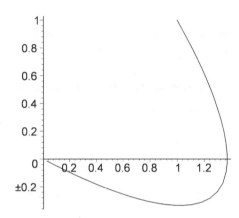

■

L–R–C Circuit with Two Loops

The differential equations that model the circuit become more difficult to derive as the number of loops in the circuit increases. For example, consider the circuit in the following figure that contains two loops.

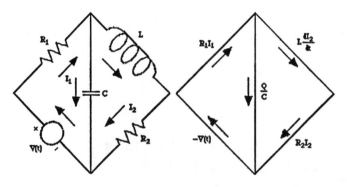

A Two-Loop Circuit

In this case, the current through the capacitor is equivalent to $I_1 - I_2$. Summing the voltage drops around each loop, we have

$$\begin{cases} R_1I_1 + \frac{1}{C}Q - V(t) = 0 \\ L\frac{dI_2}{dt} + R_2I_2 - \frac{1}{C}Q = 0 \end{cases}.$$

Solving the first equation for I_1 so that $I_1 = \frac{1}{R_1}V(t) - \frac{1}{R_1C}Q$ and using the relationship $dQ/dt = I = I_1 - I_2$ we have the following system:

$$\begin{cases} \frac{dQ}{dt} = -\frac{1}{R_1C}Q - I_2 + \frac{V(t)}{R_1} \\ \frac{dI_2}{dt} = \frac{1}{LC}Q - \frac{R_2}{L}I_2 \end{cases}.$$

EXAMPLE 2: Find $Q(t)$, $I(t)$, $I_1(t)$, and $I_2(t)$ in the L–R–C circuit with two loops given that $R_1 = R_2 = C = 1$ and $V(t) = e^{-t}$ if $Q(0) = 3$ and $I_2(0) = 1$.

SOLUTION: The nonhomogeneous system that models this circuit is

$$\begin{cases} dQ/dt = -Q - I_2 + e^{-t} \\ dI_2/dt = Q - I_2 \end{cases}$$

with initial conditions $Q(0) = 3$ and $I_2(0) = 1$. We solve the initial-value problem with dsolve, naming the result sol.

```
> q:='q':i:='i':
> sol:=dsolve({diff(q(t),t)=-q(t)-i2(t)+exp(-t),
      diff(i2(t),t)=q(t)-i2(t),q(0)=3,i2(0)=1},
         {q(t),i2(t)});
```

$$sol := \{q(t) = 3e^{(-t)}\cos(t), i2(t) = e^{(-t)}(3\sin(t) + 1)\}$$

We verify that these functions satisfy the system by substituting back into each equation and simplifying the result with simplify.

```
> assign(sol):
simplify(diff(q(t),t)-(-q(t)-i2(t)+exp(-t)));
```

$$0$$

```
> simplify(diff(i2(t),t)-(q(t)-i2(t)));
```

0

We use the relationship $dQ/dt = I$ to find $I(t)$

```
> i:=diff(q(t),t);
```

$$i := -3\mathbf{e}^{(-t)}\cos(t) - 3\mathbf{e}^{(-t)}\sin(t)$$

and then $I_1(t) = I(t) + I_2(t)$ to find $I_1(t)$.

```
> i1:=simplify(i+i2(t));
```

$$i1 := -3\mathbf{e}^{(-t)}\cos(t) + \mathbf{e}^{(-t)}$$

We graph $Q(t)$, $I(t)$, $I_1(t)$, and $I_2(t)$ with `plot` and display the result using `display` and the option `insequence = true`.

```
> p1:=plot(q(t),t=0..5,color=BLACK):
p2:=plot(i,t=0..5,color=BLACK):
p3:=plot(i1,t=0..5,color=BLACK):
p4:=plot(i2(t),t=0..5,color=BLACK):
```

```
> with(plots):
display(array([[p1,p2],[p3,p4]]),insequence=true);
```

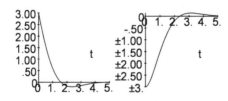

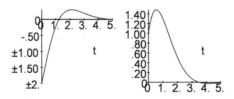

■

For nonlinear systems, solutions can often be visualized graphically by taking advantage of the DEplot and DEplot3d commands, which are contained in the **DEtools** package.

EXAMPLE 3: A certain circuit with two capacitors and an inductor in parallel and a nonlinear resistor in series with the capacitors is modeled by the *dimensionless* initial-value problem

$$\begin{cases} dx/dt = -cf(y - x) \\ dy/dt = -f(y - x) - z \\ dz/dt = y \\ x(0) = 1, y(0) = 1, z(0) = 1 \end{cases},$$

where c is a positive constant and $f(u) = -0.07u + 0.085(|u + 1| - |u - 1|)$. x and y represent the voltages across the capacitor and z is the current through the inductor.

Graph the orbits or projections of the orbits for various values of c.

SOLUTION: We begin by loading the **DEtools** package, defining $f(u) = -0.07u + 0.085(|u + 1| - |u - 1|)$,

```
> with(DEtools):
f:=u->-0.07*u+0.085*(abs(u+1)-abs(u-1));
```

$$f := u \rightarrow -.07u + .085|u + 1| - .085|u - 1|$$

and the system of equations $\begin{cases} dx/dt = -cf(y - x) \\ dy/dt = -f(y - x) - z \\ dz/dt = y \end{cases}$ in sys.

```
> sys:={diff(x(t),t)=-c*f(y(t)-x(t)),
       diff(y(t),t)=-f(y(t)-x(t))-z(t),diff(z(t),t)=y(t)};
```

$$sys := \{\frac{\partial}{\partial t}z(t) = y(t)\frac{\partial}{\partial t}x(t) =$$
$$-c(-.07y(t) + .07x(t) + .085|y(t) - x(t) + 1| - .085|y(t) - x(t) - 1|),$$
$$\frac{\partial}{\partial t}y(t) =$$
$$.07y(t) - .07x(t) - .085|y(t) - x(t) + 1| + .085|y(t) - x(t) - 1| - z(t)$$
$$\}$$

We then graph the orbit for $0 \le t \le 500$ with DEplot3d and the projection of the orbit in the xz-plane with DEplot.

```
> c:=0.1:
```

```
> c:=0.01:
```

```
> DEplot3d(sys,[x(t),y(t),z(t)],t=0..1000,
          [[x(0)=1,y(0)=1,z(0)=1]],x=1/2..3/2,y=-2..2,
       z=-2..2,linecolor=BLACK,
             thickness=1,stepsize=0.2);
```

```
> DEplot(sys,[x(t),y(t),z(t)],t=0..500,
          [[x(0)=1,y(0)=1,z(0)=1]],x=1..3/2,y=-3..3,
       z=-3..3,scene=[x(t),z(t)],linecolor=BLACK,
             thickness=1,stepsize=0.1);
```

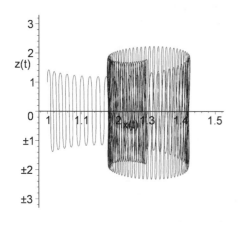
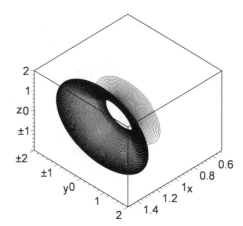

Notice how the projections of the orbits change dramatically as c increases.

```
> c:=1:
DEplot(sys,[x(t),y(t),z(t)],t=0..500,
        [[x(0)=1,y(0)=1,z(0)=1]],x=1/2..2,y=-3..3,
     z=-5/2..5/2,scene=[x(t),z(t)],linecolor=BLACK,
           thickness=1,stepsize=0.1);
```

```
> c:=8:
DEplot(sys,[x(t),y(t),z(t)],t=0..500,
      [[x(0)=1,y(0)=1,z(0)=1]],x=-2..5,y=-3..3,
      z=-7/2..7/2,scene=[x(t),z(t)],linecolor=BLACK,
          thickness=1,stepsize=0.1);
```

```
> c:=10:
DEplot(sys,[x(t),y(t),z(t)],t=0..500,
      [[x(0)=1,y(0)=1,z(0)=1]],x=-1..4,y=-5/2..5/2,
      z=-4..4,scene=[x(t),z(t)],linecolor=BLACK,
          thickness=1,stepsize=0.1);
```

```
> c:=17:
DEplot(sys,[x(t),y(t),z(t)],t=0..500,
      [[x(0)=1,y(0)=1,z(0)=1]],x=-2..5,y=-3..3,
      z=-7/2..7/2,scene=[x(t),z(t)],linecolor=BLACK,
          thickness=1,stepsize=0.1);
```

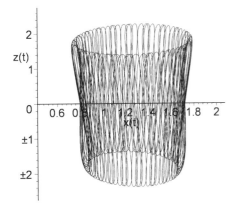
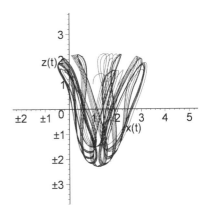

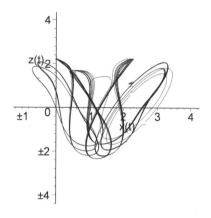

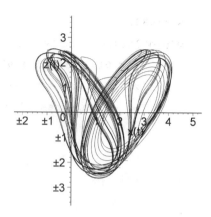

By the time $c = 34$ and 35, the result is a *strange attractor*.

```
> c:=34:
DEplot(sys,[x(t),y(t),z(t)],t=0..500,
        [[x(0)=1,y(0)=1,z(0)=1]],x=-6..6,y=-3..3,
        z=-6..6,scene=[x(t),z(t)],linecolor=BLACK,
                thickness=1,stepsize=0.1);
```

```
> c:=35:
DEplot(sys,[x(t),y(t),z(t)],t=0..500,
        [[x(0)=1,y(0)=1,z(0)=1]],x=-8..8,y=-3..3,
        z=-4..4,scene=[x(t),z(t)],linecolor=BLACK,
                thickness=1,stepsize=0.1);
```

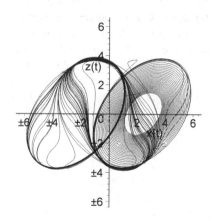

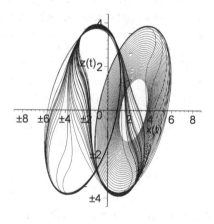

```
> c:=35:

DEplot3d(sys,[x(t),y(t),z(t)],t=0..500,

        [[x(0)=1,y(0)=1,z(0)=1]],x=-7..7,y=-4..4,

    z=-4..4,linecolor=BLACK,

            thickness=1,stepsize=0.1);
```

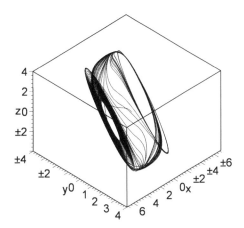

Spring–Mass Systems

The displacement of a mass attached to the end of a spring was modeled with a second-order linear differential equation with constant coefficients in Chapter 5. This situation can then be expressed as a system of first-order ordinary differential equations as well. Recall that if there is no external forcing function, then the second-order differential equation that models this situation is $md^2x/dt^2 + cdx/dt + kx = 0$, where m is the mass attached to the end of the spring, c is the damping coefficient, and k is the spring constant found with Hooke's law. This equation is easily transformed into a system of equations by letting $dx/dt = y$ so that $dy/dt = d^2x/dt^2 = -\frac{k}{m}x - \frac{c}{m}dx/dt$ and then solving the differential equation for d^2x/dt^2. After substitution, we have the system

$$\begin{cases} \frac{dx}{dt} = y \\ \frac{dy}{dt} = -\frac{k}{m}x - \frac{c}{m}y \end{cases}$$

In previous chapters, the displacement of the spring was illustrated as a function of time. However, problems of this type may also be investigated using the phase plane.

EXAMPLE 4: Solve the system of differential equations to find the displacement of the mass if $m = 1$, $c = 0$, and $k = 1$.

SOLUTION: In this case, the system is

$$\begin{cases} dx/dt = y \\ dy/dt = -x \end{cases}$$

which in matrix form is

$$\mathbf{X}' = \begin{pmatrix} 0 & 1 \\ -1 & 0 \end{pmatrix} \mathbf{X};$$

a general solution is found with `dsolve` and named `gensol` for later use.

```
> gensol:=dsolve({diff(x(t),t)=y(t),diff(y(t),t)=-x(t)},
      {x(t),y(t)});
```

$$gensol := \{x(t) = \cos(t)\_C1 + \sin(t)\_C2, y(t) = -\sin(t)\_C1 + \cos(t)\_C2\}$$

Note that this system is equivalent to the second-order differential equation $d^2x/dt^2 + x = 0$, which we solved in Chapters 4 and 5. At that time, we found a general solution to be $x(t) = c_1 \cos t + c_2 \sin t$ which is equivalent to the first component of $\mathbf{X}(t) = \begin{pmatrix} x(t) \\ y(t) \end{pmatrix}$, the result obtained with `dsolve`. Also notice that (0,

0) is the equilibrium point of the system. The eigenvalues of $\begin{pmatrix} 0 & 1 \\ -1 & 0 \end{pmatrix}$ are $\lambda = \pm i$,

```
> with(linalg):
eigenvalues([[0,1],[-1,0]]);
```

$$I, -I$$

so we classify the origin as a center. We graph several members of the phase plane and the direction field for this system with `DEplot`.

```
> with(DEtools):
> DEplot([diff(x(t),t)=y(t),diff(y(t),t)=-x(t)],
      [x(t),y(t)],t=0..2*Pi,[[x(0)=0.2,y(0)=0],
         [x(0)=0.4,y(0)=0],[x(0)=0.6,y(0)=0],
            [x(0)=0.8,y(0)=0]],x=-1..1,y=-1..1,
               color=GRAY,linecolor=BLACK);
```

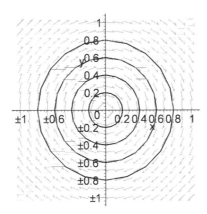

8.2 Diffusion and Population Problems with First-Order Linear Systems

Diffusion through a Membrane

Solving problems to determine the diffusion of a substance (such as glucose or salt) in a medium (such as a blood cell) also leads to systems of first-order linear ordinary differential equations. For example, suppose that two solutions of a substance are separated by a membrane where the amount of the substance that passes through the membrane is proportional to the difference in the concentrations of the solutions. The constant of proportionality is called the **permeability** P of the membrane. Therefore, if we let x and y represent the concentrations of the solutions and V_1 and V_2 represent the volumes of the solutions, the system of differential equations is given by

$$\begin{cases} \frac{dx}{dt} = \frac{P}{V_1}(y - x) \\ \frac{dy}{dt} = \frac{P}{V_2}(x - y) \end{cases}$$

where the initial concentrations of x and y are given.

EXAMPLE 1: Suppose that two salt concentrations of equal volume V are separated by a membrane of permeability P. Given that $P = V$, determine each concentration at time t if $x(0) = 2$ and $y(0) = 0$.

SOLUTION: In this case, the initial-value problem that models the situation is

$$\begin{cases} dx/dt = y - x \\ dy/dt = x - y \\ x(0) = 2, y(0) = 10 \end{cases}.$$

A general solution of the system is found with `dsolve` and named `gensol`.

```
> gensol:=dsolve({diff(x(t),t)=y(t)-x(t),
        diff(y(t),t)=x(t)-y(t)},[x(t),y(t)]);
```

$$gensol := \{ y(t) = \frac{1}{2}\_C3 - \frac{1}{2}\_C3e^{(-2t)} + \frac{1}{2}\_C4 + \frac{1}{2}\_C4e^{(-2t)},$$

$$x(t) = \frac{1}{2}\_C3 + \frac{1}{2}\_C3e^{(-2t)} + \frac{1}{2}\_C4 - \frac{1}{2}\_C4e^{(-2t)}\}$$

We then apply the initial conditions and use `solve` to determine the values of the arbitrary constants.

```
> assign(gensol):
cvals:=solve(eval(subs(t=0,{x(t)=2,y(t)=0})));
```

$$cvals := \{\_C3 = 2, \_C4 = 0\}$$

The solution is obtained by substiting these values back into the general solution.

```
> assign(cvals);
eval([x(t), y(t)]);
```

$$[e^{(-2t)} + 1, 1 - e^{(-2t)}]$$

Of course, `dsolve` can be used to solve the initial-value problem directly as well.

```
> x:='x':y:='y':
sol:=dsolve({diff(x(t),t)=y(t)-x(t),
        diff(y(t),t)=x(t)-y(t),x(0)=2,y(0)=0},
        [x(t),y(t)],method=laplace);
```

$$sol := \{x(t) = e^{(-2t)} + 1, y(t) = 1 - e^{(-2t)}\}$$

We graph this solution parametrically with `plot` and then graph $x(t)$ and $y(t)$ simultaneously. Notice that the amount of salt in each concentration approaches 1, which is the average value of the two initial amounts.

```
> assign(sol):
plot([x(t),y(t)],t=0..5,color=[GRAY,BLACK]);
plot([x(t),y(t),t=0..5],color=BLACK);
```

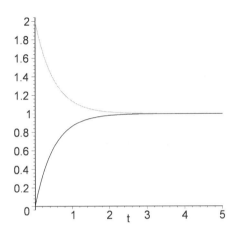 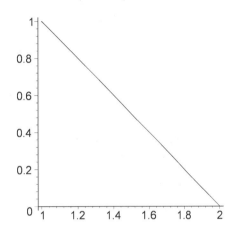

Diffusion through a Double-Walled Membrane

Next, consider the situation in which two solutions are separated by a double-walled membrane, where the inner wall has permeability P_1 and the outer wall has permeability P_2 with $0 < P_1 < P_2$. Suppose that the volume of solution within the inner wall is V_1 and that between the two walls is V_2. Let x represent the concentration of the solution within the inner wall and y the concentration between the two walls. Assuming that the concentration of the solution outside the outer wall is constantly C, we have the following system of first-order ordinary differential equations

$$\begin{cases} \frac{dx}{dt} = \frac{P_1}{V_1}(y - x) \\ \frac{dy}{dt} = \frac{1}{V_2}(P_2(C - y) + P_1(x - y)) \\ \quad x(0) = x_0, y(0) = y_0 \end{cases} .$$

EXAMPLE 2: Given that $P_1 = 3$, $P_2 = 8$, $V_1 = 2$, $V_2 = 10$, and $C = 10$, determine x and y if $x(0) = 2$ and $y(0) = 1$.

SOLUTION: In this case, we must solve the initial-value problem

$$\begin{cases} dx/dt = \frac{3}{2}(y - x) \\ dy/dt = -\frac{11}{10}y + \frac{3}{10}x + 8 \\ \quad x(0) = 2, y(0) = 1 \end{cases}.$$

A fundamental matrix of the corresponding homogeneous system is found with `matrixDE`.

```
> with(DEtools);

A:=matrix([[-3/2,3/2],[3/10,-11/10]]):

B:=matrix([[0],[8]]):

fundmat:=matrixDE(A,B,t);
```

$$\textit{fundmat} := \left[\left[\begin{matrix} \mathbf{e}^{(-2t)} & \mathbf{e}^{\left(-\frac{3}{5}t\right)} \\ -\frac{1}{3}\mathbf{e}^{(-2t)} & \frac{3}{5}\mathbf{e}^{\left(-\frac{3}{5}t\right)} \end{matrix} \right], [10, 10] \right]$$

The result indicates that a fundamental matrix for the corresponding homogeneous system is $\Phi(t) = \begin{pmatrix} e^{-2t}e^{-3t/5} \\ -\frac{1}{3}e^{-2t} \quad \frac{3}{5}e^{-3t/5} \end{pmatrix}$.

```
> Phi:=fundmat[1];
```

$$\Phi := \left[\begin{matrix} \mathbf{e}^{(-2t)} & \mathbf{e}^{\left(-\frac{3}{5}t\right)} \\ -\frac{1}{3}\mathbf{e}^{(-2t)} & \frac{3}{5}\mathbf{e}^{\left(-\frac{3}{5}t\right)} \end{matrix} \right]$$

Therefore, using the method of variation of parameters, the solution to the initial-value problem is given by

$$\mathbf{X}(t) = \Phi(t)\Phi^{-1}(0)\mathbf{X}(0) + \Phi(t) \int_0^t \Phi^{-1}(u)\mathbf{F}(u)\, du.$$

```
> sol:=simplify(evalm(Phi &*

      eval(subs(t=0,inverse(Phi)))
```

$$sol := \left[\begin{matrix} -\frac{1}{2}\left(-9\mathbf{e}^{\left(\frac{3}{5}t\right)} + 25\mathbf{e}^{(2t)} - 20\mathbf{e}^{\left(\frac{13}{5}t\right)}\right)\mathbf{e}^{\left(-\frac{13}{5}t\right)} \\ -\frac{1}{2}\left(3\mathbf{e}^{\left(\frac{3}{5}t\right)} + 15\mathbf{e}^{(2t)} - 20\mathbf{e}^{\left(\frac{13}{5}t\right)}\right)\mathbf{e}^{\left(-\frac{13}{5}t\right)} \end{matrix} \right]$$

Of course, `dsolve` can be used to solve the initial-value problem directly, as well.

```
> x:='x':y:='y':
    sol:=dsolve({diff(x(t),t)=3/2*(y(t)-x(t)),
            diff(y(t),t)=-11/10*y(t)+3/10*x(t)+8,
                x(0)=2,y(0)=1},{x(t),y(t)});
```

$$sol := \left\{ y(t) = -\frac{3}{2}e^{(-2t)} - \frac{15}{2}e^{\left(-\frac{3}{5}t\right)} + 10, x(t) = \frac{9}{2}e^{(-2t)} - \frac{25}{2}e^{\left(-\frac{3}{5}t\right)} + 10 \right\}$$

We graph this solution parametrically in addition to graphing the two functions simultaneously. Notice that initially $x(t) > y(t)$. However, the two graphs intersect at a value of t near $t = 0$ so that $y(t) > x(t)$ as the values of the two functions approach 10 (which is the concentration of the solution outside the outer wall) as t increases.

```
> assign(sol):
plot([x(t),y(t),t=0..7],color=BLACK);
plot([x(t),y(t)],t=0..7,color=[BLACK,GRAY]);
```

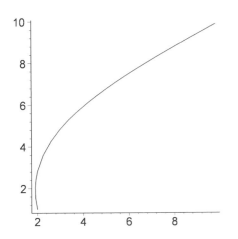
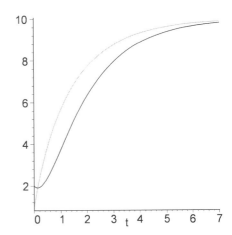

Last, we plot the solution parametrically for various initial conditions.

```
> x:='x':y:='y':
sol:=dsolve({diff(x(t),t)=3/2*(y(t)-x(t)),
        diff(y(t),t)=-11/10*y(t)+3/10*x(t)+8,
```

```
x(0)=x0,y(0)=y0},{x(t),y(t)});
```

$$sol := \left\{ x(t) = \frac{9}{14}(-10 + x0)e^{(-2t)} + \frac{5}{14}(-10 + x0)e^{\left(-\frac{3}{2}t\right)} \right.$$

$$+ \frac{15}{14}(-10 + y0)e^{\left(-\frac{3}{2}t\right)} - \frac{15}{14}(-10 + y0)e^{(-2t)} + 10, y(t) =$$

$$\frac{3}{14}(-10 + x0)e^{\left(-\frac{3}{2}t\right)} - \frac{3}{14}(-10 + x0)e^{(-2t)} + \frac{5}{14}(-10 + y0)e^{(-2t)}$$

$$\left. + \frac{9}{14}(-10 + y0)e^{\left(-\frac{3}{2}t\right)} + 10 \right\}$$

We use `seq` and construct a list of (pairs of) functions to be plotted with `plot`. The list of functions in `toplot` is then graphed with `ParametricPlot` for $0 \leq t \leq 7$.

```
> assign(sol):

vals:=seq(2*i,i=0..5):

toplot:={seq(seq(subs({x0=i,y0=j},
        [x(t),y(t),t=0..7]),i=vals),j=vals)}:
```

```
> plot(toplot,color=BLACK);
```

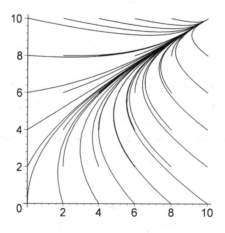

Population Problems

In Chapter 3, population problems were discussed that were based on the principle that the rate at which a population grows (or decays) is proportional to the number present in the population at any time t. Hence, if $x(t)$ represents the population at time t, $dx/dt = kx$ for some constant k. This idea can be extended to problems involving more than one population and leads to systems of ordinary differential equations. We illustrate several situations through the following examples. Note that in each problem, we determine the rate at which a population P changes with the equation

$$dP/dt = \text{(rate entering)} - \text{(rate leaving)}.$$

We begin by determining the populations in two neighboring territories. Suppose that the populations x and y of two neighboring territories depend on several factors. The birth rate of x is a_1 while that of y is b_1. The rate at which citizens of x move to y is a_2 while that at which citizens move from y to x is b_2. Finally, the mortality rate of each territory is disregarded. Determine the respective populations of these two territories for any time t.

Using the simple principles of previous examples, we have that the rate at which population x changes is

$$dx/dt = a_1 x - a_2 x + b_1 y = (a_1 - a_2)x + b_1 y$$

while the rate at which population y changes is

$$dy/dt = b_1 y - b_2 y + a_2 x = (b_1 - b_2)y + a_2 x.$$

Therefore, the system of equations that must be solved is

$$\begin{cases} \frac{dx}{dt} = (a_1 - a_2)x + b_1 y \\ \frac{dy}{dt} = a_2 x + (b_1 - b_2)y \end{cases}$$

where the initial populations of the two territories $x(0) = x_0$ and $y(0) = y_0$ are given.

EXAMPLE 3: Determine the populations $x(t)$ and $y(t)$ in each territory if $a_1 = 5$, $a_2 = 4$, $b_1 = 2$, and $b_2 = 3$ given that $x(0) = 60$ and $y(0) = 10$.

SOLUTION: In this case, the initial-value problem that models the situation is

$$\begin{cases} dx/dt = x + y \\ dy/dt = 4x - 2y \\ x(0) = 60, y(0) = 10 \end{cases},$$

which we solve directly with `dsolve`.

```
> x:='x':y:='y':
sol:=dsolve({diff(x(t),t)=x(t)+y(t),
```

```
diff(y(t),t)=4*x(t)-2*y(t),x(0)=60,y(0)=10},

    {x(t),y(t)});
```

$$sol := \{y(t) = 50e^{(2t)} - 40e^{(-3t)}, x(t) = 10e^{(-3t)} + 50e^{(2t)}\}$$

We graph these two population functions with `plot`. Notice that as t increases, the two populations are approximately the same.

```
> assign(sol):

plot([x(t),y(t)],t=0..1.2,color=[BLACK,GRAY]);
```

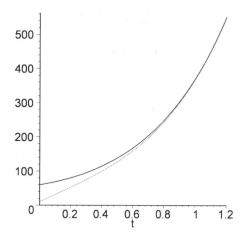

■

Population problems that involve more than two neighboring populations can be solved with a system of differential equation as well. Suppose that the populations of three neighboring territories x, y, and z depend on several factors. The birth rates of x, y, and z are a_1, b_1, and c_1, respectively. The rate at which citizens of x move to y is a_2 while that at which citizens move from x to z is a_3. Similarly, the rate at which citizens of y move to x is b_2 while that at which citizens move from y to z is b_3. Also, the rate at which citizens of z move to x is c_2 while that at which citizens move from z to y is c_3. Suppose that the mortality rate of each territory is ignored in the model.

The system of equations in this case is similar to that derived in the previous example. The rate at which population x changes is

$$dx/dt = a_1 x - a_2 x - a_3 x + b_2 y + c_2 z = (a_1 - a_2 - a_3)x + b_2 y + c_2 z,$$

while the rate at which population y changes is

$$dy/dt = b_1 y - b_2 y - b_3 y + a_2 x + c_3 z = (b_1 - b_2 - b_3)y + a_2 x + c_3 z,$$

and that of z is

$$dz/dt = c_1 z - c_2 z - c_3 z + a_3 x + b_3 y = (c_1 - c_2 - c_3)z + a_3 x + b_3 y.$$

Hence, we must solve the 3×3 system

$$\begin{cases} \frac{dx}{dt} = (a_1 - a_2 - a_3)x + b_2 y + c_2 z \\ \frac{dy}{dt} = a_2 x + (b_1 - b_2 - b_3)y + c_3 z \\ \frac{dz}{dt} = a_3 x + b_3 y + (c_1 - c_2 - c_3)z \end{cases}$$

where the initial populations $x(0) = x_0$, $y(0) = y_0$, and $z(0) = z_0$ are given.

EXAMPLE 4: Determine the populations of the three territories if $a_1 = 3$, $a_2 = 0$, $a_3 = 2$, $b_1 = 4$, $b_2 = 2$, $b_3 = 1$, $c_1 = 5$, $c_2 = 3$, and $c_3 = 0$ if $x(0) = 50$, $y(0) = 60$, and $z(0) = 25$.

SOLUTION: In this case, the system of differential equations is

$$\begin{cases} dx/dt = x + 2y + 3z \\ dy/dt = y \\ dz/dt = 2x + y + 2z \\ x(0) = 50, y(0) = 60, z(0) = 25 \end{cases},$$

which we solve directly with dsolve.

```
> x:='x':y:='y':z:='z':
sol:=dsolve({diff(x(t),t)=x(t)+2*y(t)+3*z(t),
      diff(y(t),t)=y(t),diff(z(t),t)=2*x(t)+y(t)+2*z(t),
      x(0)=50,y(0)=60,z(0)=25},{x(t),y(t),z(t)});
```

$$sol := \{x(t) = 63e^{(4t)} - 3e^{(-t)} - 10e^t, y(t) = 60e^t,$$
$$z(t) = 2e^{(-t)} + 63e^{(4t)} - 40e^t\}$$

The graphs of these three population functions are generated with plot. We notice that although y was initially greater than populations x and z, these populations increase at a much higher rate than does y.

```
> assign(sol):
plot([x(t),y(t),z(t)],t=0..0.5,
      color=[BLACK,GRAY,RED]);
```

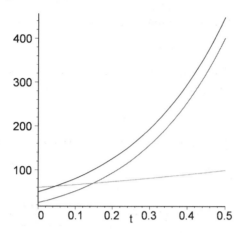

8.3 Applications Using Laplace Transforms

Coupled Spring–Mass Systems

The motion of a mass attached to the end of a spring was modeled with a second-order linear differential equation with constant coefficients in Chapter 5. Similarly, if a second spring and mass are attached to the end of the first mass, then the model becomes that of a system of second-order equations. To state the problem more precisely, let masses m_1 and m_2 be attached to the ends of springs S_1 and S_2 having spring constants k_1 and k_2, respectively. Then, spring S_2 is attached to the base of mass m_1. Suppose that $x(t)$ and $y(t)$ represent the vertical displacements from the equilibrium position of springs S_1 and S_2, respectively. Because spring S_2 undergoes both elongation and compression when the system is in motion (due to the spring S_1 and the mass m_2), then according to Hooke's law, S_2 exerts the force $k_2(y - x)$ on m_2 while S_1 exerts the force $-k_1 x$ on m_2 Therefore, the force acting on mass m_1 is the sum $-k_1 x + k_2(y - x)$ and that acting on m_2 is $-k_2(y - x)$. Hence, using Newton's second law $(F = ma)$ with each mass, we have the system

$$\begin{cases} m_1 \frac{d^2 x}{dt^2} = -k_1 x + k_2(y - x) \\ m_2 \frac{d^2 y}{dt^2} = -k_2(y - x) \end{cases}$$

The initial positions and velocities of the two masses m_1 and m_2 are given by $x(0), x'(0), y(0)\$,$ and $y'(0)$, respectively.

    If external forces $F_1(t)$ and $F_2(t)$ are applied to the masses, the system of equations becomes

$$\begin{cases} m_1 \frac{d^2 x}{dt^2} = -k_1 x + k_2(y - x) + F_1(t) \\ m_2 \frac{d^2 y}{dt^2} = -k_2(y - x) + F_2(t)) \end{cases}.$$

Therefore, the method of Laplace transforms can be used to solve problems of this type. As in Chapter 6, be sure that you have loaded the **inttrans** package before completing any of the following examples.

EXAMPLE 1: Consider the spring–mass system with $m_1 = m_2 = 1$, $k_1 = 3$, and $k_2 = 2$. Find the position functions $x(t)$ and $y(t)$ if $x(0) = 0$, $x'(0) = 1$, $y(0) = 1$, and $y'(0) = 0$. (Assume there are no external forces.)

SOLUTION: In order to find $x(t)$ and $y(t)$, we must solve the initial-value problem

$$\begin{cases} d^2 x/dt^2 = -5x + 2y \\ d^2 y/dt^2 = 2x - 2y \\ x(0) = 0, x'(0) = 1, y(0) = 1, y'(0) = 0 \end{cases}.$$

After defining the equations in Eqs, we use `laplace` to take the Laplace transform of both sides of each equation.

```
> x:='x':y:='y':
Eqs:={diff(x(t),t$2)=-5*x(t)+2*y(t),
        diff(y(t),t$2)=2*x(t)-2*y(t)}:
```

```
> step_1:=laplace(Eqs,t,s);
```

$$step\_1 := \{s(s \text{ laplace}(x(t), t, s) - x(0)) - D(x)(0) =$$
$$- 5 \text{ laplace}(x(t), t, s) + 2 \text{ laplace}(y(t), t, s),$$
$$s(s \text{ laplace}(y(t), t, s) - y(0)) - D(y)(0) =$$
$$2 \text{ laplace}(x(t), t, s) - 2 \text{ laplace}(y(t), t, s)\}$$

We then apply the initial conditions.

```
> step_2:=subs({x(0)=0,D(x)(0)=1,y(0)=1,D(y)(0)=0},
        step_1);
```

$$step\_2 := \{$$

$$s(s \text{ laplace}(y(t), t, s) - 1) = 2 \text{ laplace}(x(t), t, s) - 2 \text{ laplace}(y(t), t, s)$$

$$,$$

$$s^2 \text{ laplace}(x(t), t, s) - 1 = -5 \text{ laplace}(x(t), t, s) + 2 \text{ laplace}(y(t), t, s)$$

$$\}$$

We solve this system of algebraic equations for $X(s)$ and $Y(s)$ with `solve`.

```
> step_3:=solve(step_2,
     {laplace(x(t),t,s),laplace(y(t),t,s)});
```

$$step\_3 := \left\{ \right.$$

$$\text{laplace}(y(t), t, s) = \frac{s^3 + 2 + 5s}{s^4 + 7s^2 + 6}, \text{laplace}(x(t), t, s) = \frac{s^2 + 2s + 2}{s^4 + 7s^2 + 6}$$

$$\left. \right\}$$

Taking the inverse Laplace transform with `invlaplace` yields $x(t)$ and $y(t)$.

```
> Sol:=invlaplace(step_3,s,t);
```

$$Sol := \{y(t) = -\frac{1}{15}\sqrt{6}\sin(\sqrt{6}t) + \frac{1}{5}\cos(\sqrt{6}t) + \frac{4}{5}\cos(t) + \frac{2}{5}\sin(t),$$

$$x(t) = \frac{2}{15}\sqrt{6}\sin(\sqrt{6}t) - \frac{2}{5}\cos(\sqrt{6}t) + \frac{2}{5}\cos(t) + \frac{1}{5}\sin(t)\}$$

Note that `dsolve` is equally successful in solving the initial-value problem.

```
> Init_Val_Prob:=
      Eqs union {x(0)=0,D(x)(0)=1,y(0)=1,D(y)(0)=0}:
Sol2:=dsolve(Init_Val_Prob,{x(t),y(t)});
```

$$Sol2 := \{$$

$$y(t) = -\frac{1}{15}\sqrt{6}\sin(\sqrt{6}t) + \frac{1}{5}\cos(\sqrt{6}t) + \frac{4}{5}\cos(t) + \frac{2}{5}\sin(t),$$

$$x(t) = \frac{2}{15}\sqrt{6}\sin(\sqrt{6}t) - \frac{2}{5}\cos(\sqrt{6}t) + \frac{2}{5}\cos(t) + \frac{1}{5}\sin(t)\}$$

After naming $x(t)$ and $y(t)$ the result obtained in `Sol2` with `assign`, we graph $x(t)$ and $y(t)$ together and then parametrically. Note that $y(t)$ starts at (0,1) while $x(t)$ has initial point (0,0). Notice that this phase plane is different from those discussed in previous sections. One of the reasons for this is that the equations in the system of differential equations are second order instead of first order.

```
> assign(Sol2):
plot([x(t),y(t)],t=0..2*Pi,color=[BLACK,GRAY]);
plot([x(t),y(t),t=0..2*Pi],numpoints=200,color=BLACK);
```

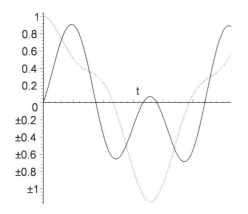
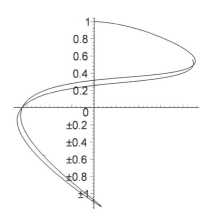

We can take advantage of Maple's animation capabilities to view the motion of the spring. We begin by defining the function `spring` in the same way that we did in Section 5.1.

```
> spring:='spring':
n:=15:
eps:=0.1:
L1:=1:
L2:=1:
spring:=proc(t0)
        local m,xt0,yt0,pts;
        xt0:=evalf(subs(t=t0,x(t)));
        yt0:=evalf(subs(t=t0,y(t)));
        pts:=[[0,yt0-L2],
            seq([eps*(-1)^m,yt0-L2+m*(xt0-yt0+L2)/n],m=1..n-1),
                [0,xt0],
```

```
          seq([eps*(-1)^m,xt0+m*(L1-xt0)/n],m=1..n-1),
               [0,L1]];
     plot(pts,view=[-1..1,-2.5..1.5],
               xtickmarks=2,ytickmarks=2);
     end:
```

We then use `seq` to generate the list of graphs spring (k) for 60 equally spaced values of k between 0 and 16, naming the resulting list of 60 graphs `to_animate`.

```
> k_vals:=seq(k*16/59,k=0..59):
to_animate:=[seq(spring(k),k=k_vals)]:
```

The list of graphs `to_animate` is animated using `display`, which is contained in the **plots** package, together with the `insequence` option. The following screen shot shows one frame from the resulting animation.

```
> with(plots):
display(to_animate,insequence=true);
```

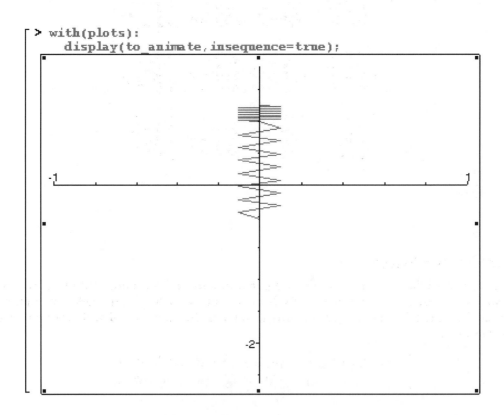

Alternatively, we use `seq` to generate the list of graphs `spring(k)` for 16 equally spaced values of k between 0 and 16, naming the resulting list of 16 graphs `to_animate`.

```
> k_vals:=seq(k*16/59,k=0..59):

to_animate:=[seq(spring(k),k=k_vals)]:
```

The list of graphs `to_animate` is displayed as a graphics array using `display` together with the `insequence` option followed by `display`.

```
> k_vals:=seq(k*16/15,k=0..15):

to_animate:=[seq(spring(k),k=k_vals)]:

p2:=display(to_animate,insequence=true):

display(p2);
```

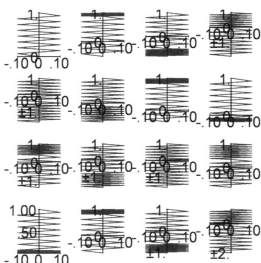

The Double Pendulum

In a method similar to that of the simple pendulum in Chapter 5 and that of the coupled spring–mass system, the motion of a double pendulum as shown in the following figure is modeled by the following system of equations using the approximation $\sin\theta \approx \theta$ for small displacements:

$$\begin{cases} (m_1+m_2)\ell_1^2\theta_1'' + m_2\ell_1\ell_2\theta_2'' + (m_1+m_2)\ell_1 g\theta_1 = 0 \\ m_2\ell_2^2\theta_2'' + m_2\ell_1\ell_2\theta_1'' + m_2\ell_2 g\theta_2 = 0 \end{cases}$$

where θ_1 represents the displacement of the upper pendulum and θ_2 that of the lower pendulum. Also, m_1 and m_2 represent the masses attached to the upper and lower pendulums, respectively, and this lengths are given by ℓ_1 and ℓ_2.

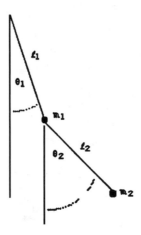

EXAMPLE 2: Suppose that $m_1 = 3$, $m_2 = 1$, and each pendulum has length 16. If $g = 32$, determine $\theta_1(t)$ and $\theta_2(t)$ if $\theta_1(0) = 1$, $\theta_1{}'(0) = 0$, $\theta_2(0) = 0$, and $\theta_2{}'(0) = -1$.

SOLUTION: In this case, the system is

$$\begin{cases} 4 \cdot 16^2\theta_1{}'' + 16^2\theta_2{}'' + 4 \cdot 16 \cdot 32\theta_1 = 0 \\ 16^2\theta_1{}'' + 16^2\theta_2{}'' + 16 \cdot 32\theta_2 = 0 \end{cases}$$

which can be simplified to obtain

$$\begin{cases} 4\theta_1{}'' + \theta_2{}'' + 8\theta_1 = 0 \\ \theta_1{}'' + \theta_2{}'' + 2\theta_2 = 0 \end{cases}.$$

After defining Eqs, we use `laplace` to compute the Laplace transform of each side of both equations, naming the result `step_1`, and then apply the initial conditions to `step_1` with `subs`, naming the resulting output `step_2`.

```
> Eqs:={4*diff(theta[1](t),t$2)+diff(theta[2](t),t$2)+
           8*theta[1](t)=0,
        diff(theta[1](t),t$2)+diff(theta[2](t),t$2)+
           2*theta[2](t)=0}:
step_1:=laplace(Eqs,t,s):
```

```
step_2:=subs({theta[1](0)=1,D(theta[1])(0)=0,
        theta[2](0)=0,D(theta[2])(0)=-1},step_1);
```

$$step\_2 := \{s(s \text{ laplace}(\theta_1(t), t, s) - 1) + 1 + s^2 \text{ laplace}(\theta_2(t), t, s)$$
$$+ 2\text{laplace}(\theta_2(t), t, s) = 0, 4s(s \text{ laplace}(\theta_1(t), t, s) - 1) + 1$$
$$+ s^2 \text{ laplace}(\theta_2(t), t, s) + 8 \text{ laplace}(\theta_1(t), t, s) = 0\}$$

We then solve the equations in `step_2` for `laplace (theta [1] (t), t, s)`, which represents the Laplace transform of $\theta_1(t)$, and `laplace (theta [2] (t), t, s)`, which represents the Laplace transform of $\theta_2(t)$, with `solve`.

```
> step_3:=solve(step_2,
        {laplace(theta[1](t),t,s),laplace(theta[2](t),t,s)});
```

$$step\_3 := \left\{ \text{laplace}(\theta_2(t), t, s) = -\frac{3s^2 - 8s + 8}{3s^4 + 16s^2 + 16}, \right.$$
$$\left. \text{laplace}(\theta_1(t), t, s) = \frac{3s^3 + 8s - 2}{3s^4 + 16s^2 + 16} \right\}$$

`invlaplace` yields the solution, named `sol`.

```
> Sol:=invlaplace(step_3,s,t);
```

$$Sol := \left\{ \theta_1(t) = \frac{1}{2}\cos(2t) + \frac{1}{8}\sin(2t) + \frac{1}{2}\cos\left(\frac{2}{3}\sqrt{3}t\right) - \frac{1}{8}\sqrt{3}\sin\left(\frac{2}{3}\sqrt{3}t\right), \right.$$
$$\left. \theta_2(t) = -\cos(2t) - \frac{1}{4}\sin(2t) - \frac{1}{4}\sqrt{3}\sin\left(\frac{2}{3}\sqrt{3}t\right) + \cos\left(\frac{2}{3}\sqrt{3}t\right) \right\}$$

As in many previous examples, we see that we obtain the same results with `dsolve`, provided that we include the option `method = laplace` in the `dsolve` command.

```
> Init_Val_Prob:=Eqs union {theta[1](0)=1,D(theta[1])(0)=0,
        theta[2](0)=0,D(theta[2])(0)=-1}:
        dsolve(Init_Val_Prob,{theta[1](t),theta[2](t)});
```

$$\left\{ \theta_1(t) = \frac{1}{2}\cos(2t) + \frac{1}{8}\sin(2t) + \frac{1}{2}\cos\left(\frac{2}{3}\sqrt{3}t\right) - \frac{1}{8}\sqrt{3}\sin\left(\frac{2}{3}\sqrt{3}t\right), \right.$$
$$\left. \theta_2(t) = -\cos(2t) - \frac{1}{4}\sin(2t) - \frac{1}{4}\sqrt{3}\sin\left(\frac{2}{3}\sqrt{3}t\right) + \cos\left(\frac{2}{3}\sqrt{3}t\right) \right\}$$

These two functions are graphed together and then parametrically with `plot` to show the solution in the phase plane.

```
> assign(Sol):
plot([theta[1](t),theta[2](t)],t=0..15,color=[BLACK,GRAY]);
plot([theta[1](t),theta[2](t),t=0..15],numpoints=200,
        color=BLACK);
```

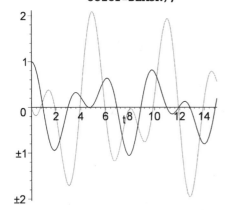 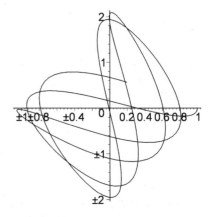

In the same way as in Section 5.5, we can animate the motion of the pendulum with Maple's animation capabilities. First, we define the function pen2. The result of entering pen2 (t, l1, l2) looks like the motion of the pendulum with lengths l1 and l2 at time t.

```
> pen2:=proc(t0,len1,len2)
        local pt1,pt2,xt0,yt0;
        xt0:=evalf(subs(t=t0,theta[1](t)));
        yt0:=evalf(subs(t=t0,theta[2](t)));
        pt1:=[len1*cos(3*Pi/2+xt0),len1*sin(3*Pi/2+xt0)];
        pt2:=[len1*cos(3*Pi/2+xt0)+len2*cos(3*Pi/2+yt0),
                len1*sin(3*Pi/2+xt0)+len2*sin(3*Pi/2+yt0)];
        plot([[0,0],pt1,pt2],xtickmarks=2,ytickmarks=2,
                view=[-32..32,-32..0]);
        end:
```

As in the previous example, we can generate several graphs and display the result as a graphics array.

```
> k_vals:=seq(k*10/15,k=0..15):
to_animate:=[seq(pen2(k,16,16),k=k_vals)]:
```

```
with(plots):

p1:=display(to_animate,insequence=true):

display(p1);
```

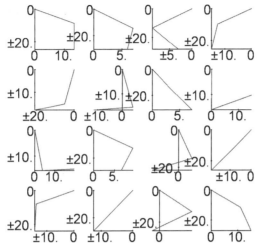

Alternatively, we can generate a list of graphs and then animate the result using the `display` function together with the `insequence = true` option.

```
> k_vals:=seq(k*10/119,k=0..119):

to_animate:=[seq(pen2(k,16,16),k=k_vals)]:

display(to_animate,insequence=true);
```

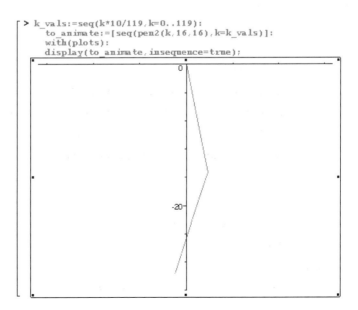

APPLICATION

Free Vibration of a Three-Story Building

If you have ever gone to the top of a tall building such as the Sears Tower, World Trade Center, or Empire State Building on a windy day, you may have been acutely aware of the *sway* of the building. In fact, all buildings sway, or vibrate, naturally. Usually, we are aware of the sway of a building only when we are in a very tall building or in a building during an event such as an earthquake. In some tall buildings, such as the John Hancock Building in Boston, the sway of the building during high winds is reduced by installing a tuned mass damper at the top of the building that oscillates at the same frequency as the building but out of phase. We will investigate the sway of a three-story building and then try to determine how we would investigate the sway of a tall building.

We make two assumptions to solve this problem. First, we assume that the mass distribution of the building can be represented by the lumped masses at the different levels. Second, we assume that the girders of the structure are infinitely rigid in comparison with the supporting columns. With these assumptions, we can determine the motion of the building by interpreting the columns as springs in parallel.

Assume that the coordinates x_1, x_2, and x_3 as well as the velocities and accelerations are positive to the right. Also assume that $x_3 > x_2 > x_1$.

In applying Newton's second law of motion, recall that we have assumed that acceleration is in the positive direction. Therefore, we sum forces in the same direction as the acceleration positively and others negatively. With this configuration, Newton's second law for each of the three masses yields the following system of differential equations.

$$\begin{cases} -k_1 x_1 + k_2(x_2 - x_1) = m_1 \frac{d^2 x_1}{dt^2} \\ -k_2(x_2 - x_1) + k_3(x_3 - x_2) = m_2 \frac{d^2 x_2}{dt^2} \\ -k_3(x_3 - x_2) = m_3 \frac{d^2 x_3}{dt^2} \end{cases}$$

which we write as

$$\begin{cases} m_1 \frac{d^2 x_1}{dt^2} + (k_1 + k_2)x_1 - k_2 x_2 = 0 \\ m_2 \frac{d^2 x_2}{dt^2} - k_2 x_1 + (k_2 + k_3)x_2 - k_3 x_3 = 0, \\ m_3 \frac{d^2 x_3}{dt^2} - k_3 x_2 + k_3 x_3 = 0 \end{cases}$$

where m_1, m_2, and m_3 represent the mass of the building on the first, second, and third levels and k_1, k_2, and k_3, corresponding to the spring constants, represent the total stiffness of the columns supporting a given floor.

If we attempt to find an exact solution with the method of Laplace transforms, we find that each denominator of $\mathscr{L}\{x_1(t)\}$, $\mathscr{L}\{x_2(t)\}$, and $\mathscr{L}\{x_3(t)\}$ is a positive function of s. Therefore, the roots are complex and solutions will involve sines and/or cosines. (Here, we use $x(t)$, $y(t)$, and $z(t)$ in the place of $x_1(t)$, $x_2(t)$, and $x_3(t)$.)

```
> x:='x':y:='y':
```

```
z:='z':k:='k':m:='m':
> eq1:=m[1]*diff(x(t),t$2)+(k[1]+k[2])*x(t)-k[2]*y(t)=0:
eq2:=m[2]*diff(y(t),t$2)-k[2]*x(t)+
        (k[2]+k[3])*y(t)-k[3]*z(t)=0:
eq3:=m[3]*diff(z(t),t$2)-k[3]*y(t)+k[3]*z(t)=0:
```

```
> with(inttrans):
step1:=laplace({eq1,eq2,eq3},t,s);
```

$$step1 := \{ m_1(s(s \, \mathrm{laplace}(\mathrm{x}(t),t,s) - \mathrm{x}(0)) - \mathrm{D}(x)(0))$$
$$+ (k_1 + k_2) \, \mathrm{laplace}(\mathrm{x}(t),t,s) - k_2 \, \mathrm{laplace}(\mathrm{y}(t),t,s) = 0,$$
$$m_2(s(s \, \mathrm{laplace}(\mathrm{y}(t),t,s) - \mathrm{y}(0)) - \mathrm{D}(y)(0)) - k_2 \, \mathrm{laplace}(\mathrm{x}(t),t,s)$$
$$+ (k_2 + k_3) \, \mathrm{laplace}(\mathrm{y}(t),t,s) - k_3 \, \mathrm{laplace}(\mathrm{z}(t),t,s) = 0,$$
$$m_3(s(s \, \mathrm{laplace}(\mathrm{z}(t),t,s) - \mathrm{z}(0)) - \mathrm{D}(z)(0)) - k_3 \, \mathrm{laplace}(\mathrm{y}(t),t,s)$$
$$+ k_3 \, \mathrm{laplace}(\mathrm{z}(t),t,s) = 0 \}$$

```
> step2:=solve(step1,{laplace(x(t),t,s),
        laplace(y(t),t,s),laplace(z(t),t,s)}):
```

Suppose that $k_1 = 3$, $k_2 = 2$, $k_3 = 1$, $m_1 = 1$, $m_2 = 2$, and $m_3 = 3$ and that the initial conditions are $x_1(0) = 0$, $x_1'(0) = 1/4$, $x_2(0) = 0$, $x_2'(0) = -1/2$, $x_3(0) = 0$, and $x_3'(0) = 1$.

```
> step3:=subs({m[1]=1,m[2]=2,m[3]=3,
        k[1]=3,k[2]=2,k[3]=1,x(0)=0,D(x)(0)=1/4,
        y(0)=0,D(y)(0)=-1/2,z(0)=0,D(z)(0)=1},step2);
```

$$step3 := \left\{ \mathrm{laplace}(\mathrm{y}(t),t,s) = \frac{\frac{21}{2} - \frac{23}{2}s^2 - 3s^4}{6 + 45s^2 + 41s^4 + 6s^6}, \right.$$
$$\mathrm{laplace}(\mathrm{x}(t),t,s) = \frac{\frac{3}{2}s^4 + \frac{9}{2} - \frac{13}{4}s^2}{6 + 45s^2 + 41s^4 + 6s^6},$$
$$\left. \mathrm{laplace}(\mathrm{z}(t),t,s) = \frac{\frac{57}{2} + 38s^2 + 6s^4}{6 + 45s^2 + 41s^4 + 6s^6} \right\}$$

For these values, we use `invlaplace` to compute $x(t)$, $y(t)$, and $z(t)$.

```
> sol:=invlaplace(step3,s,t);
```

$$sol := \left\{ y(t) = -\frac{1}{2} \left(\sum_{\_\alpha = \text{RootOf}(6\_Z^6 + 41\_Z^4 + 45\_Z^2 + 6)} \right. \right.$$

$$\left. \left(\frac{1}{222612} \_\alpha(130242\_\alpha^4 + 838183\_\alpha^2 + 663069)e^{(-\alpha t)} \right) \right), z(t)$$

$$= \frac{1}{2} \left(\sum_{\_\alpha = \text{RootOf}(6\_Z^6 + 41\_Z^4 + 45\_Z^2 + 6)} \right.$$

$$\left. \left(-\frac{1}{222612} \_\alpha(170634\_\alpha^4 + 1152535\_\alpha^2 + 1174971)e^{(-\alpha t)} \right) \right)$$

$$x(t) = \frac{1}{4} \left(\sum_{\_\alpha = \text{RootOf}(6\_Z^6 + 41\_Z^4 + 45\_Z^2 + 6)} \right.$$

$$\left. \left. \left(-\frac{1}{111306} \_\alpha(58242\_\alpha^4 + 367261\_\alpha^2 + 276876)e^{(-\alpha t)} \right) \right) \right\}$$

Note that the terms are given in terms of `RootOf`. Here is Maple's information about `RootOf`.

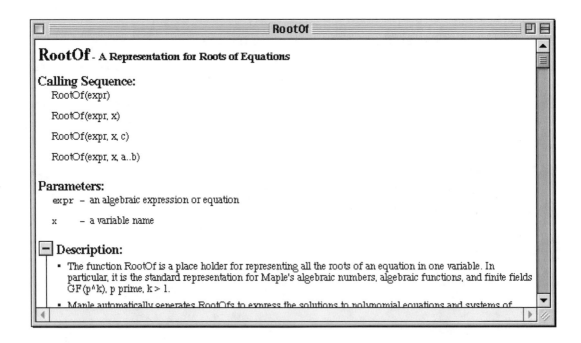

Identical results for these values of the constants are obtained using `dsolve` together with the `method = laplace` option.

```
> m[1]:=1:m[2]:=2:m[3]:=3:k[1]:=3:k[2]:=2:k[3]:=1:
```

```
> sol:=dsolve({eq1,eq2,eq3,x(0)=0,D(x)(0)=1/4,
        y(0)=0,D(y)(0)=-1/2,z(0)=0,D(z)(0)=1},
           {x(t),y(t),z(t)},method=laplace);
```

$$sol := \left\{ y(t) = -\frac{1}{2} \left(\sum_{\_\alpha = \text{RootOf}(6\_Z^6 + 41\_Z^4 + 45\_Z^2 + 6)} \right. \right.$$

$$\left. \left(\frac{1}{222612} \_\alpha (130242\_\alpha^4 + 838183\_\alpha^2 + 663069) e^{(-\alpha t)} \right) \right), z(t)$$

$$= \frac{1}{2} \left(\sum_{\_\alpha = \text{RootOf}(6\_Z^6 + 41\_Z^4 + 45\_Z^2 + 6)} \right.$$

$$\left. \left(-\frac{1}{222612} \_\alpha (170634\_\alpha^4 + 1152535\_\alpha^2 + 1174971) e^{(-\alpha t)} \right) \right)$$

$$x(t) = \frac{1}{4} \left(\sum_{\_\alpha = \text{RootOf}(6\_Z^6 + 41\_Z^4 + 45\_Z^2 + 6)} \right.$$

$$\left. \left. \left(-\frac{1}{111306} \_\alpha (58242\_\alpha^4 + 367261\_\alpha^2 + 276876) e^{(-\alpha t)} \right) \right) \right\}$$

The graphs of $x(t)$, $y(t)$, and $z(t)$ indicate that they are indeed periodic functions.

```
> assign(sol):
plot(evalf(x(t)),t=0..200);
plot(evalf(y(t)),t=0..200);
plot(evalf(z(t)),t=0..200);
```

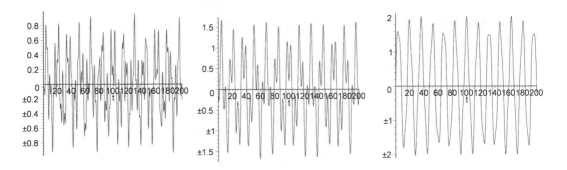

We can construct an outline of a three-story building and observe its vibration. The width and height of the floors were selected arbitrarily to be 20 and 1, respectively. Notice how we define, X, Y, and Z to be functions of t: for example, given α, X(a) returns an approximation of the value of x(t) if t = a. We include Re in each function to be sure that complex numbers are avoided in the roundoff.

```
> X:='X':Y:='Y':Z:='Z':

X:=t0->Re(evalf(subs(t=t0,x(t)))):

Y:=t0->Re(evalf(subs(t=t0,y(t)))):

Z:=t0->Re(evalf(subs(t=t0,z(t)))):
```

p1 draws the corners of each floor.

```
> p1:=t->plot([[0,0],[20,0],[X(t),1],
      [20+X(t),1],[Y(t),2],[20+Y(t),2],[Z(t),3],
      [20+Z(t),3]],style=POINT,color=BLACK,
         symbol=BOX,axes=NONE):
```

p2 draws the walls of each floor.

```
> p2:=t->plot({[[0,0],[20,0]],[[0,0],[X(t),1]],
      [[20,0],[20+X(t),1]],[[X(t),1],[Y(t),2]],
      [[20+X(t),1],[20+Y(t),2]],[[Y(t),2],[Z(t),3]],
         [[20+Y(t),2],[20+Z(t),3]],
            [[Z(t),3],[20+Z(t),3]]},color=BLACK):
```

p3 displays the results of p1 and p2 together.

```
> with(plots):
p3:=t->display({p1(t),p2(t)}):
```

In the following, we show the building at 25 equally spaced values of t between 0 and 6.

```
> tvals:=[seq(0+.25*i,i=0..24)]:
graphs:=map(p3,tvals):
graphicsarray:=display(graphs,insequence=true):
display(graphicsarray);
```

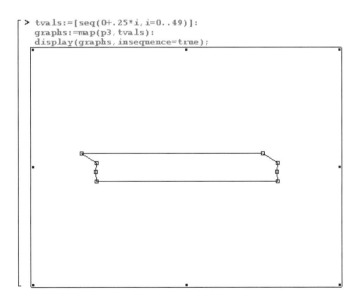

Alternatively, we can use `display` together with the `insequence = true` option to generate an animation of the building.

```
> tvals:=[seq(0+.25*i,i=0..49)]:
graphs:=map(p3,tvals):
display(graphs,insequence=true);
```

To investigate the sway of a taller building, we note that increasing the number of stories increases the size of the system of differential equations. A 5-story building corresponds to a system of 5 second-order differential equations; a 50-story building, a system of 50 second-order differential equations, and so on.

Sources: M. L. James, G. M. Smith, J. C. Wolford, P. W. Whaley, *Vibration of Mechanical and Structural Systems with Microcomputer Applications*, Harper & Row (1989), pp. 282–286. Robert K. Vierck, *Vibration Analysis*, Second Edition, HarperCollins (1979), pp. 266–290.

8.4 Applications That Lead to Nonlinear Systems

Several special equations and systems that arise in the study of many areas of applied mathematics can be solved using the techniques of Chapter 7. These include the predator–prey population dynamics problem, the Van der Pol equation that models variable damping in a spring–mass system, and the Bonhoeffer–Van der Pol (BVP) oscillator. We begin by considering the Lotka–Volterra system which models the interaction between two populations.

Biological Systems: Predator–Prey Interaction

Let $x(t)$ and $y(t)$ represent the numbers of members at time t of the prey and predator populations, respectively. (Examples of such populations include fox–rabbit and shark–seal.) Suppose that the positive constant a is the birth rate of $x(t)$ so that in the absence of the predator $dx/dt = ax$ and that c is the death rate of y, which indicates that $dy/dt = -cy$ in the absence of the prey population. In addition to these factors, the number of interactions between predator and prey affects the number of members in the two populations. Note that an interaction increases the growth of the predator population and decreases the growth of the prey population, because an interaction between the two populations indicates that a predator overtakes a member of the prey population. In order to include these interactions in the model, we assume that the number of interactions is directly proportional to the product of $x(t)$ and $y(t)$. Therefore, the rate at which $x(t)$ changes with respect to time is $dx/dt = ax - bxy$.

Similarly, the rate at which $y(t)$ changes with respect to time is $dy/dt = -cy + dxy$.

Therefore, we must solve the **Lotka–Volterra system**

$$\begin{cases} dx/dt = ax - bxy \\ dy/dt = -cy + dxy \end{cases}$$

subject to the initial populations $x(0) = x_0$ and $y(0) = y_0$.

EXAMPLE 1: Find and classify the equilibrium points of the Lotka–Volterra system.

SOLUTION: We begin by defining `eq1` and `eq2` to be $x' = a_1 x - a_2 xy$ and $y' = -b_1 y + b_2 xy$, respectively, and then solving the system of equations

$$\begin{cases} a_1 x - a_2 xy = 0 \\ -b_1 y + b_2 xy = 0 \end{cases}$$ for x and y to locate the equilibrium points.

```
> x:='x':y:='y':
```

```
rhs_eq1:=a[1]*x-a[2]*x*y:

rhs_eq2:=-b[1]*y+b[2]*x*y:

cps:=solve({rhs_eq1=0,rhs_eq2=0},{x,y});
```

$$cps := \{x = 0, y = 0\}, \{x = \frac{b_1}{b_2}, y = \frac{a_1}{a_2}\}$$

To classify the equilibrium points, we first define `linmatrix` to be the matrix

$$\begin{pmatrix} \frac{d}{dx}(a_1 x - a_2 xy) & \frac{d}{dy}(a_1 x - a_2 xy) \\ \frac{d}{dx}(-b_1 y + b_2 xy) & \frac{d}{dy}(-b_1 y + b_2 xy) \end{pmatrix}.$$

```
> linmatrix:=[[diff(rhs_eq1,x),diff(rhs_eq1,y)],

      [diff(rhs_eq2,x),diff(rhs_eq2,y)]];
```

$$linmatrix := [[a_1 - a_2 y, -a_2 x], [b_2 y, -b_1 + b_2 x]]$$

We then compute the value of `linmatrix` if $x = b_1/b_2$ and $y = a_1/a_2$,

```
> lin2:=subs(cps[2],linmatrix);
```

$$lin2 := \left[\left[0, -\frac{a_2 b_1}{b_2} \right], \left[\frac{b_2 a_1}{a_2}, 0 \right] \right]$$

and the eigenvalues of this matrix. Because the eigenvalues are complex conjugates with the real part equal to 0, we conclude that the equilibrium point $(b_1/b_2, a_1/a_2)$ is a center.

```
> with(linalg):

eigenvals(lin2);
```

$$\sqrt{-b_1 a_1}, -\sqrt{-b_1 a_1}$$

Similarly we compute the value of `linmatrix` if $x = 0$ and $y = 0$,

```
> eigenvals(subs(cps[1],linmatrix));
```

$$a_1, -b_1$$

and then the eigenvalues. Because the eigenvalues are real and have opposite signs, we conclude that the equilibrium point (0,0) is a saddle.

We use `dsolve` together with the `numeric` option to solve the system numerically if $x(0) = 1$ and $y(0) = 1$.

```
> eq1:=diff(x(t),t)=2*x(t)-x(t)*y(t):

eq2:=diff(y(t),t)=-3*y(t)+x(t)*y(t):

sol1:=dsolve({eq1,eq2,x(0)=1,y(0)=1},

        {x(t),y(t)},numeric);
```

$$sol1 := \mathbf{proc}(rkf45\_x) \ \dots \ \mathbf{end}$$

We can evaluate this result for particular values of t. For example, entering

```
> sol1(4);
```

$$[t = 4, x(t) = 5.251673011028663, y(t) = .5583196721418087]$$

shows us that $x(4) \approx 5.25167$ and $y(4) \approx 0.558341$. On the other hand, entering

```
> seq(sol1(t),t=0..5);
```

$$[0 = 0, x(0) = 1., y(0) = 1.],$$
$$[1 = 1, x(1) = 4.335622930119876, y(1) = .4448252098341622],$$
$$[2 = 2, x(2) = 2.026705097058897, y(2) = 5.503181686191604],$$
$$[3 = 3, x(3) = 1.152502750135020, y(3) = .7823187697402739],$$
$$[4 = 4, x(4) = 5.251673019498925, y(4) = .5583196719204030],$$
$$[5 = 5, x(5) = 1.366670016318617, y(5) = 4.639806357798973]$$

creates a table of values of $x(t)$ and $y(t)$ for $t = 0, 1, 2, 3, 4, 5$. We then use `odeplot` to graph $x(t)$ and $y(t)$ together and then graph the solution parametrically.

```
> with(plots):

> odeplot(sol1,[[t,x(t)],[t,y(t)]],0..10,

        color=BLACK);
```

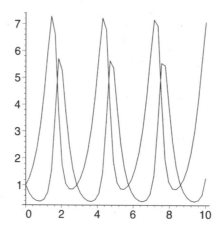

```
> odeplot(sol1,[x(t),y(t)],0..4,color=BLACK);
```

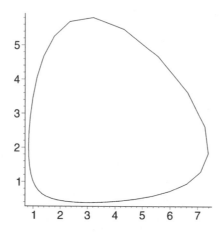

We can obtain a more accurate representation of the phase plane of the nonlinear system by graphing several solutions along with the direction field. First, we define a list of initial conditions, inits,

```
> inits:=[seq([x(0)=3*(1/8+3*i/40),
y(0)=2*(1/8+3*i/40)],i=1..10)];
```

$$inits := \left[\left[x(0) = \frac{3}{5}, y(0) = \frac{2}{5}\right], \left[x(0) = \frac{33}{40}, y(0) = \frac{11}{20}\right], \left[x(0) = \frac{21}{20}, y(0) = \frac{7}{10}\right],\right.$$

$$\left[x(0) = \frac{51}{40}, y(0) = \frac{17}{20}\right], \left[x(0) = \frac{3}{2}, y(0) = 1\right], \left[x(0) = \frac{69}{40}, y(0) = \frac{23}{20}\right],$$

$$\left[x(0) = \frac{39}{20}, y(0) = \frac{13}{10}\right], \left[x(0) = \frac{87}{40}, y(0) = \frac{29}{20}\right], \left[x(0) = \frac{12}{5}, y(0) = \frac{8}{5}\right],$$

$$\left.\left[x(0) = \frac{21}{8}, y(0) = \frac{7}{4}\right]\right]$$

and then use `phaseportrait` to graph the solutions that satisfy the initial conditions specified `inits` together with the direction field.

```
> phaseportrait([eq1,eq2],[x(t),y(t)],t=0..4,
        inits,x=0..12,y=0..12,linecolor=BLACK,
            color=GRAY,stepsize=0.05);
```

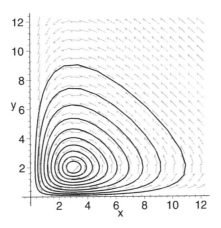

Notice that all of the solutions oscillate about the center. These solutions reveal the relationship between the two populations: prey, $x(t)$, and predator, $y(t)$. As we follow one cycle counterclockwise beginning, for example, near the point (2,0), we notice that as $x(t)$ increases, $y(t)$ increases until $y(t)$ becomes overpopulated. Then, because the prey population is too small to supply the predator population, $y(t)$ decreases, which leads to an increase in the population of $x(t)$. Because the number of predators becomes too small to control the number in the prey population, $x(t)$ becomes overpopulated and the cycle repeats itself.

■

An interesting variation of the Lotka–Volterra equations is to assume that a depends strongly on environomental factors and might be given by the differential equation

$$\frac{da}{dt} = -ax + \bar{a} + k\sin(\omega t + \phi),$$

where the term $-ax$ represents the loss of nutrients due to species x; $\bar{a}$, k, ω, and ω are constants.

Suppose that $x(0) = y(0) = a(0) = 0.5$, $b = d = 1$, $\bar{a} = 0.25$, $k = 0.125$, and $\phi = 0$. Given ω, we define the function `graph` to graph parametrically the x and y components of the solution to

$$\begin{cases} dx/dt = ax - xy \\ dy/dt = -0.5y + xy \\ da/dt = -ax + 0.25 + 0.125\sin\omega t \\ x(0) = y(0) = a(0) = 0.5 \end{cases}$$

```
> with(DEtools);

eq1:=diff(x(t),t)=a(t)*x(t)-x(t)*y(t):

eq2:=diff(y(t),t)=-0.5*y(t)+x(t)*y(t):

eq3:=omega->

        diff(a(t),t)=-a(t)*x(t)+0.25+0.125*sin(omega*t):


> graph:=omega->DEplot([eq1,eq2,eq3(omega)],

        [x(t),y(t),a(t)],t=0..40,

            [[x(0)=0.5,y(0)=0.5,a(0)=0.5]],scene=[x(t),y(t)],

            x=0..1,y=0..1,linecolor=BLACK,stepsize=0.05,

            thickness=1):
```

We then use `map` to apply `graph` to the list of numbers [0.01,0.1,0.25,0.5,0.75,1,1.25,1.5,2.5] and display the resulting graphs as a graphics array using `display` together with the `insequence = true` option.

```
> toshow:=map(graph,[0.01,0.1,0.25,0.5,0.75,

                    1,1.25,1.5,2.5]):


> with(plots):

graphicsarray:=display(toshow,insequence=true):

display(graphicsarray);
```

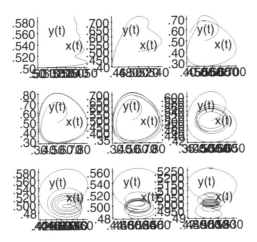

From the graphs, we see that larger values of ω appear to stabilize the populations of both species; smaller values of ω appear to cause the size of the populations to oscillate widely.

Physical Systems: Variable Damping

In some physical systems, energy is fed into the system when there are small oscillations and energy is taken from the system when there are large oscillations. This indicates that the system undergoes "negative damping" for small oscillations and "positive damping" for large oscillations. A differential equation that models this situation is **Van der Pol's equation**

$$x'' + \mu(x^2 - 1)x' + x = 0$$

where μ is a positive constant. We can transform this second-order differential equation into a system of first-order differential equations with the substitution $x' = y$. Hence, $y' = x'' = -x - \mu(x^2 - 1)x' = -x - \mu(x^2 - 1)y$, so the system of equations is

$$\begin{cases} x' = y \\ y' = -x - \mu(x^2 - 1)y \end{cases}$$

which is solved using an initial position $x(0) = x_0$ and an initial velocity $y(0) = y_0$. Notice that $\mu(x^2 - 1)$ represents the damping coefficient. This system models variable damping because $\mu(x^2 - 1) < 0$ if $-1 < x < 1$ and $\mu(x^2 - 1) > 0$ if $|x| > 1$. Therefore, damping is negative for small oscillations, $-1 < x < 1$, and positive for large oscillations, $|x| > 1$.

EXAMPLE 2: Find and classify the equilibrium points of the system of differential equations that is equivalent to Van der Pol's equation.

SOLUTION: We find these equilibrium points by solving

$$\begin{cases} y = 0 \\ -x - \mu(x^2 - 1)y = 0 \end{cases}$$

From the first equation, we see that $y = 0$. Then, substitution of $y = 0$ into the second equation yields $x = 0$ as well. Therefore, the only equilibrium point is (0,0). The Jacobian matrix for this system is

$$J(x, y) = \begin{pmatrix} 0 & 1 \\ -1 - 2\mu xy & -\mu(x^2 - 1) \end{pmatrix}.$$

The eigenvalues of $J(0,0)$ are $\lambda_{1,2} = \frac{1}{2}\left(\mu \pm \sqrt{\mu^2 - 4}\right)$.

```
> lin_mat:=array([[0,1],[-1,mu]]):
```

```
with(linalg):
```

```
eigs:=eigenvals(lin_mat);
```

$$eigs := \frac{1}{2}\mu + \frac{1}{2}\sqrt{\mu^2 - 4}, \frac{1}{2}\mu - \frac{1}{2}\sqrt{\mu^2 - 4}$$

Notice that if $\mu > 2$, both eigenvalues are positive and real. Hence, we classify (0,0) as an **unstable node.** On the other hand, if $0 < \lambda < 2$, the eigenvalues are a complex conjugate pair with a positive real part. Hence, (0,0) is an **unstable spiral.** (We omit the case $\mu = 2$ because the eigenvalues are repeated.) To graph the solutions for various values of μ and different initial conditions, we define sys.

Given μ, sys (μ) returns the system $\begin{cases} x' = y \\ y' = \mu(1 - x^2)y - x \end{cases}$.

```
sys:=mu->[diff(x(t),t)=y(t),
        diff(y(t),t)=mu*(1-x(t)^2)*y(t)-x(t)];
```

$$sys := \mu \rightarrow \left[\frac{\partial}{\partial t}x(t) = y(t), \frac{\partial}{\partial t}y(t) = \mu(1 - x(t)^2)y(t) - x(t) \right]$$

We then form various initial conditions in inits1, inits2, inits3, inits4, and inits5 and use union to join these five sets into initconds.

```
> inits1:={seq([x(0)=0.1*cos(2*Pi*i/4),
y(0)=0.1*sin(2*Pi/4)],i=0..4)};
inits2:={seq([x(0)=-5,y(0)=-5+10*i/9],i=0..9)};
inits3:={seq([x(0)=5,y(0)=-5+10*i/9],i=0..9)}:
inits4:={seq([x(0)=-5+10*i/9,y(0)=-5],i=0..9)}:
```

```
inits5:={seq([x(0)=-5+10*i/9,y(0)=5],i=0..9)}:

initconds:='union'(inits1,inits2,inits3,

     inits4,inits5):
```

We then use `phaseportrait` to graph the solution to `sys` (μ) for $\mu = 1/2, 1, 3/2,$ and 3 using the initial conditions in the set `initconds`.

```
> with(DEtools):

A:=array(1..2,1..2):

A[1,1]:=phaseportrait(sys(1/2),[x(t),y(t)],

     t=0..20,initconds,x=-5..5,y=-5..5,

          arrows=NONE,linecolor=BLACK,stepsize=0.05):

A[1,2]:=phaseportrait(sys(1),[x(t),y(t)],

     t=0..20,initconds,x=-5..5,y=-5..5,

     arrows=NONE,linecolor=BLACK,stepsize=0.05):

A[2,1]:=phaseportrait(sys(3/2),[x(t),y(t)],

     t=0..20,initconds,x=-5..5,y=-5..5,

     arrows=NONE,linecolor=BLACK,stepsize=0.05):

          A[2,2]:=phaseportrait(sys(3),[x(t),y(t)],

          t=0..20,initconds,x=-5..5,y=-5..5,

          arrows=NONE,linecolor=BLACK,stepsize=0.05):
```

These four graphics objects are displayed as an array with `display`. In each figure, we see that all of the curves approach a curve called a *limit cycle*. Physically, the fact that the system has a limit cycle indicates that for all oscillations, the motion eventually becomes periodic, which is represented by a closed curve in the phase plane.

```
> with(plots):

display(A);
```

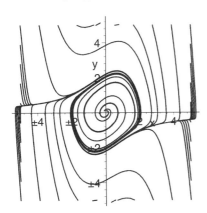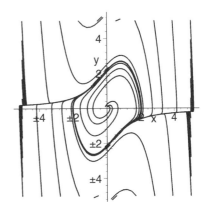

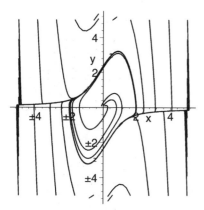
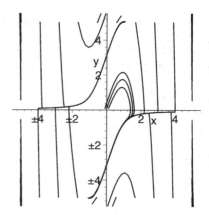

On the other hand, next we graph the solution that satisfies the initial conditions $x(0) = 1$ and $y(0) = 0$ parametrically and individually for various values of μ. Notice that for small values of μ the system more closely approximates that of the harmonic oscillator because the damping coefficient is small. The curves are more circular than those for larger values of μ.

```
> with(DEtools):

graph:=mu->DEplot([diff(x(t),t)=y(t),

        diff(y(t),t)=mu*(1-x(t)^2)*y(t)-x(t)],

            [x(t),y(t)],t=0..20,[[x(0)=1,y(0)=0]],

                x=-4..4,y=-4..4,stepsize=0.025,

                    linecolor=BLACK,color=GRAY,thickness=1):



> vals:=[seq(.2+2.8/8*i,i=0..8)];

> toshow:=map(graph,vals):

with(plots):

graphicsarray:=display(toshow,insequence=true):

display(graphicsarray);
```

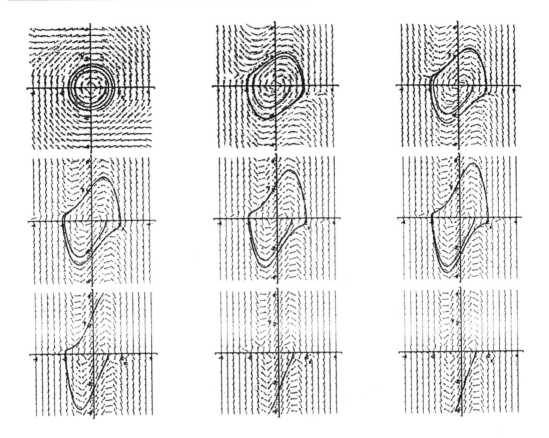

CHAPTER 9

Eigenvalue Problems and Fourier Series

In previous chapters, we have seen that many physical situations can be modeled by either ordinary differential equations or systems of ordinary differential equations. However, to understand the motion of a string at a particular location and at a particular time, the temperature in a thin wire at a particular location and a particular time, or the electrostatic potential at a point on a plate, we must solve partial differential equations as each of these quantities depends on (at least) two independent variables.

Wave equation	$c^2 u_{xx} = u_{tt}$	
Heat equation	$u_t = c^2 u_{xx}$	
Laplace's equation	$u_{xx} + u_{yy} = 0$	

In Chapter ten, we introduce a particular method for solving these partial differential equations (as well as others). In order to carry out this method, however, we introduce the necessary tools in this chapter. We begin with a discussion of boundary value problems and their solutions.

9.1 Boundary Value Problems, Eigenvalue Problems, Sturm–Liouville Problems

Boundary Value Problems

In previous sections, we have solved initial-value problems. However, at this time we will consider boundary value problems, which are solved in much the same way as initial-value problems except that the value-of the function and its derivatives are given at two values of the independent variable instead of one. A general form of a second-order (two-point) boundary value problem is

$$\begin{cases} a_2(x)y'' + a_1(x)y' + a_0(x)y = f(x), a<x<b \\ k_1 y(a) + k_2 y'(a) = \alpha, h_1 y(b) + h_2 y'(b) = \beta \end{cases}$$

where $k_1, k_2, \alpha, h_1, h_2,$ and β are constants and at least one of k_1, k_2 and at least one of h_1, h_2 are not zero.

Note that if $\alpha = \beta = 0$, then we say the problem has **homogeneous boundary conditions**. We also consider boundary value problems that include a parameter in the differential equation. We solve these problems, called **eigenvalue problems**, in order to investigate several useful properties associated with their solutions.

EXAMPLE 1: Solve $\begin{cases} y'' + y = 0, 0<x<\pi \\ y'(0) = 0, y'(\pi) = 0 \end{cases}$.

SOLUTION: Because the characteristic equation is $m^2 + 1 = 0$ with roots $m = \pm i$, a general solution is $y(x) = c_1 \cos x + c_2 \sin x$ where $y'(x) = -c_1 \sin x + c_2 \cos x$. Applying the boundary conditions, we have $y'(0) = c_2 = 0$. Then, $y(x) = c_1 \cos x$. With this solution, we have $y'(\pi) = -c_1 \sin \pi = 0$ for any value of c_1. Therefore, there are infinitely many solutions, $y(x) = c_1 \cos x$, of the boundary value problem, depending on the choice of c_1.

In this case, we are able to use dsolve to solve the boundary value problem

```
> Sol:=dsolve({diff(y(x),x$2)+y(x)=0,

       D(y)(0)=0,D(y)(Pi)=0},y(x));
```

$$Sol := y(x) = \_C2 \cos(x)$$

We confirm that the boundary conditions are satisfied for any value of _C2 by graphing several solutions with plot. After naming $y(x)$ the solution obtained in Sol, we define to_plot to be the set of functions obtained by replacing _C2 in

$y(x)$ by i for $i = -3, -2, -1, 0, 1, 2$, and 3. The set of functions `to_plot` is then graphed on the interval $[0, \pi]$ with `plot`.

```
> assign(Sol):
to_plot:={seq(subs(_C2=i,y(x)),i=-3..3)}:
plot(to_plot,x=0..Pi);
```

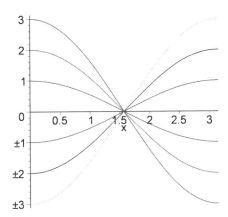

From the result of Example 1, we notice a difference between **initial-value problems** and **boundary value problems**: an initial-value problem (that meets the hypotheses of the existence and uniqueness theorem) has a unique solution whereas a boundary value problem may have more than one solution (or may have no solution).

EXAMPLE 2: Solve $\begin{cases} y'' + y = 0, 0 < x < \pi \\ y'(0) = 0, y'(\pi) = 1 \end{cases}$.

SOLUTION: Using the general solution obtained in the previous example, we have $y(x) = c_1 \cos x + c_2 \sin x$. As before, $y'(0) = c_2 = 0$, so $y(x) = c_1 \cos x$. However, because $y'(\pi) = -c_1 \sin \pi = 0 \neq 1$, the boundary conditions cannot be satisfied with any choice of c_1. Therefore, there is no solution to the boundary value problem.

As indicated in the general form of a boundary value problem, the boundary conditions in these problems can involve the function and its derivative. However, this modification of the problem does not affect the method of solution.

EXAMPLE 3: Solve $\begin{cases} y'' - y = 0, 0 < x < 1 \\ y'(0) + 3y(0) = 0, y'(1) + y(1) = 1 \end{cases}$.

SOLUTION: The characteristic equation is $m^2 - 1 = 0$ with roots $m = \pm 1$. Hence, a general solution is $y(x) = c_1 e^x + c_2 e^{-x}$ with derivative $y'(x) = c_1 e^x - c_2 e^{-x}$. Applying $y'(0) + 3y(0) = 0$ yields $y'(0) + 3y(0) = c_1 - c_2 + 3(c_1 + c_2) = 4c_1 + 2c_2 = 0$. Then,

$$y'(1) + y(1) = c_1 e^1 - c_2 e^{-1} + c_1 e^1 + c_2 e^{-1} = 2c_1 e = 1,$$

so $c_1 = \frac{1}{2e}$ and $c_2 = -\frac{1}{e}$. Thus, the boundary value problem has the unique solution $y(x) = \frac{1}{2e} e^x - \frac{1}{e} e^{-x} = \frac{1}{2} e^{x-1} - e^{-x-1}$, which we confirm with Maple.

```
> y:='y':
Sol:=dsolve(diff(y(x),x$2)-y(x)=0,y(x));
```

$$Sol := y(x) = \_C1 \sinh(x) + \_C2 \cosh(x)$$

```
> Sol:=simplify(convert(Sol,exp));
```

$$Sol := y(x) = -\frac{1}{2}(-\_C1 + \_C1 e^{(-2x)} - \_C2 - \_C2 e^{(-2x)})e^x$$

We then name $y(x)$ the result obtained in `Sol` with `assign`.

```
> assign(Sol):
```

We evaluate $y'(x) + 3y(x)$ if $x = 0$ and $y'(x) + y(x)$ if $x = 1$ with `eval` and `subs`.

```
> Eq_1:=eval(subs(x=0,diff(y(x),x)+3*y(x)=0));
Eq_2:=simplify(eval(subs(x=1,diff(y(x),x)+y(x)=1)));
```

$$Eq\_1 := \_C1 + 3\_C2 = 0$$

$$Eq\_2 := e(\_C1 + \_C2) = 1$$

We then solve these two equations for _C1 and _C2 with `solve`.

```
> c_vals:=solve({Eq_1,Eq_2});
```

$$c\_vals := \left\{ \_C1 = \frac{3}{2}\frac{1}{e}, \_C2 = -\frac{1}{2}\frac{1}{e} \right\}$$

_C1 and _C2 are then named the values obtained in c_vals with assign and we determine $y(x)$ using eval.

```
> assign(c_vals):
eval(y(x));
```

$$-\frac{1}{2}\left(-\frac{1}{e} + 2\frac{e^{(-2x)}}{e} \right) e^x$$

∎

Eigenvalue Problems

We now consider **eigenvalue problems**, boundary value problems that include a parameter. Values of the parameter for which a boundary value problem has a nontrivial solution are called **eigenvalues** of the problem. For each eigenvalue, the **nontrivial** solution that satisfies the problem is called the corresponding **eigenfunction**. (Notice that if a value of the parameter leads to the **trivial** solution, then the value is *not* considered an eigenvalue of the problem.)

EXAMPLE 4: Solve the eigenvalue problem $y'' + \lambda y = 0$, $0 < x < p$, subject to $y(0) = 0$ and $y(p) = 0$.

SOLUTION: Notice that the differential equation in this problem differs from those solved earlier because it includes the parameter λ. However, we solve it in a similar manner by solving the characteristic equation $m^2 + \lambda = 0$. Of course, the values of m depend on the value of the parameter λ. Hence, we consider the following three cases.

Case I: $(\lambda = 0)$
In this case, the characteristic equation is $m^2 = 0$ with roots $m_1 = m_2 = 0$, which indicates that a general solution is $y(x) = c_1 x + c_2$.
Application of the boundary condition $y(0) = 0$ yields $y(0) = c_1 \cdot 0 + c_2 = 0$, so

$c_2 = 0$. Then, $y(p) = c_1 p = 0$, so $c_1 = 0$ and $y(x) = 0$. Because we obtain the trivial solution, $\lambda = 0$ is *not* an eigenvalue.

Case II: $(\lambda < 0)$

To represent λ as a negative value, we let $\lambda = -k^2 < 0$. Then, the characteristic equation is $m^2 - k^2 = 0$, so $m_1 = k$ and $m_2 = -k$. A general solution is, therefore, $y(x) = c_1 e^{kx} + c_2 e^{-kx}$ (or $y(x) = c_1 \cosh kx + c_2 \sinh kx$). Substitution of the boundary condition $y(0) = 0$ yields $y(0) = c_1 + c_2 = 0$, so $c_2 = -c_1$. Because $y(p) = 0$ indicates that $y(p) = c_1 e^{kp} + c_2 e^{-kp} = 0$, substitution gives us the equation $y(p) = c_1 e^{kp} - c_1 e^{-kp} = c_1(e^{kp} - e^{-kp}) = 0$. Notice that $e^{kp} - e^{-kp} = 0$ only if $e^{kp} = e^{-kp}$, which can occur only if $k = 0$ or $p = 0$. If $k = 0$, then $\lambda = -k^2 = -0^2 = 0$, which contradicts the assumption that $\lambda < 0$. We also assumed that $p > 0$, so $e^{kp} - e^{-kp} \neq 0$. Hence, $y(p) = c_1(e^{kp} - e^{-kp}) = 0$ implies that $c_1 = 0$, so $c_2 = -c_1 = 0$ as well. Because $\lambda < 0$ leads to the trivial solution $y(x) = 0$, there are no negative eigenvalues.

Case III: $(\lambda > 0)$

In order to represent λ as a positive value, we let $\lambda = k^2 > 0$. Then, we have the characteristic equation $m^2 + k^2 = 0$ with complex conjugate roots $m_{1,2} = \pm ki$. Thus, a general solution is

$$y(x) = c_1 \cos kx + c_2 \sin kx.$$

Because $y(0) = c_1 \cos k \cdot 0 + c_2 \sin k \cdot 0 = c_1$, the boundary condition $y(0) = 0$ indicates that $c_1 = 0$. Hence, $y(x) = c_2 \sin kx$. Application of $y(p) = 0$ yields $y(p) = c_2 \sin kp = 0$, so either $c_2 = 0$ or $\sin kp = 0$. Selecting $c_2 = 0$ leads to the trivial solution that we want to avoid, so we determine the values of k that satisfy $\sin kp = 0$. Because $\sin n\pi = 0$ for integer values of n, $\sin kp = 0$ if $kp = n\pi$, $n = 1, 2, \ldots$. Solving for k, we have $k = n\pi/p$, so the eigenvalues are

$$\lambda = \lambda_n = k^2 = (n\pi/p)^2, \qquad n = 1, 2, \ldots.$$

Notice that the subscript n is used to indicate that the parameter depends on the value of n. (Notice also that we omit $n = 0$, because the value $k = 0$ was considered in case I.) For each eigenvalue, the corresponding eigenfunction is obtained by substitution into $y(x) = c_2 \sin kx$. Because c_2 is arbitrary, let $c_2 = 1$. Therefore, the eigenvalue $\lambda_n = (n\pi/p)^2$, $n = 1, 2, \ldots$, has corresponding eigenfunction

$$y(x) = y_n(x) = \sin(n\pi x/p), \qquad n = 1, 2, \ldots.$$

We did not consider negative values of n because $\sin(-n\pi x/p) = -\sin(n\pi x/p)$; the negative sign can be taken into account in the constant, and we do not obtain additional eigenvalues or eigenfunctions by using $n = -1, -2, \ldots$.

■

We will find the eigenvalues and eigenfunctions in Example 4 quite useful in future sections. The following eigenvalue problem will be useful as well.

EXAMPLE 5: Solve $y'' + \lambda y = 0$, $0 < x < p$, subject to $y'(0) = 0$ and $y'(p) = 0$.

SOLUTION: Notice that the only difference between this problem and that in Example 4 is in the boundary conditions. Again, the characteristic equation is $m^2 + \lambda = 0$, so we must consider the three cases $\lambda = 0$, $\lambda < 0$, and $\lambda > 0$. Note that a general solution in each case is the same as that obtained in Example 4. However, the final results may differ due to the boundary conditions.

Case I: $(\lambda = 0)$
Because $y(x) = c_1 x + c_2$, $y'(x) = c_1$. Therefore, $y'(0) = c_1 = 0$, so $y(x) = c_2$. Notice that this constant function satisfies $y'(p) = 0$ for all values of c_2. Hence, if we let $c_2 = 1$, then $\lambda = 0$ is an eigenvalue with corresponding eigenfunction $y(x) = y_0(x) = 1$.

Case II: $(\lambda < 0)$
If $\lambda = -k^2 < 0$, then $y(x) = c_1 e^{kx} + c_2 e^{-kx}$ and $y'(x) = c_1 k e^{-kx} - c_2 k e^{-kx}$. Applying the first condition results in $y'(0) = c_1 k - c_2 k = 0$, so $c_1 = c_2$. Therefore, $y'(p) = c_1 k e^{kp} - c_1 k e^{-kp} = 0$, which is not possible unless $c_1 = 0$, because $k \neq 0$ and $p \neq 0$. Thus, $c_1 = c_2 = 0$, so $y(x) = 0$. Because we have the trivial solution, there are no negative eigenvalues.

Case III: $(\lambda > 0)$
By letting $\lambda = k^2 > 0$, $y(x) = c_1 \cos kx + c_2 \sin kx$ and $y'(x) = -c_1 k \sin kx + c_2 k \cos kx$. Hence, $y'(0) = c_2 k = 0$, so $c_2 = 0$. Consequently, $y'(p) = -c_1 k \sin kp = 0$, which is satisfied if $kp = n\pi$, $n = 1, 2, \ldots$, Therefore, the eigenvalues are

$$\lambda = \lambda_n = (n\pi/p)^2, \qquad n = 1, 2, \ldots.$$

Note that we found $c_2 = 0$ in $y(x) = c_1 \cos kx + c_2 \sin kx$, so the corresponding eigenfunctions are

$$y(x) = y_n(x) = \cos(n\pi x/p), \qquad n = 1, 2, \ldots$$

if we let $c_1 = 1$.

■

EXAMPLE 6: Consider the eigenvalue problem $y'' + \lambda y = 0$, $y(0) = 0$, $y(1) + y'(1) = 0$.
(a) Show that the positive eigenvalues $\lambda = k^2$ satisfy the relationship $k = -\tan k$.
(b) Approximate the first eight positive eigenvalues. Notice that for larger values of k, the eigenvalues are approximately the vertical asymptotes of $y = \tan k$, so $\lambda_n \approx [(2n-1)\pi/2]^2, n = 1, 2, \ldots$.

SOLUTION: In order to solve the eigenvalue problem, we consider the three cases.

Case I: $(\lambda = 0)$
The problem $y'' = 0, y(0) = 0, y(1) + y'(1) = 0$ has the solution $y = 0$, so $\lambda = 0$ is not an eigenvalue.

Case II: $(\lambda < 0)$
Similarly, $y'' - k^2 y = 0, y(0) = 0, y(1) + y'(1) = 0$ has solution $y = 0$, so there are no negative eigenvalues.

Case III: $(\lambda > 0)$
If $\lambda = k^2 > 0$, we solve $y'' + k^2 y = 0, y(0) = 0, y(1) + y'(1) = 0$. A general solution of $y'' + k^2 y = 0$ is $y = A \cos kx + B \sin kx$. Applying $y(0) = 0$ indicates that $A = 0$, so $y = B \sin kx$. Applying $y(1) + y'(1) = 0$ where $y' = kB \cos kx$ yields $B \sin k + Bk \cos k = 0$. Because we want to avoid requiring that $B = 0$, we note that this condition is satisfied if $-\sin k = k \cos k$ or $-\tan k = k$. To approximate the first eight positive roots of this equation, we graph $y = -\tan x$ and $y = x$ simultaneously. (We only look for positive roots because $\tan(-k) = -\tan k$, meaning that no additional eigenvalues are obtained by considering negative values of k.) The eigenfunctions of this problem are $y = \sin kx$, where k satisfies $-\tan k = k$.

Sturm–Liouville Problems

Because of the importance of eigenvalue problems, we express these problems in the general form

$$a_2(x)y'' + a_1(x)y' + [a_0(x) + \lambda]y = 0, \qquad a < x < b,$$

where $a_2(x) \neq 0$ on $[a,b]$ and the boundary conditions at the endpoints $x = a$ and $x = b$ can be written as

$$k_1 y(a) + k_2 y'(a) = 0 \quad \text{and} \quad h_1 y(b) + h_2 y'(b) = 0$$

for the constants k_1, k_2, h_1, and h_2 where at least one of k_1, k_2 and at least one of h_1, h_2 are not zero. This equation can be rewritten by letting

$$p(x) = e^{\int a_1(x)/a_2(x)dx}, \quad q(x) = \frac{a_0(x)}{a_2(x)}p(x), \quad \text{and} \quad s(x) = \frac{p(x)}{a_2(x)}.$$

By making this change, we obtain the equivalent equation

$$\frac{d}{dx}\left(p(x)\frac{dy}{dx}\right) + (q(x) + \lambda s(x))y = 0$$

which is called a **Sturm–Liouville equation**, which along with boundary conditions is called a **Sturm–Liouville problem**. This particular from of the equation is known as the **self-adjoint** form. This form is of interest because of the relationship of the function $s(x)$ and the solutions of the problem.

EXAMPLE 7: Place the equation $x^2y'' + 2xy' + \lambda y = 0$, $x > 0$, in self-adjoint form.

SOLUTION: In this case, $a_2(x) = x^2$, $a_1(x) = 2x$, and $a_0(x) = 0$. Hence,

$$p(x) = e^{\int a_1(x)/a_2(x)dx} = e^{\int 2x/x^2 dx} = e^{2\ln x} = x^2, \qquad q(x) = \frac{a_0(x)}{a_2(x)}p(x) = 0, \qquad \text{and}$$

$s(x) = \frac{p(x)}{a_2(x)} = \frac{x^2}{x^2} = 1$, so the self-adjoint form of the equation is $\frac{d}{dx}\left(x^2\frac{dy}{dx}\right) + \lambda y = 0$.
We see that our result is correct by differentiating.

■

Solutions of Sturm–Liouville problems have several interesting properties, two of which are included in the following theorem.

Theorem **Linear Independence** **and Orthogonality of** **Eigenfunctions**	If $y_m(x)$ and $y_n(x)$ are eigenfunctions of the regular Sturm–Liouville problem where $m \neq n$, $y_m(x)$ and $y_n(x)$ are **linearly independent** and the **orthogonality** condition $\int_a^b s(x)y_m(x)y_n(x)dx = 0$ holds.

Because we integrate the product of the eigenfunctions with the function $s(x)$ in the orthogonality condition, we call $s(x)$ the **weighting function**.

EXAMPLE 8: Consider the eigenvalue problem $y'' + \lambda y = 0$, $0 < x < p$, subject to $y(0) = 0$ and $y(p) = 0$, which we solved in Example 4. Verify that the eigenfunctions $y_1(x) = \sin(\pi x/p)$ and $y_2(x) = \sin(2\pi x/p)$ are linearly independent. Also, verify the orthogonality condition.

SOLUTION: We can verify that $y_1(x) = \sin(\pi x/p)$ and $y_2(x) = \sin(2\pi x/p)$ are linearly independent by computing the Wronskian:

$$W(y_1, y_2) = \begin{vmatrix} \sin\frac{\pi x}{p} & \sin\frac{2\pi x}{p} \\ \frac{\pi}{p}\cos\frac{\pi x}{p} & \frac{2\pi}{p}\cos\frac{2\pi x}{p} \end{vmatrix}.$$

After defining $y_1(x) = \sin(\pi x/p)$ and $y_2(x) = \sin(2\pi x/p)$ and loading the **linalg** package, we use det and wronksian, which are both contained in the **linalg** package, to find the Wronskian.

```
> y[1]:=x->sin(Pi*x/p):

y[2]:=x->sin(2*Pi*x/p):

with(linalg):

step_1:=det(Wronskian([y[1](x),y[2](x)],x));
```

$$step\_1 := \frac{\pi\left(2\sin\left(\frac{\pi x}{p}\right)\cos\left(2\frac{\pi x}{p}\right) - \sin\left(2\frac{\pi x}{p}\right)\cos\left(\frac{\pi x}{p}\right)\right)}{p}$$

We then simplify the result obtained in step_1 using combine together with the trig option.

```
> combine(step_1,trig);
```

$$\frac{1}{2}\frac{\pi\sin\left(3\frac{\pi x}{p}\right) - 2\pi\sin\left(\frac{\pi x}{p}\right)}{p}$$

Because $W(y_1, y_2)$ is not identically zero, the two functions are linearly independent.

The equation $y'' + \lambda y = 0$ is in self-adjoint form with $s(x) = 1$. Hence, the orthogonality condition is

$$\int_0^p 1 \cdot y_1(x)y_2(x)dx = \int_0^p 1 \cdot y_1(x)y_2(x)dx = \int_0^p \sin\frac{\pi x}{p}\sin\frac{2\pi x}{p}dx$$

$$= \int_0^p 2\sin^2\frac{\pi x}{p}\cos\frac{\pi x}{p}dx = \frac{2p}{3\pi}\left[\sin^3\frac{\pi x}{p}\right]_0^p = 0,$$

which is verified with int.

```
> int(y[1](x)*y[2](x),x=0..p);
```

$$0$$

■

9.2 Fourier Sine Series and Cosine Series

Recall the eigenvalue problem $\begin{cases} y'' + \lambda y = 0 \\ y(0) = 0, y(p) = 0 \end{cases}$, which was solved in Example 4 in Section

9.1. The eigenvalues of this problem are $\lambda = \lambda_n = (n\pi/p)^2, n = 1, 2, \ldots$ with corresponding eigenfunctions $\phi_n(x) = \sin(n\pi x/p), n = 1, 2, \ldots$. We will see that for some functions f, we can find coefficients c_n so that

$$f(x) = \sum_{n=1}^{\infty} c_n \sin \frac{n\pi x}{p}.$$

A series of this form is called a **Fourier sine series**. In order to make use of these series, we must determine the coefficients c_n. We accomplish this by taking advantage of the orthogonality properties of eigenfunctions. Because the differential equation $y'' + \lambda y = 0$ is in self-adjoint form, we have that $s(x) = 1$. Therefore, the orthogonality condition is $\int_0^p \sin(n\pi x/p) \sin(m\pi x/p)dx = 0, m \neq n$. In order to use this condition, multiply both sides of $f(x) = \sum_{n=1}^{\infty} c_n \sin(n\pi x/p)$ by the eigenfunction $\sin(m\pi x/p)$ and $s(x) = 1$. Then, integrate the result from $x = 0$ to $x = p$ (because the boundary conditions of the corresponding eigenvalue problem are given at these two values of x). This yields

$$\int_0^p f(x) \sin \frac{m\pi x}{p} dx = \int_0^p \sum_{n=1}^{\infty} c_n \sin \frac{n\pi x}{p} \sin \frac{m\pi x}{p} dx.$$

Assuming that term-by-term integration is allowed on the right-hand side of the equation, we have

$$\int_0^p f(x) \sin \frac{m\pi x}{p} dx = \sum_{n=1}^{\infty} \int_0^p c_n \sin \frac{n\pi x}{p} \sin \frac{m\pi x}{p} dx = \sum_{n=1}^{\infty} c_n \int_0^p \sin \frac{n\pi x}{p} \sin \frac{m\pi x}{p} dx.$$

Recall that the eigenfunctions $\phi_n(x) = \sin(n\pi x/p), n = 1, 2, \ldots$, are orthogonal, so $\int_0^p \sin(n\pi x/p) \sin(m\pi x/p)dx = 0$ if $m \neq n$. On the other hand, if $m = n$,

$$\int_0^p \sin \frac{n\pi x}{p} \sin \frac{m\pi x}{p} dx = \int_0^p \sin^2 \frac{n\pi x}{p} dx = \int_0^p \frac{1}{2}\left(1 - \cos \frac{2n\pi x}{p}\right) dx = \frac{1}{2}\left[x - \frac{p}{2\pi n} \sin \frac{2n\pi x}{p}\right]_0^p = \frac{p}{2}.$$

```
> step1:=int(sin(n*Pi*x/p)^2,x=0..p);
```

$$step1 := \frac{1}{2}\frac{p(-\cos(n\pi)\sin(n\pi) + n\pi)}{n\pi}$$

```
> assume(n,integer);

step1;
```

$$\frac{1}{2}p$$

Therefore, each term in the sum $\sum_{n=1}^{\infty} c_n \int_0^p \sin(n\pi x/p) \sin(m\pi x/p)dx$ equals zero except when $m = n$. Hence, $\int_0^p f(x) \sin(n\pi x/p)dx = c_n \cdot p/2$, so the Fourier sine series coefficients are given by

$$c_n = \frac{2}{p} \int_0^p f(x) \sin\frac{n\pi x}{p} dx$$

where we assume that f is integrable on $[0, p]$.

EXAMPLE 1: Find the Fourier sine series for $f(x) = x, 0 \le x \le \pi$.

SOLUTION: In this case, $p = \pi$, so

$$c_n = \frac{2}{\pi} \int_0^\pi f(x) \sin\frac{n\pi x}{\pi} dx = \frac{2}{\pi} \int_0^\pi x \sin nx dx \text{ (integration by parts)}$$

$$= \frac{2}{\pi}\left[-\frac{x\cos nx}{n}\right]_0^\pi + \frac{2}{\pi}\int_0^\pi \frac{\cos nx}{n}dx = -\frac{2\cos n\pi}{n} + \frac{2}{\pi}\left[\frac{\sin nx}{n^2}\right]_0^\pi = -\frac{2\cos n\pi}{n} + \frac{2}{\pi n^2}(\sin n\pi - \sin 0)$$

$$= -\frac{2\cos n\pi}{n} + \frac{2}{\pi n^2}\cdot 0 = -\frac{2\cos n\pi}{n}.$$

```
> n:='n':
step1:=int(2*x*sin(n*x)/Pi,x=0..Pi);
```

$$step1 := -2\frac{-\sin(n\pi) + n\pi\cos(n\pi)}{\pi n^2}$$

Observe that n is an integer so $\cos n\pi = (-1)^n$. Hence, $c_n = -2(-1)^n/n = 2(-1)^{n+1}/n$,

```
> assume(n,integer);
step1;
```

$$-2\frac{(-1)^{n\sim}}{n\sim}$$

and the Fourier sine series is

$$f(x) = \sum_{n=0}^{\infty} c_n \sin\frac{n\pi x}{\pi} = \sum_{n=1}^{\infty} \frac{2(-1)^{n+1}}{n} \sin nx$$

$$= 2\sin x - \sin 2x + \frac{2}{3}\sin 3x - \frac{1}{2}\sin 4x + \ldots$$

We can use a finite number of terms of the series to obtain a trigonometric polynomial that approximates $f(x) = x, 0 \le x \le \pi$ as follows. Let $f_k(x) = \sum_{n=1}^{k} \frac{2(-1)^{n+1}}{n} \sin nx$. Then, $f_k(x) = f_{k-1}(x) + \frac{2(-1)^{k+1}}{k} \sin kx$. Thus, to calculate the kth partial sum of the Fourier sine series, we need only add $\frac{2(-1)^{k+1}}{k} \sin kx$ to the $(k-1)$st partial sum: we need not recompute all k terms of the kth partial sum if we know the $(k-1)$st partial sum. Using this observation, we define the recursively defined function f to return the kth partial sum of the series. We use the option `remember` when defining f so that Maple "remembers" each $f_k(x)$ that is computed. The advantage of doing so is that Maple need not recompute $f_k(x)$ to compute $f_{k+1}(x)$.

```
> f:=proc(k) option remember;
        f(k-1)+subs(n=k,step1)*sin(k*x)
     end:
f(1):=subs(n=1,step1)*sin(x):
```

and then use `approx` to compute the 3rd, 6th, 9th, and 12th partial sums.

```
> k_vals:=3,6,9,12:
array([seq([k,f(k)],k=k_vals)]);
```

$$\left\lfloor 3, 2\sin(x) - \sin(2x) + \frac{2}{3}\sin(3x) \right\rfloor$$

$$\left[6, 2\sin(x) - \sin(2x) + \frac{2}{3}\sin(3x) - \frac{1}{2}\sin(4x) + \frac{2}{5}\sin(5x) - \frac{1}{3}\sin(6x) \right]$$

$$\left[9, 2\sin(x) - \sin(2x) + \frac{2}{3}\sin(3x) - \frac{1}{2}\sin(4x) + \frac{2}{5}\sin(5x) - \frac{1}{3}\sin(6x) \right.$$

$$\left. + \frac{2}{7}\sin(7x) - \frac{1}{4}\sin(8x) + \frac{2}{9}\sin(9x) \right]$$

$$\left[12, 2\sin(x) - \sin(2x) + \frac{2}{3}\sin(3x) - \frac{1}{2}\sin(4x) + \frac{2}{5}\sin(5x) - \frac{1}{3}\sin(6x) \right.$$

$$+ \frac{2}{7}\sin(7x) - \frac{1}{4}\sin(8x) + \frac{2}{9}\sin(9x) - \frac{1}{5}\sin(10x) + \frac{2}{11}\sin(11x)$$

$$\left. - \frac{1}{6}\sin(12x) \right]$$

We then use a `for` loop to graph $f(x) = x$ and the kth partial sum on the interval $[0, \pi]$ for $k = 3, 6, 9,$ and 12. Similarly, we graph the absolute value of the difference of $f(x) = x$ and the kth partial sum. The results are shown side by side.

```
> for k in k_vals do plot({x,f(k)},x=0..Pi) od;
```

```
for k in k_vals do plot(abs(x-f(k)),x=0..Pi) od;
```

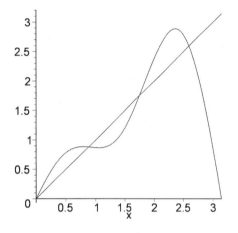

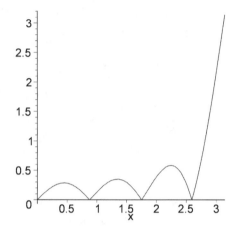

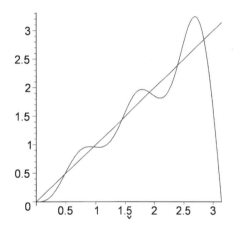

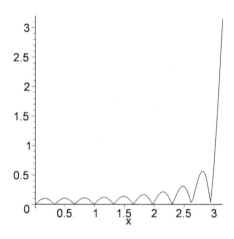

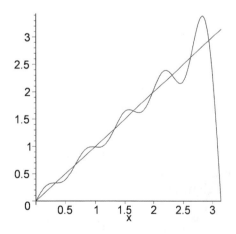

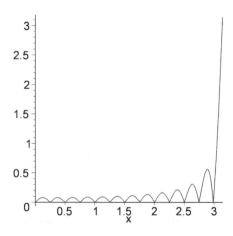

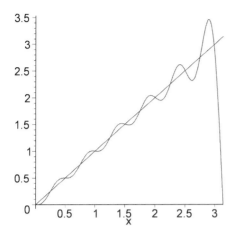

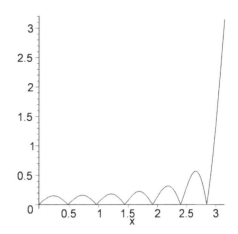

As we increase the number of terms used in approximating f, we improve the accuracy. Notice from the graphs that none of the polynomials attain the value of $f(\pi) = \pi$ at $x = \pi$. This is due to the fact that at $x = \pi$, each of the polynomials yields a value of 0. Hence, our approximation can be reliable only on the interval $0 < x < \pi$. In general, however, we are assured of accuracy only at points of continuity of f on the open interval.

■

EXAMPLE 2: Find the Fourier sine series for $f(x) = \begin{cases} 1, 0 \leq x < 1 \\ -1, 1 \leq x \leq 2 \end{cases}$.

SOLUTION: Because f is defined on $0 \leq x \leq 2$, $p = 2$.

```
> n:='n':
```
```
f:=x->1-2*Heaviside(x-1);
```

$$f := x \rightarrow 1 - 2\,\text{Heaviside}(x-1)$$

```
> plot(f(x),x=0..2);
```

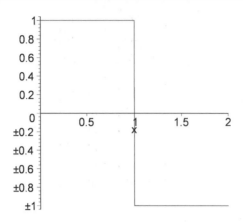

Hence,

$$c_n = \frac{2}{2}\int_0^2 f(x)\sin\frac{n\pi x}{2}\,dx = \frac{2}{n\pi}\left(-2\cos\frac{n\pi}{2} + \cos n\pi + 1\right)$$

```
> c:=n->int(f(x)*sin(n*Pi*x/2),x=0..2):
c(n);
```

$$2\frac{\cos(n\pi) - 2\cos\left(\frac{1}{2}n\pi\right)}{n\pi} + 2\frac{1}{n\pi}$$

```
> assume(n,integer):
c(n);
```

$$2\frac{(-1)^{n\sim} - 2\cos\left(\frac{1}{2}n\sim\pi\right)}{n\sim\pi} + 2\frac{1}{n\sim\pi}$$

We use seq to calculate a few of the c_n's

```
> array([seq([n,c(n)],n=1..10)),
      array([seq([n,c(n)],n=11..20)]);
```

$$
\begin{bmatrix}
1 & 0 \\
2 & 4\frac{1}{\pi} \\
3 & 0 \\
4 & 0 \\
5 & 0 \\
6 & 4\frac{1}{3\pi} \\
7 & 0 \\
8 & 0 \\
9 & 0 \\
10 & 4\frac{1}{5\pi}
\end{bmatrix},
\begin{bmatrix}
11 & 0 \\
12 & 0 \\
13 & 0 \\
14 & 4\frac{1}{7\pi} \\
15 & 0 \\
16 & 0 \\
17 & 0 \\
18 & 4\frac{1}{9\pi} \\
19 & 0 \\
20 & 0
\end{bmatrix}
$$

As we can see, most of the coefficients are zero. In fact, only those c_n's where n is an odd multiple of 2 yield a nonzero value. For example, $c_6 = c_{2 \cdot 3} = \frac{2}{6\pi} \cdot 4 = \frac{4}{3\pi}$, $c_{10} = c_{2 \cdot 5} = \frac{2}{10\pi} \cdot 4 = \frac{4}{5\pi}$, ..., $c_2(2n-1) = \frac{4}{(2n-1)\pi}$, $n = 1, 2, \ldots$, so we have the series

$$
f(x) = \sum_{n=1}^{\infty} \frac{4}{(2n-1)\pi} \sin\frac{2(2n-1)\pi x}{2} = \sum_{n=1}^{\infty} \frac{4}{(2n-1)\pi} \sin(2n-1)\pi x
$$

$$
= \frac{4}{\pi} \sin \pi x + \frac{4}{3\pi} \sin 3\pi x + \frac{4}{5\pi} \sin 5\pi x + \ldots
$$

As in Example 1, we graph $f(x)$ with several polynomial approximations.

```
> approx:='approx':
approx:=proc(k) option remember;
        approx(k-1)+c(k)*sin(k*Pi*x/2)
        end:
approx(1):=0:
```

```
> k_vals:=10,18,26,34:
array([seq([k,approx(k)],k=k_vals)]);
```

$$\left\lfloor 10, 4\frac{\sin(\pi x)}{\pi} + \frac{4}{3}\frac{\sin(3\pi x)}{\pi} + \frac{4}{5}\frac{\sin(5\pi x)}{\pi} \right\rfloor$$

$$\left[18, \right.$$

$$\left. 4\frac{\sin(\pi x)}{\pi} + \frac{4}{3}\frac{\sin(3\pi x)}{\pi} + \frac{4}{5}\frac{\sin(5\pi x)}{\pi} + \frac{4}{7}\frac{\sin(7\pi x)}{\pi} + \frac{4}{9}\frac{\sin(9\pi x)}{\pi} \right]$$

$$\left[26, 4\frac{\sin(\pi x)}{\pi} + \frac{4}{3}\frac{\sin(3\pi x)}{\pi} + \frac{4}{5}\frac{\sin(5\pi x)}{\pi} + \frac{4}{7}\frac{\sin(7\pi x)}{\pi} \right.$$

$$\left. + \frac{4}{9}\frac{\sin(9\pi x)}{\pi} + \frac{4}{11}\frac{\sin(11\pi x)}{\pi} + \frac{4}{13}\frac{\sin(13\pi x)}{\pi} \right]$$

$$\left[34, 4\frac{\sin(\pi x)}{\pi} + \frac{4}{3}\frac{\sin(3\pi x)}{\pi} + \frac{4}{5}\frac{\sin(5\pi x)}{\pi} + \frac{4}{7}\frac{\sin(7\pi x)}{\pi} \right.$$

$$+ \frac{4}{9}\frac{\sin(9\pi x)}{\pi} + \frac{4}{11}\frac{\sin(11\pi x)}{\pi} + \frac{4}{13}\frac{\sin(13\pi x)}{\pi} + \frac{4}{15}\frac{\sin(15\pi x)}{\pi}$$

$$\left. + \frac{4}{17}\frac{\sin(17\pi x)}{\pi} \right]$$

Notice that with a large number of terms the approximation is quite good at values of x where f is continuous.

```
> for k in k_vals do plot({f(x),approx(k)},x=0..2) od;

for k in k_vals do plot(abs(f(x)-approx(k)),x=0..2) od;
```

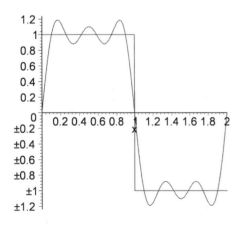

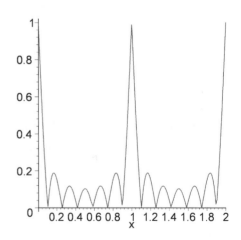

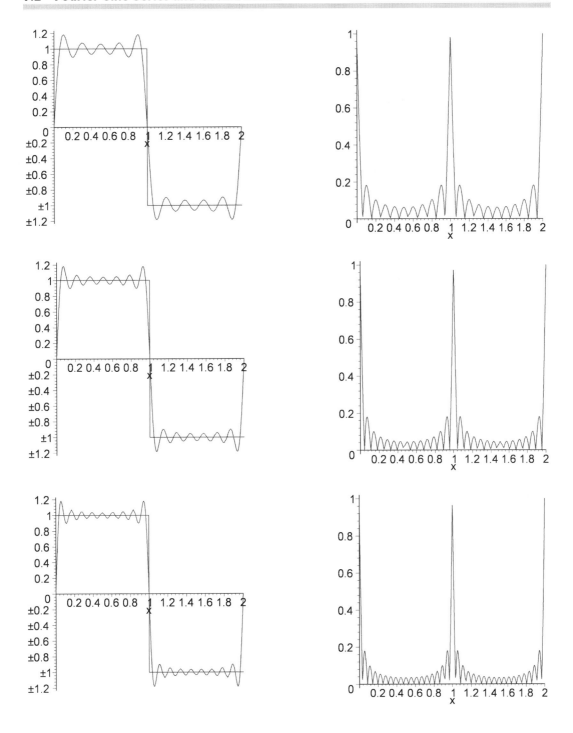

The behavior of the series near points of discontinuity such that the approximation overshoots the function is called the **Gibbs phenomenon**. The approximation continues to "miss" the function even though more and more terms from the series are used!

■

Fourier Cosine Series

Another important eigenvalue problem that has useful eigenfunctions is

$$\begin{cases} y'' + \lambda y = 0 \\ y'(0) = 0, y'(p) = 0 \end{cases}$$

which has eigenvalues and eigenfunctions given by

$$\lambda_n = \begin{cases} 0, n = 0 \\ (n\pi/p)^2, n = 1, 2, \ldots \end{cases} \quad \text{and} \quad y_n(x) = \begin{cases} 1, n = 0 \\ \cos(n\pi x/p), n = 1, 2, \ldots \end{cases}$$

Therefore, for some functions $f(x)$, we can find a series expansion of the form

$$f(x) = \frac{1}{2}a_0 + \sum_{n=1}^{\infty} a_n \cos\frac{n\pi x}{p}.$$

We call this expansion a **Fourier cosine series**, where in the first term (associated with $\lambda_0 = 0$), the constant $\frac{1}{2}a_0$ is written in this form for convenience in finding the formula for the coefficients a_n, $n = 0, 1, 2, \ldots$. We find these coefficients in a manner similar to that followed to find the coefficients in the Fourier sine series. Notice that in this case, the orthogonality condition is $\int_0^p \cos(n\pi x/p)\cos(m\pi x/p)dx = 0, m \neq n$. We use this condition by multiplying both sides of the series expansion by $\cos(m\pi x/p)$ and integrating from $x = 0$ to $x = p$. This yields

$$\int_0^p f(x)\cos\frac{m\pi x}{p}\,dx = \int_0^p \frac{1}{2}a_0 \cos\frac{m\pi x}{p}\,dx + \int_0^p \sum_{n=1}^{\infty} a_n \cos\frac{n\pi x}{p}\cos\frac{m\pi x}{p}\,dx.$$

Assuming that term-by-term integration is allowed,

$$\int_0^p f(x)\cos\frac{m\pi x}{p}\,dx = \int_0^p \frac{1}{2}a_0 \cos\frac{m\pi x}{p}\,dx + \sum_{n=1}^{\infty}\int_0^p a_n \cos\frac{n\pi x}{p}\cos\frac{m\pi x}{p}\,dx.$$

If $m = 0$, then this expression reduces to

$$\int_0^p f(x)dx = \int_0^p \frac{1}{2}a_0 dx + \sum_{n=1}^{\infty}\int_0^p a_n \cos\frac{n\pi x}{p}\,dx$$

where $\int_0^p \cos(n\pi x/p)dx = 0$ and $\int_0^p \frac{1}{2}a_0 dx = pa_0/2$. Therefore, $\int_0^p f(x)dx = pa_0/2$, so

$$a_0 = \frac{2}{p} \int_0^p f(x)dx.$$

If $m > 0$, we note that by the orthogonality property, $\int_0^p \cos(n\pi x/p)\cos(m\pi x/p)dx = 0, m \neq n$. We also note that $\int_0^p \frac{1}{2}a_0 \cos(m\pi x/p)dx = 0$ and $\int_0^p \cos^2(n\pi x/p)dx = p/2$. Hence, $\int_0^p f(x)\cos(n\pi x/p)dx = 0 + a_n \cdot p/2$. Solving for a_n, we have

$$a_n = \frac{2}{p} \int_0^p f(x)\cos\frac{n\pi x}{p}dx, \qquad n = 1, 2, \ldots$$

Notice that this formula also works for $n = 0$ because $\cos\frac{0\cdot\pi x}{a} = \cos 0 = 1$.

EXAMPLE 3: Find the Fourier cosine series for $f(x) = x, 0 \leq x \leq \pi$.

SOLUTION: In this case, $p = \pi$. Hence,

$$a_0 = \frac{2}{\pi} \int_0^\pi x dx = \frac{2}{\pi} \left[\frac{x^2}{2}\right]_0^\pi = \pi$$

and

$$a_n = \frac{2}{\pi} \int_0^\pi x \cos\frac{n\pi x}{\pi}dx = \frac{2}{\pi} \int_0^\pi x \cos nx \, dx \quad \text{(integration by parts)}$$

$$= \frac{2}{\pi} \left\{ \left[\frac{x\sin nx}{n}\right]_0^\pi - \int_0^\pi \frac{\sin nx}{n}dx \right\} = \frac{2}{\pi} \left[\frac{\cos nx}{n^2}\right]_0^\pi = \frac{2}{\pi n^2}(\cos n\pi - 1) = \frac{2}{\pi n^2}[(-1)^n - 1].$$

```
> a:='a':n:='n':

a:=proc(n) option remember;

        2/Pi*int(x*cos(n*x),x=0..Pi)

        end:

a(n);
```

$$2\frac{\cos(n\pi) + n\pi \sin(n\pi) - 1}{\pi n^2}$$

```
> assume(n,integer):

a(n);
```

$$2\frac{(-1)^{n\sim} - 1}{\pi n\sim^2}$$

Notice that for even values of n, $(-1)^n - 1 = 0$. Therefore, $a_n = 0$ if n is even. On the other hand, if n is odd, $(-1)^n - 1 = -2$.

Hence, $a_1 = -\frac{4}{\pi}, a_3 = -\frac{4}{9\pi}, a_5 = -\frac{4}{25\pi}, \ldots, a_{2n-1} = -\frac{4}{(2n-1)^2\pi}$ so the Fourier cosine series is

$$f(x) = \frac{1}{2} \cdot \pi - \sum_{n=1}^{\infty} \frac{1}{n^2\pi} \cos\frac{2n\pi x}{\pi} = \frac{\pi}{2} - \frac{4}{\pi} \sum_{n=1}^{\infty} \frac{1}{(2n-1)^2} \cos(2n-1)x.$$

We plot the function with several terms of the series. Compare these results with those obtained when approximating this function with a sine series. Which series yields the better approximation with the fewer terms?

```
> approx:='approx':

approx:=proc(k) option remember;

        approx(k-1)+a(k)*cos(k*x)

        end:

approx(0):=Pi/2:
```

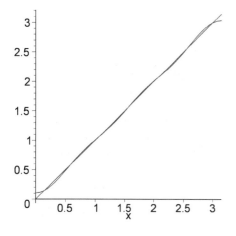

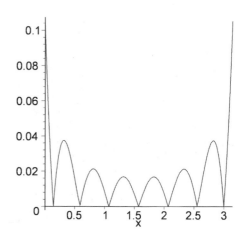

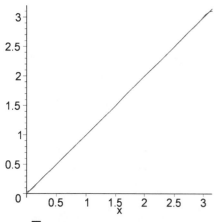

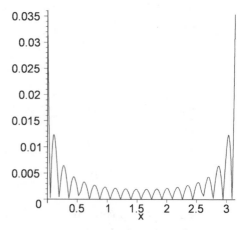

9.3 Fourier Series

Fourier Series

The eigenvalue problem

$$\begin{cases} y'' + \lambda y = 0, -p < x < p \\ y(-p) = y(p), y'(-p) = y'(p) \end{cases}$$

has eigenvalues

$$\lambda_n = \begin{cases} 0, n = 0 \\ (n\pi/p)^2, n = 1, 2, \dots \end{cases}$$

and eigenfunctions

$$y_n(x) = \begin{cases} 1, n = 0 \\ a_n \cos \frac{n\pi x}{p} + b_n \sin \frac{n\pi x}{p}, n = 1, 2, \dots \end{cases},$$

so we can consider a series made up of these functions. Hence, we write

$$f(x) = \frac{1}{2}a_0 + \sum_{n=1}^{\infty} \left(a_n \cos \frac{n\pi x}{p} + b_n \sin \frac{n\pi x}{p} \right),$$

which is called a **Fourier series.** As was the case with Fourier sine and Fourier cosine series, we must determine the coefficients $a_0, a_n (n = 1, 2, \dots)$, and $b_n (n = 1, 2, \dots)$. Because we use a method similar to that used to find the coefficients in Section 9.2, we give the values a of several integrals in the following table.

$\int_{-p}^{p} \cos \frac{n\pi x}{p} dx = 0$	$\int_{-p}^{p} \sin \frac{n\pi x}{p} dx = 0$	$\int_{-p}^{p} \cos \frac{m\pi x}{p} \sin \frac{n\pi x}{p} dx = 0$
$\int_{-p}^{p} \cos \frac{m\pi x}{p} \cos \frac{n\pi x}{p} dx = \begin{cases} 0, m \neq n \\ p, m = n \end{cases}$	$\int_{-p}^{p} \sin \frac{m\pi x}{p} \sin \frac{n\pi x}{p} dx = \begin{cases} 0, m \neq n \\ p, m = n \end{cases}$	

We begin by finding a_0 and $a_n (n = 1, 2, \dots)$. Multiplying both sides of

$$f(x) = \frac{1}{2}a_0 + \sum_{n=1}^{\infty} \left(a_n \cos \frac{n\pi x}{p} + b_n \sin \frac{n\pi x}{p} \right) \text{ by } \cos \frac{m\pi x}{p} \text{ and integrating from } x = -p \text{ to } x = p$$

(because of the boundary conditions) yields

$$\int_{-p}^{p} f(x) \cos \frac{m\pi x}{p} dx = \int_{-p}^{p} \frac{1}{2} a_0 \cos \frac{m\pi x}{p} dx + \int_{-p}^{p} \sum_{n=1}^{\infty} \left(a_n \cos \frac{n\pi x}{p} \cos \frac{m\pi x}{p} + b_n \sin \frac{n\pi x}{p} \cos \frac{m\pi x}{p} \right) dx$$

$$= \int_{-p}^{p} \frac{1}{2} a_0 \cos \frac{m\pi x}{a} dx + \sum_{n=1}^{\infty} \left(\int_{-p}^{p} a_n \cos \frac{n\pi x}{p} \cos \frac{m\pi x}{p} dx + \int_{-p}^{p} a_n \sin \frac{n\pi x}{p} \cos \frac{m\pi x}{p} dx \right).$$

If $m = 0$, we notice that all of the integrals that we are summing have the value zero. Thus, this expression simplifies to

$$\int_{-p}^{p} f(x)dx = \int_{-p}^{p} \frac{1}{2} a_0 dx$$

$$\int_{-p}^{p} f(x)dx = \frac{1}{2} a_0 \cdot 2p$$

$$a_0 = \frac{1}{p} \int_{-p}^{p} f(x)dx.$$

If $m \neq 0$, only one of the integrals on the right-hand side of the expression yields a value other than zero and this occurs with $\int_{-p}^{p} \cos\frac{m\pi x}{p} \cos\frac{n\pi x}{p} dx = \begin{cases} 0, m \neq n \\ p, m = n \end{cases}$ if $m = n$. Hence,

$$\int_{-p}^{p} f(x) \cos\frac{n\pi x}{p} dx = p \cdot a_n$$

$$a_n = \frac{1}{p} \int_{-a}^{a} f(x) \cos\frac{n\pi x}{p} dx, \qquad n = 1, 2, \ldots$$

We find $b_n (n = 1, 2, \ldots)$ by multiplying the series by $\sin\frac{m\pi x}{p}$ and integrating from $x = -p$ to $x = p$. This yields

$$\int_{-p}^{p} f(x) \sin\frac{m\pi x}{p} dx = \int_{-p}^{p} \frac{1}{2} a_0 \sin\frac{m\pi x}{p} dx + \int_{-p}^{p} \sum_{n=1}^{\infty} \left(a_n \cos\frac{n\pi x}{p} \sin\frac{m\pi x}{p} + b_n \sin\frac{n\pi x}{p} \sin\frac{m\pi x}{p} \right) dx$$

$$= \int_{-p}^{p} \frac{1}{2} a_0 \sin\frac{m\pi x}{p} dx + \sum_{n=1}^{\infty} \left(\int_{-p}^{p} a_n \cos\frac{n\pi x}{p} \sin\frac{m\pi x}{p} dx + \int_{-p}^{p} a_n \sin\frac{n\pi x}{p} \sin\frac{m\pi x}{p} dx \right).$$

Again, we note that only one of the integrals on the right-hand side is not zero. In this case, we use $\int_{-p}^{p} \sin\frac{m\pi x}{p} \sin\frac{n\pi x}{p} dx = \begin{cases} 0, m \neq n \\ p, m = n \end{cases}$ to obtain

$$\int_{-p}^{p} f(x) \sin\frac{n\pi x}{p} dx = p \cdot b_n$$

$$b_n = \frac{1}{p} \int_{-p}^{p} f(x) \sin\frac{n\pi x}{p} dx, \qquad n = 1, 2, \ldots$$

Definition
Fourier Series

Suppose that f is defined on $-p < x < p$. The **Fourier series** for f is

$$f(x) = \frac{1}{2}a_0 + \sum_{n=1}^{\infty}\left(a_n \cos\frac{n\pi x}{p} + b_n \sin\frac{n\pi x}{p}\right)$$

where

$$a_0 = \frac{1}{p}\int_{-p}^{p} f(x)dx$$

$$a_n = \frac{1}{p}\int_{-p}^{p} f(x)\cos\frac{n\pi x}{p}dx, n = 1, 2, \ldots \quad \text{and}$$

$$b_n = \frac{1}{p}\int_{-p}^{p} f(x)\sin\frac{n\pi x}{p}dx, n = 1, 2, \ldots$$

The following theorem tells us that the Fourier series for any function converges to the function except at points of discontinuity.

Theorem
Convergence of Fourier
Series

Suppose that f and f' are piecewise continuous functions on $-p < x < p$. Then the Fourier series for f on $-p < x < p$ converges to $f(x)$ at every x where f is continuous. If f is discontinuous at $x = x_0$, the Fourier series converges to the average

$$\frac{1}{2}(f(x_0^{+}) + f(x_0^{-})),$$

where $f(x_0^{+}) = \lim_{x \to x_0^{+}} f(x)$ and $f(x_0^{-}) = \lim_{x \to x_0^{-}} f(x)$.

EXAMPLE 1: Find the Fourier series for $f(x) = \begin{cases} 1, -2 \le x < 0 \\ 2, 0 \le x < 2 \end{cases}$, where $f(x + 4) = f(x)$.

SOLUTION: In this case, $p = 2$.

```
> f:=x->1+Heaviside(x):
plot(f(x),x=-2..2,view=[0..2,-2..2]);
```

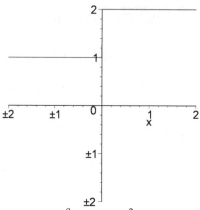

First we find $a_0 = \frac{1}{2}\int_{-2}^{2} f(x)dx = \frac{1}{2}\int_{-2}^{0} 1dx + \frac{1}{2}\int_{0}^{2} 2dx = 3,$

```
> a:='a':
a(0):=1/2*int(f(x),x=-2..2);
```

$$a(0) := 3$$

$$a_n = \frac{1}{2}\int_{-2}^{2} f(x)\cos\frac{n\pi x}{2}dx = \frac{1}{2}\int_{-2}^{0}\cos\frac{n\pi x}{2}dx + \frac{1}{2}\int_{0}^{2} 2\cos\frac{n\pi x}{2}dx = 0, \text{ and}$$

```
> assume(n,integer):
a:=proc(n) option remember;
        1/2*int(f(x)*cos(n*Pi*x/2),x=-2..2);
        end:
a(n);
```

$$0$$

$$b_n = \frac{1}{2}\int_{-2}^{2} f(x)\sin\frac{n\pi x}{2}dx = \frac{1}{2}\int_{-2}^{0}\sin\frac{n\pi x}{2}dx + \frac{1}{2}\int_{0}^{2} 2\sin\frac{n\pi x}{2}dx = \frac{1}{n\pi}(1-\cos n\pi) = \frac{1}{n\pi}(1-(-1)^n).$$

```
> assume(n,integer):
b:='b':
b:=proc(n) option remember;
        1/2*int(f(x)*sin(n*Pi*x/2),x=-2..2);
        end:
b(n);
```

$$-\frac{2(-1)^{n\sim}-1}{n\sim\pi} + \frac{(-1)^{n\sim}}{n\sim\pi}$$

Therefore,

$$f(x) = \frac{3}{2} + \sum_{n=1}^{\infty} (1 - (-1)^n) \frac{1}{n\pi} \sin \frac{n\pi x}{2} = \frac{3}{2} + \frac{2}{\pi} \sin \frac{\pi x}{2} + \frac{2}{3\pi} \sin \frac{3\pi x}{2} + \frac{2}{5\pi} \sin \frac{5\pi x}{2} + \dots$$

We now graph $f(x)$ with the 3rd, 5th, 11th, and 15th partial sums of the series. First, we define $pk(x)$ to be the k partial sum of the Fourier series

```
> p:=proc(k) option remember;

        p(k-1)+b(k)*sin(k*Pi*x/2)

        end:

p(0):=3/2:
```

and then `toplot` to be a list of the 3rd, 5th, 11th, and 15th partial sums of the series.

```
> toplot:=[p(3),p(5),p(11),p(15)];
```

$$toplot := \left[\frac{3}{2} + 2 \frac{\sin\left(\frac{1}{2}\pi x\right)}{\pi} + \frac{2}{3} \frac{\sin\left(\frac{3}{2}\pi x\right)}{\pi}, \right.$$

$$\frac{3}{2} + 2 \frac{\sin\left(\frac{1}{2}\pi x\right)}{\pi} + \frac{2}{3} \frac{\sin\left(\frac{3}{2}\pi x\right)}{\pi} + \frac{2}{5} \frac{\sin\left(\frac{5}{2}\pi x\right)}{\pi}, \frac{3}{2} + 2 \frac{\sin\left(\frac{1}{2}\pi x\right)}{\pi}$$

$$+ \frac{2}{3} \frac{\sin\left(\frac{3}{2}\pi x\right)}{\pi} + \frac{2}{5} \frac{\sin\left(\frac{5}{2}\pi x\right)}{\pi} + \frac{2}{7} \frac{\sin\left(\frac{7}{2}\pi x\right)}{\pi} + \frac{2}{9} \frac{\sin\left(\frac{9}{2}\pi x\right)}{\pi}$$

$$+ \frac{2}{11} \frac{\sin\left(\frac{11}{2}\pi x\right)}{\pi}, \frac{3}{2} + 2 \frac{\sin\left(\frac{1}{2}\pi x\right)}{\pi} + \frac{2}{3} \frac{\sin\left(\frac{3}{2}\pi x\right)}{\pi} + \frac{2}{5} \frac{\sin\left(\frac{5}{2}\pi x\right)}{\pi}$$

$$+ \frac{2}{7} \frac{\sin\left(\frac{7}{2}\pi x\right)}{\pi} + \frac{2}{9} \frac{\sin\left(\frac{9}{2}\pi x\right)}{\pi} + \frac{2}{11} \frac{\sin\left(\frac{11}{2}\pi x\right)}{\pi} + \frac{2}{13} \frac{\sin\left(\frac{13}{2}\pi x\right)}{\pi}$$

$$\left. + \frac{2}{15} \frac{\sin\left(\frac{15}{2}\pi x\right)}{\pi} \right]$$

We also redefine $f(x)$.

```
> f:='f':

f:=proc(x) option remember;

        if x >=-2 and x < 0 then 1 elif

        x >=0 and x < 2 then 2 else

        f(x-4) fi end:
```

We graph $f(x)$ and the kth partial sum for $k = 3, 5, 11$, and 15 on the interval [0,8] using the following `for` loop.

```
> for q in toplot
    do plot({'f(x)',q},'x'=0..8,0..2.5) od;
```

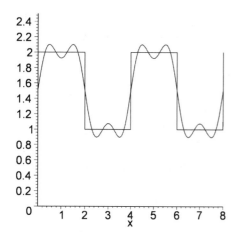
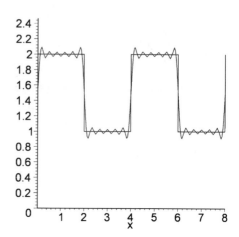
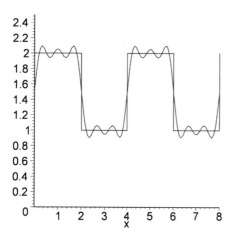
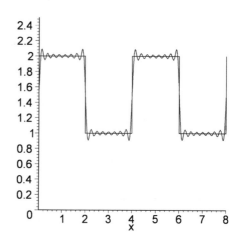

We graph the error by graphing the absolute value of the difference of $f(x)$ and the kth partial sum using the following `for` loop.

```
> for q in toplot do plot(abs('f(x)'-q),'x'=0..8) od;
```

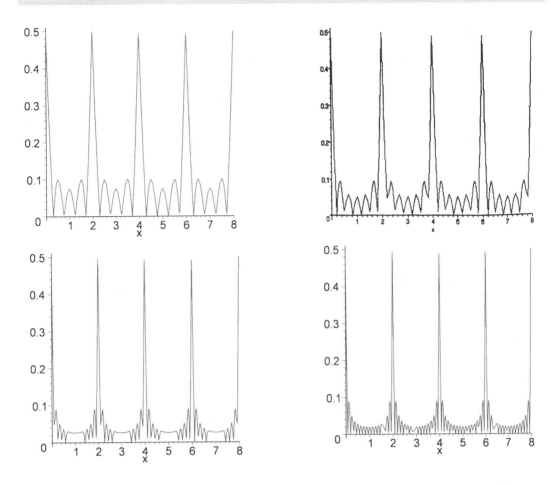

The graphs show that if we extend f over more periods, then the approximation by the Fourier series carries over to these intervals.

∎

EXAMPLE 2: Find the Fourier series for $f(x) = \begin{cases} 0, -1 \leq x < 0 \\ \sin \pi x, 0 \leq x < 1 \end{cases}$ where $f(x+2) = f(x)$.

SOLUTION: In this case, $p = 1$, so $a_0 = \frac{1}{1}\int_{-1}^{1} f(x)dx = \int_0^1 \sin \pi x dx = \frac{2}{\pi}$ and

```
> a:='a':b:='b':n:='n':

a(0):=int(sin(Pi*x),x=0..1);
```

$$a(0) := 2\frac{1}{\pi}$$

$a_n = \int_0^1 \sin \pi x \cos n\pi x \, dx = \int_0^1 \sin \pi x \cos n\pi x \, dx$. The value of this integral depends on the value of n. If $n = 1$, we have $a_1 = \int_0^1 \sin \pi x \cos \pi x \, dx = \frac{1}{2}\int_0^1 \sin 2\pi x \, dx = 0$, where we use the identity $\sin A \cos A = \frac{1}{2}\sin 2A$.

```
> assume(n,integer):

a:=proc(n) option remember;

        int(sin(Pi*x)*cos(n*Pi*x),x=0..1)

        end:

> a(1);
```

$$0$$

If $n \neq 1$, we use the identity $\sin A \cos B = \frac{1}{2}[\sin(A - B) + \sin(A + B)]$ to obtain

$$a_n = \frac{1}{2}\int_0^1 [\sin(1-n)\pi x + \sin(1+n)\pi x]dx = -\frac{1}{2}\left[\frac{\cos(1-n)\pi x}{(1-n)\pi} + \frac{\cos(1+n)\pi x}{(1+n)\pi}\right]_0^1$$

$$= -\frac{1}{2}\left\{\left[\frac{\cos(1-n)\pi}{(1-n)\pi} + \frac{\cos(1+n)\pi}{(1+n)\pi}\right] - \left[\frac{1}{(1-n)\pi} + \frac{1}{(1+n)\pi}\right]\right\}.$$

```
> a(n);
```

$$-\frac{(-1)^{n\sim} + 1}{\pi(1 + n\sim)(-1 + n\sim)}$$

Notice that if n is odd, both $1 - n$ and $1 + n$ are even. Hence, $\cos(1 - n)\pi x = \cos(1 + n)\pi x = 1$, so

$$a_n = -\frac{1}{2}\left\{\left[\frac{1}{(1-n)\pi} + \frac{1}{(1+n)\pi}\right] - \left[\frac{1}{(1-n)\pi} + \frac{1}{(1+n)\pi}\right]\right\} = 0$$

if n is odd. On the other hand, if n is even, $1 - n$ and $1 + n$ are odd. Therefore, $\cos(1 - n)\pi x = \cos(1 + n)\pi x = -1$, so

$$a_n = -\frac{1}{2}\left\{\left[\frac{-1}{(1-n)\pi} + \frac{-1}{(1+n)\pi}\right] - \left[\frac{1}{(1-n)\pi} + \frac{1}{(1+n)\pi}\right]\right\}$$

$$= \frac{1}{(1-n)\pi} + \frac{1}{(1+n)\pi} = \frac{2}{(1-n)(1+n)\pi} = -\frac{2}{(n-1)(1+n)\pi},$$

if n is even. We confirm this observation by computing several coefficients.

```
> array([seq([n,a(n)],n=0..10)]);
```

$$\begin{bmatrix} 0 & 2\frac{1}{\pi} \\ 1 & 0 \\ 2 & -\frac{2}{3}\frac{1}{\pi} \\ 3 & 0 \\ 4 & -\frac{2}{15}\frac{1}{\pi} \\ 5 & 0 \\ 6 & -\frac{2}{35}\frac{1}{\pi} \\ 7 & 0 \\ 8 & -\frac{2}{63}\frac{1}{\pi} \\ 9 & 0 \\ 10 & -\frac{2}{99}\frac{1}{\pi} \end{bmatrix}$$

Putting this information together, we can write the coefficients as

$$a_{2n} = -\frac{2}{(2n-1)(1+2n)\pi}, \qquad n = 1, 2, \ldots .$$

Similarly,

$$b_n = \frac{1}{1}\int_{-1}^{1} f(x) \sin n\pi x \, dx = \int_0^1 \sin \pi x \sin n\pi x \, dx,$$

so if $n = 1$, $b_1 = \int_0^1 \sin^2 \pi x \, dx = \int_0^1 \frac{1}{2}(1 - \cos 2\pi x) dx = \frac{1}{2}\left[x - \frac{\sin 2\pi x}{2\pi}\right]_0^1 = \frac{1}{2}.$

```
> b:=proc(n) option remember;
      int(sin(Pi*x)*sin(n*Pi*x),x=0..1)
  end:
```

```
> b(1);
```

$$\frac{1}{2}$$

If $n \neq 1$, we use $\sin A \sin B = \frac{1}{2}[\cos(A - B) - \cos(A + B)]$. Hence,

$$b_n = \frac{1}{2}\int_0^1 [\cos(1-n)\pi x - \cos(1+n)\pi x] dx = \frac{1}{2}\left[\frac{\sin(1-n)\pi x}{(1-n)\pi} - \frac{\sin(1+n)\pi x}{(1+n)\pi}\right]_0^1 = 0, \qquad n = 2, 3, \ldots .$$

```
> b(n);
```

$$0$$

Therefore, we write the Fourier series as

$$f(x) = \frac{1}{\pi} + \frac{1}{2}\sin \pi x - \frac{2}{\pi}\sum_{n=1}^{\infty} \frac{1}{(2n-1)(1+2n)}\cos 2n\pi x.$$

We graph f along with several approximations using this series in the same way as in Section 9.2. Let $pk(x) = \frac{1}{\pi} + \frac{1}{2}\sin \pi x - \frac{2}{\pi}\sum_{n=1}^{k}\frac{1}{(2n-1)(1+2n)}\cos 2n\pi x$ denote the kth partial sum of the Fourier series. Note that

$$pk(x) = \frac{1}{\pi} + \frac{1}{2}\sin \pi x - \frac{2}{\pi}\sum_{n=1}^{k-1}\frac{1}{(2n-1)(1+2n)}\cos 2n\pi x - \frac{2}{\pi}\frac{1}{(2k-1)(1+2k)}\cos 2k\pi x$$

$$= pk - 1(x) - \frac{2}{\pi}\frac{1}{(2k-1)(1+2k)}\cos 2k\pi x.$$

Thus, to calculate the kth partial sum of the Fourier series, we need only subtract $\frac{2}{\pi}\frac{1}{(2k-1)(1+2k)}\cos 2k\pi x.$ from the $(k-1)$ st partial sum: we need not recompute all k terms of the kth partial sum if we know the $(k-1)$ st partial sum. Using this observation, we define the recursively defined function p to return the kth partial sum of the series.

```
> p:=proc(k) option remember;
        p(k-1)+a(k)*cos(k*Pi*x)
        end:
p(0):=1/Pi+1/2*sin(Pi*x):


> f:='f':
f:=proc(x) option remember;
        if x >=-1 and x < 0 then 0 elif
        x >=0 and x < 1 then sin(Pi*x) else
        f(x-2) fi end:
```

We graph f along with the 2nd, 6th, and 10th partial sums of the series.

```
> k_vals:=2,6,10:
for k in k_vals do
        plot({'f(x)',p(k)},'x'=0..4,-.5..1.5) od;
```

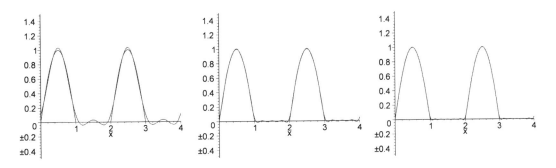

The corresponding errors are graphed as well.

```
> for k in k_vals do
      plot(abs('f(x)'-p(k)),'x'=0..4) od;
```

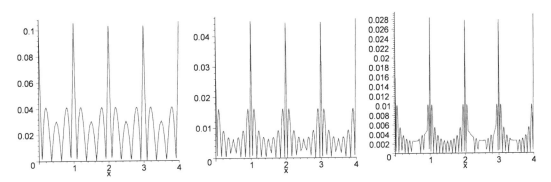

■

Even, Odd, and Periodic Extensions

In the discussion so far in this section, we have assumed that the f was defined on the interval $-p < x < p$. However, this is not always the case. Sometimes, we must take a function that is defined on the interval $0 < x < p$ and represent it in terms of trigonometric functions. Three ways to accomplish this task are to extend f to obtain (1) an **even** function on $-p < x < p$, (2) an **odd** function on $-p < x < p$, or (3) a **periodic** function on $-p < x < p$.

We can notice some interesting properties associated with the Fourier series in each of these three cases by noting the properties of even and odd functions. If f is an even function and g is an odd function, then the product fg is an odd function. Similarly, if f is an even function and g is an even function, then fg is an even function, and if f is an odd function and g is an odd function, then fg is an even function. Recall from integral calculus that if f is odd on $-p<x<p$, then $\int_{-p}^{p} f(x)dx = 0$, while if g is even on $-p<x<p$, then $\int_{-p}^{p} g(x)dx = 2\int_{0}^{p} g(x)dx$. These properties are useful in determining the coefficients in the Fourier series for the even, odd, and periodic extensions of a function, because $\cos(n\pi x/p)$ and $\sin(n\pi x/p)$ are even and odd periodic functions, respectively, on $-p<x<p$.

The **even extension** f_{even} of f is an even function. Therefore,

$$a_0 = \frac{1}{p}\int_{-p}^{p} f_{\text{even}}(x)dx = \frac{2}{p}\int_{0}^{p} f(x)dx, a_n = \frac{1}{p}\int_{-p}^{p} f_{\text{even}}(x)\cos\frac{n\pi x}{p}dx = \frac{2}{p}\int_{0}^{p} f(x)\cos\frac{n\pi x}{p}dx \ (n = 1, 2, \ldots),$$

and $b_n = \frac{1}{p}\int_{-p}^{p} f_{\text{even}}(x)\sin\frac{n\pi x}{p}dx = 0(n = 1, 2, \ldots)$.

The **odd extension** f_{odd} of f is an odd function, so $a_0 = \frac{1}{p}\int_{-p}^{p} f_{\text{odd}}(x)dx = 0$,
$a_n = \frac{1}{p}\int_{-p}^{p} f_{\text{odd}}(x)\cos\frac{n\pi x}{p}dx = 0(n = 1, 2, \ldots)$, and $b_n = \frac{1}{p}\int_{-p}^{p} f_{\text{odd}}(x)\sin\frac{n\pi x}{a}dx$
$= \frac{2}{p}\int_{0}^{p} f(x)\sin\frac{n\pi x}{p}dx(n = 1, 2, \ldots)$.

The **periodic extension** f_p has period p. Because half of the period is $p/2$,

$$a_0 = \frac{2}{p}\int_{0}^{p} f(x)dx, a_n = \frac{2}{p}\int_{0}^{p} f(x)\cos\frac{2n\pi x}{p}dx(n = 1, 2, \ldots), \text{ and } b_n = \frac{2}{p}\int_{0}^{p} f(x)\sin\frac{2n\pi x}{p}dx(n = 1, 2, \ldots).$$

EXAMPLE 3: Let $f(x) = x$ on $(0, 1)$. Find the Fourier series for (a) the even extension of f; (b) the odd extension of f; (c) the periodic extension of f.

SOLUTION: (a) Here $p = 1$, so $a_0 = 2\int_{0}^{1} xdx = 1$,

```
> a(0):=2*int(x,x=0..1);
```

$$a(0) := 1$$

$$a_n = 2\int_{0}^{1} x\cos n\pi x dx = \frac{2}{n^2\pi^2}(\cos n\pi - 1) = \frac{2}{n^2\pi^2}[(-1)^n - 1](n = 1, 2, \ldots)., \text{ and } b_n = 0(n = 1, 2, \ldots).$$

```
> a:='a':n:='n':

assume(n,integer):

a:=n->2*int(x*cos(n*Pi*x),x=0..1):

a(n);
```

$$2\frac{(-1)^{n\sim} - 1}{n\sim^2\pi^2}$$

Because $a_n = 0$ if n is even,

```
> a(2*n);
```

$$0$$

we can represent the coefficients with odd subscripts as $a_{2n-1} = -\frac{4}{(2n-1)^2\pi^2}$.

```
> a(2*n-1);
```

$$-4\frac{1}{(2n\sim -1)^2\pi^2}$$

Therefore, the Fourier cosine series is

$$f_{\text{even}}(x) = \frac{1}{2} - \sum_{n=1}^{\infty} \frac{4}{(2n-1)^2\pi^2}\cos(2n-1)\pi x.$$

We graph the even extension with several terms of the Fourier cosine series by first defining f to be the even extension of $f(x) = x$ on $(0,1)$

```
> f:=proc(x) option remember;
      if x > =-1 and x < 0 then -x elif
      x > =0 and x < 1 then x elif
      x > =1 then f(x-2) elif
      x < =-1 then f(x+2) fi end:
```

and $p_k(x) = \frac{1}{2} - \sum_{n=1}^{k} \frac{4}{(2n-1)^2\pi^2}\cos(2n-1)\pi x.$

```
> p:=(x,k)->a(0)/2+sum(a(2*n-1)*cos((2*n-1)*Pi*x),
      n=1..k);
```

$$p := (x,k) \rightarrow \frac{1}{2}a(0) + \left(\sum_{n=1}^{k} a(2n-1)\cos((2n-1)\pi x)\right)$$

We then graph f together with $p1$ and $p5$.

```
> plot({'f(x)',p(x,1)},'x'=-2..2);
plot({'f(x)',p(x,5)},'x'=-2..2);
```

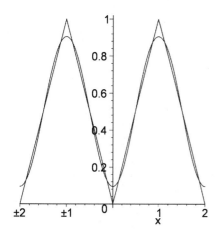 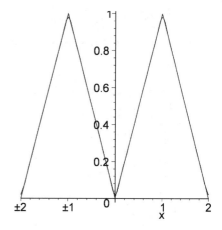

(b) For the odd extension f_{odd}. We note that $a_0 = 0$, $a_n = 0 (n = 1, 2, \ldots).$, and $b_n = 2 \int_0^1 x \sin \frac{n\pi x}{a} dx = -\frac{2}{n\pi} \cos n\pi = \frac{2(-1)^{n+1}}{n\pi} (n = 1, 2, \ldots).$

```
> b:=n->2*int(x*sin(n*Pi*x),x=0..1):
b(n);
```

$$-2\frac{(-1)^{n\sim}}{n\sim\pi}$$

Hence, the Fourier sine series is

$$f_{\text{odd}}(x) = \sum_{n=1}^{\infty} \frac{2(-1)^{n+1}}{n\pi} \sin n\pi x.$$

We graph the odd extension along with several terms of the Fourier sine series in the same manner as in (a).

```
> f:=proc(x) option remember;
        if x>=-1 and x<1 then x elif
        x>=1 then f(x-2) elif
        x<=-1 then f(x+2) fi end:
```

```
> p:=(x,k)->sum(b(n)*sin(n*Pi*x),n=1..k);
```

$$p := (x, k) \rightarrow \sum_{n=1}^{k} b(n) \sin(n\pi x)$$

```
> plot({'f(x)',p(x,5)},'x'=-2..2);
plot({'f(x)',p(x,10)},'x'=-2..2);
```

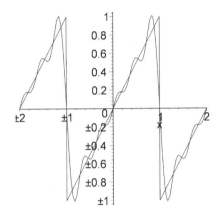
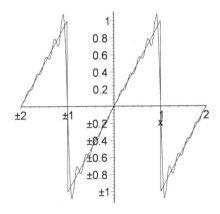

(c) The periodic extension has period $2p = 1$, so $p = 1/2$. Thus, $a_0 = \frac{1}{1/2} \int_0^1 x\,dx = 2 \int_0^1 x\,dx = 1$

```
> a(0):=2*int(x,x=0..1);
```

$$a(0) := 1$$

$$a_n = 2 \int_0^1 x \cos 2n\pi x\,dx = \frac{1}{2} \frac{\cos 2n\pi + 2n\pi \sin 2n\pi}{n^2\pi^2} - \frac{1}{2n^2\pi^2} = 0 \ (n = 1, 2, \ldots), \text{ and}$$

```
> a:=n->2*int(x*cos(2*n*Pi*x),x=0..1):
a(n);
```

$$0$$

$$b_n = 2 \int_0^1 x \sin 2n\pi x\,dx = -\frac{1}{2} \frac{1 - \sin 2n\pi + 2n\pi \cos 2n\pi}{n^2\pi^2} = -\frac{2n\pi}{2n^2\pi^2} = -\frac{1}{n\pi} \ (n = 1, 2, \ldots).$$

```
> b:=n->2*int(x*sin(2*n*Pi*x),x=0..1):
b(n);
```

$$-\frac{1}{n \sim \pi}$$

Hence, the Fourier series for the periodic extension is

$$f_p(x) = \frac{1}{2} - \sum_{n=1}^{\infty} \frac{1}{n\pi} \sin 2n\pi x.$$

We graph the periodic extension with several terms of the Fourier series in the same way as in (a) and (b).

```
> f:=proc(x) option remember;
        if x >=-1 and x < 0 then x+1 elif
        x >=0 and x < 1 then x elif
        x >=1 then f(x-2) elif
        x <=-1 then f(x+2) fi end:
```

```
> p:=(x,k)->a(0)/2+sum(b(n)*sin(2*n*Pi*x),n=1..k);
```

$$p := (x,k) \rightarrow \frac{1}{2}a(0) + \left(\sum_{n=1}^{k} b(n) \sin(2n\pi x) \right)$$

```
> plot({'f(x)',p(x,5)},'x'=-2..2);
plot({'f(x)',p(x,10)},'x'=-2..2);
```

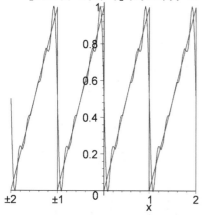
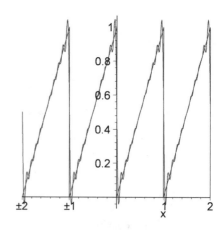

■

Differentiation and Integration of Fourier Series

Definition	A function $f(x)$, $-p < x < p$, is **piecewise smooth** if f and all of its
Piecewise Smooth	derivatives are piecewise continuous.

Theorem	Let $f(x)$, $-p < x < p$, be a continuous piecewise smooth function
Term-by-Term	with Fourier series $\frac{1}{2}a_0 + \sum\limits_{n=1}^{\infty}\left(a_n \cos\frac{n\pi x}{p} + b_n \sin\frac{n\pi x}{p}\right)$. Then,
Differentiation	$f'(x)$, $-p < x < p$ has Fourier series
	$$\sum_{n=1}^{\infty}\frac{n\pi}{p}\left(-a_n \sin\frac{n\pi x}{p} + b_n \cos\frac{n\pi x}{p}\right).$$

In other words, we differentiate the Fourier series for f term by term to obtain the Fourier series for f'.

Theorem	Let $f(x)$, $-p < x < p$, be a piecewise smooth function with Fourier
Term-by-Term	series $\frac{1}{2}a_0 + \sum\limits_{n=1}^{\infty}\left(a_n \cos\frac{n\pi x}{p} + b_n \sin\frac{n\pi x}{p}\right)$. Then, the Fourier series of
Integration	an antiderivative of f can be found by integrating the Fourier series of f term by term.

EXAMPLE 4: Use the Fourier series for $f(x) = \frac{1}{12}x(\pi^2 - x^2)$, $-\pi < x < \pi$ to show how term-by-term differentiation and term-by-term integration can be used to find the Fourier series of $g(x) = \frac{1}{12}\pi^2 - \frac{1}{4}x^2$, $-\pi < x < \pi$ and $h(x) = \frac{1}{24}\pi^2 x^2\left(1 - \frac{1}{2}x^2\right)$, $-\pi < x < \pi$.

SOLUTION: After defining f and the substitutions in rule to simplify our results, we calculate a_0, a_n, and b_n. (Because f is an odd function, $a_n = 0$, $n \geq 0$.)

```
> f:=x->1/12*x*(Pi^2-x^2):

> a:='a':b:='b':n:='n':

assume(n,integer):
```

```
a(0):=1/Pi*int(f(x),x=-Pi..Pi);
```

$$a(0) := 0$$

```
> a:=n->1/Pi*int(f(x)*cos(n*x),x=-Pi..Pi):
a(n);
```

$$0$$

```
> b:=n->1/Pi*int(f(x)*sin(n*x),x=-Pi..Pi):
b(n);
```

$$-\frac{(-1)^{n\sim}}{n\sim^3}$$

We define the nth term in $\sum_{n=1}^{\infty}\left(a_n\cos\frac{n\pi x}{p}+b_n\sin\frac{n\pi x}{p}\right)$ in `f s (n)` and the finite sum $\frac{1}{2}a_0+\sum_{n=1}^{k}\left(a_n\cos\frac{n\pi x}{p}+b_n\sin\frac{n\pi x}{p}\right)$ in `fourier (k)`.

```
> fs:=n->a(n)*cos(n*x)+b(n)*sin(n*x):
fourier:=k->fourier(k-1)+fs(k):
fourier(0):=a(0)/2:
```

We see how quickly the Fourier series converges to f by graphing $f(x)$, $-\pi < x < \pi$, together with `fourier (1)` and `fourier (3)`.

```
> plot({f(x),fourier(1)},x=-Pi..Pi);
plot({f(x),fourier(3)},x=-Pi..Pi);
```

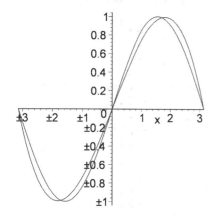
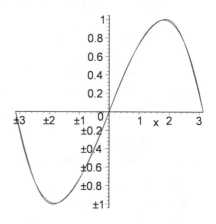

Notice that $g(x) = \frac{1}{12}\pi^2 - \frac{1}{4}x^2, -\pi < x < \pi$, is the derivative of $f(x), -\pi < x < \pi$. Of course, we could compute the Fourier series of $f'(x), -\pi < x < \pi$ directly by applying the integral formulas with $g(x), -\pi < x < \pi$, to find the Fourier series coefficients. However, the objective here is to illustrate how term-by-term differentiation of the Fourier series for $f(x), -\pi < x < \pi$, gives us the Fourier series for $f'(x), -\pi < x < \pi$. We calculate the derivative of $f(x)$ in df in order to make graphical comparisons. In dfs (n), we determine the derivative of the nth term of the Fourier series for $f(x), -\pi < x < \pi$, found before, and in dfourier (k), we calculate the kth partial sum of the Fourier series for $f'(x), -\pi < x < \pi$. (Notice that this series does not include a constant term because the derivative of $\frac{1}{2}a_0$ is zero.)

```
> df:=diff(f(x),x);
```

$$df := \frac{1}{12}\pi^2 - \frac{1}{4}x^2$$

```
> dfs:=n->diff(fs(n),x):

dfs(n);
```

$$-\frac{(-1)^{n\sim}\cos(n\sim x)}{n\sim^2}$$

```
> dfourier:=k->dfourier(k-1)+dfs(k):

dfourier(0):=0:
```

Next, we graph $f'(x), -\pi < x < \pi$, simultaneously with dfourier (1) and dfourier (3). Again, the convergence of the Fourier series approximations to $f'(x), -\pi < x < \pi$ is quick.

```
> plot({df,dfourier(1)},x=-Pi..Pi);

plot({df,dfourier(3)},x=-Pi..Pi);
```

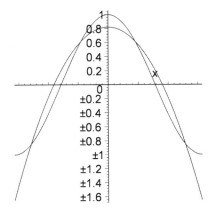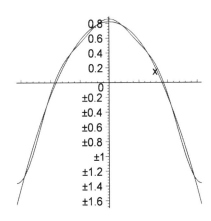

Notice that $h(x)$, $-\pi < x < \pi$, is an antiderivative of $f(x)$, $-\pi < x < \pi$. We calculate this antiderivative in `intf`. Of course, this is the antiderivative of f with zero constant of integration because Maple does not include an integration constant. When we integrate the terms of the Fourier series of $f(x)$, $-\pi < x < \pi$, a constant term is not included. However, the Fourier series of the even function $h(x) = \frac{1}{24}\pi^2 x^2\left(1 - \frac{1}{2}x^2\right)$, $-\pi < x < \pi$, should include the constant term $\frac{1}{2}\tilde{a}_0$. We calculate the value of $\tilde{a}_0$ in `inta(0)` with the integral formula $\frac{1}{\pi}\int_{-\pi}^{\pi} h(x)\,dx$.

```
> intf:=int(f(x),x);
```

$$intf := -\frac{1}{48}x^4 + \frac{1}{24}\pi^2 x^2$$

```
> inta(0):=1/Pi*int(intf,x=-Pi..Pi);
```

$$\text{inta}(0) := \frac{7}{360}\pi^4$$

In `intfs (n)`, we integrate the nth term of the Fourier series of $f(x)$, $-\pi < x < \pi$, found before to determine the coefficients of $\cos nx$ and $\sin nx$ in the Fourier series of $h(x)$, $-\pi < x < \pi$. In `intfourier (k)`, we determine the sum of the first k terms of the Fourier series of $h(x)$, $-\pi < x < \pi$ obtained by adding $\frac{1}{2}\tilde{a}_0$ to the expression obtained through term-by-term integration of the Fourier series of $f(x)$, $-\pi < x < \pi$.

```
> intfs:=n->int(fs(n),x):
intfs(n);
```

$$\frac{(-1)^{n\sim}\cos(n\sim x)}{n\sim^4}$$

```
> intfourier:=k->intfourier(k-1)+intfs(k):
intfourier(0):=inta(0)/2:
```

By graphing $h(x)$, $-\pi < x < \pi$, simultaneously with the approximation in `intfourier (k)` for $n = 1$ and 3, we see how the graphs of Fourier series approximations obtained through term-by-term integration converge to the graph of $h(x)$, $-\pi < x < \pi$.

```
> plot({intf,intfourier(1)},x=-Pi..Pi);
plot({intf,intfourier(3)},x=-Pi..Pi);
```

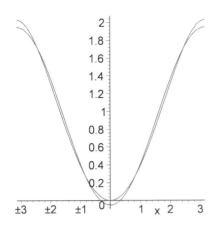

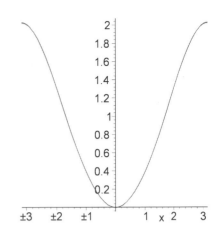

■

Parseval's Equality

Let $f(x)$, $-p < x < p$, be a piecewise smooth function with Fourier series

$$\frac{1}{2}a_0 + \sum_{n=1}^{\infty}\left(a_n \cos\frac{n\pi x}{p} + b_n \sin\frac{n\pi x}{p}\right).$$

Parseval's equality states that

$$\frac{1}{p}\int_{-p}^{p}[f(x)]^2 dx = \frac{1}{2}a_0^2 + \sum_{n=1}^{\infty}(a_n^2 + b_n^2) = 2A_0^2 + \sum_{n=1}^{\infty}(a_n^2 + b_n^2)$$

where $A_0 = \frac{1}{2}a_0$, the constant term in the Fourier series.

EXAMPLE 5: Verify Parseval's equality for $f(x) = \frac{1}{12}x(\pi^2 - x^2)$, $-\pi < x < \pi$.

SOLUTION: Notice that the function $f(x) = \frac{1}{12}x(\pi^2 - x^2)$, $-\pi < x < \pi$, is odd as we see from its graph.

```
> f:='f':a:='a':b:='b':n:='n':

f:=x->1/12*x*(Pi^2-x^2):
```

```
> plot(f(x),x=-Pi..Pi);
```

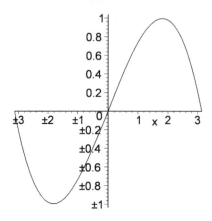

Therefore, the only nonzero coefficients in the Fourier series of f are found in b_n.

```
> assume(n,integer):
b:=n->2/Pi*int(f(x)*sin(n*x),x=0..Pi):
b(n);
```

$$-\frac{(-1)^{n\sim}}{n\sim^3}$$

Next, we evaluate $\frac{1}{\pi}\int_{-\pi}^{\pi}[f(x)]^2 dx$.

```
> 1/Pi*int(f(x)^2,x=-Pi..Pi);
```

$$\frac{1}{945}\pi^6$$

```
> evalf(%);
```

$$1.017343063$$

We compare this result with the value of $\sum\limits_{n=1}^{\infty} b_n^2$ by calculating $\sum\limits_{n=1}^{k} b_n^2$ for $k = 1, 2, \ldots, 20$. Notice that this sequence of partial sums converges quickly to 1.01734, the numerical approximation of $\pi^6/945$.

```
> seq(evalf(sum(b(n)^2,n=1..j)),j=1..20);
```

$$1., 1.015625000, 1.016996742, 1.017240883, 1.017304883,$$

$$1.017326316, 1.017334816, 1.017338631, 1.017340512,$$

$$1.017341512, 1.017342077, 1.017342412, 1.017342619,$$

$$1.017342752, 1.017342840, 1.017342899, 1.017342941,$$

$$1.017342970, 1.017342991, 1.017343007$$

Therefore, the sum of the infinite convergent p-series

$$\sum_{n=1}^{\infty} b_n^2 = \sum_{n=1}^{\infty} \left(\frac{-(-1)^n}{n^3} \right)^2 = \sum_{n=1}^{\infty} \frac{1}{n^6} = 1 + \frac{1}{2^6} + \frac{1}{3^6} + \dots \text{ is } \pi^6/945.$$

■

9.4 Generalized Fourier Series

In addition to the trigonometric eigenfunctions that were used to form the Fourier series in Sections 9.2 and 9.3, the eigenfunctions of other eigenvalue problems can be used to form what we call **generalized Fourier series.** We will find that these series will assist in solving problems in applied mathematics that involve physical phenomena that cannot be modeled with trigonometric functions.

Recall **Bessel's equation of order zero**

$$x^2 y'' + xy' + \lambda^2 x^2 y = 0.$$

If we require that the solutions of this differential equation satisfy the boundary conditions $|y(0)| < \infty$ (meaning that the solution is bounded at $x = 0$) and $y(p) = 0$, we can find the eigenvalues of the boundary value problem

$$\begin{cases} x^2 y'' + xy' + \lambda^2 x^2 y = 0, 0 < x < p \\ |y(0)| < \infty, y(p) = 0 \end{cases}$$

A general solution of Bessel's equation of order zero is $y = c_1 J_0(\lambda x) + c_2 Y_0(\lambda x)$. Because $|y(0)| < \infty$, we must choose $c_2 = 0$ because $\lim_{x \to 0^+} Y_0(\lambda x) = -\infty$. Hence, $y(p) = c_1 J_0(\lambda p) = 0$. Just as we did with the eigenvalue problems solved earlier in Section 9.1, we want to avoid choosing $c_1 = 0$, so we must select λ so that $J_0(\lambda p) = 0$. Let α_n represent the nth zero of the Bessel function of order zero, $J_0(x)$, where $n = 1, 2, \dots$.

Unfortunately, the values of x where $J_0(\lambda p) = 0$ are not as easily expressed as they are with trigonometric functions. From our study of Bessel functions in Section 4.7, we know that this function intersects the x-axis in infinitely many places. In this case, we let α_n represent the

nth zero of the Bessel function of order zero, J_0, where $n = 1, 2, \ldots$. To approximate the α_n's, we begin by graphing $J_0(x)$ with `plot`.

```
> plot(BesselJ(0,x),x=0..40);
```

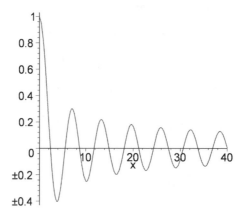

The command `fsolve (equation, x, a..b)` attempts to locate an approximation of the solution to `equation`, which represents an equation in x, in the interval `(a,b)`. For example, from the graph we see that the first zero of the Bessel function of order zero occurs in the interval (2,3). Thus, entering

```
> fsolve(BesselJ(0,x)=0,x,2..3);
```
$$2.404825558$$

returns an approximation of the solution to $J_0(x) = 0$ in the interval (2,3). We interpret the result to mean $\alpha_1 \approx 2.4048$. To approximate the first 10 zeros, we see from the graph that the second zero occurs in the interval (5,6), the third in (8,9), the fourth in (11,12), the fifth in (14,15), the sixth in (18,19), the seventh in (21,22), the eighth in (24,25), the ninth in (27,28), and the tenth in (30,31). These 10 ordered pairs are entered in the list guesses using `..` to indicate that they represent intervals.

```
> guesses:=[2..3,5..6,8..9,11..12,14..15,18..19,
      21..22,24..25,27..28,30..31]:
```

Next, we define the function `approx`. Given guess, where guess is of the form `a..b`, `approx (guess)` attempts to find a numerical solution to $J_0(x) = 0$ in the interval guess.

```
> approx:=proc(guess)
      fsolve(BesselJ(0,x)=0,x,guess)
   end:
```

We then use `map` to apply the function `approx` to the list `guesses`, naming the resulting list `alpha`.

> `alpha:=map(approx,guesses);`

$$\alpha := [2.404825558, 5.520078110, 8.653727913, 11.79153444,$$
$$14.93091771, 18.07106397, 21.21163663, 24.35247153, 27.49347913,$$
$$30.63460647]$$

Thus, entering

> `alpha[5];`

$$14.93091771$$

returns an approximation of the fifth zero of $J_0(x)$: $\alpha_5 \approx 14.9309$.

Therefore, in trying to find the eigenvalues, we must solve $J_0(\lambda p) = 0$. From our definition of α_n, this equation is satisfied if $\lambda p = \alpha_n, n = 1, 2, \ldots$. Hence, the eigenvalues are $\lambda = \lambda_n = \alpha_n/p, n = 1, 2, \ldots$, and the corresponding eigenfunctions are $y(x) = y_n(x) = J_0(\lambda_n x) = J_0(\alpha_n x/p), n = 1, 2, \ldots$.

As with the trigonometric eigenfunctions that we found in Section 9.2 and 9.3, $J_0(\alpha_n x/p)$ can be used to build an eigenfunction series expansion of the form

$$f(x) = \sum_{n=1}^{\infty} c_n J_0\left(\frac{\alpha_n x}{p}\right)$$

(called a **Bessel–Fourier series**), where we use the orthogonality properties of $J_0(\alpha_n x/p)$ to find the coefficients c_n.

We determine the orthogonality condition by placing Bessel's equation of order zero in the self-adjoint form

$$\frac{d}{dx}(xy') + \lambda^2 xy = 0.$$

Because the weighting function is $s(x) = x$, the orthogonality condition is

$$\int_0^p xJ_0\left(\frac{\alpha_n x}{p}\right)J_0\left(\frac{\alpha_m x}{p}\right)dx = 0, \qquad n \neq m.$$

Multiplying $f(x) = \sum_{n=1}^{\infty} c_n J_0(\frac{\alpha_n x}{p})$ by $xJ_0(\frac{\alpha_m x}{p})$ and integrating from $x = 0$ to $x = p$ yields

$$\int_0^p xf(x)J_0(\frac{\alpha_m x}{p})dx = \int_0^p \sum_{n=1}^{\infty} c_n x J_0\left(\frac{\alpha_n x}{p}\right)J_0(\frac{\alpha_m x}{p})dx$$

$$= \sum_{n=1}^{\infty} c_n \int_0^p x J_0\left(\frac{\alpha_n x}{p}\right)J_0(\frac{\alpha_m x}{p})dx.$$

However, by the orthogonality condition, each of the integrals on the right-hand side of the equation equals zero except for $m = n$. Therefore,

$$c_n = \frac{\int_0^p xf(x)J_0(\frac{\alpha_n x}{p})dx}{\int_0^p x[J_0(\frac{\alpha_n x}{p})]^2 dx}, \qquad n = 1, 2, \ldots.$$

The value of the integral in the denominator can be found through the use of several of the identities associated with the Bessel functions. Because $\lambda_n = \alpha_n/p, n = 1, 2, \ldots$, the function $J_0(\alpha_n x/p) = J_0(\lambda_n x)$ satisfies Bessel's equation of order zero:

$$\frac{d}{dx}\left(x\frac{d}{dx}J_0(\lambda_n x)\right) + \lambda_n^2 x J_0(\lambda_n x) = 0.$$

Multiplying by the factor $2x\frac{d}{dx}J_0(\lambda_n x)$, we can write this equation as

$$\frac{d}{dx}\left(x\frac{d}{dx}J_0(\lambda_n x)\right)^2 + \lambda_n^2 x^2 \frac{d}{dx}J_0(\lambda_n x)^2 = 0.$$

Integrating this from $x = 0$ to $x = p$, we have

$$2\lambda_n^2 \int_0^p x(J_0(\lambda_n x))^2 dx = \lambda_n^2 p^2 (J_0'(\lambda_n p))^2 + \lambda_n^2 p^2 (J_0(\lambda_n p))^2.$$

Because $\lambda_n p = \alpha_n$, we make the following substitutions:

$$2\lambda_n^2 \int_0^p x(J_0(\lambda_n x))^2 dx = \lambda_n^2 p^2 (J_0'(\alpha_n))^2 + \lambda_n^2 p^2 (J_0(\alpha_n))^2.$$

Now, $J_0(\alpha_n) = 0$, because α_n is the nth zero of J_0. Also, with $n = 0$, the identity $\frac{d}{dx}(x^{-n}J_n(x)) = -x^{-n}J_{n+1}(x)$ indicates that $J_0'(\alpha_n) = -J_1(\alpha_n)$. Therefore,

$$2\lambda_n^2 \int_0^p x(J_0(\lambda_n x))^2 dx = \lambda_n^2 p^2 (-J_1(\alpha_n))^2 + \lambda_n^2 p^2 \cdot 0$$

$$\int_0^p x(J_0(\lambda_n x))^2 dx = \frac{p^2}{2}(J_1(\alpha_n))^2.$$

Therefore, the series coefficients are found with

$$c_n = \frac{2}{p^2(J_1(\alpha_n))^2}\int_0^p xf(x)J_0\left(\frac{\alpha_n x}{p}\right)dx, \qquad n = 1, 2, \ldots.$$

EXAMPLE 1: Find the Bessel–Fourier series for $f(x) = 1 - x^2$ on $0 < x < 1$.

SOLUTION: In this case, $p = 1$, so

$$c_n = \frac{2}{(J_1(\alpha_n))^2} \int_0^1 x(1 - x^2)J_0(\alpha_n x)dx = \frac{2}{(J_1(\alpha_n))^2} \left\{ \int_0^1 xJ_0(\alpha_n x)dx - \int_0^1 x^3 J_0(\alpha_n x)dx \right\}.$$

Using the formula, $\frac{d}{dx}(x^n J_n(x)) = -x^n J_{n-1}(x)$ with $n = 1$ yields

$$\int_0^1 xJ_0(\alpha_n x)dx = \left[\frac{1}{\alpha_n} xJ_1(\alpha_n x) \right]_0^1 = \frac{1}{\alpha_n} J_1(\alpha_n).$$

Note that the factor $1/\alpha_n$ is due to the chain rule for differentiating the argument of $J_1(\alpha_n x)$. We use integration by parts with $u = x^2$ and $dv = xJ_0(\alpha_n x)$ to evaluate $\int_0^1 x^3 J_0(\alpha_n x)dx$. As in the first integral, we obtain $v = -\frac{x}{\alpha_n}J_1(\alpha_n x)$. Then, because $du = 2xdx$, we have

$$\int_0^1 x^3 J_0(\alpha_n x)dx = \left[\frac{1}{\alpha_n} x^3 J_1(\alpha_n x) \right]_0^1 - \frac{2}{\alpha_n} \int_0^1 x^2 J_1(\alpha_n x)dx$$

$$= \frac{1}{\alpha_n} J_1(\alpha_n) - \frac{2}{\alpha_n} \left[\frac{1}{\alpha_n} x^2 J_2(\alpha_n x) \right]_0^1 = \frac{1}{\alpha_n} J_1(\alpha_n) - \frac{2}{\alpha_n^2} J_2(\alpha_n).$$

Thus, the coefficients are

$$c_n = \frac{2}{[J_1(\alpha_n)]^2} \int_0^1 x(1 - x^2)J_0(\alpha_n x)dx = \frac{2}{[J_1(\alpha_n)]^2} \left\{ \int_0^1 xJ_0(\alpha_n x)dx - \int_0^1 x^3 J_0(\alpha_n x)dx \right\}$$

$$= \frac{2}{[J_1(\alpha_n)]^2} \left[\frac{1}{\alpha_n} J_1(\alpha_n) - \left(\frac{1}{\alpha_n} J_1(\alpha_n) - \frac{2}{\alpha_n^2} J_2(\alpha_n) \right) \right] = \frac{4J_2(\alpha_n)}{\alpha_n^2 [J_1(\alpha_n)]^2}, \quad n = 1, 2, \ldots$$

so that the Bessel–Fourier series is

$$f(x) = \sum_{n=1}^{\infty} \frac{4J_2(\alpha_n)}{\alpha_n^2 [J_1(\alpha_n)]^2} J_0(\alpha_n x).$$

To graph f along with several terms of the series, we first define the function c and then compute c (n) for $n = 1, 2, 3, 4,$ and 5.

```
> readlib('evalf/int'):

c:=proc(n) option remember;
        2/BesselJ(1,alpha[n])^2*
                'evalf/int'(x*(1-x^2)*BesselJ(0,alpha[n]*x),x=0..1)
        end:
array([seq([n,c(n)],n=1..5)]);
```

$$\begin{bmatrix} 1 & 1.108022261 \\ 2 & -.1397775052 \\ 3 & .04547647070 \\ 4 & -.02099090194 \\ 5 & .01163624312 \end{bmatrix}$$

Then, in the same manner as in Sections 9.2 and 9.3, we define `approx(k)` to be the kth partial sum of the series.

```
> approx:=proc(k) option remember;
        approx(k-1)+c(k)*BesselJ(0,alpha[k]*x)
      end:

approx(1):=c(1)*BesselJ(0,alpha[1]*x):
```

We graph f along with several terms of the series. Notice that the polynomial with two terms yields an accurate approximation of f. When using four or six terms, the graphs are practically indistinguishable.

```
> f:=x->1-x^2:

k_vals:=2,4,6:

for k in k_vals do plot({f(x),approx(k)},x=0..1) od;
```

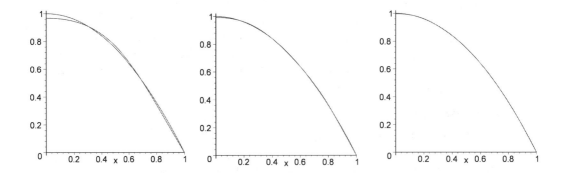

We verify that the error is "small" by graphing the absolute value of the difference between $f(x)$ and the kth partial sum for $k = 2, 4,$ and 6.

```
> for k in k_vals do plot(abs(f(x)-approx(k)),x=0..1) od;
```

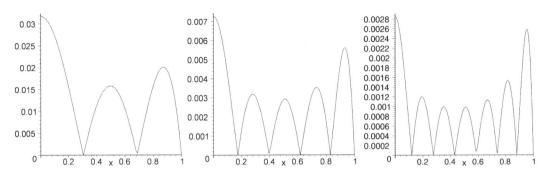

Application

Constructing a Table of Zeros of Bessel Functions

The zeros of the Bessel functions play an important role in the generalized Fourier series involving Bessel functions. Use Maple to find the first eight zeros of the Bessel functions of the first kind, $J_n(x)$, of order $n = 0, 1, 2, \ldots, 6$.

SOLUTION: The **Bessel function of the first kind of order** n, $J_n(x)$, is represented by `BesselJ(n,x)`. We graph the Bessel functions of the first kind of order n for $n = 0, 1, \ldots, 8$ on the interval [0,40] with `animate` and then display the graphs as an array with `display`.

```
> with(plots):

    A:=animate(BesselJ(i,x),x=0..40,i=0..8,

    frames=9,color=BLACK,numpoints=100):

display(A);
```

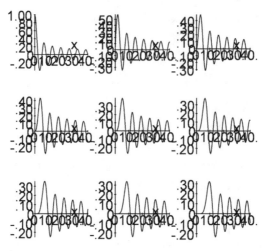

In order to approximate the zeros of the Bessel functions, we will use the command `fsolve`. Recall that `fsolve (equation, x, a..b)` attempts to locate an approximation of the solution to `equation`, which represents an equation in x, on the interval (a,b). We use the proceeding graphs to approximate (open) intervals containing the zeros to be approximated. For example, we see that the first zero of the Bessel function of order zero occurs in the interval (2,3), the second in (5,6), the third in (8,9), the fourth in (11,12), the fifth in (14,15), the sixth in (18,19), the seventh in (21,22), and the eighth in (24,25). Hence, these initial intervals are entered in the array `guesses`. Note that `guesses [1]` corresponds to the list of eight intervals containing the first eight zeros of the Bessel function of order 0. Hence, `guesses [i]` corresponds to the list of eight intervals containing the first eight zeros of the Bessel function of order i-1; `guesses [i,j]` corresponds to the interval containing the jth zero of the Bessel function of order i-1.

```
> guesses:=array(
            [[2..3,5..6,8..9,11..12,14..15,
                18..19,21..22,24..25],
            [3..4,7..8,10..11,13..14,16..17,
                19..20,22..23,25..26],
            [5..6,8..9,11..12,14..15,17..18,
                21..22,24..25,27..28],
            [6..7,9..10,13..14,16..17,19..20,
                22..23,25..26,28..29],
            [7..8,11..12,14..15,17..18,20..21,
                24..25,27..28,30..31],
            [8..9,12..13,15..16,18..19,22..23,
                25..26,28..29,31..32],
```

$$[9..10,13..14,17..18,20..21,23..24,$$

$$26..27,30..31,33..34]]):$$

The function `alpha` uses `fsolve` and the intervals in `guesses` to approximate the zeros of the Bessel functions. In the following two `for` loops, i corresponds to the Bessel function of order i and j represents the jth zero. For example, for $i=0$, the first eight zeros of the Bessel function of order 0 are computed by using `fsolve` to approximate the zero on each of the intervals in the first list in the array `guesses`. This is carried out for $i = 0$ to $i = 6$ to yield a table of zeros of the Bessel functions of order 0 to order 6. This table is then printed. Notice that each zero corresponds to an orderd pair that represents the jth zero of the Bessel function of order i.

```
> i:='i':j:='j':

alpha:=table():

for i from 0 to 6 do

        for j to 8 do

                alpha[i,j]:=

                        fsolve(BesselJ(i,x)=0,x,guesses[i+1,j])

                od od:
```

```
> print(alpha);
```

table([	$(2,1) = 5.135622302$
$(0,6) = 18.07106397$	$(4,2) = 11.06470949$
$(2,6) = 21.11699705$	$(1,1) = 3.831705970$
$(4,7) = 27.19908777$	$(0,1) = 2.404825558$
$(6,5) = 23.58608444$	$(3,2) = 9.761023130$
$(1,6) = 19.61585851$	$(5,3) = 15.70017408$
$(3,7) = 25.74816670$	$(2,2) = 8.417244140$
$(0,8) = 24.35247153$	$(4,3) = 14.37253667$
$(5,5) = 22.21779990$	$(6,1) = 9.936109524$
$(2,7) = 24.27011231$	$(1,2) = 7.015586670$
$(4,8) = 30.37100767$	$(3,3) = 13.01520072$
$(6,6) = 26.82015198$	$(0,2) = 5.520078110$
$(1,7) = 22.76008438$	$(2,3) = 11.61984117$
$(3,8) = 28.90835078$	$(4,4) = 17.61596605$
$(5,6) = 25.43034115$	$(6,2) = 13.58929017$
$(2,8) = 27.42057355$	$(1,3) = 10.17346814$
$(6,7) = 30.03372239$	$(3,4) = 16.22346616$
$(1,8) = 25.90367209$	$(0,7) = 21.21163663$
$(5,7) = 28.62661831$	$(2,4) = 14.79595178$
$(0,5) = 14.93091771$	$(4,5) = 20.82693296$
$(6,8) = 33.23304176$	$(6,3) = 17.00381967$

$$(5,8) = 31.81171672 \qquad (0,4) = 11.79153444$$
$$(5,1) = 8.771483816 \qquad (1,4) = 13.32369194$$
$$(4,1) = 7.588342435 \qquad (3,5) = 19.40941523$$
$$(5,4) = 18.98013388 \qquad (2,5) = 17.95981950$$
$$(3,1) = 6.380161896 \qquad (4,6) = 24.01901952$$
$$(5,2) = 12.33860420 \qquad (6,4) = 20.32078921$$
$$(1,5) = 16.47063005$$
$$(3,6) = 22.58272959$$
$$(0,3) = 8.653727913$$
$$])$$

```
> ALPHA:=array(1..7,1..8):

> for i from 1 to 7 do

      for j from 1 to 8 do

          ALPHA[i,j]:=evalf(alpha[i-1,j],5) od od:

print(ALPHA);
```

$$\begin{bmatrix}
2.4048 & 5.5201 & 8.6537 & 11.792 & 14.931 & 18.071 & 21.212 & 24.352 \\
3.8317 & 7.0156 & 10.173 & 13.324 & 16.471 & 19.616 & 22.760 & 25.904 \\
5.1356 & 8.4172 & 11.620 & 14.796 & 17.960 & 21.117 & 24.270 & 27.421 \\
6.3802 & 9.7610 & 13.015 & 16.223 & 19.409 & 22.583 & 25.748 & 28.908 \\
7.5883 & 11.065 & 14.373 & 17.616 & 20.827 & 24.019 & 27.199 & 30.371 \\
8.7715 & 12.339 & 15.700 & 18.980 & 22.218 & 25.430 & 28.627 & 31.812 \\
9.9361 & 13.589 & 17.004 & 20.321 & 23.586 & 26.820 & 30.034 & 33.233
\end{bmatrix}$$

We then save this table of numbers, for later use, in a text editor and name the resulting file **besselzeros**. In this particular case, a word processor was used to manipulate the list so that the result is an array of numbers. The array of numbers is displayed in the following screen shot. Note that the first row corresponds to the zeros of the Bessel function of order zero, the second row corresponds to the zeros of the Bessel function of order one, and so on. An alternative approach would be to save a Maple scratchpad as **besselzeros**.

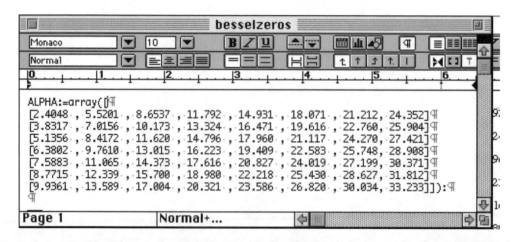

In order to use this file, we first open the file in Maple, enter `ALPHA`, and then define `alpha(i, j) = α_{ij}`, the jth zero of the Bessel function of order i, with the following command.

```
> alpha:=(i,j)->ALPHA[i+1][j]:
```

■

As was the case with Fourier series, we can make a statement about the convergence of the Bessel–Fourier series.

Theorem **Convergence of Bessel–** **Fourier Series**	Suppose that f and f' are piecewise continuous functions on $0 < x < p$. Then the Bessel–Fourier series for f on $-p < x < p$ converges to $f(x)$ at every x where f is continuous. If f is discontinuous at $x = x_0$, the Bessel–Fourier series converges to the average $$\frac{1}{2}(f(x_0{}^+) + f(x_0{}^-)).$$

Series involving the eigenfunctions of other eigenvalue problems can be formed as well.

EXAMPLE 2: The problem $\begin{cases} y'' + 2y' - (\lambda - 1)y = 0 \\ y(0) = 0, y(2) = 0 \end{cases}$ has eigenvalues $\lambda_n = -(n\pi/2)^2$, $n = 1, 2, \ldots$, and eigenfunctions $y_n(x) = e^{-x}\sin(n\pi x/2)$. Use these eigenfunctions to approximate $f(x) = e^{-x}$ on $[0,2]$.

SOLUTION: In order to approximate f, we need the orthogonality condition for these eigenfunctions. We obtain this condition by placing the differential equation in self-adjoint form using the formulas given in Section 9.1. In the general equation, $a_2(x) = 1$, $a_1(x) = 2$, and $a_0(x) = 0$. Therefore, $p(x) = e^{\int 2\,dx} = e^{2x}$ and $s(x) = p(x)/a_2(x) = e^{2x}$, so the equation is $\frac{d}{dx}\left(e^{2x}\frac{dy}{dx}\right) - (\lambda - 1)e^{2x}y = 0$. This means that the orthogonality condition, $\int_a^b s(x)y_n(x)y_m(x)\,dx = 0 \ (m \neq n)$, is

$$\int_0^2 e^{2x}e^{-x}\sin\frac{m\pi x}{2}e^{-x}\sin\frac{n\pi x}{2}\,dx = \int_0^2 \sin\frac{m\pi x}{2}\sin\frac{n\pi x}{2}\,dx = 0 \qquad (m \neq n).$$

We use this condition to determine the coefficients in the eigenfunction expansion

$$f(x) = \sum_{n=1}^{\infty} c_n y_n(x) = \sum_{n=1}^{\infty} c_n e^{-x}\sin\frac{n\pi x}{2}.$$

Multiplying both sides of this equation by $y_m(x) = e^{-x} \sin(m\pi x/2)$ and $s(x) = e^{2x}$ and then integrating from $x = 0$ to $x = 2$ yield

$$\int_0^2 f(x) e^{2x} e^{-x} \sin\frac{m\pi x}{2}\,dx = \int_0^2 \sum_{n=1}^{\infty} c_n e^{-x} \sin\frac{n\pi x}{2} e^{2x} e^{-x} \sin\frac{m\pi x}{2}\,dx$$

$$\int_0^2 f(x) e^x \sin\frac{m\pi x}{2}\,dx = \sum_{n=1}^{\infty} \int_0^2 c_n \sin\frac{n\pi x}{2} \sin\frac{m\pi x}{2}\,dx.$$

Each integral in the sum on the right-hand side of the equation is zero except if $m = n$. In this case, $\int_0^2 \sin^2\frac{n\pi x}{2}\,dx = 1$. Therefore,

$$c_n = \int_0^2 f(x) e^x \sin\frac{n\pi x}{2}\,dx.$$

For $f(x) = e^{-x}$,

$$c_n = \int_0^2 e^x e^{-x} \sin\frac{n\pi x}{2}\,dx = \int_0^2 \sin\frac{n\pi x}{2}\,dx = -\frac{2}{n\pi}(\cos n\pi - 1).$$

```
> assume(n,integer):
c:=n->int(sin(n*Pi*x/2),x=0..2):
c(n);
```

$$-2\frac{(-1)^{n\sim} - 1}{n\sim\pi}$$

Because $\cos n\pi = (-1)^n$, we can write the eigenfunction expansion of f as

$$f(x) = \sum_{n=1}^{\infty} -\frac{2}{n\pi}((-1)^n - 1)e^{-x} \sin\frac{n\pi x}{2} = \frac{4}{\pi} e^{-x} \sin\frac{\pi x}{2} + \frac{4}{3\pi} e^{-x} \sin\frac{3\pi x}{2} + \frac{4}{5\pi} e^{-x} \sin\frac{5\pi x}{2} + \cdots.$$

We graph $f(x) = e^{-x}$ together with $pk(x) = \sum_{n=1}^{k} -\frac{2}{n\pi}((-1)^n - 1)e^{-x} \sin\frac{n\pi x}{2}$ for $k = 6$, 10, and 14.

```
> f:=exp(-x):
p:=(x,k)->sum(c(n)*exp(-x)*sin(n*Pi*x/2),n=1..k):


> k_vals:=6,10,14:
for k in k_vals do plot({f(x),p(x,k)},x=0..1) od;
```

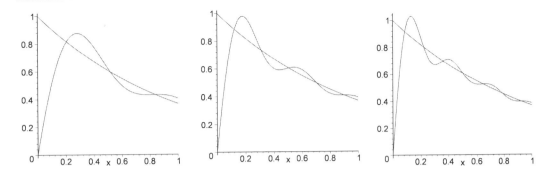

■

EXAMPLE 3: Use the eigenvalues and eigenfunctions of the eigenvalue problem $\begin{cases} y'' + \lambda y = 0 \\ y(0) = 0, y(1) + y'(1) = 0 \end{cases}$ (found in Example 6 of Section 9.1) to obtain a generalized Fourier series for $f(x) = x(1-x), 0 < x < 1$.

SOLUTION: The eigenvalues of this problem $\lambda = k^2$ were shown to satisfy the relationship $k = -\tan k$ in Example 6 of Section 9.1. We approximated the first eight roots of this equation by 2.02876, 4.91318, 7.97867, 11.0855, 14.2074, 17.3364, 20.4692, and 23.6043 entered in `kvals`.

```
> kvals:=[seq(fsolve(-tan(x)=x,x,
        (2*n-1)*Pi/2..(2*n-1)*Pi/2+0.5),n=1..8)];
```

$$kvals := [2.028757838, 4.913180439, 7.978665712, 11.08553841,$$
$$14.20743673, 17.33637792, 20.46916740, 23.60428477]$$

Let k_n represent the nth positive root of $k = -\tan k$. Therefore, the eigenfunctions of $\begin{cases} y'' + \lambda y = 0 \\ y(0) = 0, y(1) + y'(1) = 00 \end{cases}$ are $y = \sin k_n x$. Because of the orthogonality of the eigenfunctions, we have the orthogonality condition $\int_0^1 \sin k_n x \sin k_m x \, dx = 0$, $m \neq n$. If $k = m$, we have $\int_0^1 \sin^2 k_n x \, dx = \frac{1}{2} - \frac{\sin 2k}{4k}$.

```
> k:='k':
int(sin(k*x)^2,x=0..1);
```

$$\frac{1}{2} - \frac{\cos(k)\sin(k) + k}{k}$$

Therefore, $\int_0^1 \sin^2 k_n x \, dx = \frac{2k_n - \sin 2k_n}{4k_n} = \frac{2k_n - 2\sin k_n \cos k_n}{4k_n}$. With the condition $k_n = -\tan k_n$ or $\sin k_n = -k_n \cos k_n$ from the eigenvalue problem, we have $\int_0^1 \sin^2 k_n x \, dx = \frac{k_n - \sin k_n \cos k_n}{2k_n} = \frac{k_n - (-k_n \cos k_n) \cos k_n}{2k_n} = \frac{1 + \cos^2 k_n}{2}$. To determine the coefficients c_n in the generalized Fourier series $f(x) = \sum_{n=1}^{\infty} c_n \sin k_n x$ using the eigenfunctions $y = \sin k_n x$ of the eigenvalue problem $y'' + \lambda y = 0, y(0) = 0, y(1) + y'(1) = 0$, we multiply both sides of $f(x) = \sum_{n=1}^{\infty} c_n \sin k_n x$ by $\sin k_m x$ and integrate from 0 to 1. This yields $\int_0^1 f(x) \sin k_n x \, dx = \int_0^1 \sum_{n=1}^{\infty} c_k \sin k_n x \sin k_m x \, dx$. Assuming uniform convergence of the series, we have $\int_0^1 f(x) \sin k_n x \, dx = \sum_{n=1}^{\infty} c_n \int_0^1 \sin k_n x \sin k_m x \, dx$. All terms on the right are zero except if $m = n$. In this case, we have $\int_0^1 f(x) \sin k_n x \, dx = c_n \int_0^1 \sin^2 k_n x \, dx = c_n \left(\frac{1 + \cos^2 k_n}{2}\right)$ so that $c_n = \frac{2}{1 + \cos^2 k_n} \int_0^1 f(x) \sin k_n x \, dx$. We approximate the value of c_n for $n = 1, 2, \ldots 8$ in cvals using the values of k in kvals.

```
> f:=x->x*(1-x):
```

```
> cvals:=[seq(evalf(2*int(f(x)*sin(kvals[i]*x),x=0..1)/
        (1+cos(kvals[i])^2)),i=1..8)];
```

$$cvals := [.2132851322, .1040487897, -.02197880768, .01873025456,$$
$$- .008349935820, .007342554332, -.004268411122, .003870742708]$$

We define the sum of the first j terms of $\sum_{n=1}^{\infty} c_n \sin k_n x$ for $f(x) = x(1 - x), 0 < x < 1$, with fapprox (x,n) and then create a table of fapprox (x,n) for $n = 1, 2, \ldots, 8$ in funcs.

```
> fapprox:=(x,j)->sum(cvals[n]*sin(kvals[n]*x),n=1..j);
```

$$fapprox := (x,j) \rightarrow \sum_{n=1}^{j} cvals_n \sin(kvals_n x)$$

```
> funcs:=seq(fapprox(x,k),k=1..8);
```

$funcs := .2132851322 \sin(2.028757838x),$

$\qquad .2132851322 \sin(2.028757838x) + .1040487897 \sin(4.913180439x),$

$\qquad .2132851322 \sin(2.028757838x) + .1040487897 \sin(4.913180439x)$

$\qquad - .02197880768 \sin(7.978665712x), .2132851322 \sin(2.028757838x)$

$\qquad + .1040487897 \sin(4.913180439x)$

$\qquad - .02197880768 \sin(7.978665712x)$

$\qquad + .01873025456 \sin(11.08553841x), .2132851322 \sin(2.028757838x)$

$\qquad + .1040487897 \sin(4.913180439x)$

$\qquad - .02197880768 \sin(7.978665712x)$

$\qquad + .01873025456 \sin(11.08553841x)$

$\qquad - .008349935820 \sin(14.20743673x),$

$\qquad .2132851322 \sin(2.028757838x) + .1040487897 \sin(4.913180439x)$

$\qquad - .02197880768 \sin(7.978665712x)$

$\qquad + .01873025456 \sin(11.08553841x)$

$\qquad - .008349935820 \sin(14.20743673x)$

$\qquad + .007342554332 \sin(17.33637792x),$

$\qquad .2132851322 \sin(2.028757838x) + .1040487897 \sin(4.913180439x)$

$\qquad - .02197880768 \sin(7.978665712x)$

$\qquad + .01873025456 \sin(11.08553841x)$

$\qquad - .008349935820 \sin(14.20743673x)$

$\qquad + .007342554332 \sin(17.33637792x)$

$\qquad - .004268411122 \sin(20.46916740x)$

$\qquad .2132851322 \sin(2.028757838x) + .1040487897 \sin(4.913180439x)$

$\qquad - .02197880768 \sin(7.978665712x)$

$\qquad + .01873025456 \sin(11.08553841x)$

$\qquad - .008349935820 \sin(14.20743673x)$

$\qquad + .007342554332 \sin(17.33637792x)$

$\qquad - .004268411122 \sin(20.46916740x)$

$\qquad + .003870742708 \sin(23.60428477x)$

We graph $f(x) = x(1 - x), 0 < x < 1$, simultaneously with the first term of the generalized Fourier series to observe the accuracy.

```
> plot({fapprox(x,1),f(x)},x=0..1);
```

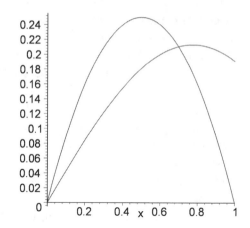

Next, we plot the approximation using the sum of the first 2, 4, 6, and 8 terms of the generalized Fourier series and display the results. We see that the approximation improves as the number of terms increases.

```
> k_vals:=2,4,6,8:

for k in k_vals do plot({f(x),fapprox(x,k)},x=0..1) od;
```

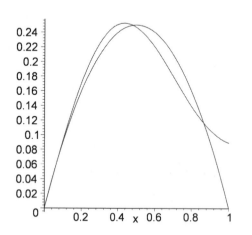

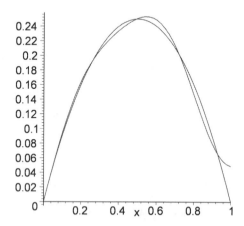

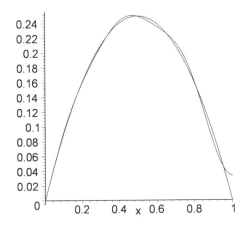 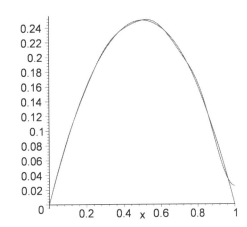

CHAPTER 10

Partial Differential Equations

10.1 Introduction to Partial Differential Equations and Separation of Variables

We begin our study of partial differential equations with an introduction of some of the terminology associated with the topic. A **linear second-order partial differential equation (PDE)** in the two independent variables x and y has the form

$$A(x,y)u_{xx} + B(x,y)u_{xy} + C(x,y)u_{yy} + D(x,y)u_x + E(x,y)u_y + F(x,y)u = G(x,y)$$

where the solution is $u(x,y)$. If $G(x,y) = 0$ for all x and y, we say that the equation is **homogeneous.** Otherwise, the equation is **nonhomogeneous.**

EXAMPLE 1: Classify the following partial differential equations: (a) $u_{xx} + u_{yy} = u$; (b) $u_{xy} + uu_x = x$.

SOLUTION: (a) This equation satisfies the form of the linear second-order partial differential equation with $A = C = 1$, $F = -1$, and $B = D = E = 0$. Because $G(x,y) = 0$, the equation is homogeneous. (b) This equation is nonlinear, because the coefficient of u_x is a function of u. It is also nonhomogeneous because $G(x,y) = x$.

■

Definition	A **solution of a partial differential equation** in some region R of
Solution of a Partial	the space of the independent variables is a function that
Differential Equation	possesses all of the partial derivatives that are present in the
	PDE in some region containing R and satisfies the PDE
	everywhere in R.

EXAMPLE 2: Show that $u(x,y) = y^2 - x^2$ and $u(x,y) = e^y \sin x$ are solutions to Laplace's equation $u_{xx} + u_{yy} = 0$.

SOLUTION: For $u(x,y) = y^2 - x^2$, $u_x(x,y) = -2x$, $u_y(x,y) = 2y$, $u_{xx}(x,y) = -2$, and $u_{yy}(x,y) = 2$, so we have that $u_{xx} + u_{yy} = (-2) + 2 = 0$, which we quickly verify with Maple.

```
> u:=(x,y)->y^2-x^2:
diff(u(x,y),x$2)+diff(u(x,y),y$2);
```

$$0$$

Similarly, for $u(x,y) = e^y \sin x$, we have $u_x = e^y \cos x$, $u_y = e^y \cos x$, $u_{xx} = -e^y \sin x$, and $u_{yy} = e^y \sin x$. Therefore, $u_{xx} + u_{yy} = (-e^y \sin x) + e^y \sin x = 0$, so the equation is satisfied for both functions.

```
> u:=(x,y)->exp(y)*sin(x):
diff(u(x,y),x$2)+diff(u(x,y),y$2);
```

$$0$$

We notice that the solutions to Laplace's equation differ in form. This is unlike solutions to homogeneous linear ordinary differential equations. There, we found that solutions were similar in form. (Recall, all solutions could be generated from a general solution.)

■

Some of the techniques used in constructing solutions of homogeneous linear ordinary differential equations can be extended to the study of partial differential equations as we see with the following theorem.

Theorem	If $u_1, u_2, \ldots, u_m$ are solutions to a linear homogeneous partial differential equation in a region R, then
Principle if Superposition	

$$c_1 u_1 + c_2 u_2 + \cdots + c_m u_m = \sum_{k=1}^{m} c_k u_k$$

where $c_1, c_2, \ldots, c_m$ are constants is also a solution in R.

The principle of superposition will be used in solving partial differential equations throughout the rest of the chapter. In fact, we will find that equations can have an infinite set of solutions so that we construct another solution in the form of an infinite series.

Separation of Variables

A method that can be used to solve linear partial differential equations is called **separation of variables** (or the **product method**). Generally, the goal of the method of separation of variables is to transform the partial differential equation into a system of ordinary differential equations each of which depends on only one of the functions in the product form of the solution. Suppose that the function $u(x,y)$ is a solution of a partial differential equation in the independent variables x and y. In separating variables, we assume that u can be written as the product of a function of x and a function of y. Hence,

$$u(x, y) = X(x)Y(y),$$

and we substitute this product into the partial differential equation to determine $X(x)$ and $Y(y)$. Of course, in order to substitute into the differential equation, we must be able to differentiate this product. However, this is accomplished by following the differentiation rules of multivariate calculus:

$$u_x = X'Y, u_{xx} = X''Y, u_{xy} = X'Y', u_y = XY', \text{ and } u_{yy} = XY'',$$

where X' represents dX/dx and Y' represents dY/dy. After these substitutions are made and if the equation is separable, we can obtain an ordinary differential equation for X and an ordinary differential equation for Y. These two equations are then solved to find $X(x)$ and $Y(y)$.

EXAMPLE 3: Use separation of variables to find a solution of $xu_x = u_y$.

SOLUTION: If $u(x, y) = X(x)Y(y)$, then $u_x = X'Y$ and $u_y = XY'$. The equation then becomes

$$xX'Y = XY',$$

which can be written as the separated equation

$$\frac{xX'}{X} = \frac{Y'}{Y}.$$

Notice that the left-hand side of the equation is a function of x while the right-hand side is a function of y. Hence, the only way that this situation can be true is for xX'/X and Y'/Y both to be constant. Therefore,

$$\frac{xX'}{X} = \frac{Y'}{Y} = k,$$

so we obtain the ordinary differential equations $xX' - kX = 0$ and $Y' - kY = 0$. We find X first.

$$xX' - kX = 0$$

$$x\frac{dX}{dx} = kX$$

$$\frac{dX}{X} = \frac{k}{x}dx$$

$$\ln|X| = k\ln|x| + c_1$$

$$X(x) = e^{c_1}x^k = C_1 x^k.$$

Similarly, we find

$$Y' - kY = 0$$

$$Y' = kY$$

$$\frac{dY}{dy} = kY$$

$$\frac{dY}{Y} = kdy$$

$$\ln|Y| = ky + c_2$$

$$Y(y) = e^{c_2}e^{ky} = C_2 e^{ky}.$$

Therefore, a solution is $u(x,y) = X(x)Y(y) = (C_1 x^k)(C_2 e^{ky}) = C_3 x^k e^{ky}$, where k and C_3 are arbitrary constants. (Notice that the first equation is a Cauchy–Euler equation, so we could have used the techniques covered in Section 4.6 to solve it. Similarly, we could have used a characteristic equation to solve the second equation.) We can also use the pdsolve command to find a solution of this partial differential equation as well. Notice that the syntax for the pdsolve command is nearly identical to the syntax of the dsolve command.

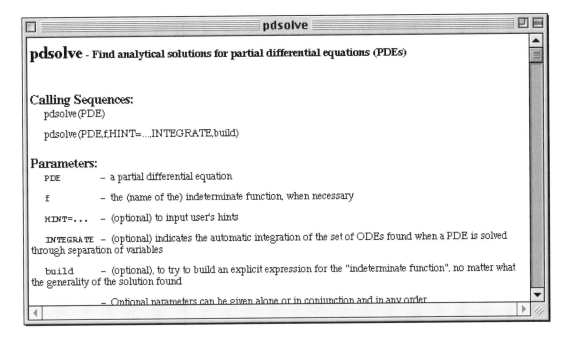

```
> u:='u':x:='x':y:='y':
sol:=pdsolve(x*diff(u(x,y),x)=diff(u(x,y),y));
```

$$sol := \mathrm{u}(x,y) = \_F1(y + \ln(x))$$

In this result, the symbol _F1 represents an arbitrary differentiable *function*. That is, if f is a differentiable function of a single variable, $u(x,y) = f(y + \ln x)$ is a solution to $xu_x = u_y$, which we verify by substituting this result into the partial differential equation.

```
> assign(sol):
evalb(x*diff(u(x,y),x)=diff(u(x,y),y));
```

$$true$$

■

10.2 The One-Dimensional Heat Equation

One of the more important differential equations is the **heat equation,**

$$u_t = c^2 u_{xx}.$$

In one spatial dimension, the solution of the heat equation represents the temperature (at any position x and any time t) in a thin rod or wire of length p. Because the rate at which heat flows through the rod depends on the material that makes up the rod, the constant c^2, which is related to the thermal diffusivity of the material, is included in the heat equation. Several different situations can be considered when determining the temperature in the rod. The ends of the wire can be held at a constant temperature, the ends may be insulated, or there can be a combination of these situations.

The Heat Equation with Homogeneous Boundary Conditions

The first problem that we investigate is the situation in which the temperature at the ends of the rod is constantly kept at zero and the initial temperature distribution in the rod is represented as the given function $f(x)$. Hence, the fixed end zero temperature is given in the boundary conditions

$$u(0, t) = u(p, t) = 0$$

while the initial temperature distribution is given by

$$u(x, 0) = f(x).$$

Because the temperature is zero at the endpoints, we say that the problem has **homogeneous boundary conditions** (which are important in finding a solution with separation of variables). We call problems of this type **initial–boundary value problems** (IBVPs), because they include initial as well as boundary conditions. Thus, the problem is summarized as

$$\begin{cases} u_t = c^2 u_{xx}, 0 < x < p, t > 0 \\ u(0, t) = 0, u(p, t) = 0, t > 0. \\ \quad u(x, 0) = f(x), 0 < x < p \end{cases}$$

We solve this problem through separation of variables by assuming that

$$u(x, t) = X(x)T(t).$$

Substitution into the differential equation yields

$$\frac{T'}{c^2 T} = \frac{X''}{X} = -\lambda$$

where $-\lambda$ is the separation constant. (Note that we selected this constant in order to obtain an eigenvalue problem that was solved in Section 10.1). Separating the variables, we have the two equations

$$T' + c^2\lambda T = 0 \quad \text{and} \quad X'' + \lambda X = 0.$$

Now that we have successfully separated the variables, we turn our attention to the homogeneous boundary conditions. In terms of the functions $X(x)$ and $T(t)$, these boundary conditions become

$$u(0,t) = X(0)T(t) = 0 \quad \text{and} \quad u(p,t) = X(p)T(t) = 0.$$

In each case, we must avoid setting $T(t) = 0$ for all t, because if this were the case, our solution would be the trivial solution $u(x,t) = X(x)T(t) = 0$. Therefore, we have the boundary conditions

$$X(0) = 0 \quad \text{and} \quad X(p) = 0,$$

so we solve the eigenvalue problem

$$\begin{cases} X'' + \lambda X = 0 \\ X(0) = 0, X(p) = 0 \end{cases}.$$

The eigenvalues of this problem are $\lambda_n = (n\pi/p)^2$, $n = 1, 2, \ldots$ with corresponding eigenfunctions $X_n(x) = \sin(n\pi x/p)$, $n = 1, 2, \ldots$ (see Example 4 in Section 9.1).

Similarly, a general solution of $T' + c^2\lambda_n T = 0$ is $T_n(t) = Ae^{-c^2\lambda_n t}$, where A is an arbitrary constant and $\lambda_n = (n\pi/p)^2$, $n = 1, 2, \ldots$

```
> dsolve(diff(T(t),t)+c^2*lambda[n]*T(t)=0,T(t));
```

$$T(t) = \_C1e^{(-c^2\lambda_n^t)}$$

Because $X(x)$ and $T(t)$ both depend on n, the solution $u(x,t) = X(x)T(t)$ does as well. Hence,

$$u_n(x,t) = X_n(x)T_n(t) = c_n \sin\frac{n\pi x}{p} e^{-c^2\lambda_n t}$$

where we have replaced the constant A by one that depends on n. In order to find the value of c_n, we apply the initial condition $u(x,0) = f(x)$. Notice that

$$u_n(x,0) = c_n \sin\frac{n\pi x}{p} e^{-c^2\lambda_n \cdot 0} = c_n \sin\frac{n\pi x}{p}$$

is satisfied only by functions of the form $\sin(\pi x/p)$, $\sin(2\pi x/p)$, $\sin(3\pi x/p),\ldots$ (which, in general, is not the case). Therefore, we use the principle of superposition to state that

$$u(x,t) = \sum_{n=1}^{\infty} u_n(x,t) = \sum_{n=1}^{\infty} c_n \sin\frac{n\pi x}{p} e^{-c^2\lambda_n t}$$

is also a solution of the problem, because this solution satisfies the heat equation as well as the boundary conditions. Then, when we apply the initial condition $u(x,0) = f(x)$, we find that

$$u(x,0) = \sum_{n=1}^{\infty} c_n \sin\frac{n\pi x}{p} e^{-c^2\lambda_n \cdot 0} = \sum_{n=1}^{\infty} c_n \sin\frac{n\pi x}{p} = f(x).$$

Therefore, c_n represents the Fourier sine series coefficients for $f(x)$, which are given by

$$c_n = \frac{2}{p}\int_0^p f(x)\sin\frac{n\pi x}{p}\,dx, \qquad n = 1, 2, \ldots$$

Maple knows to use separation of variables to solve this equation.

```
> u:='u':x:='x':t:='t':
sol:=pdsolve(diff(u(x,t),t)=c^2*diff(u(x,t),x$2),u(x,t));
```

$$sol := (u(x,t) = \_F2(x)\_F3(t))$$

$$\left[\left\{\frac{\partial^2}{\partial x^2}\_F2(x) = \_c1\_F2(x), \frac{\partial}{\partial t}\_F3(t) = c^2\_c1\_F3(t)\right\}\right]$$

We can then use `build`, which is contained in the **PDEtools** package, to determine an explicit form for u.

```
> with(PDEtools);
sol2:=build(sol);
```

$$sol2 := u(x,t) = \_C3e^{(c^2\_c1t)}\_C1\sinh(\sqrt{\_c1}x) + \_C3e^{(c^2\_c1t)}\_C2\cosh(\sqrt{\_c1}x)$$

EXAMPLE 1: Solve $\begin{cases} u_t = u_{xx}, 0<x<1, t>0 \\ u(0,t) = 0, u(1,t) = 0, t>0. \\ u(x,0) = 50, 0<x<1 \end{cases}$

SOLUTION: In this case, $c = 1$, $p = 1$, and $f(x) = 50$. Hence,

$$u(x,t) = \sum_{n=1}^{\infty} c_n \sin n\pi x e^{-\lambda_n t}$$

where

$$c_n = \frac{2}{1}\int_0^1 50\sin n\pi x\,dx = -\frac{100}{n\pi}(\cos n\pi - 1) = -\frac{100}{n\pi}[(-1)^n - 1], \qquad n = 1, 2, \ldots$$

and $\lambda_n = (n\pi)^2$.

```
> assume(n,integer):
c:=n->100*int(sin(n*Pi*x),x=0..1):
```

```
c(n);
```

$$-100\frac{(-1)^{n^{\sim}}-1}{n\sim\pi}$$

```
> lambda:=n->(n*Pi)^2;
```

$$\lambda := n \rightarrow n^2\pi^2$$

Therefore, because $c_n = 0$ if n is even, we write $u(x,t)$ as

$$u(x,t) = \sum_{n=1}^{\infty}\frac{200}{(2n-1)\pi}\sin(2n-1)\pi x e^{-(2n-1)^2\pi^2 t}.$$

We graph an approximation of u at various times by graphing
$u_k(x,t) = \sum_{n=1}^{k}\frac{200}{(2n-1)\pi}\sin(2n-1)\pi x e^{-(2n-1)^2\pi^2 t}$ if $k = 10$.

```
> approx:='approx':t:='t':x:='x':

approx:=proc(k) option remember;

        approx(k-1)+200/((2*k-1)*Pi)*sin((2*k-1)*Pi*x)*

                exp(-(2*k-1)^2*Pi^2*t)

        end:

approx(1):=200/Pi*sin(Pi*x)*exp(-Pi^2*t):
```

```
> tvals:=seq(0.07/20*i,i=0..20):

toplot:=[seq(subs(t=tval,approx(10)),tval=tvals)]:

plot(toplot,x=0..1);
```

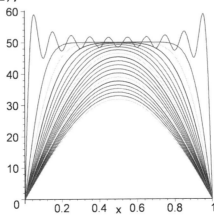

An alternative approach to visualizing the solution is to generate a density plot of $u(x,t)$ with densityplot, which is contained in the **plots** package.

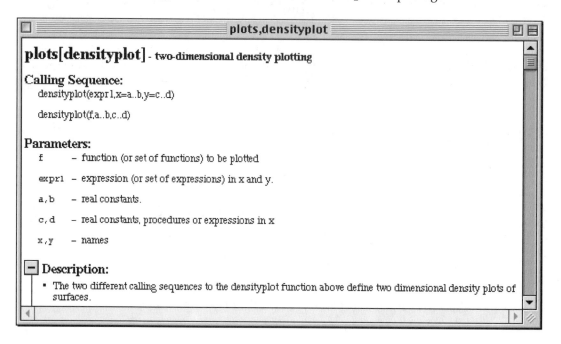

plots,densityplot

plots[densityplot] - two-dimensional density plotting

Calling Sequence:
densityplot(expr1,x=a..b,y=c..d)

densityplot(f,a..b,c..d)

Parameters:
f   – function (or set of functions) to be plotted

expr1 – expression (or set of expressions) in x and y.

a,b   – real constants.

c,d   – real constants, procedures or expressions in x

x,y   – names

Description:
- The two different calling sequences to the densityplot function above define two dimensional density plots of surfaces.

```
> with(plots):
densityplot(approx(10),x=0..1,t=0..0.25,grid=[25,25]);
```

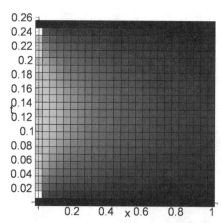

Notice that as t increases, the temperature throughout the rod approaches zero.

Nonhomogeneous Boundary Conditions

The ability to apply the method of separation of variables depends on the presence of homogeneous boundary conditions as we just saw in the previous problem. However, with the heat equation, the temperature at the endpoints may not be held constantly at zero. Instead, consider the case in which the temperature at the left-hand endpoint is $T_0 \neq 0$ and at the right-hand endpoint is $T_1 \neq 0$. Mathematically, we state these **nonhomogeneous boundary conditions** as

$$u(0,t) = T_0 \quad \text{and} \quad u(p,t) = T_1,$$

so we are faced with solving the problem

$$\begin{cases} u_t = c^2 u_{xx}, 0 < x < p, t > 0 \\ u(0,t) = T_0, u(p,t) = T_1, t > 0 \\ u(x,0) = f(x), 0 < x < p \end{cases}$$

In this case, we must modify the problem in order to introduce homogeneous boundary conditions to the problem. We do this by using the physical observation that as $t \to \infty$, the temperature in the wire does not depend on t. Hence,

$$\lim_{t \to \infty} u(x,t) = S(x)$$

where we call $S(x)$ the **steady-state temperature**. Therefore, we let

$$u(x,t) = v(x,t) + S(x)$$

where $v(x,t)$ is called the **transient temperature**. We use these two functions to obtain two problems that we can solve. In order to substitute $u(x,t)$ into the heat equation $u_t = c^2 u_{xx}$, we calculate the derivatives

$$u_t(x,t) = v_t(x,t) + 0 \quad \text{and} \quad u_{xx}(x,t) = v_{xx}(x,t) + S''(x).$$

Substitution into the heat equation yields

$$u_t = c^2 u_{xx}$$

$$v_t = c^2 v_{xx} + c^2 S''$$

so we have the two equations $v_t = c^2 v_{xx}$ and $S'' = 0$. We then consider the boundary conditions. Because

$$u(0,t) = v(0,t) + S(0) = T_0 \quad \text{and} \quad u(p,t) = v(p,t) + S(p) = T_1,$$

we can choose the boundary conditions for S to be the nonhomogeneous conditions

$$S(0) = T_0 \quad \text{and} \quad S(p) = T_1$$

and the boundary conditions for $v(x,t)$ to be the homogeneous conditions

$$v(0,t) = 0 \quad \text{and} \quad v(p,t) = 0.$$

Of course, we have failed to include the initial temperature. Applying this condition, we have $u(x,0) = v(x,0) + S(x) = f(x)$, so the initial condition for v is

$$v(x,0) = f(x) - S(x).$$

Therefore, we have two problems, one for v with homogeneous boundary conditions and one for S that has nonhomogeneous boundary conditions:

$$\begin{cases} S'' = 0, 0 < x < p \\ S(0) = T_0, S(p) = T_1 \end{cases} \quad \text{and} \quad \begin{cases} v_t = c^2 v_{xx}, 0 < x < p, t > 0 \\ v(0,t) = 0, v(p,t) = 0, t > 0 \\ v(x,0) = f(x) - S(x), 0 < x < p \end{cases}$$

Because S is needed in the determination of v, we begin by finding the steady-state temperature and obtain $S(x) = T_0 + \frac{T_1 - T_0}{p}x$.

> **dsolve({diff(s(x),x$2)=0,s(0)=T[0],s(p)=T[1]},s(x));**

$$s(x) = \frac{(-T_0 + T_1)x}{p} + T_0$$

We are now able to find $v(x,t)$ by solving the heat equation with homogeneous boundary conditions for v. Because we solved this problem at the beginning of this section, we do not need to go through the separation of variables procedure. Instead, we use the formula that we derived there using the initial temperature $f(x) - S(x)$. Therefore,

$$v(x,t) = \sum_{n=1}^{\infty} c_n \sin\frac{n\pi x}{p} e^{-c^2 \lambda_n t}$$

where $v(x,0) = \sum_{n=1}^{\infty} c_n \sin\frac{n\pi x}{p} = f(x) - S(x)$. This means that c_n represents the Fourier sine series coefficients for the function $f(x) - S(x)$ given by

$$c_n = \frac{2}{p} \int_0^p (f(x) - S(x)) \sin\frac{n\pi x}{p} dx, \qquad n = 1, 2, \ldots$$

EXAMPLE 2: Solve $\begin{cases} u_t = u_{xx}, 0 < x < 1, t > 0 \\ u(0,t) = 10, u(1,t) = 60, t > 0. \\ u(x,0) = 10, 0 < x < 1 \end{cases}$

SOLUTION: In this case, $c = 1, p = 1, T_0 = 10, T_1 = 60$, and $f(x) = 10$. Therefore, the steady-state solution is

$$S(x) = T_0 + \frac{T_1 - T_0}{p}x = 10 + \frac{60 - 10}{1}x = 10 + 50x$$

Then, the initial transient temperature is

$$v(x, 0) = 10 - (10 + 50x) = -50x$$

so that the series coefficients in the solution $v(x, t) = \sum_{n=1}^{\infty} c_n \sin n\pi x e^{-n^2\pi^2 t}$ are

$$c_n = \frac{2}{1}\int_0^1 -50x \sin n\pi x\, dx = -100\int_0^1 x \sin n\pi x\, dx = \frac{100\cos n\pi}{n\pi} = \frac{100(-1)^n}{n\pi}, \qquad n = 1, 2, \ldots$$

```
> assume(n,integer):
c:=n->2*int(-50*x*sin(n*Pi*x),x=0..1):
c(n);
```

$$100\frac{(-1)^{n~}}{n \sim \pi}$$

```
> lambda:=n->(n*Pi)^2:
```

so the transient temperature is

$$v(x, t) = \sum_{n=1}^{\infty} c_n \sin\frac{n\pi x}{p}e^{-c^2\lambda_n t} = \sum_{n=1}^{\infty} \frac{100(-1)^n}{n\pi}\sin n\pi x e^{-n^2\pi^2 t}$$

and

$$u(x, t) = v(x, t) + S(x) = 10 + 50x + \sum_{n=1}^{\infty} \frac{100(-1)^n}{n\pi}\sin n\pi x e^{-n^2\pi^2 t}.$$

We graph an approximation of $u(x,t)$ for several values of t by graphing $10 +$

$50x + \sum_{n=1}^{30} \frac{100(-1)^n}{n\pi}\sin n\pi x e^{-n^2\pi^2 t}$. Notice that as $t \to \infty$, $u(x, t) \to S(x)$

```
> uapprox:=(x,t)->10+50*x+
    sum(c(n)*sin(n*Pi*x)*exp(-lambda(n)^2*t),n=1..30);
```

$$uapprox := (x, t) \to 10 + 50x + \left(\sum_{n=1}^{30} c(n)\sin(n\pi x)e^{(-\lambda(n)^2 t)}\right)$$

```
> tvals:=seq(0.07/20*i,i=0..20):
toplot:=[seq(uapprox(x,t),t=tvals)]:
```

```
> plot(toplot,x=0..1);
```

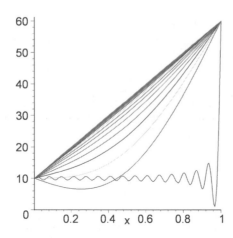

We generate a density plot of this function as well.

```
> with(plots):
densityplot(uapprox(x,t),x=0..1,t=0..0.5);
```

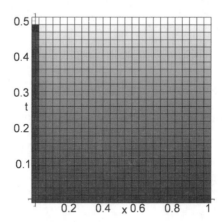

Notice that the temperature throughout the bar approaches the steady-state temperature as t increases.

Insulated Boundary

Another important situation concerning the flow of heat in a wire involves insulated ends. In this case, heat is not allowed to escape from the ends of the wire. Mathematically, we express these boundary conditions as

$$u_x(0, t) = 0 \quad \text{and} \quad u_x(p, t) = 0,$$

because the rate at which the heat changes along the x-axis at the endpoints $x = 0$ and $x = p$ is zero. Therefore, if we want to determine the temperature in a wire of length p with insulated ends, we solve the initial–boundary value problem

$$\begin{cases} u_t = c^2 u_{xx}, 0 < x < p, t > 0 \\ u_x(0, t) = 0, u_x(p, t) = 0, t > 0 \\ u(x, 0) = f(x), 0 < x < p \end{cases}$$

Notice that the boundary conditions are homogeneous, so we can use separation of variables to find $u(x, t) = X(x)T(t)$. By following the steps taken in the solution of the problem with homogeneous boundary conditions, we obtain the ordinary differential equations

$$T' + c^2 \lambda T = 0 \quad \text{and} \quad X'' + \lambda X = 0.$$

However, when we consider the boundary conditions

$$u_x(0, t) = X'(0)T(t) = 0 \quad \text{and} \quad u_x(p, t) = X'(p)T(t) = 0,$$

we wish to avoid letting $T(t) = 0$ for all t (which leads to the trivial solution), and we have the homogeneous boundary conditions

$$X'(0) = 0 \quad \text{and} \quad X'(p) = 0,$$

Therefore, we solve the problem

$$\begin{cases} X'' + \lambda X = 0, 0 < x < p \\ X'(0) = 0, X'(p) = 0 \end{cases}$$

to find $X(x)$. The eigenvalues and corresponding eigenfunctions of this problem are (see Example 5 in Section 9.1)

$$\lambda_n = \begin{cases} 0, n = 0 \\ (n\pi/p)^2, n = 1, 2, \ldots \end{cases} \quad \text{and} \quad X_n(x) = \begin{cases} 1, n = 0 \\ \cos(n\pi x/p), n = 1, 2, \ldots \end{cases}$$

Next, we solve the equation $T' + c^2 \lambda_n T = 0$. First, for $\lambda_0 = 0$, we have the equation $T' = 0$, which has the solution $T(t) = A_0$ where A_0 is a constant. Therefore, for $\lambda_0 = 0$, the solution is the product

$$u_0(x, t) = X_0(x)T_0(t) = A_0.$$

For $\lambda_n = n\pi/p$, $T' + c^2 \lambda_n T = 0$ has general solution $T_n(t) = a_n e^{-c^2 \lambda_n t}$. For these eigenvalues, we have the solution

$$u_n(x,t) = X_n(x)T_n(t) = a_n \cos\frac{n\pi x}{p}e^{-c^2\lambda_n t}.$$

Therefore, by the principle of superposition, the solution is

$$u(x,t) = A_0 + \sum_{n=1}^{\infty} a_n \cos\frac{n\pi x}{p}e^{-c^2\lambda_n t}.$$

Application of the initial temperature yields

$$u(x,0) = A_0 + \sum_{n=1}^{\infty} a_n \cos\frac{n\pi x}{p} = f(x),$$

which is the Fourier cosine series for $f(x)$ where the coefficient A_0 is equivalent to $a_0/2$ in the original Fourier series given in Section 9.2. Therefore,

$$A_0 = \frac{1}{2}a_0 = \frac{1}{2}\frac{2}{p}\int_0^p f(x)dx = \frac{1}{p}\int_0^p f(x)dx \quad \text{and} \quad a_n = \frac{2}{p}\int_0^p f(x)\cos\frac{n\pi x}{p}dx, \quad n = 1,2,\ldots.$$

EXAMPLE 3: Solve $\begin{cases} u_t = u_{xx}, 0 < x < \pi, t > 0 \\ u_x(0,t) = 0, u_x(\pi,t) = 0, t > 0. \\ u(x,0) = x, 0 < x < \pi \end{cases}$

SOLUTION: In this case, $p = \pi$ and $c = 1$. The Fourier cosine series coefficients for $f(x) = x$ are given by (see Example 3 in Section 9.2)

$$A_0 = \frac{1}{2}a_0 = \frac{1}{\pi}\int_0^\pi xdx = \frac{\pi}{2}$$

and

$$a_n = \frac{2}{\pi}\int_0^\pi x\cos\frac{n\pi x}{\pi}dx = \frac{2}{\pi n^2}[(-1)^n, -1], \qquad n = 1,2,\ldots.$$

Therefore, the solution is

$$u(x,t) = \frac{\pi}{2} - \sum_{n=1}^{\infty}\frac{4}{(2n-1)^2\pi}\cos((2n-1)x)e^{-(2n-1)^2 t}$$

where we have used the fact that $a_n = 0$ if n is even. We graph an approximation of u by graphing $u(x,t) = \frac{\pi}{2} - \sum_{n=1}^{40}\frac{4}{(2n-1)^2\pi}\cos((2n-1)x)e^{-(2n-1)^2 t}$.

```
> a:=n->4/((2*n-1)^2*Pi):

uapprox:=(x,t)->Pi/2-

      sum(a(n)*cos((2*n-1)*x)*exp(-(2*n-1)^2*t),n=1..40);
```

$$uapprox := (x, t) \rightarrow \frac{1}{2}\pi - \left(\sum_{n=1}^{40} a(n) \cos((2n-1)x) e^{(-(2n-1)^2 t)} \right)$$

```
> tvals:=seq(i/20,i=0..20):
toplot:=[seq(uapprox(x,t),t=tvals)]:
plot(toplot,x=0..Pi);
```

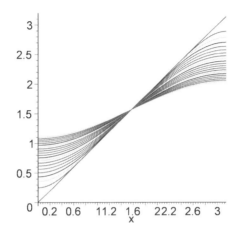

```
> with(plots):
densityplot(uapprox(x,t),x=0..Pi,t=0..0.5);
```

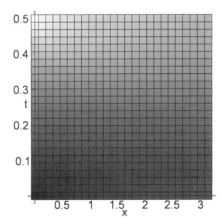

Notice that the temperature eventually becomes $A_0 = \pi/2$ throughout the wire. Temperatures to the left of $x = \pi/2$ increase while those to the right decrease.

10.3 The One-Dimensional Wave Equation

The one-dimensional **wave equation** is important in solving an interesting problem. Suppose that we pluck a string (such as a guitar or violin string) of length p and constant mass density that is fixed at each end. A question that we might ask is: "What is the position of the string at a particular instant of time?" We answer this question by modeling the physical situation with a partial differential equation, namely the **wave equation** in one spatial variable. We will not go through this derivation as we did with the heat equation, but we point out that it is based on determining the forces that act on a small segment of the string and applying Newton's second law of motion. The partial differential equation that is found is

$$c^2 u_{xx} = u_{tt},$$

which is called the (one-dimensional) **wave equation.** In this equation $c^2 = T/\rho$, where T is the tension of the string and ρ is the constant mass of the string per unit length. The solution $u(x,t)$ represents the displacement of the string from the x-axis at time t. In order to determine u we must describe the boundary and initial conditions that model the physical situation. At the ends of the string, the displacement from the x-axis is fixed at zero, so we use the homogeneous boundary conditions

$$u(0,t) = 0 \quad \text{and} \quad u(p,t) = 0 \quad \text{for } t > 0$$

The motion of the string also depends on the displacement and the velocity at each point of the string at $t = 0$. If the initial displacement is given by $f(x)$ and the initial velocity by $g(x)$, we have the initial conditions

$$u(x,0) = f(x) \quad \text{and} \quad u_t(x,0) = g(x) \quad \text{for } 0 < x < p.$$

Therefore, we determine the displacement of the string with the initial–boundary value problem

$$\begin{cases} c^2 u_{xx} = u_{tt}, 0 < x < p, t > 0 \\ u(0,t) = 0, u(p,t) = 0, t > 0 \\ u(x,0) = f(x), u_t(x,0) = g(x), 0 < x < p \end{cases}$$

(Notice that the wave equation requires two initial conditions where the heat equation needed only one. This is due to the fact that there is a second derivative with respect to t, whereas there is only one derivative with respect to t in the heat equation.)

This problem is solved through separation of variables by assuming that $u(x,t) = X(x)T(t)$.

```
> sol1:=pdsolve(c^2*diff(u(x,t),x$2)=diff(u(x,t),t$2),

    HINT='*');
```

$$sol1 := (u(x,t) = \_F1(x)\_F2(t))$$

$$\left[\left\{\frac{\partial^2}{\partial t^2}\_F2(t) = c^2\_c_1\_F2(t), \frac{\partial^2}{\partial x^2}\_F1(x) = \_c_1\_F1(x)\right\}\right]$$

Substitution into the wave equation yields

$$c^2 X'' T = X T''$$

$$\frac{X''}{X} = \frac{T''}{c^2 T} = -\lambda$$

so we obtain the two second-order ordinary differential equations

$$X'' + \lambda X = 0 \quad \text{and} \quad T'' + c^2 \lambda T = 0.$$

At this point, we solve the equation that involves the homogeneous boundary conditions. As was the case with the heat equation, the boundary conditions in terms of $u(x,t) = X(x)T(t)$ are $u(0,t) = X(0)T(t) = 0$ and $u(p,t) = X(p)T(t) = 0$, so we have

$$X(0) = 0 \quad \text{and} \quad X(p) = 0.$$

Therefore, we determine $X(x)$ by solving the eigenvalue problem

$$\begin{cases} X'' + \lambda X = 0 \\ X(0) = 0, X(p) = 0 \end{cases}$$

which we encountered when solving the heat equation and solved in Section 9.2. The eigenvalues of this problem are

$$\lambda_n = (n\pi/p)^2, \qquad n = 1, 2, \ldots.$$

with corresponding eigenfunctions

$$X_n(x) = \sin(n\pi x/p), \qquad n = 1, 2, \ldots.$$

Next, we solve the equation $T'' + c^2 \lambda_n T = 0$. A general solution is

$$T_n(t) = a_n \cos(c\sqrt{\lambda_n}t) + b_n \sin(c\sqrt{\lambda_n}t) = a_n \cos\frac{cn\pi t}{p} + b_n \sin\frac{cn\pi t}{p}$$

where the coefficients a_n and b_n must be determined. Putting this information together, we obtain

$$u_n(x,t) = \left(a_n \cos\frac{cn\pi t}{p} + b_n \sin\frac{cn\pi t}{p}\right)\sin\frac{n\pi x}{p},$$

so by the principle of superposition, we have

$$u(x,t) = \sum_{n=1}^{\infty}\left(a_n \cos\frac{cn\pi t}{p} + b_n \sin\frac{cn\pi t}{p}\right)\sin\frac{n\pi x}{p}.$$

Applying the initial position yields

$$u(x, 0) = \sum_{n=1}^{\infty} a_n \sin \frac{n\pi x}{p} = f(x),$$

so a_n is the Fourier sine series coefficient for $f(x)$, which is given by

$$a_n = \frac{2}{p} \int_0^p f(x) \sin \frac{n\pi x}{p} dx, \qquad n = 1, 2, \ldots.$$

In order to determine b_n, we must use the initial velocity. Therefore, we compute

$$u_t(x, t) = \sum_{n=1}^{\infty} \left(-a_n \frac{cn\pi}{p} \sin \frac{cn\pi t}{p} + b_n \frac{cn\pi}{p} \cos \frac{cn\pi}{p} \right) \sin \frac{n\pi x}{p}.$$

Then,

$$u_t(x, 0) = \sum_{n=1}^{\infty} b_n \frac{cn\pi}{p} \sin \frac{n\pi x}{p} = g(x),$$

so b_n represents the Fourier sine series coefficient for $g(x)$, which means that

$$b_n = \frac{p}{cn\pi} \frac{2}{p} \int_0^p g(x) \sin \frac{n\pi x}{p} dx = \frac{2}{cn\pi} \int_0^p g(x) \sin \frac{n\pi x}{p} dx, \qquad n = 1, 2, \ldots.$$

EXAMPLE 1: Solve $\begin{cases} u_{xx} = u_{tt}, 0 < x < 1, t > 0 \\ u(0, t) = 0, u(1, t) = 0, t > 0 \\ u(x, 0) = x(1 - x), u_t(x, 0) = 0, 0 < x < 1 \end{cases}$

SOLUTION: In this case, $c = p = 1$, $f(x) = x(1 - x)$, and $g(x) = 0$. From this information, we compute

$$a_n = \frac{2}{1} \int_0^1 x(1 - x) \sin n\pi x dx = -\frac{4 \cos n\pi}{n^3 \pi^3} + \frac{4}{n^3 \pi^3} = \frac{4}{n^3 \pi^3} (1 - (-1)^n), \qquad n = 1, 2, \ldots.$$

```
> a:='a':n:='n':

assume(n,integer):

a:=n->2*int(x*(1-x)*sin(n*Pi*x),x=0..1):

a(n);
```

$$-4 \frac{(-1)^{n\sim} - 1}{n\sim^3 \pi^3}$$

Because $g(x) = 0$, the coefficients $b_n = 0$ for all n. Using the fact that $a_n = 0$ for even values of n,

```
> a(2*n);
```

$$0$$

```
> a(2*n-1);
```

$$8 \frac{1}{(2n \sim -1)^3 \pi^3}$$

the solution is

$$u(x,t) = \sum_{n=1}^{\infty} \frac{8}{(2n-1)^3 \pi^3} \cos((2n-1)\pi t) \sin((2n-1)\pi x).$$

We illustrate the motion of the string by graphing $u_k(x,t) = \sum_{n=1}^{k} \frac{8}{(2n-1)^3\pi^3} \cos((2n-1)\pi t) \sin((2n-1)\pi x)$ using $k = 10$ for various values of t between 0 and 1.

```
> u:=(x,t)->sum(a(2*n-1)*cos((2*n-1)*Pi*t)*
      sin((2*n-1)*Pi*x),n=1..10);
```

$$u := (x,t) \rightarrow \sum_{n=1}^{10} a(2n-1)\cos((2n-1)\pi t)\sin((2n-1)\pi x)$$

```
> with(plots):
anarray:=animate(u(x,t),x=0..1,t=0..1,
      color=BLACK,frames=16):
display(anarray);
```

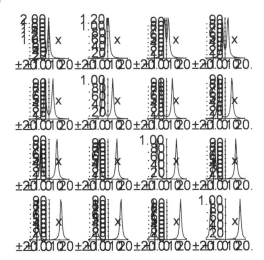

To see the motion of the string, we use `animate` to generate a sequence of graphs and animate the result. Several frames from an animation are shown here.

```
> animate(u(x,t),x=0..1,t=0..1,frames=60);
```

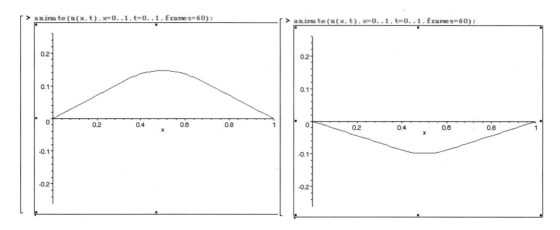

■

EXAMPLE 2: Solve $\begin{cases} u_{xx} = u_{tt}, 0<x<1, t>0 \\ u(0,t) = 0, u(1,t) = 0, t>0 \\ u(x,0) = \sin \pi x, u_t(x,0) = 3x+1, 0<x<1 \end{cases}$

SOLUTION: The appropriate parameters and initial conditions are defined first.

```
> f:=x->sin(Pi*x):

g:=x->3*x+1:
```

Next, the functions to determine the coefficients a_n and b_n in the series approximation of the solution $u(x, t)$ are defined.

```
> a:='a':b:='b':n:='n':

assume(n,integer):

a(1):=2*int(f(x)*sin(Pi*x),x=0..1);
```

$$a(1) := 1$$

```
> a:=n->2*int(f(x)*sin(n*Pi*x),x=0..1):
```

```
a(n);
```

$$0$$

```
> b:=n->2*int(g(x)*sin(n*Pi*x),x=0..1)/(n*Pi):
b(n);
```

$$-2\frac{4(-1)^{n^\sim} - 1}{n\sim^2\pi^2}$$

The function u defined next computes the *n*th term in the series expansion. Hence, approx determines the approximation of order *k* by summing the first *k* terms of the expansion as illustrated with approx (2).

```
> u:=n->(a(n)*cos(n*Pi*t)+b(n)*sin(n*Pi*t))*sin(n*Pi*x):
approx:=proc(k) option remember;
        approx(k-1)+u(k)
        end:
approx(1):=u(1):
```

```
> approx(2);
```

$$\left(\cos(\pi t) + 10\frac{\sin(\pi t)}{\pi^2}\right)\sin(\pi x) - \frac{3}{2}\frac{\sin(2\pi t)\sin(2\pi x)}{\pi^2}$$

To illustrate the motion of the string, we graph uapprox(10), the 10th partial sum of the series, on the interval [0,1] for 16 equally spaced values of *t* between 0 and 2 using animate together with display.

```
> with(plots):
graphicsarray:=animate(approx(10),x=0..1,t=0..2,
        frames=16):
display(graphicsarray);
```

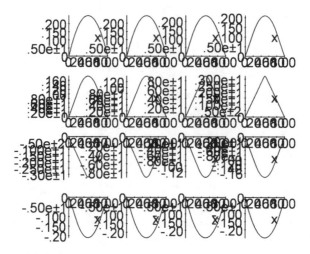

To *see* the motion of the string, we use `approx(10)`, the 10th partial sum of the series, to approximate the motion of the spring together with `animate` to graph the solution on the interval [0,1] for 30 equally spaced values of *t* between 0 and 2. Several frames from the resulting animation are shown here.

```
> animate(approx(10),x=0..1,t=0..2,color=BLACK,frames=30);
```

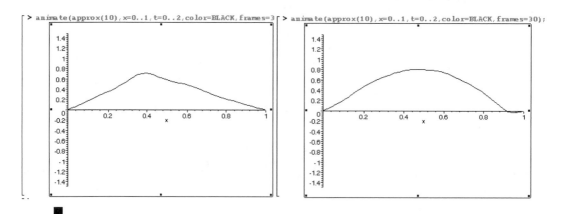

D'Alembert's Solution

An interesting version of the wave equation is illustrated by considering a string of infinite length. Therefore, the boundary conditions are no longer of importance. Instead, we simply work with the wave equation with the initial position and velocity functions. In order to solve the problem

$$\begin{cases} c^2 u_{xx} = u_{tt}, -\infty < x < \infty, t > 0 \\ u(x,0) = f(x), u_t(x,0) = g(x) \end{cases}$$

we use the change of variables $r = x + ct$ and $s = x - ct$. Using the chain rule, we compute the derivatives u_{xx} and u_{tt} in terms of the variables r and s:

$$u_x = u_r r_x + u_s s_x = u_r + u_s,$$

$$u_{xx} = (u_r + u_s)_r r_x + (u_r + u_s)_s s_x = u_{rr} + 2u_{rs} + u_{ss},$$

$$u_t = u_r r_t + u_s s_t = cu_r - cu_s = c(u_r - u_s),$$

and

$$u_{tt} = c\big[(u_r - u_s)_r r_t + (u_r - u_s)_s s_t\big] = c^2\big[u_{rr} - 2u_{rs} + u_{ss}\big].$$

Substitution into the wave equation yields

$$c^2 u_{xx} = u_{tt}$$

$$c^2\big[u_{rr} + 2u_{rs} + u_{ss}\big] = c^2\big[u_{rr} - 2u_{rs} + u_{ss}\big]$$

$$4c^2 u_{rs} = 0$$

$$u_{rs} = 0.$$

The partial differential equation $u_{rs} = 0$ can be solved by first integrating with respect s to obtain

$$u_r = f(r)$$

where $f(r)$ is an arbitrary function of r. Then, integrating with respect to s, we have

$$u(r,s) = F(r) + G(s)$$

where F is an antiderivative of f and G is an arbitrary function of s. Returning to our original variables then gives us

$$u(x,t) = F(x + ct) + G(x - ct).$$

We see that this is the solution that `pdsolve` returns as well, when we include the option HINT = ' + '. (Note that _F1 and _F2 represent the arbitrary functions F and G.)

```
> u:='u':x:='x':t:='t':
pdsolve(c^2*diff(u(x,t),x$2)=diff(u(x,t),t$2),HINT='+');
```

$$\left.\left(u(x,t) = \_F1(x) + \_F2(t)\right) \text{ where } \left|\left\{\frac{\partial^2}{\partial x^2}\_F1(x) = \frac{\_c_2}{c^2}, \frac{\partial^2}{\partial t^2}\_F2(t) = \_c_2\right\}\right.\right|$$

The functions F and G are determined by the initial conditions, which indicate that

$$u(x,0) = F(x) + G(x) = f(x)$$

and

$$u_t(x,0) = cF'(x) - cG'(x) = g(x).$$

We can rewrite the second equation by integrating to obtain

$$F'(x) - G'(x) = \frac{g(x)}{c}$$

$$F(x) - G(x) = \frac{1}{c}\int_0^x g(v)dv.$$

Therefore, we solve the system

$$F(x) + G(x) = f(x)$$

$$F(x) - G(x) = \frac{1}{c}\int_0^x g(v)dv.$$

Adding these equations yields

$$F(x) = \frac{1}{2}\left[f(x) + \frac{1}{c}\int_0^x g(v)dv\right],$$

and subtracting gives us

$$G(x) = \frac{1}{2}\left[f(x) - \frac{1}{c}\int_0^x g(v)dv\right]$$

Therefore,

$$F(x+ct) = \frac{1}{2}\left[f(x+ct) + \frac{1}{c}\int_0^{x+ct} g(v)dv\right] \quad \text{and} \quad G(x-ct) = \frac{1}{2}\left[f(x-ct) - \frac{1}{c}\int_0^{x-ct} g(v)dv\right],$$

so the solution is

$$u(x,t) = \frac{1}{2}(f(x+ct) + f(x-ct)) + \frac{1}{2c}\int_{x-ct}^{x+ct} g(v)dv.$$

EXAMPLE 3: Solve $\begin{cases} u_{xx} = u_{tt}, -\infty < x < \infty, t > 0 \\ u(x,0) = \frac{2}{1+x^2}, u_t(x,0) = 0. \end{cases}$

SOLUTION: Because $c = 1$, $f(x) = \frac{1}{1+x^2}$, and $g(x) = 0$, we have the solution

$$u(x, t) = \frac{1}{2}[f(x + t) + f(x - t)] = \frac{1}{2}\left[\frac{2}{1 + (x + t)^2} + \frac{2}{1 + (x - t)^2}\right]$$

$$= \frac{1}{1 + (x + t)^2} + \frac{1}{1 + (x - t)^2}.$$

We plot the solution for $t = 0$ to $t = 15$ to illustrate the motion of the string of infinite length.

```
> u:=(x,t)->1/(1+(x+t)^2)+1/(1+(x-t)^2):
with(plots):
graphicsarray:=animate(u(x,t),x=-20..20,t=0..15,
      color=BLACK,frames=16):
display(graphicsarray);
```

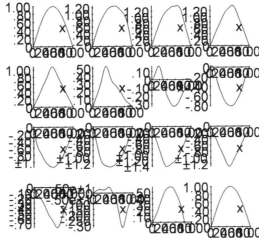

Alternatively, we can generate several graphs and animate the results to see the motion of the string. Here we show several frames from the resulting animation.

```
> animate(u(x,t),x=-20..20,t=0..15,frames=60);
```

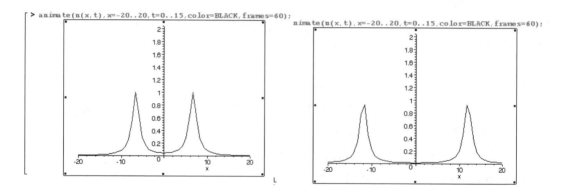

D'Alembert's solution is sometimes referred to as the **travelling wave solution** because of the behaviour of its graph. The waves appear to move in opposite directions along the x-axis as t increases, as we can see in the graphs.

■

10.4 Problems in Two Dimensions: Laplace's Equation

Laplace's Equation

Laplace's equation, often called the **potential equation**, is given by

$$u_{xx} + u_{yy} = 0$$

in rectangular coordinates and is one of the most useful partial differential equations in that it arises in many fields of study. These include fluid flows as well as electrostatic and gravitational potential. Because the potential $u(x,y)$ does not depend on time, no initial condition is required, so we are faced with solving a pure boundary value problem when working with Laplace's equation. The boundary conditions can be stated in different forms. If the value of the solution is given around the boundary of the region, then the boundary value problem is called a **Dirichlet problem**, whereas if the normal derivative of the solution is given around the boundary, the problem is known as a **Neumann problem**. We now investigate the solutions to Laplace's equation in a rectangular region by first stating the general form of the Dirichlet problem:

$$\begin{cases} u_{xx} + u_{yy} = 0, 0<x<a, 0<y<b \\ u(x,0) = f_1(x), u(x,b) = f_2(x), 0<x<a. \\ u(0,y) = g_1(y), u(a,y) = g_2(y), 0<y<b \end{cases}$$

This boundary-value problem is solved through separation of variables. We begin by considering the problem

$$\begin{cases} u_{xx} + u_{yy} = 0, 0 < x < a, 0 < y < b \\ u(x,0) = 0, u(x,b) = f(x), 0 < x < a \\ u(0,y) = 0, u(a,y) = 0, 0 < y < b \end{cases}$$

In this case, we assume that

$$u(x,y) = X(x)Y(y)$$

so substitution into Laplace's equation yields

$$X''Y + XY'' = 0$$

$$\frac{X''}{X} = -\frac{Y''}{Y} = -\lambda,$$

where $-\lambda$ is the separation constant. Therefore, we have the ordinary differential equations $X'' + \lambda X = 0$ and $Y'' - \lambda Y = 0$. Notice that the boundary conditions along the lines $x = 0$ and $x = a$ are homogeneous. In fact, because $u(0,y) = X(0)Y(y) = 0$ and $u(a,y) = X(a)Y(y) = 0$, we have $X(0) = 0$ and $X(a) = 0$. Therefore, we first solve the eigenvalue problem

$$\begin{cases} X'' + \lambda X = 0 \\ X(0) = 0, X(a) = 0 \end{cases}$$

which was solved with a=p in Section 9.1. There, we found the eigenvalues and corresponding eigenfunctions to be $\lambda_n = (n\pi/a)^2$, $n = 1, 2, \ldots$, and $X_n(x) = \sin(n\pi x/a)$, $n = 1, 2, \ldots$. We then solve the equation $Y'' - \lambda_n^2 Y = 0$. From our experience with second-order equations, we know that $Y_n(y) = a_n e^{\lambda_n y} + b_n e^{-\lambda_n y}$, which can be written in terms of the hyperbolic trigonometric functions as

$$Y_n(y) = A_n \cosh \lambda_n y + B_n \sinh \lambda_n y = A_n \cosh \frac{n\pi y}{a} + B_n \sinh \frac{n\pi y}{a}.$$

Then, using the homogeneous boundary condition $u(x,0) = X(x)Y(0) = 0$, which indicates that $Y(0) = 0$, we have

$$Y(0) = Y_n(0) = A_n \cosh 0 + B_n \sinh 0 = A_n = 0,$$

so $A_n = 0$ for all n. Therefore, $Y_n(y) = B_n \sinh \lambda_n y$, and a solution is

$$u_n(x,y) = B_n \sinh \frac{n\pi y}{a} \sin \frac{n\pi x}{a},$$

so by the principle of superposition,

$$u(x,y) = \sum_{n=1}^{\infty} B_n \sinh \frac{n\pi y}{a} \sin \frac{n\pi x}{a}$$

is also a solution, where the coefficients are determined with the boundary condition $u(x,b) = f(x)$. Substitution into the solution yields

$$u(x,b) = \sum_{n=1}^{\infty} B_n \sinh \frac{n\pi b}{a} \sin \frac{n\pi x}{a} = f(x)$$

where $B_n \sinh(n\pi b/a)$ represents the Fourier sine series coefficients given by

$$B_n \sinh \frac{n\pi b}{a} = \frac{2}{a} \int_0^a f(x) \sin \frac{n\pi x}{a} dx$$

$$B_n = \frac{2}{a \sinh \frac{n\pi b}{a}} \int_0^a f(x) \sin \frac{n\pi x}{a} dx.$$

EXAMPLE 1: Solve $\begin{cases} u_{xx} + u_{yy} = 0, 0<x<1, 0<y<2; \\ u(x,0) = 0, u(x,2) = x(1-x), 0<x<1 \\ u(0,y) = 0, u(1,y) = 0, 0<y<2 \end{cases}$

SOLUTION: In this case, $a = 1$, $b = 2$, and $f(x) = x(1-x)$. Therefore,

$$B_n = \frac{2}{\sinh 2n\pi} \int_0^1 x(1-x) \sin n\pi x dx = \frac{1}{\sinh 2n\pi} \left(-\frac{4\cos n\pi}{n^3\pi^3} + \frac{4}{n^3\pi^3} \right)$$

$$= \frac{4}{n^3\pi^3 \sinh 2n\pi} (1 - (-1)^n), \qquad n = 1, 2, \ldots$$

```
> B:='B':
assume(n,integer):
B:=n->2/sinh(2*n*Pi)*int(x*(1-x)*sin(n*Pi*x),x=0..1):
B(n);
```

$$-4 \frac{(-1)^{n\sim} - 1}{\sinh(2\pi n\sim)\pi^3 n\sim^3}$$

so the solution is

$$u(x,y) = \sum_{n=1}^{\infty} B_n \sinh n\pi y \sin n\pi x = \sum_{n=1}^{\infty} \frac{8 \sinh(2n-1)\pi y \sin(2n-1)\pi x}{(2n-1)^3 \pi^3 \sinh 2(2n-1)\pi}$$

We plot $u(x,y)$ using the first 15 terms of the series solution.

```
> u:='u':
u:=(x,y)->sum(B(n)*sinh(n*Pi*y)*sin(n*Pi*x),n=1..15);
```

```
> plot3d(u(x,y),x=0..1,y=0..2,axes=BOXED);
```

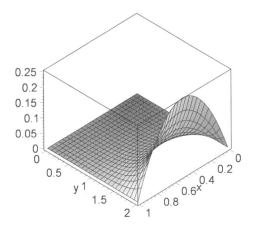

Alternatively, we can generate a contour or density plot of $u(x, y)$.

```
> with(plots):
contourplot(u(x,y),x=0..1,y=0..2);
densityplot(u(x,y),x=0..1,y=0..2,grid=[25,25]);
```

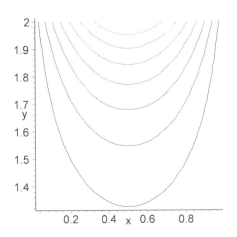

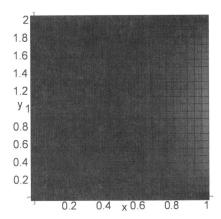

■

Any version of Laplace's equation on a rectangular region can be solved through separation of variables as long as we have a pair of homogeneous boundary conditions in the same variable.

EXAMPLE 2: Solve
$$\begin{cases} u_{xx} + u_{yy} = 0, 0 < x < \pi, 0 < y < 1 \\ u(x,0) = 0, u(x,1) = 0, 0 < x < \pi \\ u(0,y) = \sin 2\pi y, u(\pi,y) = 4, 0 < y < 1 \end{cases}$$

SOLUTION: As we did in the previous problem, we assume that $u(x,y) = X(x)Y(y)$. Notice that this problem differs from the previous problem in that the homogeneous boundary conditions are in terms of the variable y. Hence, when we separate variables, we use a different constant of separation. This yields

$$X''Y + XY'' = 0$$

$$\frac{X''}{X} = -\frac{Y''}{Y} = \lambda,$$

so we have the ordinary differential equations $X'' - \lambda X = 0$ and $Y'' + \lambda Y = 0$. Therefore, with the homogeneous boundary conditions $u(x,0) = X(x)Y(0) = 0$ and $u(x,1) = X(x)Y(1) = 0$, we have $Y(0) = 0$ and $Y(1) = 0$. The eigenvalue problem

$$\begin{cases} Y'' + \lambda Y = 0 \\ Y(0) = 0, Y(1) = 0 \end{cases}$$

has eigenvalues $\lambda_n = (n\pi/1)^2 = n^2\pi^2$, $n = 1,2,\ldots$, and eigenfunctions $Y_n(y) = \sin n\pi y$, $n = 1,2,\ldots$. We then solve the equation $X'' - \lambda_n X = 0$ obtaining $X_n(x) = a_n e^{n\pi x} + b_n e^{-n\pi x}$, which can be written in terms of the hyperbolic trigonometric functions as

$$X_n(x) = A_n \cosh n\pi x + B_n \sinh n\pi x.$$

Now, because the boundary conditions on the boundaries $x = 0$ and $x = \pi$ are nonhomogeneous, we use the principle of superposition to obtain

$$u(x,y) = \sum_{n=1}^{\infty}(A_n \cosh n\pi x + B_n \sinh n\pi x)\sin n\pi y$$

Therefore,

$$u(0,y) = \sum_{n=1}^{\infty} A_n \sin n\pi y = \sin 2\pi y,$$

so $A_2 = 1$ and $A_n = 0$ for $n \neq 2$. Similarly,

$$u(\pi,y) = A_2 \cosh 2\pi^2 + \sum_{n=1}^{\infty} B_n \sinh n\pi^2 \sin n\pi y = 4$$

which indicates that $\sum_{n=1}^{\infty} B_n \sinh n\pi^2 \sin n\pi y = 4 - \cosh 2\pi^2$. Then, $B_n \sinh n\pi^2$ are the

Fourier sine series coefficients for the constant function $4 - \cosh 2\pi^2$, which are given by

$$B_n \sinh n\pi^2 = \frac{2}{1} \int_0^1 \left(4 - \cosh 2\pi^2\right) \sin n\pi y\, dy = -2\left(4 - \cosh 2\pi^2\right) \left[\frac{\cos n\pi y}{n\pi}\right]_0^1$$

$$= \frac{-2(4 - \cosh 2\pi^2)}{n\pi} \left[(-1)^n - 1\right], \qquad n = 1, 2, \ldots.$$

From this formula, we see that $B_n = 0$ if n is even. Therefore, we express these coefficients as

$$B_{2n-1} = \frac{4(4 - \cosh 2\pi^2)}{(2n-1)\pi}, \qquad n = 1, 2, \ldots$$

so that the solution is

$$u(x, y) = \cosh 2\pi x + \sum_{n=1}^{\infty} \frac{4(4 - \cosh 2\pi^2)}{(2n-1)\pi} \sin(2n-1)\pi y$$

As in Example 1, we generate contour and density plots of an approximation of u.

```
> u:='u':
u:=(x,y)->cosh(2*Pi*x)+
        sum(4*(4-cosh(2*Pi^2))*sin((2*n-1)*Pi*y)/((2*n-1)*Pi),
                n=1..30);
u := (x, y) -> cosh(2 Pi x)
```

$$u := (x, y) \rightarrow \cosh(2\pi x) + \left(\sum_{n=1}^{30} \left(4 \frac{(4 - \cosh(2\pi^2)) \sin((2n-1)\pi y)}{(2n-1)\pi}\right)\right)$$

```
> with(plots):
contourplot(u(x,y),x=0..Pi,y=0..1);
densityplot(u(x,y),x=0..Pi,y=0..1,grid=[25,25]);
```

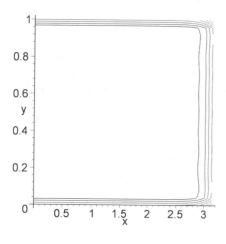

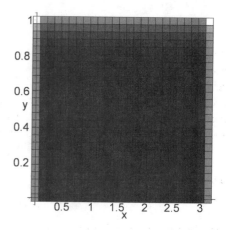

10.5 Two-Dimensional Problems in a Circular Region

In some situations, the region on which we solve a boundary value problem or an initial boundary, value problem is not rectangular in shape. For example, we usually do not have rectangular-shaped drumheads and the heating elements on top of the stove are not square. Instead, these objects are typically circular in shape, so we find the use of polar coordinates convenient. In this section, we discuss problems of this type by presenting two important problems solved on circular regions: Laplace's equation, which is related to the steady-state temperature, and the wave equation, which is used to find the displacement of a drumhead.

Laplace's Equation in a Circular Region

In calculus, we found that polar coordinates are useful in solving many problems. The same can be said for solving boundary value problems in a circular region. With the change of variables

$$\begin{cases} x = r\cos\theta \\ y = r\sin\theta, \end{cases}$$

we transform Laplace's equation in rectangular coordinates, $u_{xx} + u_{yy} = 0$, to polar coordinates

$$u_{rr} + \frac{1}{r}u_r + \frac{1}{r^2}u_{\theta\theta} = 0, \qquad 0 < r < \rho, -\pi < \theta < \pi.$$

Recall that for the solution of Laplace's equation in a rectangular region, we had to specify a boundary condition on each of the four boundaries of the rectangle. However, in the case of a circle, there are not four sides, so we must alter the boundary conditions. Because in polar coordinates the points (r, π) and $(r, -\pi)$ are equivalent for the same value of r, we want our solution and its derivative with respect to θ to match at these points (so that the solution is smooth). Therefore, two of the boundary conditions are

$$u(r, -\pi) = u(r, \pi) \quad \text{and} \quad u_\theta(r, -\pi) = u_\theta(r, \pi) \quad \text{for } 0 < r < R.$$

Also, we want our solution to be bounded at $r = 0$, so another boundary condition is

$$|u(0, \theta)| < \infty \quad \text{for } -\pi < \theta < \pi.$$

Finally, we can specify the value of the solution around the boundary of the circle. This is given by

$$u(R, \theta) = f(\theta) \quad \text{for } -\pi < \theta < \pi.$$

Therefore, we solve the following boundary value problem to solve Laplace's equation (the Dirichlet problem) in a circular region of radius ρ.

$$\begin{cases} u_{rr} + \frac{1}{r}u_r + \frac{1}{r^2}u_{\theta\theta} = 0, 0 < r < p, -\pi < \theta < \pi \\ u(r, -\pi) = u(r, -\pi), u_\theta(r, -\pi) = u_\theta(r, -\pi), 0 < r < p \\ |u(0, \theta)| < \infty, u(p, \theta) = f(\theta), -\pi < \theta < \pi \end{cases}$$

Using separation of variables, we assume that $u(r, \theta) = R(r)H(\theta)$. Substitution into Laplace's equation yields

$$R''H + \frac{1}{r}R'H + RH'' = 0$$

$$R''H + \frac{1}{r}R'H = -RH''$$

$$\frac{rR'' + R'}{rR} = -\frac{H''}{H} = \lambda$$

Therefore, we have the ordinary differential equations

$$H'' + \lambda H = 0 \quad \text{and} \quad r^2 R'' + rR' - \lambda R = 0$$

Notice that the boundary conditions

$$u(r, -\pi) = u(r, \pi) \qquad \text{and} \qquad u_\theta(r, -\pi) = u_\theta(r, \pi)$$

imply that

$$R(r)H(-\pi) = R(r)H(\pi) \quad \text{and} \quad R(r)H'(-\pi) = R(r)H'(\pi)$$

so that

$$H(-\pi) = H(\pi) \quad \text{and} \quad H'(-\pi) = H'(\pi)$$

This means that we begin by solving the eigenvalue problem

$$\begin{cases} H'' + \lambda H = 0 \\ H(-\pi) = H(\pi), H'(-\pi) = H'(\pi) \end{cases}$$

The eigen values and corresponding eigenfunctions of this problem are

$$\lambda_n = \begin{cases} 0, n = 0 \\ n^2, n = 1, 2, \ldots \end{cases} \quad \text{and} \quad H_n(\theta) = \begin{cases} 1, n = 0 \\ a_n \cos n\theta + b_n \sin n\theta, n = 1, 2, \ldots \end{cases}$$

Because $r^2 R'' + rR' - \lambda_n{}^2 R = 0$ is a Cauchy–Euler equation, we assume that $R(r) = r^m$:

$$m(m - 1)r^2 r^{m-2} + mrr^{m-1} - \lambda_n r^m = 0$$

$$r^m[m(m - 1) + m - \lambda_n] = 0$$

Therefore,

$$m^2 - \lambda_n{}^2 = 0$$

$$m = \pm \lambda_n$$

If $\lambda_0 = 0$, then $R_0(r) = c_1 + c_2 \ln r$. However, because we require that the solution be bounded near $r = 0$ and $\lim_{r \to 0^+} \ln r = -\infty$, we must choose $c_2 = 0$. Therefore, $R_0(r) = c_1$. On the other hand, if $\lambda_n = n^2, n = 1, 2, \ldots$, then $R_n(r) = c_3 r^n + c_4 r^{-n}$. Similarly, because $\lim_{r \to 0^+} r^{-n} = \infty$, we must let $c_4 = 0$, so $R_n(r) = c_3 r^n$. By the principle of superposition, we have the solution

$$u(r, \theta) = A_0 + \sum_{n=1}^{\infty} r^n (A_n \cos n\theta + B_n \sin n\theta)$$

where $A_0 = c_1, A_n = c_3 a_n$, and $B_n = c_3 b_n$. We find these coefficients by applying the boundary condition $u(p, \theta) = f(\theta)$. This yields

$$u(p, \theta) = A_0 + \sum_{n=1}^{\infty} p^n (A_n \cos n\theta + B_n \sin n\theta) = f(\theta)$$

so A_0, A_n, and B_n are related to the Fourier series coefficients in the following way:

$$A_0 = \frac{1}{2\pi} \int_{-\pi}^{\pi} f(\theta) d\theta,$$

$$A_n = \frac{1}{\pi p^n} \int_{-\pi}^{\pi} f(\theta) \cos n\theta d\theta,$$

and

$$B_n = \frac{1}{\pi p^n} \int_{-\pi}^{\pi} f(\theta) \sin n\theta d\theta, \qquad n = 1, 2, \ldots$$

.

EXAMPLE 1: Solve

$$\begin{cases} u_{rr} + \frac{1}{r} u_r + \frac{1}{r^2} u_{\theta\theta} = 0, 0 < r < 2, -\pi < \theta < \pi \\ u(r, -\pi) = u(r, \pi), u_\theta(r, -\pi) = u_\theta(r, \pi), 0 < r < 2 \\ |u(0, \theta)| < \infty, u(2, \theta) = |\theta|, -\pi < \theta < \pi \end{cases}$$

SOLUTION: Notice that $f(\theta) = |\theta|$ is an even function on $-\pi < \theta < \pi$. Therefore, $B_n = 0$ for $n = 1, 2, \ldots$,

$$A_0 = \frac{1}{2\pi} \int_{-\pi}^{\pi} |\theta| d\theta = \frac{1}{\pi} \int_0^{\pi} \theta d\theta = \frac{\pi}{2},$$

```
> A:='A':B:='B':
A(0):=1/Pi*int(theta,theta=0..Pi);
```

$$A(0) := \frac{1}{2}\pi$$

and

$$A_n = \frac{1}{\pi 2^n} \int_{-\pi}^{\pi} |\theta| \cos n\theta d\theta = \frac{1}{\pi 2^{n-1}} \int_0^{\pi} \theta \cos n\theta d\theta = \frac{1}{\pi 2^n n^2} (\cos n\pi - 1) = \frac{1}{\pi 2^n n^2} ((-1)^n - 1), \qquad n = 1, 2, \ldots$$

```
> assume(n,integer):
A:=n->1/(Pi*2^(n-1))*int(theta*cos(n*theta),theta=0..Pi):
A(n);
```

$$\frac{(-1)^{n\sim} - 1}{\pi 2^{(n\sim-1)} n\sim^2}$$

Notice that $A_{2n} = 0, n = 1, 2, \ldots$

```
> A(2*n);
```

$$0$$

while $A_{2n-1} = \frac{-2}{\pi 2^{2n-1}(2n-1)^2}, n = 1, 2, \ldots,$

```
> A(2*n-1);
```

$$-2\,\frac{1}{\pi 2^{(2n\sim-2)}(2n\sim-1)^2}$$

so the solution is

$$u(r,\theta) = \frac{\pi}{2} - \sum_{n=1}^{\infty} \frac{r^{2n-1}}{\pi 2^{2n-2}(2n-1)^2}\cos(2n-1)\theta$$

In the same manner as in previous examples, we graph an approximation of this solution.

```
> u:='u':
u:=(r,theta)->Pi/2-
      sum(A(2*n-1)*r^(2*n-1)*cos((2*n-1)*theta),n=1..20);
```

$$u := (r,\theta) \rightarrow \frac{1}{2}\pi - (\sum_{n=1}^{20} A(2n-1)r^{(2n-1)}\cos((2n-1)\theta))$$

```
> plot3d([r*cos(theta),r*sin(theta),u(r,theta)],r=0..2,
      theta=-Pi..Pi,axes=FRAME);
```

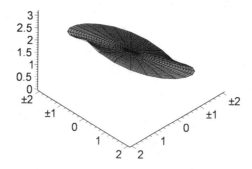

The Wave Equation in a Circular Region

One of the more interesting problems involving two spatial dimensions (x and y) is the wave equation.

$$c^2(u_{xx} + u_{yy}) = u_{tt}$$

This is due to the fact that the solution to this problem represents something with which we are all familiar, the displacement of a drumhead. Because most drumheads are circular in shape, we investigate the solution of the wave equation in a circular region. Therefore, we transform the wave equation into polar coordinates. In Section 10.4, we found that

$$u_{xx} + u_{yy} = u_{rr} + \frac{1}{r}u_r + \frac{1}{r^2}u_{\theta\theta}$$

Then, the wave equation in polar coordinates becomes

$$c^2\left(u_{rr} + \frac{1}{r}u_r + \frac{1}{r^2}u_{\theta\theta}\right) = u_{tt}$$

If we assume that the displacement of the drumhead from the xy-plane at time t is the same at equal distances from the origin, we say that the solution u is **radially symmetric.** (In other words, the value of u does not depend on the angle θ.) Therefore, $u_{\theta\theta} = 0$, so the wave equation can be expressed in terms of r and t only as

$$c^2\left(u_{rr} + \frac{1}{r}u_r\right) = u_{tt}$$

Of course, to find $u(r,t)$ we need the appropriate boundary and initial conditions. Because the circular boundary of the drumhead must be fixed so that it doesn't move, we say that

$$u(\rho, t) = 0 \text{ for } t > 0.$$

Then, as we had in Laplace's equation on a circular region, we require that the solution u be bounded near the origin, so we have the condition

$$|u(0, t)| < \infty \quad \text{for } t > 0.$$

The initial position and initial velocity functions are given as functions of r as

$$u(r, 0) = f(r) \quad \text{and} \quad u_t(r, 0) = g(r) \quad \text{for } 0 < r < \rho.$$

Therefore, the initial–boundary value problem to find the displacement u of a circular drumhead (of radius ρ is given by

$$\begin{cases} c^2(u_{rr} + \frac{1}{r}u_r) = u_{tt} \\ u(\rho, t) = 0, |u(0, t)| < \infty, t > 0 \\ u(r, 0) = f(r), u_t(r, 0) = g(r), 0 < r < \rho \end{cases}$$

As with other problems, we are able to use separation of variables to find u by assuming that $u(r, t) = R(r)T(t)$. Substitution into the wave equation yields

$$c^2\left(R''T + \frac{1}{r}R'T\right) = RT''$$

$$\frac{rR'' + R'}{rR} = \frac{T''}{c^2T} = -k^2,$$

where $-k^2$ is the separation constant. Separating the variables, we have the ordinary differential equations

$$r^2R'' + rR' + k^2r^2R = 0 \quad \text{and} \quad T'' + c^2k^2T = 0$$

We recognize the equation $r^2R'' + rR' + k^2r^2R = 0$ as Bessel's equation of order zero that has solution

$$R(r) = c_1J_0(kr) + c_2Y_0(kr),$$

where J_0 and Y_0 are the Bessel functions of order zero of the first and second kind, respectively. In terms of R, we express the boundary condition $|u(0,t)| < \infty$ as

$$|R(0)| < \infty.$$

Therefore, because $\lim\limits_{r \to 0^+} Y_0(kr) = -\infty$, we must choose $c_2 = 0$. Applying the other boundary condition $R(\rho) = 0$, we have

$$R(\rho) = c_1J_0(k\rho) = 0,$$

so to avoid the trivial solution with $c_1 = 0$, we have

$$k\rho = \alpha_n.$$

where α_n is the nth zero of J_0 that was discussed in Sections 4.7 and 9.4. Because k depends on n, we write

$$k_n = \alpha_n/\rho$$

The solution of $T'' + c^2k_n^2T = 0$ is

$$T_n(t) = A_n \cos ck_nt + B_n \sin ck_nt,$$

so with the principle of superposition, we have

$$u(r,t) = \sum_{n=1}^{\infty}(A_n \cos ck_nt + B_n \sin ck_nt)J_0(k_nr)$$

where the coefficients A_n and B_n are found through application of the initial position and velocity functions. With $u(r,0) = \sum_{n=1}^{\infty} A_nJ_0(k_nr) = f(r)$ and the orthogonality conditions of the Bessel functions, we find that

$$A_n = \frac{\int_0^\rho rf(r)J_0(k_nr)dr}{\int_0^\rho r[J_0(k_nr)]^2dr} = \frac{2}{[J_1(\alpha_n)]^2}\int_0^\rho rf(r)J_0(k_nr)dr, \qquad n = 1, 2, \ldots,$$

where we calculated the Bessel–Fourier series coefficients in Section 9.4. Similarly, because

$$u_t(r,t) = \sum_{n=1}^{\infty}(-ck_nA_n \sin ck_nt + ck_nB_n \cos ck_nt)J_0(k_nr),$$

we have $u_t(r,0) = \sum_{n=1}^{\infty} ck_nB_nJ_0(k_nr) = g(r)$. Therefore,

$$B_n = \frac{\int_0^{\rho} rg(r)J_0(k_nr)dr}{ck_n \int_0^{\rho} r[J_0(k_nr)]^2dr} = \frac{2}{ck_n[J_1(\alpha_n)]^2} \int_0^{\rho} rg(r)J_0(k_nr)dr, \qquad n = 1, 2, \dots$$

EXAMPLE 2: Solve $\begin{cases} u_{rr} + \frac{1}{r}u_{rr} = u_{tt} \\ u(1,t) = 0, |u(0,t)| < \infty, t > 0 \\ u(r,0) = r(r-1), u_t(r,0) = \sin \pi r, 0 < r < 1 \end{cases}$.

SOLUTION: In this case, $\rho = 1$, $f(r) = r(r-1)$, and $g(r) = \sin \pi r$. To calculate the coefficients, we will need to have approximations of the zeros of the Bessel functions, so we load the table of zeros that were found earlier in Chapter 9 and saved as `besseltable`. Then, for $1 \leq n \leq 8$, α_n is the nth zero of J_0.

```
> alpha:=array([2.4048, 5.5201, 8.6537, 11.792,
                14.931, 18.071, 21.212, 24.352]):
```

Next, we define the constants ρ and c and the functions $f(r) = r(r-1)$, $g(r) = \sin \pi r$, and $k_n = \alpha_n/\rho$.

```
> c:=1:
rho:=1:
f:=r->r*(r-1):
g:=r->sin(Pi*r):
k:=n->alpha[n]/rho:
```

The formulas for the coefficients A_n and B_n are then defined so that an approximate solution may be determined. Note that we use `evalf` together with `int` to approximate the coefficients and avoid the difficulties in integration associated with the presence of the Bessel function of order zero.

```
> a:=proc(n) option remember;
      2/BesselJ(1,alpha[n])^2*
          evalf(Int(r*f(r)*BesselJ(0,k(n)*r),
              r=0..rho))
      end:
```

```
b:=proc(n) option remember;

2/(c*k(n)*BesselJ(1,alpha[n])^2)*

    evalf(Int(r*g(r)*BesselJ(0,k(n)*r),

        r=0..rho))

end:
```

We now compute the first eight values of A_n and B_n. Because a and b are defined as procedures with `proc` using the option `remember`, Maple "remembers" these values for later use.

```
> array([seq([n,a(n),b(n)],n=1..8)]);
```

$$\begin{bmatrix} 1 & -.3235010276 & .5211819702 \\ 2 & .2084692034 & -.1457773395 \\ 3 & .007640292444 & -.01342290349 \\ 4 & .03838004574 & -.008330225220 \\ 5 & .005341000922 & -.002504216150 \\ 6 & .01503575901 & -.002082788164 \\ 7 & .003340078854 & -.0008805687932 \\ 8 & .007857367112 & -.0008134612340 \end{bmatrix}$$

The nth term of the series solution is defined in u. Then, an approximate solution is obtained in uapprox by summing the first eight terms of u.

```
> u:='u':n:='n':
u:=(n,r,t)->(a(n)*cos(c*k(n)*t)+b(n)*sin(c*k(n)*t))*

    BesselJ(0,k(n)*r):
```

```
> uapprox:=sum('u(n,r,t)','n'=1..8):
```

We graph uapprox for several values of t using `animate3d`.

```
> with(plots):
drumhead:=animate3d([r*cos(theta),r*sin(theta),

    uapprox],r=0..1,theta=-Pi..Pi,

    t=0..1.5,frames=9):

display(drumhead);
```

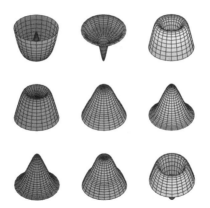

In order to actually watch the drumhead move, we can use `animate3d` to generate several graphs and animate the result. Be aware, however, that generating many three-dimensional graphics and then animating the results uses a great deal of memory and can take considerable time, even on a relatively powerful computer. We show one frame from the following animation.

```
> animate3d([r*cos(theta),r*sin(theta),
        uapprox],r=0..1,theta=-Pi..Pi,
        t=0..1.5,frames=9);
```

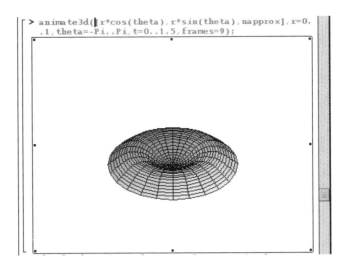

The problem that depends on the angle θ is more complicated to solve. Due to the presence of $u_{\theta\theta}$ we must include two more boundary conditions in order to solve the initial boundary value problem. So that the solution is a smooth function, we require the "artificial" boundary conditions

$$u(r, \pi, t) = u(r, -\pi, t) \quad \text{and} \quad u_\theta(r, \pi, t) = u_\theta(r, -\pi, t) \quad \text{for} \quad 0 < r < \rho \text{ and } t > 0..$$

Therefore, we solve the problem

$$\begin{cases} c^2(u_{rr} + \frac{1}{r}u_r + \frac{1}{r^2}u_{\theta\theta}) = u_{tt} \\ u(\rho, \theta, t) = 0, |u(0, \theta, t)| < \infty, -\pi < \theta < \pi, t > 0 \\ u(r, \pi, t) = u(r, -\pi, t), u_\theta(r, \pi, t) = u_\theta(r, -\pi, t), 0 < r < \rho, t > 0 \\ u(r, \theta, 0) = f(r, \theta), u_t(r, \theta, 0) = g(r, \theta), 0 < r < \rho, -\pi < \theta < \pi \end{cases}$$

if the displacement of the drumhead is not radially symmetric. Assuming that $u(r, \theta, t) = R(r)H(\theta)T(t)$ and using separation of variables yields

$$c^2\left(R''HT + \frac{1}{r}R'HT + \frac{1}{r^2}RH''T\right) = RHT''$$

Then, division by RHT gives us

$$c^2\left(\frac{R''}{R} + \frac{1}{r}\frac{R'}{R} + \frac{1}{r^2}\frac{H''}{H}\right) = \frac{T''}{T} = -\lambda^2$$

where $-\lambda^2$ is the constant of separation. Separating variables, we obtain

$$T'' + \lambda^2 c^2 T = 0 \quad \text{and} \quad \frac{R''}{R} + \frac{1}{r}\frac{R'}{R} + \frac{1}{r^2}\frac{H''}{H} = -\lambda^2.$$

Separating variables in the second equation yields

$$r^2\frac{R''}{R} + r\frac{R'}{R} + r^2\lambda^2 = \frac{-H''}{H} = \mu^2$$

where μ^2 is the constant of separation. Therefore, we have two more ordinary differential equations

$$H'' + \mu^2 H = 0 \quad \text{and} \quad r^2 R'' + rR' + (r^2\lambda^2 - \mu^2)R = 0$$

The boundary conditions in terms of H and R become $R(\rho) = 0, |R(0)| < \infty, H(-\pi) = H(\pi)$, and $H'(-\pi) = H'(\pi)$. Recall that the problem

$$\begin{cases} H'' + \mu^2 H = 0 \\ H(-\pi) = H(\pi), H'(-\pi) = H'(\pi) \end{cases}$$

has solutions

$$H_n(x) = \begin{cases} 1, n = 0 \\ a_n \cos n\theta + b_n \sin n\theta \end{cases}$$

which correspond to the eigenvalues $\mu_n^2 = \begin{cases} 0, n = 0 \\ n^2, n = 1, 2, \ldots \end{cases}$

The corresponding solutions of $r^2R'' + rR' + (r^2\lambda^2 - \mu^2)R = 0$, which we recognize as Bessel's equation of order n, are $R_n(r) = c_nJ_n(\lambda r) + d_nY_n(\lambda r)$. Because $|R(0)| < \infty, d_n = 0$. Then, because $R(\rho) = 0, J_n(\lambda\rho) = 0$, so $\lambda_{mn} = \alpha_{mn}/\rho$, where α_{mn} denotes the nth zero of the Bessel function J_m. Then, the solution is

$$u(r, \theta, t) = \sum_n a_{0n}J_0(\lambda_{0n}r)\cos(\lambda_{0n}ct) + \sum_{m,n} a_{mn}J_m(\lambda_{mn}r)\cos(m\theta)\cos(\lambda_{mn}ct)$$

$$+ \sum_{m,n} b_{mn}J_m(\lambda_{mn}r)\sin(m\theta)\cos(\lambda_{mn}ct) + \sum_n A_{0n}J_0(\lambda_{0n}r)\sin(\lambda_{0n}ct)$$

$$+ \sum_{m,n} A_{mn}J_m(\lambda_{mn}r)\cos(m\theta)\sin(\lambda_{mn}ct)$$

$$+ \sum_{m,n} B_{mn}J_m(\lambda_{mn}r)\sin(m\theta)\sin(\lambda_{mn}ct).$$

where

$$a_{0n} = \frac{\int_0^{2\pi}\int_0^{\rho} f(r, \theta)J_0(\lambda_{0n}r)r\,dr\,d\theta}{2\pi\int_0^{\rho}[J_0(\lambda_{0n}r)]^2 r\,dr},$$

$$a_{mn} = \frac{\int_0^{2\pi}\int_0^{\rho} f(r, \theta)J_m(\lambda_{mn}r)\cos(m\theta)r\,dr\,d\theta}{\pi\int_0^{\rho}[J_m(\lambda_{mn}r)]^2 r\,dr},$$

$$b_{mn} = \frac{\int_0^{2\pi}\int_0^{\rho} f(r, \theta)J_m(\lambda_{mn}r)\sin(m\theta)r\,dr\,d\theta}{\pi\int_0^{\rho}[J_m(\lambda_{mn}r)]^2 r\,dr},$$

$$A_{0n} = \frac{\int_0^{2\pi}\int_0^{\rho} g(r, \theta)J_0(\lambda_{0n}r)r\,dr\,d\theta}{2\pi\lambda_{0n}c\int_0^{\rho}[J_0(\lambda_{0n}r)]^2 r\,dr},$$

$$A_{mn} = \frac{\int_0^{2\pi}\int_0^{\rho} g(r, \theta)J_m(\lambda_{mn}r)\cos(m\theta)r\,dr\,d\theta}{\pi\lambda_{mn}c\int_0^{\rho}[J_m(\lambda_{mn}r)]^2 r\,dr},$$

$$B_{mn} = \frac{\int_0^{2\pi}\int_0^{\rho} g(r, \theta)J_m(\lambda_{mn}r)\sin(m\theta)r\,dr\,d\theta}{\pi\lambda_{mn}c\int_0^{\rho}[J_m(\lambda_{mn}r)]^2 r\,dr}.$$

EXAMPLE 3: Solve $\begin{cases} 10^2\left(u_{rr} + \frac{1}{r}u_r + \frac{1}{r^2}u_{\theta\theta}\right) = u_{tt} \\ u(1, \theta, t) = 0, |u(0, \theta, t)| < \infty, -\pi < \theta < \pi, t > 0 \\ u(r, \pi, t) = u(r, -\pi, t), u_\theta(r, \pi, t) = u_\theta(r, -\pi, t), 0 < r < 1, t > 0 \\ u(r, \theta, 0) = \cos\left(\frac{\pi r}{2}\right)\sin\theta, u_t(r, \theta, 0) = (r - 1)\cos\left(\frac{\pi\theta}{2}\right) \end{cases}$

SOLUTION: The table of zeros that were found earlier in Chapter 9 and saved as `besseltable` are read in and called `getzeros`. A function `alpha` is then defined so that these zeros of the Bessel functions can be more easily obtained from the list.

```
> ALPHA:=array([

        [2.4048, 5.5201, 8.6537, 11.792,

            14.931, 18.071, 21.212, 24.352],

        [3.8317, 7.0156, 10.173, 13.324,

            16.471, 19.616, 22.760, 25.904],
```

```
        [5.1356, 8.4172, 11.620, 14.796,

            17.960, 21.117, 24.270, 27.421],

    [6.3802, 9.7610, 13.015, 16.223,

            19.409, 22.583, 25.748, 28.908],

    [7.5883, 11.065, 14.373, 17.616,

            20.827, 24.019, 27.199, 30.371],

    [8.7715, 12.339, 15.700, 18.980,

            22.218, 25.430, 28.627, 31.812],

    [9.9361, 13.589, 17.004, 20.321,

            23.586, 26.820, 30.034, 33.233]]):
```

```
> alpha:=table():
for i from 0 to 6 do
        for j from 1 to 8 do
                alpha[i,j]:=ALPHA[i+1,j] od od:
```

The appropriate parameter values as well as the initial condition functions are defined as follows. Notice that the functions describing the initial displacement and velocity are defined as the product of functions. This enables the subsequent calculations to be carried out using `evalf` and `int`.

```
> c:=10:
rho:=1:f:='f':
f1:=r->cos(Pi*r/2):
f2:=theta->sin(theta):
f:=proc(r,theta) option remember;
        f1(r)*f2(theta)
        end:
```

```
> g1:=r->r-1:
g2:=theta->cos(Pi*theta/2):
g:=proc(r,theta) option remember;
        g1(r)*g2(theta)
        end:
```

The coefficients a_{0n} are determined with the function a0.

```
a0:=proc(n) option remember;

    evalf(Int(f1(r)*BesselJ(0,alpha[0,n]*r)*r,

         r=0..rho)*

    Int(f2(t),t=0..2*Pi)/

    (2*Pi*Int(r*BesselJ(0,alpha[0,n]*r)^2,

         r=0..rho)))

end:
```

We use `seq` to generate the first five values of a0. Because a0 was defined as a procedure using the option `remember`, Maple will not need to recompute these values when they are called later. Notice that we interpret these values to be all zero.

```
> seq(a0(n),n=1..5);
```

$$.128649336710^{-16}, -.769800602510^{-18}, .230864886310^{-18},$$

$$- .104530467110^{-18}, .570856148010^{-19}$$

Because the denominator of each integral formula used to find a_{mn} and b_{mn} is the same, the function `bjmn` that, computes this value is defined next. A table of nine values of this coefficient is then determined.

```
> bjmn:=proc(m,n) option remember;

    evalf(Int(r*BesselJ(m,alpha[m,n]*r)^2,

         r=0..rho))

end:
```

```
> seq(seq(bjmn(m,n),m=1..3),n=1..3);
```

$$.08110781816, .05768792844, .04448295755, .04503456132,$$

$$.03682464114, .03110451636, .03117913218, .02701416557,$$

$$.02382360434$$

We also note that in evaluating the numerators of a_{mn} and b_{mn}, we must compute $\int_0^\rho r f_1(r) J_m(\alpha_{mn}r)dr$. This integral is defined in `fbjmn` and the corresponding values are found for $n = 1, 2, 3$ and $m = 1, 2, 3$.

```
> fbjmn:=proc(m,n) option remember;
```

```
        evalf(Int(f1(r)*BesselJ(m,alpha[m,n]*r)*r,

            r=0..rho))

    end:
```

```
> seq(seq(fbjmn(m,n),m=1..3),n=1..3);
```

.1035741366, .07909488454, .06289255442, .02051373335,

.02755702171, .02907659382, .01039761263, .01503815332,

.01720000672

The formula to compute a_{mn} is then defined and uses the information calculated in fbjmn and bjmn. As in the previous calculation, the coefficient values for $n = 1, 2, 3$ and $m = 1, 2, 3$ are determined.

```
> a:=proc(m,n) option remember;
        evalf(fbjmn(m,n)*Int(f2(t)*cos(m*t),
            t=0..2*Pi)/(Pi*bjmn(m,n)))
    end:
```

```
> seq(seq(a(m,n),m=1..3),n=1..3);
```

$-.189151065210^{-15}, -.626177991710^{-16}, .100752847110^{-15},$

$-.674712813310^{-16}, -.341765407110^{-16}, .666149681110^{-16},$

$-.493957744410^{-16}, -.254236146910^{-16}, .514485212210^{-16}$

A similar formula is then defined for the computation of b_{mn}.

```
> b:=proc(m,n) option remember;
evalf(fbjmn(m,n)*Int(f2(t)*sin(m*t),
            t=0..2*Pi)/(Pi*bjmn(m,n)))
    end:
```

```
> seq(seq(b(m,n),m=1..3),n=1..3);
```

$$1.276993254, .676536393810^{-16}, -.697643081110^{-16}, .4555108954,$$

$$.369250818610^{-16}, -.461262117610^{-16}, .3334798599,$$

$$.274682292110^{-16}, -.356245068010^{-16}$$

The values of A_{0n} are found similarly to those of a_{0n}. After defining the function capa0 to calculate these coefficients, a table of values is then found.

```
> capa0:=proc(n) option remember;
     evalf(Int(g1(r)*BesselJ(0,alpha[0,n]*r)*r,
          r=0..rho)*
     Int(g2(t),t=0..2*Pi)/
     (2*Pi*c*alpha[0,n]*
          Int(r*BesselJ(0,alpha[0,n]*r)^2,
          r=0..rho)))
     end:
```

```
> seq(capa0(n),n=1..6);
```

$$.001422305064, .00005424800235, .00002675839648,$$

$$.640860642910^{-5}, .495941750710^{-5}, .188636410710^{-5}$$

The value of the integral of the component of g, g1, which depends on r and the appropriate Bessel functions, is defined as gbjmn.

```
> gbjmn:=proc(m,n) option remember;
     evalf(Int(g1(r)*BesselJ(m,alpha[m,n]*r)*r,r=0..rho))
          end:
```

```
> seq(seq(gbjmn(m,n),m=1..3),n=1..3);
```

$$-.07439063140, -.05543795341, -.04336144966, -.01949086495,$$

$$-.02279800021, -.02267782604, -.009892715342, -.01303894855,$$

$$-.01416851159$$

Then, A_{mn} is found by taking the product of integrals, `gbjmn` depending on r and one depending on θ. A table of coefficient values is generated in this case as well.

```
> capa:=proc(m,n) option remember;
      evalf(gbjmn(m,n)*Int(g2(t)*cos(m*t),
            t=0..2*Pi)/(Pi*alpha[m,n]*c*bjmn(m,n)))
   end:
```

```
> seq(seq(capa(m,n),m=1..3),n=1..3);
```

$.003509601162, -.002626909580, -.0005031904977, .0009045115035,$

$-.001032534693, -.0002460029071, .0004572947161, -.0005831213673,$

$-.0001504974617$

Similarly, the B_{mn} are determined.

```
> capb:=proc(m,n) option remember;
      evalf(gbjmn(m,n)*Int(g2(t)*sin(m*t),
            t=0..2*Pi)/(Pi*alpha[m,n]*c*bjmn(m,n)))
   end:
```

```
> seq(seq(capb(m,n),m=1..3),n=1..3);
```

$.009879439153, -.01478936902, -.004249405850, .002546177172,$

$-.005813118470, -.002077476020, .001287273144, -.003282944015,$

$-.001270939727$

Now that the necessary coefficients have been found, we must construct the approximate solution to the wave equation by using our results. In the following, `term1` represents the sum of the first five terms of the expansion involving a_{0n}, `term2` the sum of the first nine terms involving a_{mn}, `term3` the sum of the first nine terms involving b_{mn}, `term4` the sum of the first five terms involving A_{0n}, `term5` the sum of the first nine terms involving A_{mn}, and `term6` the sum of the first nine terms involving B_{mn}. We interprety `term1` and `term2` to be zero.

```
> term1:=sum('a0(n)*BesselJ(0,alpha[0,n]*r)*
            cos(alpha[0,n]*c*t)',n=1..5);
```

$$term1 := .128649336710^{-16}BesselJ(0, 2.4048r)\cos(24.0480t)$$

$$- .769800602510^{-18}BesselJ(0, 5.5201r)\cos(55.2010t)$$

$$+ .230864886310^{-18}BesselJ(0, 8.6537r)\cos(86.5370t)$$

$$- .104530467110^{-18}BesselJ(0, 11.792r)\cos(117.920t)$$

$$+ .570856148010^{-19}BesselJ(0, 14.931r)\cos(149.310t)$$

```
> n:='n':m:='m':
term2:=sum('sum('a(m,n)*BesselJ(m,alpha[m,n]*r)*
      cos(m*theta)*cos(alpha[m,n]*c*t)',n=1..3)',m=1..3);
```

$$term2 :=$$

$$- .189151065210^{-15}BesselJ(1, 3.8317r)\cos(\theta)\cos(38.3170t)$$

$$- .674712813310^{-16}BesselJ(1, 7.0156r)\cos(\theta)\cos(70.1560t)$$

$$- .493957744410^{-16}BesselJ(1, 10.173r)\cos(\theta)\cos(101.730t)$$

$$- .626177991710^{-16}BesselJ(2, 5.1356r)\cos(2\theta)\cos(51.3560t)$$

$$- .341765407110^{-16}BesselJ(2, 8.4172r)\cos(2\theta)\cos(84.1720t)$$

$$- .254236146910^{-16}BesselJ(2, 11.620r)\cos(2\theta)\cos(116.200t)$$

$$+ .100752847110^{-15}BesselJ(3, 6.3802r)\cos(3\theta)\cos(63.8020t)$$

$$+ .666149681110^{-16}BesselJ(3, 9.7610r)\cos(3\theta)\cos(97.6100t)$$

$$+ .514485212210^{-16}BesselJ(3, 13.015r)\cos(3\theta)\cos(130.150t)$$

```
> n:='n':m:='m':
term3:=sum('sum('b(m,n)*BesselJ(m,alpha[m,n]*r)*
      sin(m*theta)*cos(alpha[m,n]*c*t)',n=1..3)',m=1..3);
```

```
> n:='n':
term4:=sum('capa0(n)*BesselJ(0,alpha[0,n]*r)*
      sin(alpha[0,n]*c*t)',n=1..5);
```

```
> n:='n':m:='m':
term5:=sum('sum('capa(m,n)*BesselJ(m,alpha[m,n]*r)*
      cos(m*theta)*sin(alpha[m,n]*c*t)',n=1..3)',m=1..3);
```

```
> n:='n':m:='m':
```

```
term6:=sum('sum('capb(m,n)*BesselJ(m,alpha[m,n]*r)*
              sin(m*theta)*sin(alpha[m,n]*c*t)',n=1..3)',m=1..3);
```

An approximate solution is given as the sum of these terms as computed in `u`. Notice that we do not include `term1` and `term2` because they are zero.

```
> u:=term3+term4+term5+term6:
```

The position of the waves on the circular region can be viewed with Maple through the use of polar coordinates (in the *xy*-plane) and `animate3d`. We plot the membrane for $t = 0$ to $t = 1$ using increments of $1/8$. Hence, a sequence of nine three-dimensional plots is the result.

```
> with(plots):
somegraphs:=animate3d([r*cos(theta),r*sin(theta),u],
       r=0..1,theta=-Pi..Pi,t=0..1,frames=9):
display(somegraphs);
```

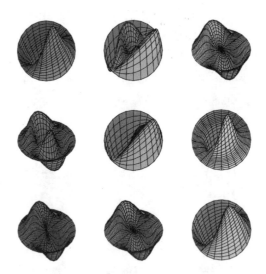

Of course, we can generate many graphs with `animate3d` and animate the result as in Example 2. Be aware, however, that generating many three-dimensional graphics and then animating the results uses a great deal of memory and can take considerable time, even on a relatively powerful computer.

■

Getting Started

Introduction to Maple V

Maple V is one of the most powerful and reliable systems for doing mathematics by computer currently available. Editions of Maple and Maple V have been available for more then 10 years. The over 100,000 users of Maple V consist of scientists, engineers, mathematicians, teachers, and students. Maple V is easy to use, accurate, and expandable. Although Maple V has modest memory requirements, with the more than 2500 math routines available, Maple V can perform symbolic and numeric mathematics, generate two- and three-dimensional graphics and animations, display output using standard mathematical notation, and provide a high-level programming language similar to Pascal.

Maple V is available for nearly every operating system including Macintosh, MS Windows, MS DOS, UNIX, VMS, NeXT, Ultrix, and UNICOS. Because Maple V worksheets share a common format they can be taken from one platform to another.

A Note Regarding Different Versions of Maple

With Release 5 of Maple V, many new functions and features have been added to Maple V. Particular examples of features available with Release 5 but not some previous releases include standard mathematical notation for output, contour and implicit plots, animation of graphics, and the interactive help browser. We encourage users of earlier releases to update to Release 5 as soon as they can. All examples in *Differential Equations with Maple V*, second edition, were done using Release 5. In most cases, the same results will be obtained if you are using Release 4. Occasionally, however, if particular features of Release 5 are illustrated in an example, of course, these features are not available with earlier releases. If you are using an earlier or later version of Maple, your results may not appear in a form identical to those found in this book: some commands found in Release 5 are not available in earlier versions of Maple; in later versions some commands will certainly be changed, new commands added, and obsolete commands removed.

In this text, we assume that Maple V has been correctly installed on the computer you are using. If you need to install Maple V on your computer, please refer to the documentation that came with the Maple V software package.

Getting Started with Maple V

After the Maple V program has been properly installed, a user can access Maple V. Here we briefly describe methods of starting Maple V on several platforms.

Macintosh: In the same manner as folders are opened, the Maple application can be started by selecting the Maple icon, going to **File,** and selecting **Open** or by simply clicking twice on the Maple icon. Of course, opening an existing Maple document also opens the Maple program.

Windows: To start Maple V for Windows, double-click on the Maple V for Windows application icon. This operation opens Maple V for Windows with a blank worksheet.

DOS: To run Maple, type the command **MAPLE** at the DOS prompt. After a few seconds, the screen will clear, the Maple logo will appear briefly, a prompt will appear at the bottom of the screen above the status line, and you will be in the Maple session. You can now begin typing Maple commands.

UNIX: To start Maple V, type **xmaple** at the UNIX prompt. A new window will open, containing Maple V Release 5. You can now begin typing Maple commands.

If you are using a text-based interface (such as UNIX), Maple V is started with the operating system command `maple` or `xmaple`. If you are using a notebook interface (such as Macintosh, Windows, or NeXT), Maple V is started by selecting the Maple V icon and double-clicking or selecting the Maple V icon and selecting **Open** from the **File** menu.

Once Maple V has been started, computations can be carried out immediately. Maple V commands are typed to the right of the prompt, a semicolon is placed at the end of the command, and the command is then evaluated by pressing **ENTER.** Generally, when a colon is placed at the end of the command, the resulting output is not displayed. Note that pressing **ENTER** evaluates commands and pressing **RETURN** yields a new line. Output is displayed below input. We illustrate some of the typical steps involved in working with Maple V in the following calculations. In each case, we type the command, place a semicolon at the end of the command, and press **ENTER.** Maple V evaluates the command, displays the result, and inserts a new prompt. For example, entering

```
> evalf(Pi,50);
```

$$3.1415926535897932384626433832795028841971693993751$$

returns a 50-digit approximation of π.

The next calculation can then be typed and entered in the same manner as the first. For example, entering

```
> solve(x^3-2*x+1=0);
```

$$1, -\frac{1}{2} + \frac{1}{2}\sqrt{5}, -\frac{1}{2} - \frac{1}{2}\sqrt{5}$$

solves the equation $x^3 - 2x + 1 = 0$ for x. Subsequent calculations are entered in the same way. For example, entering

```
> plot({sin(x),2*cos(2*x)},x=0..3*Pi);
```

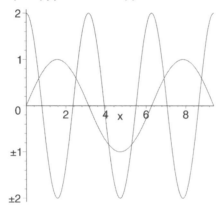

graphs the functions $\sin x$ and $2 \cos 2x$ on the interval $[0, 3\pi]$. Similarly, entering

```
> plot3d(sin(x+cos(y)),x=0..4*Pi,y=0..4*Pi);
```

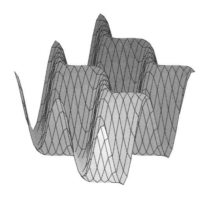

graphs the function $\sin(x + \cos y)$ on the rectangle $[0, 4\pi] \times [0, 4\pi]$.

Notice that you can change how your input and output appear in your Maple notebook by going to **Options** under the menu and selecting **Input Display**

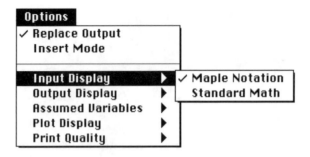

or **Output Display**.

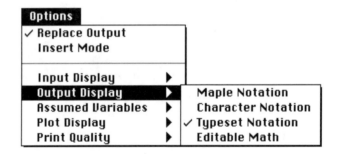

In this command, both the input and the output are in **Maple Notation**.

```
Int(x^2*sin(x)^2,x)=int(x^2*sin(x)^2,x);
                    Int(x^2*sin(x)^2,x)=

        x^2*(-1/2*cos(x)*sin(x)+1/2*x)-

1/2*x*cos(x)^2+1/4*cos(x)*sin(x)+1/4*x-1/3*x^3
```

In this case, the input is in **Maple Notation** and the output is in **Character Notation**.

```
Int(x^2*sin(x)^2,x)=int(x^2*sin(x)^2,x);
```

```
         /

        | 2       2       2

        | x  sin(x)  dx  =  x  (- 1/2 cos(x)  sin(x)  + 1/2 x)

        |

         /

                  2                                           3
        - 1/2 x cos(x)  + 1/4 cos(x)  sin(x)  + 1/4 x  - 1/3 x
```

Here, the output is in **Typeset Notation.**

```
Int(x^2*sin(x)^2,x)=int(x^2*sin(x)^2,x);
```

$$\int x^2 \sin(x)^2 dx =$$

$$x^2\left(-\frac{1}{2}\cos(x)\sin(x)+\frac{1}{2}x\right)-\frac{1}{2}x\cos(x)^2+\frac{1}{4}\cos(x)\sin(x)+\frac{1}{4}x-\frac{1}{3}x^3$$

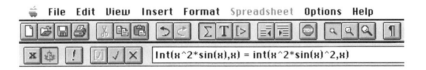

$$\int x^2 \sin(x)^2 dx = \int x^2 \sin(x)^2 dx$$

$$\int x^2 \sin(x)^2 dx = x^2\left(-\frac{1}{2}\cos(x)\sin(x)+\frac{1}{2}x\right)-\frac{1}{2}x\cos(x)^2$$

$$+\frac{1}{4}\cos(x)\sin(x)+\frac{1}{4}x-\frac{1}{3}x^3$$

Although **Editable Math** notation looks the same as **Typeset Notation,** portions of **Editable Math** notation can be selected, copied, and pasted elsewhere in your Maple notebook.

This book includes real input and output from Maple V Release 5. Appearances of input and output may vary depending on the version of Maple used, the fonts used to display input and output, the quality of the monitor, and the resolution and type of printer used to print the Maple worksheet: the results displayed on your computer may not be physically identical to those shown here.

Maple V sessions are terminated by entering quit, done, or stop. On several platforms with notebook interfaces (such as Macintosh, Windows, and NeXT), Maple V sessions are ended by selecting **Quit** from the **File** menu or by using the keyboard shortcut ⌘ **Q,** as with other applications. They can be saved by referring to ⌘ **S** from the **File** menu.

On these platforms, input and text regions in notebook interfaces can be edited. Editing input can create a notebook in which the mathematical output does not make sense in the sequence in which it appears. It is also possible simply to go into a notebook and alter input without doing any recalculation. This also creates misleading notebooks. Hence, common sense and caution should be used when editing the input regions of notebooks. Recalculating all commands in the notebook will clarify any confusion.

In order for the Maple V user to take full advantage of the capabilities of this software, an understanding of its syntax is imperative. Although all of the rules of Maple V syntax are far too numerous to list here, knowledge of the following five rules equips the beginner with the necessary tools to start using the Maple V program with little trouble.

Five Basic Rules of Maple V Syntax

1. The arguments of functions are given in parentheses (...).
2. A semicolon (;) or colon (:) must be included at the end of each command.
3. Multiplication is represented by a *.
4. Powers are denoted by a^.
5. If you get no response or an incorrect response, you may have entered or executed the command incorrectly. In some cases, the amount of memory allocated to Maple V can cause a crash; like people, Maple V is not perfect and some errors can occur.

Loading Miscellaneous Library Functions and Packages

Maple V's modularity, which gives Maple a great deal of flexibility, helps minimize Maple's memory requirements. Nevertheless, although Maple contains many built-in functions that are loaded immediately when called, some other functions are contained in packages that must be loaded separately. Other functions, miscellaneous library functions, must also be loaded separately.

Loading Miscellaneous Library Functions

Miscellaneous library commands must be loaded with the command readlib (command) before using them during a Maple session. We show this with the following example by first listing the miscellaneous library functions.

We use Maple's help facility to view a list of the miscellaneous library functions. Entering the command

> ?index[misc]

displays the following help window, which lists the miscellaneous library functions.

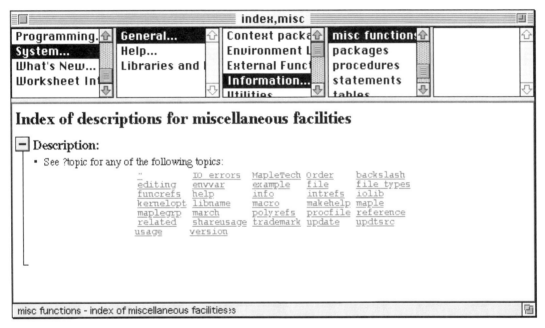

For example, we let $f(z) = \frac{5z-2}{z^2-z}$. Then, by the residue theorem, the value of the integral $\int_C f(z)dz$, where C is the circle $|z| = 2$, is $2\pi i(res(f, 0) + res(f, 1))$, where $res(f,0)$ and $res(f,1)$ denote the residue of $f(z)$ at $z = 0$ and $z = 1$, respectively. We can use Maple to compute these values by loading the command residue, defining f, and then evaluating residue (f, z = 0) and residue (f, z = 1), which compute the residue of $f(z)$ at $z = 0$ and $z = 1$, respectively.

We first load the command residue with

readlib(residue)

and then define *f*.

```
> readlib(residue):
f:=(5*z-2)/(z^2-z);
```

$$f := \frac{5z - 2}{z^2 - z}$$

Next, we compute the desired residues.

```
> r0:=residue(f,z=0);
r1:=residue(f,z=1);
```

$$r0$$
$$r1 := 3$$

Finally, we apply the residue theorem to evaluate the integral.

```
> 2*Pi*I*(r0+r1);
```

$$10I\pi$$

The following command calculates the residue of $\frac{\cot z}{z^4}$ if $z = 0$.

```
> residue(cot(z)/z^4,z=0);
```

$$\frac{-1}{45}$$

Loading Packages

In addition to the standard library functions and miscellaneous library functions just described, a tremendous number of additional commands are available in various packages. We use Maple's help facility by entering ?packages to view a list of the available packages. Entering the command

```
> ?index[packages]
```

displays the following window, which lists the available packages.

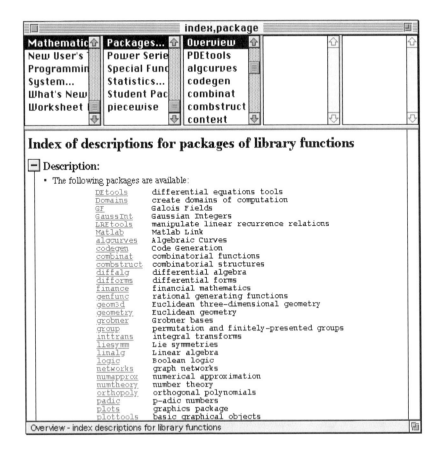

Information about particular packages is obtained with `?packagename`. For example, with the command `?student` we obtain information about the `student` calculus package, including a list of the commands contained in the package.

```
> ?student
```

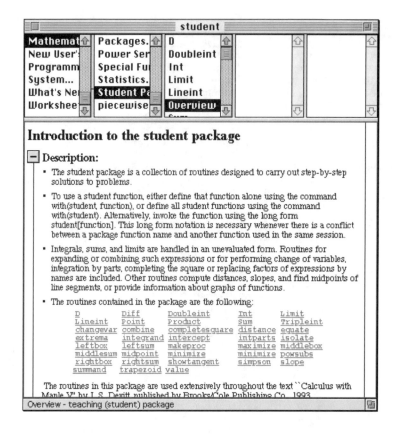

A typical example of a package is the linear algebra package, which is called **linalg**. For example, to compute the determinant of the matrix $\mathbf{A} = \begin{pmatrix} 10 & -6 & -9 \\ 6 & -5 & -7 \\ -10 & 9 & 12 \end{pmatrix}$, we must first define the matrix and then select a Maple command, or define a new one, which computes the determinant of a square matrix.

In this case, we first define **A** and then use the command linalg [det] to compute the determinant of **A**. Note that the command det is contained in the package **linalg**.

```
> A:=matrix([[10,-6,-9],[6,-5,-7],[-10,9,12]]);
```

$$A := \begin{bmatrix} 10 & -6 & -9 \\ 6 & -5 & -7 \\ -10 & 9 & 12 \end{bmatrix}$$

```
> linalg[det](A);
```

$$6$$

However, we can use the with command to load a package. After a package has been loaded, subsequent calculations involving commands contained in the package can be entered directly.

In the following, we use with to load the package **linalg**. The commands contained in the linalg package are displayed after the package is loaded. If a colon were used at the end of the command instead of a semicolon, then the package commands would be loaded but not listed.

```
> with(linalg);
```

[BlockDiagonal, GramSchmidt, JordanBlock, LUdecomp, QRdecomp, Wronskian, addcol, addrow, adj, adjoint, angle, augment, backsub, band, basis, bezout, blockmatrix, charmat, charpoly, cholesky, col, coldim, colspace, colspan, companion, concat, cond, copyinto, crossprod, curl, definite, delcols, delrows, det, diag, diverge, dotprod, eigenvals, eigenvalues, eigenvectors, eigenvects, entermatrix, equal, exponential, extend, ffgausselim, fibonacci, forwardsub, frobenius, gausselim, gaussjord, geneqns, genmatrix, grad, hadamard, hermite, hessian, hilbert, htranspose, ihermite, indexfunc, innerprod, intbasis, inverse, ismith, issimilar, iszero, jacobian, jordan, kernel, laplacian, leastsqrs, linsolve, matadd, matrix, minor, minpoly, mulcol, mulrow, multiply, norm, normalize, nullspace, orthog, permanent, pivot, potential, randmatrix, randvector, rank, ratform, row, rowdim, rowspace, rowspan, rref, scalarmul, singularvals, smith, stackmatrix, submatrix, subvector, sumbasis, swapcol, swaprow, sylvester, toeplitz, trace, transpose, vandermonde, vecpotent, vectdim, vector, wronskian]

We can then compute the determinant of **A** using the command det instead of linalg[det].

```
> det(A);
```

$$6$$

Getting Help from Maple V

Help Commands

Becoming competent with Maple can take a serious investment of time. Hopefully, messages that result from syntax errors are viewed lightheartedly. Ideally, instead of becoming frustrated, beginning Maple users will find it challenging and fun to locate the source of errors. Frequently, Maple's error messages indicate where the error(s) has (have) occurred. In this process, it is natural that one will become more proficient with Maple.

One way to obtain information about commands and functions is with the commands ? and help. In general, ?f and help(f) give information on the Maple function f. This information appears in a separate window.

EXAMPLE 1: Use ? to obtain information about the command `help`.

SOLUTION:

> ?help

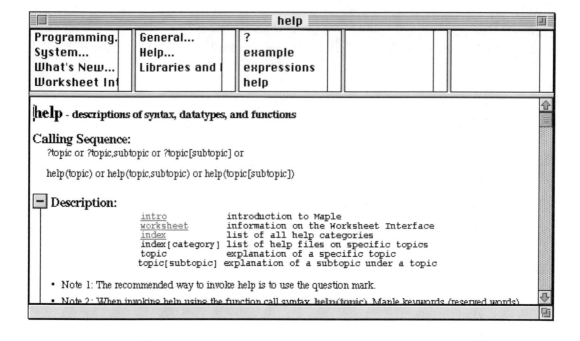

Alternatively, Maple's help facility can be accessed from the Maple menu. In this case, we select **Using Help**

which presents a window that describes Maple's online help facility.

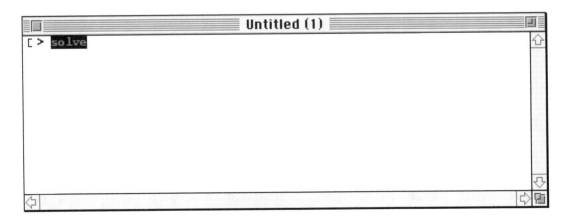

How to Use the Help System

Maple V contains a complete online help system you can use to find information on a specific topic easily and to explore the wide range of commands available:

- Explore help topics using the Help Browser.
- Locate information on Maple's commands and features by using the Topic Search feature.
- Locate help pages that contain specific words of your choice using the Full Text Search feature.
- Follow *hyperlinks* on help pages to refer to related topics easily.

[+] Browsing the Help System

[+] Searching for Help Topics

[+] Searching for Phrases Contained in Help Pages

[+] Going Back to Previously Viewed Topics

Additional Ways of Obtaining Help from Maple V

The Maple Menu offers other ways of obtaining help. For example, if a command is selected, and then **Help for Context** is selected from the **Windows** menu, Maple displays the help window for that command. This is illustrated have with the solve command. We begin by typing and selecting the solve command.

Untitled (1)

[> solve

We then select **Help for Context** from the **Windows** menu. Notice that in this case the word **Context** is replaced by the word **solve**.

The resulting help window is displayed next.

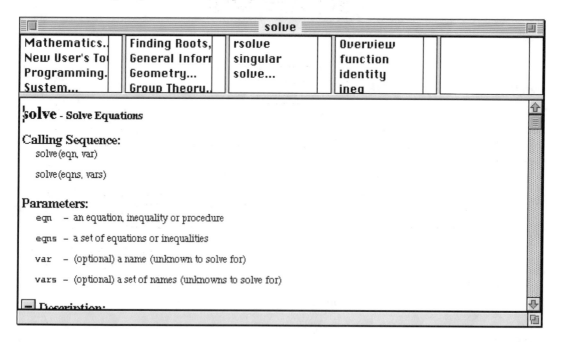

In this particular case, we wish to solve the literal equation $v = \frac{1}{3}\pi h(r^2 + rR + R^2)$ for R. We scroll through the help window and locate an example similar to the example we wish to solve. We then complete the typing of the command and enter the result. The output corresponds to the solution of $v = \frac{1}{3}\pi h(r^2 + rR + R^2)$ for R.

```
> solve(v=1/3*Pi*h*(r^2+r*R+R^2),R);
```

$$\frac{1}{2}\frac{-\pi hr + \sqrt{-3\pi^2 h^2 r^2 + 12\pi hv}}{\pi h}, \frac{1}{2}\frac{-\pi hr - \sqrt{-3\pi^2 h^2 r^2 + 12\pi hv}}{\pi h}$$

Assistance can also be obtained by selecting **Topic Search...** or **Full Text Search...** from the **Help** menu.

Selecting **Topic Search...** yields the following help window:

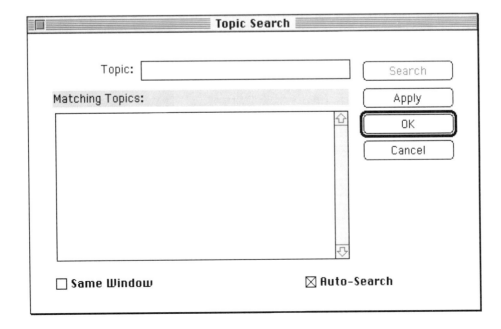

Selected References

Abbasian, R. and Ionescu, A., *Vector Calculus with Maple*, McGraw-Hill, 1996.

Abell, Martha L. and Braselton, James P., *Maple V by Example*, Second Edition, Academic Press, 1999.

Abell, Martha L. and Braselton, James P., *Modern Differential Equations: Theory, Applications, Technology*, Saunders College Publishing, 1996.

Abell, M. and Braselton, J., *Student Resource Manual for Modern Differential Equations: Theory, Applications, Technology*, Saunders College Publishing, 1996.

Adams, S., *Maple Talk*, Prentice Hall, 1996.

Andersson, G., *Applied Mathematics with Maple*, Chartwell-Bratt, 1997.

Articolo, G., *Partial Differential Equations and Boundary Value Problems with Maple V*, Academic Press, 1998.

Bauldry, W., Evans, B., and Johnson, J., *Linear Algebra with Maple*, John Wiley & Sons, 1995.

Baylis, W., *Theoretical Methods in the Physical Sciences: An Introduction to Problem Solving Using Maple V*, Birkhäuser, 1994.

Blachman, N. and Mossinghoff, M., *Maple V Quick Reference*, Brooks/Cole, 1995.

Borreli, Robert L. and Coleman, Courtney S., *Differential Equations: A Modeling Perspective*, Preliminary Edition, John Wiley & Sons, 1996.

Carlson, J. and Johnson, J., *Multivariable Mathematics with Maple: Linear Algebra, Vector Calculus and Differential Equations*, Prentice Hall, 1997.

Cheney, Ward and Kincaid, David, *Numerical Mathematics and Computing*, Second Edition, Brooks/Cole, 1985.

Churchill, Ruel V., *Operational Mathematics*, Third Edition, McGraw-Hill, 1972.

Coombes, K., Hunt, B., Lipsman, R., Osborn, J., and Stuck, G., *Differential Equations with Maple*, Second Edition, John Wiley & Sons, 1997.

Corless, R., *Essential Maple: An Introduction for Scientific Programmers*, Springer-Verlag, 1995.

Powers, David L., *Boundary Value Problems*, Second Edition, Academic Press, 1979.

Robertson, J., *Engineering Mathematics with Maple*, McGraw-Hill, 1996.

Rosen, K. H., *Exploring Discrete Mathematics with Maple*, McGraw-Hill, 1997.

Strang, Gilbert, *Linear Algebra and Its Applications*, Third Edition, Harcourt Brace Jovanovich, 1988.

Zwillinger, Daniel, *Handbook of Differential Equations*, Second Edition, Academic Press, 1992.

Note that an extensive list of books and other publications, including works in progress, devoted to Maple V can be found at

http://www.maplesoft.com/books.html

For information, including purchasing information, about Maple V contact:
Corporate Headquarters:

Waterloo Maple, Inc.
450 Phillip Street
Waterloo, ON, Canada N2L 5J2
telephone: (519) 747-2373
fax: (519) 747-5284
email: info@maplesoft.com
web: http://www.maplesoft.com/

Index